ByDesign

A Journey to Excellence through Science

SEVENTH-DAY ADVENTIST® CHURCH

Kendall Hunt Religious Publishing
A Division of Kendall Hunt Publishing

Developed in Collaboration with the
Seventh-day Adventist NAD Office of Education and
Kendall Hunt Religious Publishing
A Division of Kendall Hunt Publishing Company

Acknowledgments

SDA/NAD Office of Education

Larry Blackmer, Executive Editor/Vice-President of Education
Carol Campbell, Director of Elementary Education
Debra Fryson, Director of Education, Southern Union

Program Development and Publisher

Kendall Hunt Religious Publishing, a Division of Kendall Hunt Publishing

Program Consultants Lead Team I

Dan Wyrick, Director of Nature by Design
Lee Davidson, Associate Professor of Teacher Education/Dept. Chair, Andrews University
Jerrell Gilkeson, Associate Director of Education, Atlantic Union
Ed Zinke, Biblical Researcher
Tim Standish, Geoscience Research Institute

Associates Representing Lead Team II

Betty Bayer, Associate Director of Education, SDA Church in Canada
Diane Ruff, Associate Director of Education, Southern Union
Ileana Espinosa, Associate Director of Education, Columbia Union
James Martz, Associate Director of Education, Lake Union
Keith Waters, Associate Director of Education, North Pacific Union
LouAnn Howard, Associate Director of Education, Mid-America Union
Martha Havens, Associate Director of Education, Pacific Union
Mike Furr, Associate Director of Education, Southwestern Union
Patti Revolinski, Associate Director of Education, North Pacific Union

Contributing Writers and Reviewers

Carol Raney, Freelance Writer
Katia Reinert, NAD Director of Health Ministries
Alayne Thorpe, President of Griggs University
Rick Norskov, MD, Southern Adventist University
Lucinda Hill, MD, Southern Adventist University
Raul Esperante, PhD, Geoscience Research Institute
Mitch Menzmer, PhD, Southern Adventist University

Contributing Multi-Grade Classroom Writers and Reviewers

Karen Reinke, Freelance Writer
Mark Mirek
Heidi Jorgenson
LaVona Gillham
Cindy French Puterbaugh

www.kendallhunt.com
Send all inquiries to:
4050 Westmark Drive
Dubuque, IA 52004-1840
1-800-542-6657

ISBN 978-1-4652-0116-4

Printed in the United States of America

1 2 3 4 5 6 7 8 9 10 19 18 17 16 15 14

Science Classroom and Lab Safety

SAFETY:

It is important to be safe when you perform hands-on investigations. Always look for the safety symbol in your textbook. It tells you how to be safe as you perform your investigations. Below are some important safety rules to follow for most investigations. Remember to be alert and listen to your teacher for any additional safety rules.

Prepare for laboratory work

- Review classroom laboratory procedures with your teacher.
- Never perform unauthorized experiments.
- Keep your lab work area organized and free of clutter.
- Know the location of the safety fire blanket and first aid kit.

Dress for laboratory work

- Tie back long hair.
- Do not wear loose sleeves.
- Wear shoes that completely cover your feet.
- Wear lab apron, goggles, and gloves as required.

Avoid contact with chemicals

- Never taste or sniff chemicals.
- Never draw materials in a pipette with your mouth.
- When heating substances in a test tube, point the mouth away from people.
- Never carry dangerous chemicals or hot equipment near other people.

Avoid hazards

- Keep combustibles away from open flames.
- Use caution when handling hot glassware.
- When diluting acid, always add acid slowly to water. Never add water to acid.
- Turn off burners when not in use.
- Keep lids on reagent containers. Never switch lids.
- Use sharp equipment and objects as they were designed to be used and in a safe manner. For example, always cut away from yourself rather than toward yourself, and be careful to study which edges and points of an object are the sharp ones.

Clean Up

- Consult the teacher for proper disposal of materials.
- Wash hands thoroughly following experiments.
- Leave laboratory work area clean and neat.

In case of accident

- Report all accidents and spills immediately.
- Place broken glass in designated containers.
- Wash all acids and bases from your skin immediately with plenty of running water.
- If chemicals get in your eyes, wash them for at least 15 minutes with an eyewash or clean water.

Scavenger Hunt

Start Use this Scavenger Hunt to find where things are in your book.

Where will you find the **Essential Question** for each lesson?

What is the first **Inquiry Kick-Off** Engage about?

Write the sentence in which the first **vocabulary** word appears.

What is the first word that appears in the Index in the back of your book?

What is the title of the first **Structured Inquiry**?

What does the first page of the **Inquiry Handbook** talk about?

What is your first **SCIENCE SAFETY** rule?

What are the vocabulary words for Chapter 1, Lesson 1?

What question does the first **Check for Understanding** ask?

What is the scripture reference for the first **Scripture Spotlight**?

What do each of these symbols stand for?

What does the first **Lesson Activity** ask you to think about?

What question does the first **Explore-a-Lab** ask?

What is the solution to the first **Math in Science**?

What is the first question of the first **Concept Check** **Assess/Reflect**?

How many questions are in the first **Chapter Review**?

ByDesign

Table of Contents

Science

How would you define science? Is it all the knowledge accumulated from disciplines such as biology, chemistry, and physics? What other disciplines might it include? Does it include the latest technology? Science includes all of these things, but we usually focus most on science as a process of discovery—a systematic search for understanding of the natural world.

In order to systematically search for the truth, a scientist would have to be true about the natural world. What are some truths about the natural world that a scientist would consider? Can you think of any other assumptions that would need to be true for scientists to proceed with this process?

Scientists make several assumptions that provide the foundation for science:

- Living things and the physical world can be studied and understood.
- The processes of nature follow predictable laws.
- By observation and experimentation, we can learn what these laws are.

Where do you think scientists got the idea that nature follows predictable laws?

Sometimes we distinguish between empirical science and historical science.

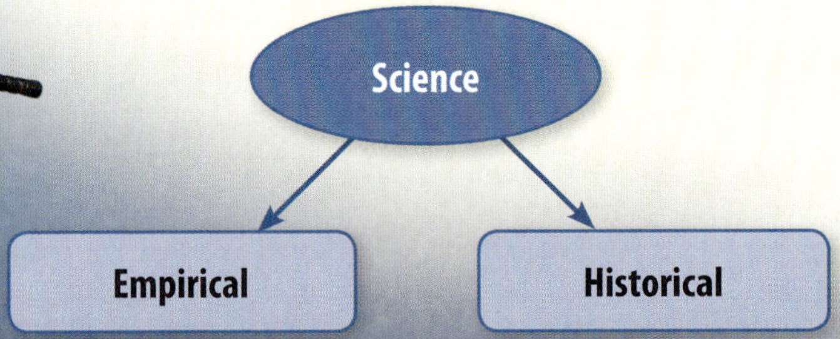

Empirical Science

You are probably most familiar with empirical science, which is the basic scientific process that you have been learning about for several years.

Process of Empirical Science	
Observe	Use your senses to make observations.
Question	Ask a question that you would like to have answered.
Hypothesize	Provide an explanation for an observation or problem that can be tested by experimentation. Hypotheses are also known as educated guesses.
Experiment	Design a test that will confirm or falsify your hypothesis.
Record data	Carefully gather and record data. (Data might include pH, time, temperature, distance, growth, or other measurements.)
Analyze data and draw conclusions	Determine whether the data confirms or falsifies your hypothesis. This may lead you to ask more questions and start the process all over again.

Scientists do not always follow these processes in the same order. New questions may be raised at every step in the process, additional observations along the way may lead to changes in the experimental procedure, and multiple experiments may need to be done before a hypothesis can be confirmed or rejected.

Testable

It is important to emphasize that in empirical science, a hypothesis must be testable. This means that a scientist must be able to design an experiment that can test whether a hypothesis is true.

Repeatable

It is also important for an experiment to be repeatable. If one scientist performs an experiment and leaves a detailed description of the exact procedure followed and the data gathered, another scientist should be able to repeat the experiment and verify the results.

Science Limitations

The success of science has encouraged a tendency to believe whatever science claims, but recognize that science has limitations.

Sample Size

It may surprise you to learn that science cannot provide absolute answers. One reason for this is that we never have all the data. Say, for example, that we want to research the number of viruses contained in a human mouth. We could never realistically check every human on Earth, so we have to settle for checking a relatively small number of humans. Even then, we could not check every human mouth completely, so we would have to take a small amount of saliva from each person. Of the small sample, we would still only count viruses from a portion of it and estimate the total contained in the complete sample. Our conclusions would have to be based on a much smaller sample of data than we might like. The possibility always exists that our conclusions might be different if we had more data. What other scientific data might be limited by sample size?

Experimental Design

It is important to carefully design experiments. An old joke illustrates how easy it is to design a faulty experiment that limits what we can learn from the scientific process. Imagine that we want to find out where the ears are located on a trained flea named Herman. Since some insects have ears on their legs, abdomen, or antennae, we cannot assume that his ears are where we would expect them to be. When we say, "Jump, Herman," Herman jumps, so we can assume that he heard us. To test whether his ears are on his antennae, we remove the antennae and tell Herman to jump. Since he jumps, we can assume he heard us, and therefore his ears are not on his antennae. To determine whether his ears are on his front legs, we remove his front legs and again tell Herman to jump. Since he jumps, we assume his ears are not on his front legs. Since insects often have ears on their back legs, we remove the back legs and again tell Herman to jump. When he does not jump this time, we assume that his ears were, in fact, on his back legs. Of course, you see the problem! What other explanation might explain why Herman didn't jump? Did we really learn the location of Herman's ears?

Example A	Example B
If I am out of gas, my car won't start. My car won't start. Therefore, I am out of gas.	If I am out of gas, my car won't start. My car does start. Therefore, I am not out of gas.

This example illustrates a real problem in logic that can occur in science.

What is the problem with Example A? Do you think it is easier to prove or disprove a hypothesis? How does this relate to how a scientist should state a hypothesis?

Historical Science

The scientific process is also limited in what it can do to help us answer questions of historical science, which deals with events that have occurred in the past. When we are trying to explain how or why something happened in the past—especially a one-time event such as the origin of the Universe or life on Earth—we are limited by the fact that the event was not observed and by the fact that we cannot design an experiment to repeat it. Scientists approach these kinds of questions in a different way, gathering as much evidence as possible and suggesting possible explanations based on what we know about cause-and-effect relationships. They often examine multiple possibilities to see which one explains the most data. While reasonable suggestions can be made, it is impossible to verify any option scientifically. What examples of historical science can you list?

Bias

Another important factor that can limit the effectiveness of the scientific process if we are not careful is personal bias. A bias is a tendency to look at things in a certain way or a prejudice that affects how a scientist sees a particular idea or situation. For example, a scientist was looking for fossilized footprints on a sloped area but was not planning to look for them on a flat area nearby because he had reasons to believe there would not be any tracks on the flat area. A graduate student with him did not have the preconceived idea about where footprints would and would not be found. Looking on the flat area, the graduate student found footprints that the scientist would not have even looked for. How was the graduate student's lack of knowledge an advantage in this situation? How might a scientist guard against bias? Although it may not be possible to eliminate all biases, it is important to recognize that they exist and to try to correct for them. When scientists share their findings and critique one another's ideas, it helps to identify and correct biases.

Worldview

Although scientists try to identify and eliminate certain kinds of biases, there is a very important kind of bias that is difficult to eliminate. It is impossible to do science without *any* worldview, so it is important for scientists to intentionally choose the worldview that will be the starting point from which they will work. This chart summarizes the assumptions shared by scientists from biblical and naturalistic worldviews as well as the assumptions on which they would disagree.

Assumptions		
Biblical Worldview Only	**Both Worldviews**	**Naturalistic Worldview Only**
God does exist.		God may or may not exist. If He exists, He did not create and does not interfere with nature.
Divine revelation may reveal information about nature that human beings might never discover by themselves.	Living things and the physical world can be studied and understood.	Human beings are capable of understanding nature without divine revelation.
God created the Universe by a process we don't understand and sustains nature through the laws He has established.	The processes of nature follow predictable laws. By observation and experimentation, we can learn what these laws are.	Nature operates on its own according to these laws.
	Change has occurred in organisms and in the physical Universe.	
Scientists are willing to consider the possibility of intervention by God at various times in history.	Scientists cannot do experiments to test the supernatural. Scientific hypotheses must be testable using only criteria accessible to the five senses.	Scientists are only willing to consider theories that do not imply or require any supernatural activity at any time in history.

Although the possibility of bias from one's worldview cannot be eliminated, it is important for every scientist to recognize the influence his or her worldview creates. This influence does not have to impede the scientific process. It can, in fact, even be beneficial. It makes sense that scientists with one worldview will notice things that other scientists with a differing worldview might not.

God and Science

You will notice in the chart on the previous page that scientists with a naturalistic worldview prefer to keep God separate from science. If He exists at all, they believe He does not interfere with nature, and they prefer to avoid any theory that even implies any intervention by God at any time in history.

Contrast that with a biblical worldview. Far from wanting to keep God separate from science, we recognize that God, as the Creator and Sustainer of the Universe, is inseparable from science. Much of what we study in science—living organisms, Earth itself, and the laws of nature—was created by God. Even the curiosity, intelligence, and creativity that allow us to do science came from God when we were created in His image.

In their book *The Privileged Planet*, Guillermo Gonzalez and Jay Wesley Richards suggest that not only was the Universe designed to support life, but it was also designed for discovery. According to the evidence they have analyzed, our Earth is in the best possible place to study the Universe. This is consistent with what the Bible says about us being able to learn about God's glory by studying the Universe (**Psalm 19**), and that the things He created can teach us about His invisible attributes, eternal power, and divine nature (**Romans 1:20**).

When scientists seek to understand God better through a study of the natural world, God is glorified. What effect might result from excluding the Creator from something He created?

The Bible and Science

Although nature can reveal to us God's glory, invisible attributes, eternal power, and divine nature, it cannot answer all our questions. The Bible is a revelation from God that supplements what we can know about Him through nature alone. Without the knowledge revealed in the Bible, we could easily be confused about why parasites exist, why catastrophes happen, and why all living things eventually die. Knowing what the Bible says about the entrance of sin into the Universe explains a lot about what we see in the natural world.

Since nature and the Bible came from the same source, it would make sense that when we correctly understand both, science and the Bible should agree with each other. When it appears that they don't, that tells us that we don't understand one or the other of them correctly yet.

Take origins, for example. The Bible teaches that the Universe, Earth, and living things were created by God. The theory of evolution suggests that we came from a single-celled common ancestor after millions of years of evolution. Both stories cannot be true. When science and the Bible disagree, how should we respond?

This year you will study the evidence for both ideas. You will see how worldview affects the interpretation of the data. You will see that some evidence is more consistent with Creation than with evolution. You will also see that there are questions for which we don't have answers yet, and that is OK.

Those who have chosen a biblical worldview recognize the Bible as the standard by which everything else should be evaluated. Even when there are questions we can't answer, it is appropriate to hold on to what the Bible teaches and continue to look for explanations of the data that are scientifically sound and consistent with Scripture. It is important to remember that science is a Creation of God and should never be elevated above God.

Gathering Data

Suppose your parents are interested in determining which diapers are the most absorbent to use for a new baby about to join your family—cloth or disposable diapers. You might like to learn how disposable diapers work. You could conduct an investigation to test the leading brands against cloth diapers. You would want to keep certain variables constant, such as the amount and kind of liquid, the temperature, and the amount of time. You would need to gather data about how much liquid is present before and after and subtract to find out how much is absorbed by the test samples.

This is not much different from gathering data using the scientific process. Once a scientist has defined a question, formulated a hypothesis, and researched relevant information, he or she begins to experiment and make observations. Gathering or collecting data takes place during experimentation. Data can be collected in a variety of ways, such as through observing, measuring, and recording. Why do you think the process of gathering data is a critical part of the scientific process?

This researcher is testing water samples to learn about water quality.

What variables might she test to gather data for a study of the effects of pollutants on organisms living in the water?

Regardless of the techniques used when gathering data, scientists must be sure to record all information that could affect the answer to their hypothesis. So, how do they know which data to focus on? Here are some questions to guide you when gathering data:

- Have I recorded all relevant data?
- Have I recorded the data in a way that can be read and understood?
- How am I keeping track of what was done at each step?
- Can someone else understand the data?
- Is my data trustworthy and reliable?

It is important to remember that scientists often build on one another's work. If your experiment is based upon information gathered from another scientist's work and his or her data collection method was flawed, how could this affect your outcome?

Using a process that is consistent when collecting data and making sure your records are detailed helps others understand and use the data collected. This builds confidence in your conclusions and the conclusions of those who use your research.

Interpreting Data

Once all of your data has been systematically and accurately gathered, it's time to step back and ask, "What do my findings tell me?" When all of the evidence is considered, you can evaluate and draw conclusions by assessing the data. Data may be interpreted in different ways based on a person's background knowledge and perceptions. One scientist may see patterns in the data that another scientist does not.

Let's go back to the diaper data in our earlier example. Gathering data is only the first step. Part of the process of interpreting data is to display and present it in a meaningful way so you can share your findings with others. What would be a good way to display your comparison of the absorbency of different diaper brands? The bar graph shown below is one way you could have chosen to represent the data. Which disposable diapers led the pack during the three trials? Which diaper of those tested would you recommend to your parents? Why?

To interpret data you must look at it critically. These questions can guide you when interpreting data:

- What conclusion can you reach from your results?
- Do the results answer the original questions?
- Are the results different from what you expected?
- Are there other hypotheses to explain the results?
- Did you think about all relevant data, even data that seemed odd?
- Did the data collection method change the results in any way?
- What are the implications of your conclusions?

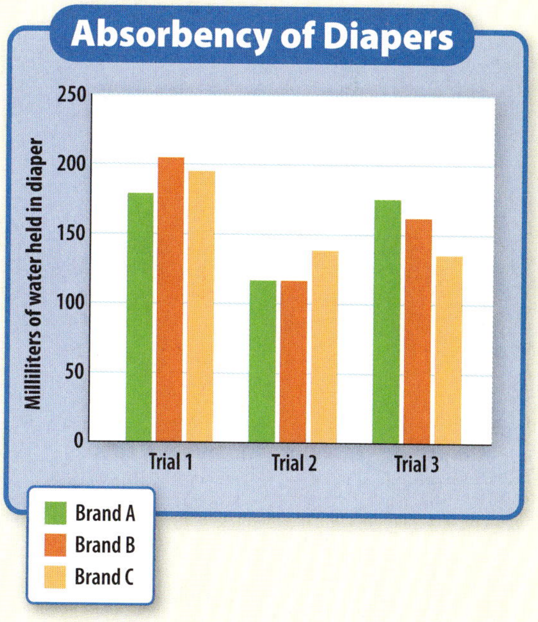

Can you think of any additional questions that can guide you when interpreting data?

Sometimes interpreting data results in an error or incorrect conclusions. The error can be systematic throughout the testing and have great effects, or it can be the result of random events and have little to no effect on the outcome. What if the measuring instrument you used was inaccurate? How would that affect the results? What can scientists do to prevent such errors and inaccuracies?

Although using carefully planned methods can reduce inaccuracy, errors still occur. Scientists indicate their level of confidence when interpreting data using a margin of error. When there is conflict about the interpretation of data, experts in that field will come to a consensus. For example, you would communicate the results of your diaper tests with another group that performed the same tests and see whether you could reach a consensus on the best brand of diaper. The process of coming to consensus may require further experimentation and additional evidence.

The Scientific Process

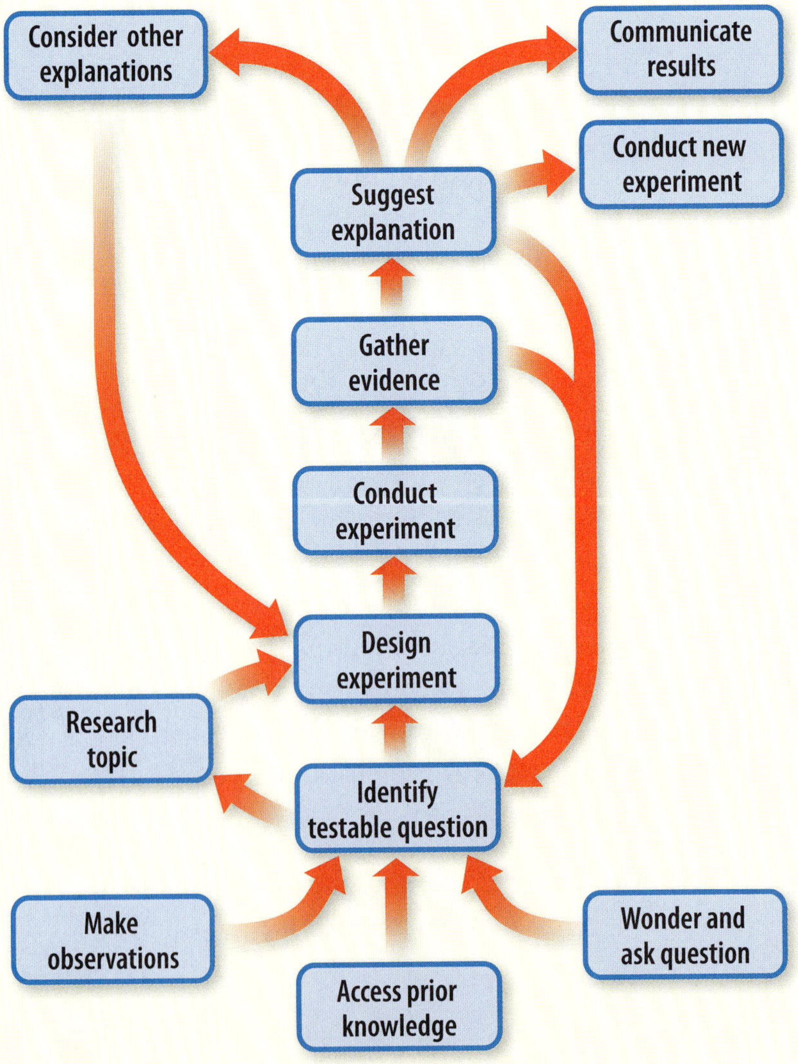

Evaluating the Claims of Science

It is important to recognize that science is not one large bundle of information to be accepted or rejected as a whole. Specific claims of science should be analyzed and evaluated separately. Recognizing both the value and the limitations of the scientific process, how does one evaluate the claims of science?

1. Evaluate the reliability of the source of information. Did you read it in a reputable scientific journal, see it on the news, or read about it in a tabloid paper in the checkout line at the grocery store?

2. Know the difference between data and interpretation. Data is information gathered using your senses—things you can see, hear, touch, count, or measure. Interpretation is the meaning you attach to the data. Recognizing which part of any scientific claim is data and which is interpretation can help you evaluate the validity of the claim.

3. Remember that *all* scientists have a worldview. Look for the influence of a naturalistic or biblical worldview as you evaluate any scientific claim.

4. Ask how the claim relates to God's divine revelation in Scripture. Is anything related to the claim mentioned in Scripture? Is the claim consistent with the Bible, or does it conflict with what Scripture teaches?

5. Show respect for scientists and their ideas, even if you don't agree with them.

Which one of these suggestions do you think is the most important? Why? Which do you think would be easiest for you to do? Which one do you think you might struggle with? Throughout this year, you will have opportunities to practice these skills as you learn about various claims made about the natural world by scientists from differing worldviews.

Unit 1

Life Science

Some regions of Earth remain relatively untouched by humans. Scientists use these regions to study about the origins of life on Earth and the survival of plant and animal species.

Unit Overview

How have living things on Earth changed over time? Are there forces that cause changes that are passed on to offspring? How are living things classified? Do we classify them according to how their structures look, how they function, or both? If they change, how would that affect our classifications?

In this unit, you will learn about theories surrounding the origins of life and theories about how life has changed, or evolved. What evidence is used to support scientific assumptions? How do you think these compare with Scripture? Scientists organize their findings and classify the organisms they study.

As you continue your studies in this unit, you will focus on the plant kingdom. This large division of life consists of vascular and nonvascular plants. How do scientists classify the plants? What structures and functions are compared? You will develop an understanding of plant reproduction and how botanists use this knowledge to classify the organisms.

You will also learn about plant and animal survival by analyzing relationships, examining processes, and distinguishing between factors that promote or deter survival. How do you think plants and animals adapt to changing environments? What impact do humans have on plant and animal species? You will conclude this unit by analyzing the impact of human behavior on ecosystems.

Your teacher may assign an Open Inquiry lab and a Lifestyle Challenge activity. Use your *Science Journal* to record your work.

Chapter 1

How Life on Earth Has Changed

Scripture Spotlight

Various theories of origins exist, some of which conflict with Creation as described in **Genesis**. Learn more about origins and grow your faith in God as Creator in the following passages in this chapter.

Isaiah 45:18 (p. 17)
Psalm 139:13–16 (p. 26)
Exodus 19:4 (p. 31)
Psalm 55:6 (p. 31)
Matthew 23:37 (p. 31)
Matthew 12:46–50 (p. 41)

Horseshoe crabs look exactly like the fossils of the animals that lived long ago. There is little evidence that they have changed over time.

The Big Idea

There are many ideas about the origins of life. Several kinds of scientific evidence are consistent with the biblical story of Creation by God.

What scientific evidence are you aware of that supports the story written in **Genesis 1–2**?

Inquiry Kick-Off

Engage

How do you think the horseshoe crab survived so long without much change? Why have some populations become extinct, while others survive? Is the survival of species random or is something more involved? In your ***Science Journal***, you will observe that what might be perceived as random isn't so random after all.

Lesson 1

Essential Question

What Are Theories of Origins?

Objectives

- Compare and contrast the theories of origins that are consistent with the biblical worldview and those that are consistent with the naturalistic worldview.
- Describe the Miller-Urey experiment.
- Explain Darwin's contribution to the theory of evolution.
- Compare and contrast microevolution with macroevolution.

Vocabulary

origins
creationism
Informed Interventionism
abiogenesis
spontaneous generation
macroevolution
primordial soup
exogenesis
panspermia
fixity of species
artificial selection
natural selection
microevolution
mutation
Intelligent Design

Do you know the two primary theories of origins—how life began on Earth? What do these two theories have in common? What are their differences? What have you learned before that can help you answer this question? Is it possible to use the scientific method to determine the origins of life? Why or why not?

Origins Explain

You may recall that **origins** means "beginnings." In the study of origins, scientists develop theories that try to explain where the world came from and how life on Earth began. How do they find evidence that helps them explain past events in Earth's history? They face many obstacles in their study of origins. The scientific process, which works so well when observations are made and experiments are repeated in the present, is limited in its ability to help with historical science.

You have already learned that all scientists are influenced by their worldviews. This is especially true when they are studying historical science. Recall what you learned in the ***Inquiry Handbook*** about the assumptions of the biblical and naturalistic worldviews. Scientists from both worldviews agree that supernatural events—like Creation by God—cannot be tested scientifically. For naturalistic scientists, this assumption prevents them from considering the possibility that God is the Creator, but for scientists with a biblical worldview, it is the starting point from which they do their research and experiments. See the chart on the next page for a comparison of the two approaches to studying the origins of Earth.

Theories of Origins

Although there are multiple variations of each theory, there are two basic ideas about where life came from: Creation by God or the origins of life through evolutionary processes. Study the chart to compare the two theories.

Creation and Informed Intervention

Creationism is the idea that God created the Universe and all life, as described in **Genesis 1–2**. **Informed Interventionism** is a broader concept that also acknowledges God's intervention at other times in history, such as the biblical Flood.

Abiogenesis

Abiogenesis, the idea that life came from nonliving things, goes back to the time of the Greek philosopher Aristotle in the fourth century B.C. Aristotle observed that no plant or animal survived when a pond dried up. When rain refilled the pond, he saw tiny fish. He concluded that the fish must have generated from the sand of the pond through **spontaneous generation**, the idea that life can arise from nonliving materials.

Fast-forward to the 1600s. Observers reported what they thought were examples of spontaneous generation. Jean-Baptiste van Helmont explained that mice would form if wheat grains were put in a pot with a dirty shirt. Many people observed maggots on rotting meat. They assumed the maggots formed from the meat.

Scripture Spotlight

What does **Isaiah 45:18** say about God's purpose in Creation?

	Worldview	
	Biblical Worldview	Naturalistic Worldview
	Theory	
	Creationism and Informed Interventionism	Abiogenesis and Evolution
Origins	Living and nonliving matter were created by God. (**Genesis 1–2**)	**Abiogenesis**: Life arose spontaneously from nonliving matter many millions of years ago.
Explanation for what we see around us now	Sin has caused change. (**Genesis 3**) Change is limited and not progressive.	**Macroevolution**: All present life forms descended from a simple, one-celled organism.

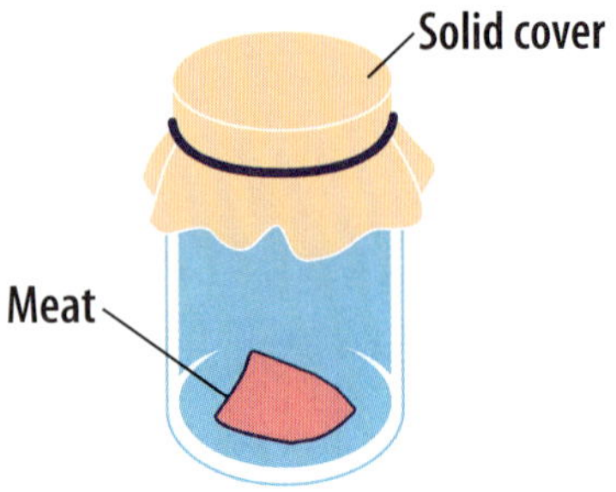

Jar Completely Sealed

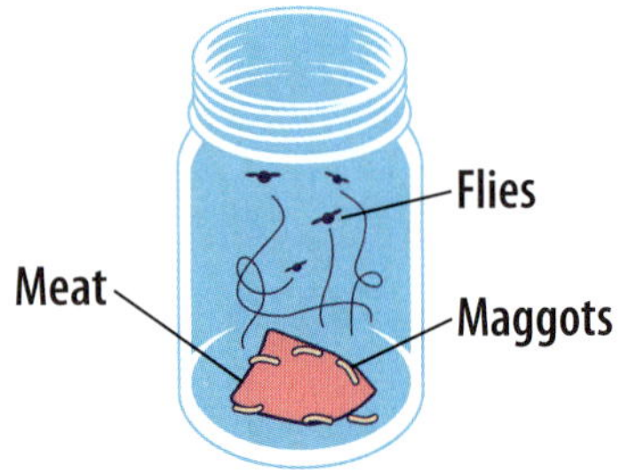

Open Jar

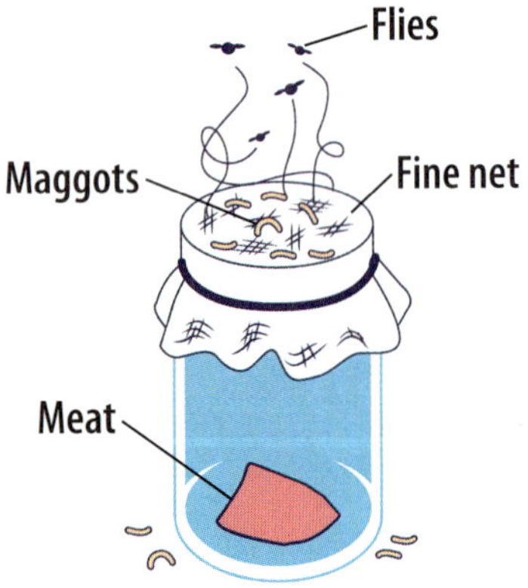

Jar Covered with Fine Net

Redi suspected that flies laid eggs too small to be seen with the naked eye on the meat.

How does Redi's experiment support his hypothesis?

In 1668, Francesco Redi, an Italian physician and biologist, designed and carried out a controlled experiment to test spontaneous generation. He placed equal samples of meat into several jars. He left some open to the air. He covered others with a fine net. Other jars were sealed completely. Redi observed flies entering and maggots forming in the open jars. In the other jars, the meat decayed, but there were no maggots and no new flies. What was an important variable in Redi's experiment? You have already learned how Louis Pasteur's experiments in the 1800s, which were built on Redi's work, finally disproved the theory of spontaneous generation. What other method could Redi have used to test spontaneous generation?

In the twentieth century, scientists investigating the origins of life shifted their attention away from whole organisms. They began looking at the organic molecules that make up living things and wondering how they originated. In 1924, the idea of abiogenesis was resurrected when Alexander Oparin proposed that life on Earth began in the oceans. In spite of the facts that Redi and Pasteur had already demonstrated that the spontaneous generation of life is not possible, Oparin suggested that it had been possible at some time in the past when conditions on Earth were different. It was hypothesized that at some previous time, the oceans were filled with carbon dioxide, methane, ammonia, and water. This is sometimes referred to as **primordial soup**. Next, it was suggested that the energy needed to change inorganic compounds into organic chemicals of life could have come from the Sun's ultraviolet radiation. It was further hypothesized that those organic compounds formed into short segments of DNA that then linked together to form longer chains. From there, the hypothesis suggested that the first actual life forms, thought by scientists to be prokaryotes, formed.

You may recall that in 1953, Stanley Miller and Harold Urey designed experiments that attempted to simulate this process and create organic molecules, such as amino acids, in the laboratory. They simulated an atmosphere made up of water, methane, ammonia, and hydrogen, shot electrical charges through it to simulate lightning, and let it bubble for a week to simulate Earth's hot environment. How did Miller and Urey know the composition of Earth's atmosphere?

When Miller and Urey analyzed the mixture, they found several different amino acids in it, which caused great optimism among scientists. Their excitement turned out to be premature, however, because of several issues.

Issues with Miller-Urey Experiment
• Many of the molecules produced were not useful for creating life and were sometimes even destructive.
• Some processes responsible for generating the amino acids and molecules could break them down even faster.
• Opinions about what Earth's atmosphere was actually like continue to change.
• Except for glycine, amino acids are either left handed or right handed, which means they are mirror images of each other. The amino acids found in living things are all left handed. The amino acids produced in the Miller-Urey experiments were a mixture of left and right handed. How would that affect the probability of the correct left-handed amino acid sequences forming naturally?

Left handed

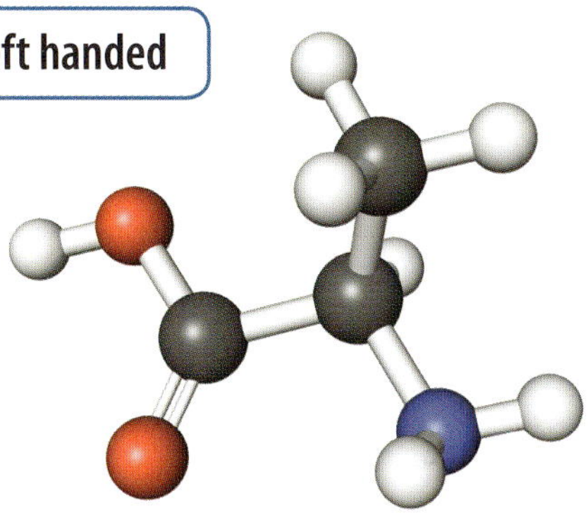

Alanine can be found as a right-handed or a left-handed amino acid.

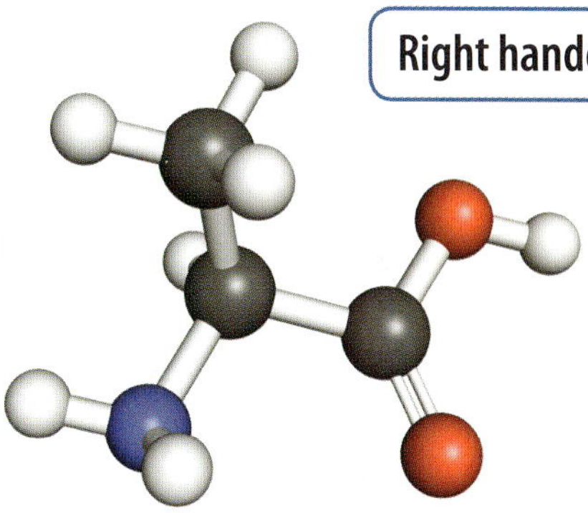

Right handed

Despite the initial optimism, the work of Miller and Urey and modified experiments by other researchers since then have come nowhere near producing life. This recognition has caused scientists to redirect their attention to other possibilities. Researchers have since investigated an idea called *RNA world,* which suggests that life began when self-replicating RNA formed an intermediate, "pre-cellular" life form with similar results as the Miller-Urey experiments.

As scientists recognize how impossible it would be for life to have sprung from nonliving material here on Earth, some have suggested a hypothesis called **exogenesis**, which suggests that life began somewhere else. Because organic molecules have been found in comets and meteorites, it has been suggested that the earliest living things traveled to Earth on an asteroid or inside a comet. In what way does this help naturalistic scientists show how life originated on Earth?

A similar idea, known as **panspermia**, suggests that life is actually widespread throughout the Universe. Many scientists reject this idea. Does suggesting that the origins of life happened somewhere else in the Universe solve any of the problems associated with its origins here?

Scientific research continues to discover the staggering complexity of cells, DNA, the proteins from which they are made, and the information they contain. As a result, the spontaneous origins of life from nonliving matter is recognized as an extremely improbable event.

Structured Inquiry

Discover

Record your work for this inquiry. Your teacher may also assign the related Guided Inquiry.

Science Journal

Natural Selection in 15 Seconds

How does natural selection work?

Materials
- paper cup
- meterstick
- paper (black, white, newspaper)
- paper punch
- stopwatch
- targets (black, white, black and white newspaper)

Procedure

1. Make three targets that measure 1 m × 1 m in size. Make one target from black paper, one from white paper, and one from newspaper.
2. Use the paper punch to make 10 paper dots each from black paper, white paper, and newspaper.
3. Mix the dots together in the paper cup.
4. Place the black paper target on the floor. Without letting your partner watch, scatter the mixed dots from the cup over the black paper target. Be sure the dots are not stacked on each other and that all are on the target.
5. See how many dots your partner can pick up in 15 s. **Record** the number of each color picked up. Repeat Steps 4 and 5 two more times.
6. Repeat Steps 3–5, first using the white paper target and then the newspaper target. **Calculate** the average number of dots picked up on each target.
7. Then switch roles with your partner and repeat Steps 3–6.

Analyze Results

Graph the average number of dots picked up on each color target.

Create Explanations

1. How does natural selection work?
2. How does this activity model natural selection?
3. Suppose you used a sheet of newspaper that was filled with color advertising. Which color dot do you think would be picked up the most? Why?
4. What are the limitations of this model? Explain.
5. Do you think that natural selection provides a complete explanation of the variety of life on Earth? Explain.

Explore-a-Lab

Structured Inquiry

What might cause scientists to have different explanations even though they have access to the same data?

Scientists gather information and make hypotheses based on what they find. Oftentimes, as more data is gathered, hypotheses will change. With your group, you will also make and alter hypotheses.

1. Your group will be given an envelope with word cards in it. Randomly choose 5 cards from the envelope and place them face up on the table.
2. Discuss with your group what story you think the words form. Write your hypothesis on a sheet of drawing paper.
3. Randomly choose 5 more word cards and place them with the previous 5. Discuss with your group what the story could be with the 10 words.
4. Write a revised hypothesis on your paper.
5. Choose another 5 word cards randomly, combine them with the previous 10 cards, and write one more revised hypothesis with your group.
6. Did your group's hypothesis change with each addition of data? Each group will share its final hypothesis about what the complete story is about. Compare the groups' hypotheses. Your teacher will tell you the final story. How does it compare with your hypothesis?

Microevolution

For many centuries, people thought that animals and plants never changed. They believed that life on Earth was exactly the same as what was created by God during the Creation week. This idea was called the **fixity of species**. In the seventeenth century, however, Swedish botanist Carolus Linnaeus observed that plants could be the result of hybridization, or the crossing of two different species. As breeders purposely selected plants or animals with desirable characteristics to become parents of the next generation, the desirable traits were passed on to the offspring and changes in the species occurred over time. This process is known as **artificial selection**, and it was influential in the transition from a belief in the fixity of species to the acceptance of a new idea, the theory of evolution.

Evolution is often defined simply as "change over time." Remember that this change does not happen to individuals, but to populations, or species, which are groups of organisms consisting of similar individuals that interbreed in nature.

Chinese shar-pei dogs have a history dating back to 200 B.C. Their uniquely wrinkled skin is a result of hundreds of years of artificial selection.

A number of different scientists over many years thought about the idea of evolution. When Charles Darwin published his book *On the Origin of Species by Means of Natural Selection* in 1859, his major contribution was his theory that the evolution of new species occurred as the result of natural selection. He suggested that a process similar to artificial selection is at work in nature.

Darwin's theory is summarized by these four points:

- **Competition** Each species produces more offspring than can actually survive. This leads to competition between the offspring for food, water, and space. What are some examples of competition among individual animals or species?
- **Variation** Each individual within a species has a slightly different genetic make-up, so its characteristics are different, producing variety within a species. How do dogs of the same breed differ? How do students in your classroom differ?
- **Adaptation** Because some variations are more advantageous than others, certain individuals within a species are better suited for surviving in their environment. What are some characteristics of individual polar bears that might make them better suited than other polar bears to survive in an Arctic environment?
- **Selection** Those that survive and reproduce pass their characteristics on to their offspring, making them better equipped to survive.

This pattern of variation, greater survival, and successful reproduction is called **natural selection**. You have already learned one of the most well-known examples of natural selection—the finches found by Darwin in the Galápagos Islands and studied later by Peter and Rosemary Grant. Over time, changes in the beak size of the finch population happened in response to changing weather conditions. During drought years, plants with smaller seeds did not grow well, which meant the small-beaked birds that ate them did not survive well.

Science Journal

Check out your *Science Journal* for a Structured Inquiry that explores variation in three different species.

Extend

As a result, fewer survived to reproduce, and the number of small-beaked birds in the population declined. During years with large amounts of rain, the process reversed. Although no individual beak sizes changed, the frequency at which certain beak sizes appeared in the population changed over time.

Another common example that illustrates natural selection involves the peppered moth and its history over the last two hundred years in England. Originally, the light-colored moths were camouflaged against the trees and flourished as a result. The darker moths stood out more and were eaten more often by predators. During the Industrial Revolution, however, pollution became widespread and the trees were often covered with soot. Which kind of moth would be better camouflaged then? Which kind would be more likely to be eaten by predators? How do you think that affected the population of peppered moths? What do you think happened when improved environmental standards improved the pollution problem?

Small changes over time, such as the size of bird beaks changing a little from generation to generation in the finch population or the color changing within the peppered moth population, are examples of **microevolution**. Scientists from both worldviews acknowledge the presence of microevolution in nature. Where scientists from different worldviews disagree is in the discussion about macroevolution.

Peppered moth (*Biston betularia*)

How do changes in the coloring of peppered moths explain the concept of microevolution?

Macroevolution

Macroevolution involves much larger changes—the kind of changes that would have been necessary to change a microscopic cell to a fish or a reptile to a bird. What kind of new structures and physiological processes would have to evolve to change a prokaryote to a eukaryote, a single-celled organism to a multicelled one, an aquatic creature to one that lives on land, or an earthbound creature to one that can fly? Recall the evolutionary claim that all present life forms descended from a simple, one-celled organism and imagine the amount of change necessary to change a microscopic cell into a human being.

How Much Change?

Although most people believe that some change has happened throughout history, the real question is, *how much change?* Many evolutionary scientists claim that macroevolution is nothing more than a lot of microevolutionary changes accumulated over billions of years. Tiny changes are introduced by genetic mutation. Those that improve an organism's chance for survival and reproduction are passed on to the offspring. Over billions of years, these tiny changes accumulate to produce major changes in body structures and physiological processes. **Mutations** are changes that occur when a gene is altered, damaged, or lost. Very few mutations have been shown to be positive. Most are harmful or even lethal. For this reason, many scientists, even some who believe in evolution, doubt that microevolutionary changes caused by mutations are capable of producing macroevolution.

The chart below summarizes the differences between microevolution and macroevolution. Scientific evidence for microevolution is abundant, but evolutionists and creationists disagree over what is considered evidence for macroevolution.

Microevolution	Macroevolution
Changes that occur among plants and animals of a population	Major changes in structure that result in changes from one kind of plant or animal to another
Numerous examples of microevolution found in nature	Macroevolution has not been shown to occur. While many scientists believe there is evidence for macroevolution, other scientists have other explanations and believe that there is considerable evidence that contradicts the theory that macroevolution has occurred.
Accepted by evolutionists and creationists	Accepted by evolutionists but rejected by most creationists

From cities to forests and mountains to coastlines, rat snakes are found in all types of habitats across the eastern and midwestern United States and parts of Canada. They come in a wide range of colors, from iridescent black to orange striped.

How do these color variations help the survival of the species?

Theories of Origins—Variations

Recall from earlier in this lesson that there are two main theories of origins: Creation and evolution. There are variations of these two main theories. Several are summarized here.

Theories of Origins	
Name	**Description**
Young-Earth Creation	God created Earth and life on it in six literal 24-hour days several thousand years ago.
Creation Passive Gap Theory	God created the raw materials of Earth sometime in the past (maybe millions of years ago) and created life on Earth in six, literal 24-hour days several thousand years ago.
Creation Active Gap Theory	God created life, destroyed it, and created life again. Fossils are from the first Creation.
Progressive Creation	God performed several Creations at various times over long periods of Earth's history. Each Creation included more complex life forms. The days of Creation represent vast periods of time.
Theistic Evolution	God initially created life and then allowed the evolutionary process to progress on its own.
Evolution (Microevolution-Macroevolution)	As a result of lots of microevolution over billions of years, current life forms evolved from a common ancestor—the first single-celled living organism. God, if He exists, had nothing to do with it.
Punctuated Equilibrium	Evolution occurred in short, rapid bursts followed by long periods with little or no change.
Exogenesis or Panspermia	Life originated somewhere else in the Universe and was carried to Earth here in debris from space.

Scripture Spotlight

Read **Psalm 139:13–16** and identify verses that sound like Intelligent Design and complexity.

Intelligent Design

Intelligent Design is the relatively recent theory that certain features of the Universe and of living things are best explained by an intelligent cause, not an undirected process such as natural selection. Although some of the scientists who work within the Intelligent Design movement are Christians who believe the Bible, the Intelligent Design movement itself claims to rely on scientific evidence, not scripture. Although it makes no claims about biblical chronology and does not identify the Designer as God, the Intelligent Design movement has generated considerable controversy in the scientific community. Intelligent Design does not see complexity alone as evidence of a designer because complex patterns can sometimes occur by chance. Instead, it focuses on two specific kinds of complexity.

Complex Patterns

Type of Complexity	Description
Irreducible complexity	This is a single system composed of several well-matched, interacting parts that contribute to the basic function. The removal of any one of the parts makes the system unable to function. Example: a mousetrap Because an irreducibly complex system cannot be produced through one small change at a time, it is evidence of a designer. Within every cell in your body, there are thousands of tiny molecular machines that are much more complex than a mousetrap. These molecular machines are evidence of a designer.
Specified complexity	This pattern of letters and spaces is complex but not specific: *Slek owjkel silejks wwezjkl.* Because it has meaning, this sequence is an example of specified complexity: *Please pass the lemonade.* Things that do not occur because of natural laws but are both complex and specific are evidence of design. The information contained in the billions of proteins (made up of chains of amino acids in specific orders that fold to specific 3-D shapes) is evidence of design.

Whether scientists believe in evolution or Creation, they do so because of certain basic assumptions they have chosen to accept as true. While science cannot prove whether God intervened in history, scientific evidence can be evaluated with each worldview in mind. In the next lesson, you will study scientific evidence that is consistent with the biblical story of Creation.

Concept Check Assess/Reflect

Summary: What are theories of origins? Although there are variations of each, two basic ideas about our origins exist: Creation and evolution. Creationists believe in Creation by God. Believers in Intelligent Design recognize that some things are best explained by a designer, but they do not attempt to identify the designer. Evolution claims that life arose spontaneously from nonliving matter and that all present life forms descended from a single-celled common ancestor. Microevolution involves small changes that can be documented, and all scientists agree that it happens. Macroevolution involves much larger changes, and not all scientists agree that the evidence supports it.

1. What evidence supports the hypothesis of exogenesis? Do you find it convincing?
2. Summarize Redi's evidence that helped lead to the rejection of the hypothesis of spontaneous generation.
3. Describe fixity of species. How was this idea disproved?
4. Think of the wide variety of dogs in the world. How did this variation come about? How does it compare to natural selection?
5. Compare microevolution with macroevolution.

Lesson 2

Objectives

- Assess the evidence that scientists use to support theories of origins.
- Compare and contrast creationist and naturalistic interpretations of observable data.
- Define vestigial structure.
- Compare homologous structures with analogous structures.

Vocabulary

vestigial structure
homologous structures
analogous structures
conservation of design

Essential Question

What Does the Evidence Show?

With so many different ideas about the origins of life on Earth, how is it possible to know whether any of them are correct? If origins falls into the category of historical science, what kind of data is available on which to base conclusions? Scientists from both worldviews are interested in how scientific evidence matches up with their theories of origins. This lesson will describe several lines of observable data—embryology, vestigial organs, genetic similarities, the fossil record, homology and analogy, and design—and how evolutionists and creationists interpret them.

Embryology Explain

In the late 1800s, Ernst Haeckel drew pictures that compared the embryonic development of lizards, chicks, rabbits, and humans. What similarities do you see in the embryos? What differences are evident? After observing that embryos of these vertebrates were much more similar than the adults, he theorized that during its development in the womb, the embryo actually passed through the different stages of its evolutionary history. Although it is true that embryos of many vertebrates are similar, Haeckel altered his drawings to support his theory, which did not hold up to modern genetic evidence.

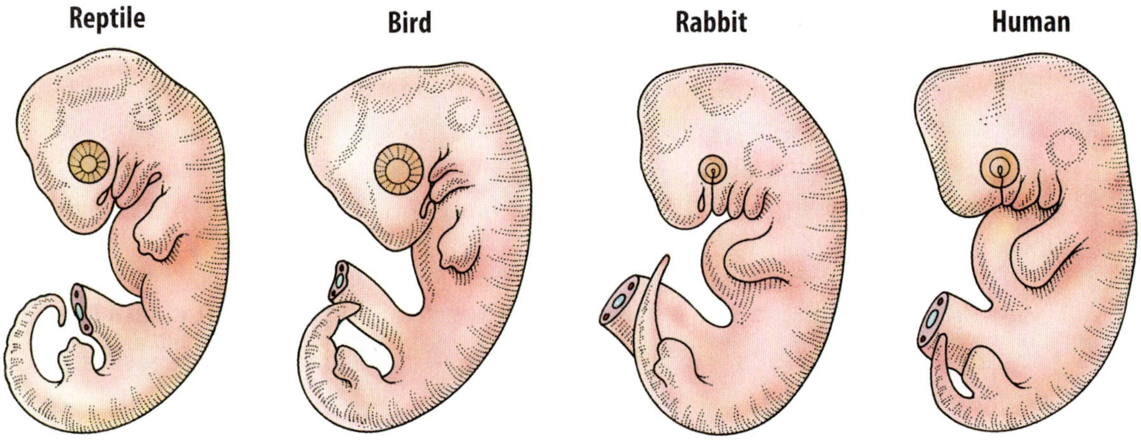

In early stages of development, vertebrate animals may be very similar.

Observable Data — Embryology	
Embryos of different animals appear to be similar, although not as similar as originally portrayed by Haeckel's drawings shown on the previous page.	
Creationist Interpretation	**Evolutionist Interpretation**
Because the basic structure of vertebrates is similar, similarities in organisms are visible in the early stages of development. The Designer created an efficient plan to make basic structures first and to then specialize as the process continued.	An embryo resembles the embryonic stages of its ancestors because of evolutionary descent from a common ancestor.

Vestigial Structures

According to the theory of macroevolution, as species evolve new structures, certain older structures that are no longer necessary will slowly disappear. Until a structure disappears completely, a remnant, or vestige, of it may remain. Structures that do not appear to have any use are often presumed to be "leftovers" and are called **vestigial structures**. In 1895, a long list of vestigial organs in mammals was compiled that was considered convincing evidence for macroevolution. More than 80 structures were on that list, including the thyroid, pituitary glands, and the middle ear. Scientists now know that these structures have very specific and important functions. As scientists continue to gather evidence, the list continues to shrink.

Some organs, however, seem to be truly vestigial. Some salamanders that live in dark caves have sightless eyes. Populations of birds found living on islands without predators sometimes have small wings and have lost the ability to fly. It is important to notice that these changes are examples of microevolution, not macroevolution. Also, instead of evolving new structures, these animals lost certain traits.

Observable Data — Vestigial Organs	
Some body parts, called vestigial organs, appear to have no function. Examples include the human coccyx (tailbone), small leg bones in some snakes, eyes in blind amphibians, and wings in flightless birds.	
Creationist Interpretation	**Evolutionist Interpretation**
Just because we don't understand a structure's purpose doesn't mean it doesn't have one. Scientists have discovered the purpose of many structures that used to be considered vestigial. Some vestigial structures may be the result of microevolution that has happened since Creation.	Vestigial organs are remains of an animal's evolutionary history.

In spite of many genetic similarities, humans and chimps are very different.

What are some physical differences between you and a chimpanzee?

Genetic Similarities **Explain**

Has anyone ever told you that the human and chimpanzee genomes are 98 percent identical? When the chimpanzee genome was actually sequenced, the ratio dropped to 96 percent. Although the published numbers may vary, understanding what these numbers mean can be tricky. Clearly, the human and chimpanzee genomes are more like one another than either is to the banana genome (about half the genes in bananas are similar to human genes). However, there are some very important differences. In particular, the *Y*, or male sex chromosome, is very different. Although the proteins are quite similar, the regulatory genes that control how these proteins are used are different. This is like very different houses being put together with the same types of nails and lumber. Can you think of other examples that illustrate this concept?

Observable Data — Genetic Similarities	
Humans, chimpanzees, and bananas have many similar, but not necessarily identical, genes.	
Creationist Interpretation	**Evolutionist Interpretation**
When the Creator designed different organisms, He used similar parts to build them, just as a builder may use similar bricks to design different buildings. Bananas have relatively few genes in common with humans and chimpanzees, because bananas differ more from either of them than they do from each other. Where necessary, God made different genes for the unique needs of bananas, chimps, and humans.	Similar genes are inherited from a common ancestor. The common ancestor of bananas and humans existed long before the common ancestor of humans and chimpanzees. Therefore, humans and chimps share fewer genes with bananas than they do with each other. Unique genes in each organism evolved by mutation and natural selection. Because this requires random unguided changes, the process takes long periods of time.

Fossil Record

The fossil record is often used to support the theory of evolution. Evidence about the fossil record will be presented in Chapter 10, but a short summary is included in the chart below.

Observable Data — Fossil Record	
Beginning with a sudden profusion of diverse life forms, the sequence of fossils (from bottom to top of the geologic column) includes invertebrates, fish, amphibians, reptiles, mammals, and humans. New organisms appear abruptly with relatively few examples of gradual "transitions."	
Creationist Interpretation	**Evolutionist Interpretation**
The fossil record is a record of the sequence in which the fossils would have been buried by the Flood and later geologic activity.	The fossil record is a record of the evolution of life forms from common ancestors.

Homology and Analogy Explain

Scientists have noticed that many animals share similar body structures. For example, the human arm and hand, a whale's flipper, a bat's wing, and a lizard's leg have roughly the same sort of bones in the same order. But despite how similar these appendages appear, their functions are very different. Humans can grasp things, whales can swim, birds can fly, and lizards can run using this same basic arrangement of bones. Structures that are similar in construction but that have different functions are called **homologous structures**.

Scripture Spotlight

What types of wings are mentioned in these verses: **Exodus 19:4**, **Psalm 55:6**, and **Matthew 23:37**?

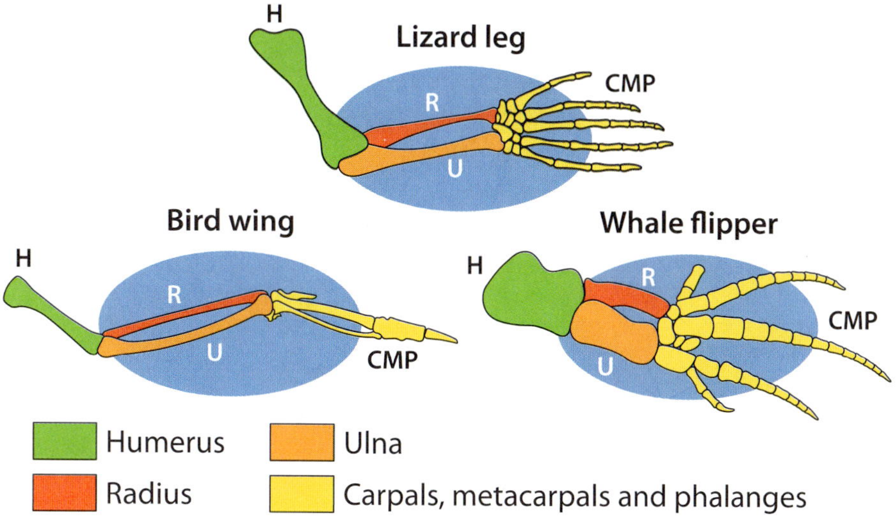

Similar structures that have different functions are called homologous structures.

Which bones are homologous in these three animals?

Notice also how different the structure is in the wings of the butterfly, bird, bat, and pterosaur on the next page. Yet, each of these different structures still results in the ability to fly. Different structures that have the same function are called **analogous structures**. What other homologous and analogous structures can you describe in nature?

Lesson Activity

Think about the many kinds of winged animals. With a partner, make a list of at least 10 different winged animals. Are all wings the same? Comparing the wings of animals can reveal more differences than you may have considered. Choose two different types of winged animals from your list and research their structures. Organize your research into a visual presentation. Include photographs, diagrams, or your own drawings and models in your presentation. Be sure to add captions and labels as needed. Then create a Venn diagram to record the differences between the wings of the two types of winged animals you researched. Share your visuals and research with the rest of the class.

What differences can you identify relating to the structure and function of wings?

Observable Data — Homology and Analogy

Homology

	Human hand/arm	Whale flipper	Bat wing	Dog leg
Same structure	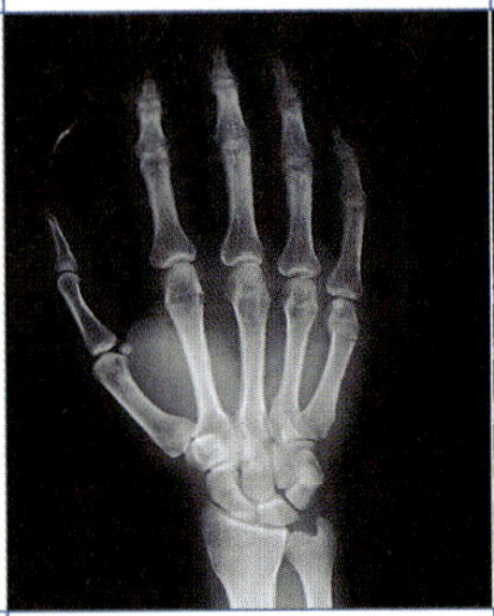	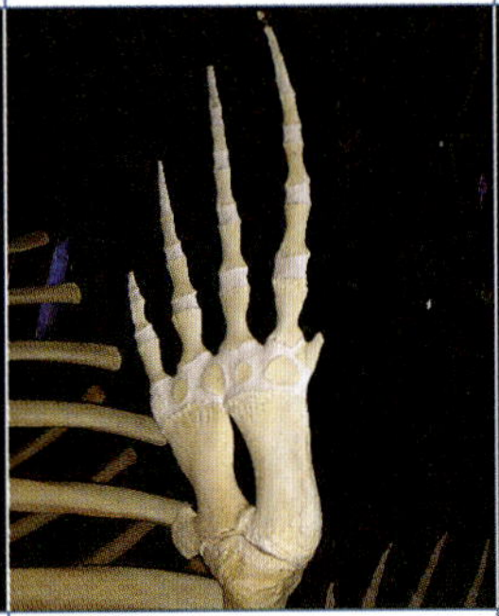		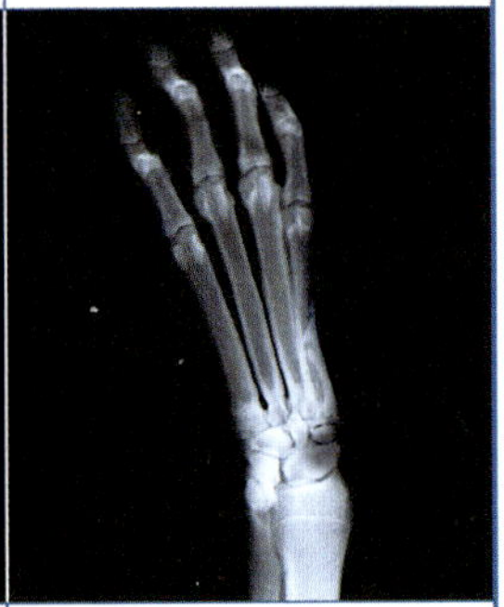
Different functions	Fine manipulations	Swimming	Flying	Running fast

Analogy

	Butterfly wing	Bird wing	Bat wing	Pterosaur wing
Different structure				
Same functions	No bones at all	Finger bones reduced; feathers are the main flight surface	A thin membrane stretches between the digits	Fourth finger alone supports the wing membrane

Creationist Interpretation

An intelligent Creator designed a flexible general plan, which could be altered for different animals. It might be like a computer programmer writing a basic program that could be modified and reused efficiently in other more complex programs. Sometimes this is called the **conservation of design**.

Evolutionist Interpretation

Homology

Similar structures are said to suggest common ancestry. In other words, these traits were passed down to these animals from a common ancestor.

Analogy

Even though animals with analogous structures did have common ancestors at some point in the past, these structures are thought to have evolved sometime since the last common ancestor. That means those traits would have to have evolved more than once, each in a separate evolutionary line.

Structured Inquiry

Discover

Science Journal

Record your work for this inquiry. Your teacher may also assign the related Guided Inquiry.

When You've Seen One, You've Seen Them All

Does similar design mean a common ancestry?

Materials

- a variety of cars

Procedure

1. As a class, visit your school parking lot. Select three cars that are obviously different.
2. **Observe** the cars and identify these common characteristics: headlights, taillights, and side-view mirrors.
3. Then choose two more characteristics that the cars have in common.
4. Create a five-column chart and **record** one common characteristic at the top of each column. Describe or draw the design of the five characteristics for each of the cars. Record your observations on the chart.

Analyze Results

Compare the common characteristics of the cars. Based on your observations, how does the design of the common characteristics of the cars differ? What are the similarities of the design?

Create Explanations

1. Does similar design mean a common ancestry?
2. Why do different makes of cars share the same features?
3. How are similarities and differences in cars similar to what is found in nature?
4. How does this activity relate to the conservation of design?

Design **Explain**

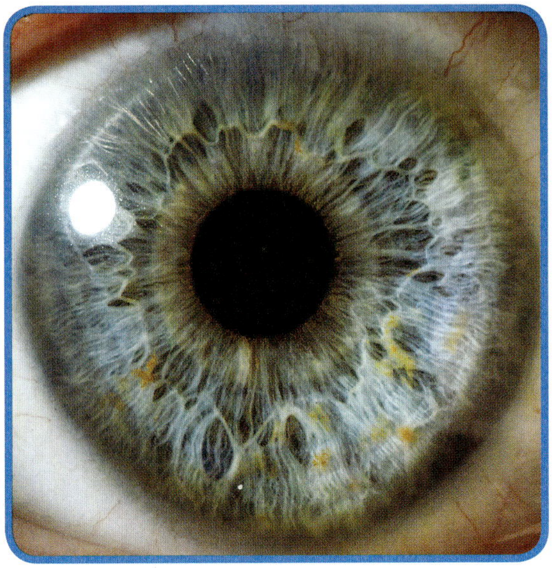

If you were walking in the woods and found a watch on the ground, where would you assume it came from? Back in the eighteenth century, a British philosopher and clergyman named William Paley used this question to illustrate the idea that when we see evidence of planning, craftsmanship, beauty, and usefulness, we recognize that those things come from a designer. Since that time, scientists have continued to gather data about the stunning precision of universal constants, the staggering complexity of the molecular machines inside our bodies, and the fascinating processes found in the natural world.

Although most people would immediately recognize design in this data, scientists wanted to find a way to be more scientifically objective about the process of identifying design. William Dembski developed something called the *explanatory filter* to help determine whether direct intervention by a designer should be inferred. If it can be determined that the object could have existed in more than one format, is complex enough that it could not have happened by chance, and is not random but specific, then design can be inferred.

Observable Data — Design		
We live in a finely tuned Universe with dozens of factors, such as gravity, that have to be exactly the way they are for life to exist.	Humans and other animals are made of complex systems, which are made of organs, which are made of tissues, which are made of cells, which are made of organelles, which have hundreds of tiny molecular machines, which are themselves irreducibly complex and work together to accomplish cell processes.	Caterpillars that crawl, eat leaves, and spin silk cocoons but that cannot reproduce melt into a jellylike substance before reassembling into butterflies that fly, drink nectar, and can reproduce caterpillars.

Creationist Interpretation	Evolutionist Interpretation
These are examples of organisms designed, created, and maintained by God.	These organisms arose from random evolutionary processes, directed by environmental constraints, but they have the appearance of conscious design.

Evaluating the Evidence

Remember that two parts of the scientific process are gathering and interpreting data. As more data is gathered, scientists evaluate it and make adjustments to their theories. As you have seen, scientists with different worldviews can interpret the same data in completely different ways. Both creationists and evolutionists show strong commitments to their worldviews and the assumptions that go with them. As you evaluate the evidence presented here, keep the following points in mind:

- Each theory is ultimately accepted by faith.
- Each theory depends on assumptions that cannot be proved.
- Each theory is supported by scientific evidence.
- Each theory is unable to explain all the scientific evidence.

Moths' eyes do not reflect light. Instead, the light is absorbed and deflected. This antiglare quality helps keep the moth safe from predators.

Concept Check Assess/Reflect

Summary: What does the evidence show? Evolutionists and creationists interpret the same observable data in different ways. Evidence used frequently to debate life's origins includes embryological similarities, vestigial organs, genome similarities, the fossil record, homologous and analogous structures, and design. Ultimately, neither evolutionism nor creationism is a provable theory. Both are based on scientific evidence, but not all is explainable. They are accepted by faith.

1. In your opinion, which is more useful to biologists when studying ancestry: homologous structures or analogous structures? Explain.
2. Would scientists classify the wings of a dragonfly and the wings of a pigeon as homologous or analogous structures? Explain.
3. How might an evolutionist and a creationist describe the evidence that wings on flightless birds give for the origins of life?
4. How would an evolutionist say similarities in genetic make-up suggest that humans and chimpanzees evolved from a common ancestor?
5. Give an example and explanation of something that appears to have been created with design in mind, rather than falling together randomly.

Lesson 3

Essential Question

What Is Phylogenetics?

Objectives

- Describe characteristics used to classify organisms.
- Explain the process of phylogenetic analysis.
- Interpret phylogenetic trees, or cladograms.

Vocabulary

systematics
phylogenetic analysis
phylogenetic tree
cladogram
parsimony
cladistics

When you sort your room or organize this class, does your view of where things came from affect how you classify the things? How does the scientist's worldview change the classification of life? Does the assumption of where something comes from change the classification of science? Can you think of an example?

Looking for evidence about the assumption that organisms evolved from common ancestors is the foundation for the work of many taxonomists. You have already learned quite a bit about classification. What do you recall about the reasons for classifying organisms, the scientists who began this work, and the categories into which organisms are classified?

How Classification Has Changed Explain

Classifying organisms today is different in at least two important ways from how it was done back in the 1700s when Carolus Linnaeus began his well-known classification work. The first has to do with the characteristics used to group the organisms.

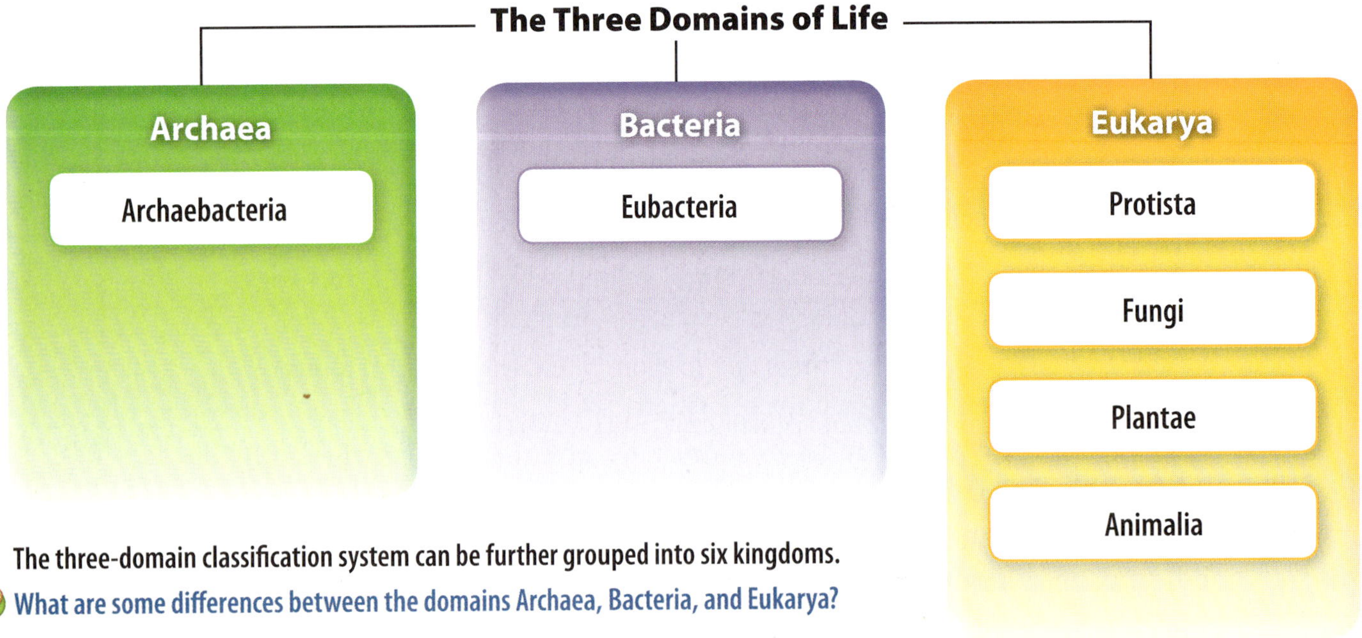

The three-domain classification system can be further grouped into six kingdoms.

What are some differences between the domains Archaea, Bacteria, and Eukarya?

In the early years of classification, organisms were placed in groups based on obvious physical characteristics (such as wings, backbones, or fur) and body processes (such as laying eggs, nursing their young, or keeping warm in different ways). As new technologies emerged, powerful microscopes and other tools enabled scientists to look at an organism's cell structure and DNA. Discoveries using this molecular analysis led to the establishment of the three domains above the kingdom level.

Scientists use the term **systematics** to refer to the classification system in use today. It uses all the evidence known about organisms—the obvious physical characteristics and body processes that have been known for centuries, as well as the new molecular data discovered more recently—to classify organisms.

The second significant way that classifying organisms has changed over the last several decades involves the assumptions accepted by taxonomists. Remember that an assumption is a statement that is assumed to be true even though it cannot be proved. Linnaeus accepted the truth of the Creation account in the Bible and attempted to classify organisms according to his belief that God had created basic "kinds" in the beginning. Today, the assumption that organisms evolved from common ancestors is the foundation for the work of many taxonomists.

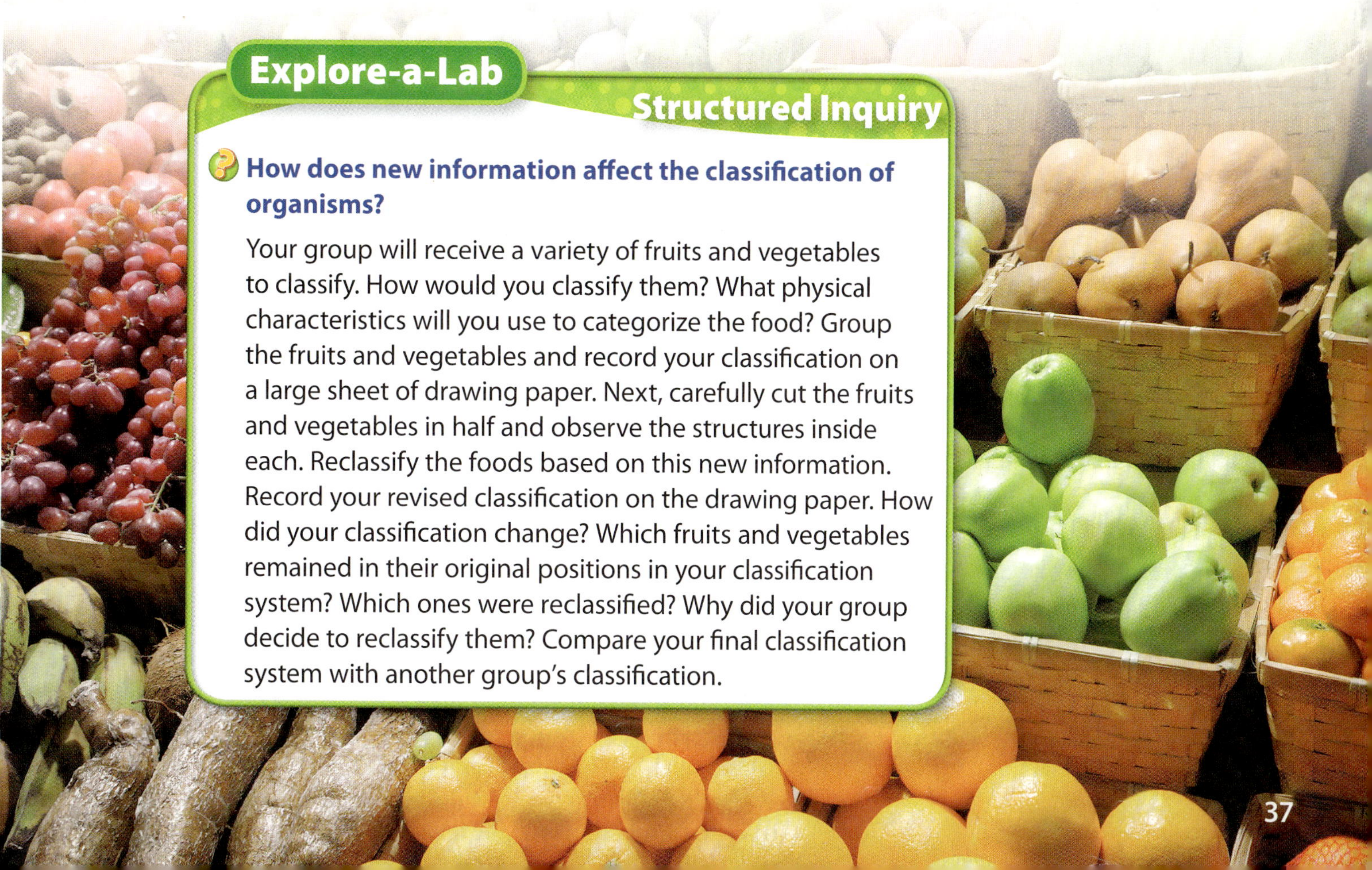

Explore-a-Lab

Structured Inquiry

How does new information affect the classification of organisms?

Your group will receive a variety of fruits and vegetables to classify. How would you classify them? What physical characteristics will you use to categorize the food? Group the fruits and vegetables and record your classification on a large sheet of drawing paper. Next, carefully cut the fruits and vegetables in half and observe the structures inside each. Reclassify the foods based on this new information. Record your revised classification on the drawing paper. How did your classification change? Which fruits and vegetables remained in their original positions in your classification system? Which ones were reclassified? Why did your group decide to reclassify them? Compare your final classification system with another group's classification.

Structured Inquiry

Discover

Record your work for this inquiry. Your teacher may also assign the related Guided Inquiry.

Science Journal

Using a Dichotomous Key

How can dichotomous keys help scientists identify organisms?

Procedure

1. Select a legume from the sample. **Observe** the bean and use the dichotomous key in your ***Science Journal*** to determine the name of the bean. If you cannot name the bean using the key, set it aside for later consideration.
2. Repeat the procedure until you have **classified** all the beans in your sample.
3. If there are any beans that you set aside because you could not classify them, revise the key so you can use it for classification.

Materials

- a sample from a bag of bean soup mix

Analyze Results

How does the key use paired comparisons to help you find the name of an organism (bean)?

Create Explanations

1. How can dichotomous keys help scientists identify organisms?
2. What characteristics of beans were used to make this key?
3. Could other characteristics have been used? Explain your answer.
4. What changes did you make to the key to help you identify the unknown beans?

Phylogenetic Analysis Explain

Phylogenetic analysis is the process of grouping organisms based on multiple shared characteristics (which scientists refer to as characters) or traits that were not shared with organisms thought to have been their ancestors. It is based on the assumption that evolution has occurred. The results of this process are displayed in graphics called **phylogenetic trees**.

Study the phylogenetic tree below. Each node, labeled A–H, represents the presence of a new trait that the presumed ancestors did not have. For example, node B represents the presence of jaws. No animal to the left of the perch on the diagram has a jaw, but every animal to the right of the perch has a jaw.

So you can look at how close different organisms are to each other on these trees and draw conclusions about how similar they are to each other. On this tree, scientists would infer that the salamander shares more characters with the lizard than with the chimpanzee. From an evolutionary perspective, it would be inferred that the shared characters result from a shared evolutionary history. From a creationist perspective, the tree simply shows which organisms are designed to be more similar and which traits are shared by multiple creatures.

Phylogenetic Tree

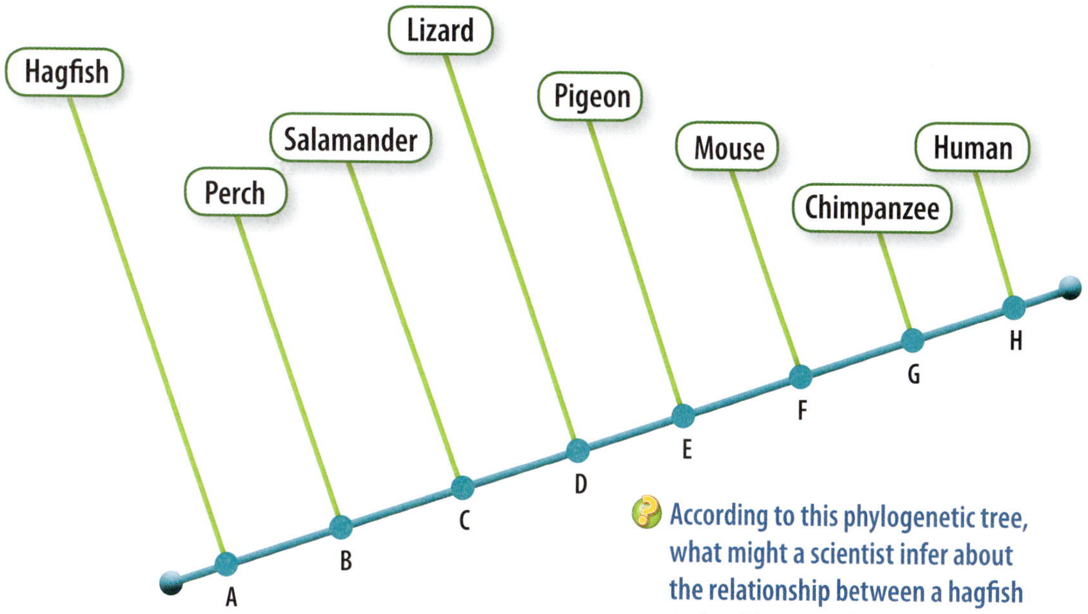

According to this phylogenetic tree, what might a scientist infer about the relationship between a hagfish and a chimpanzee?

As you learn about how the process of phylogenetic analysis works, keep in mind that scientists who are evolutionists begin with the assumption that organisms evolved. Their purpose is to try to figure out which organisms are most closely related to other ones based on the characteristics they share. Their study may include living organisms as well as fossils.

In the first step of this process, scientists collect data about specific traits that similar species of organisms have. Examples of the kinds of data they collect include visible things such as bone structure, as well as molecular data about cell structure and DNA. Sometimes they gather very detailed data, such as how many cusps there are on a tooth, tooth size (which may vary by only a millimeter or less), or the order of the base pairs in a gene. Often more than 100 different characteristics may be gathered for each organism.

Physical characteristics are one type of evidence used to place organisms within a phylogenetic tree.

How are these mammals' teeth similar? How are they different? Would a scientist place these mammals on a phylogenetic tree based solely on teeth structure? Explain.

Once scientists have gathered the data, it is entered into a computer database. A computer program analyzes the data, looks for similarities, and generates phylogenetic trees, which are also called **cladograms**. A phylogenetic tree is based on presumed relatedness that supports evolutionary progressive development, an idea that is inconsistent with a biblical worldview. The phylogenetic trees are inferred from the similarities shared by various organisms. It is easy to imagine how, depending on which characteristics are entered into the computer, different relationships might be suggested.

To illustrate this, let A, B, and C stand for three imaginary organisms. Imagine a scientist gathering data about dozens of different characteristics, or traits, of all three creatures. When the computer analyzes the data, it suggests that according to several of the traits, A is more closely related to B than to C.

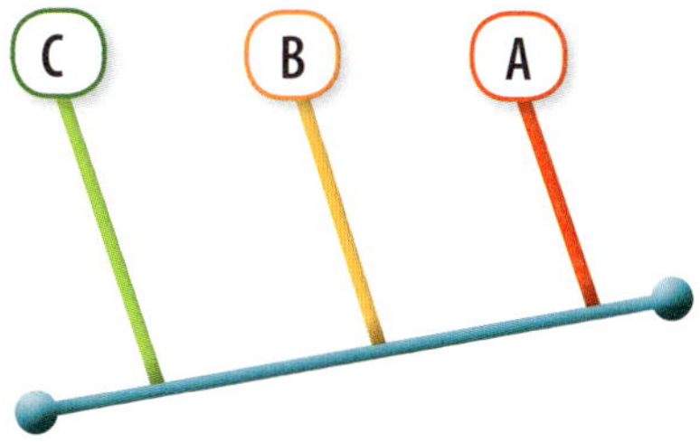

Data from several other characteristics suggests that C is more closely related to B than to A.

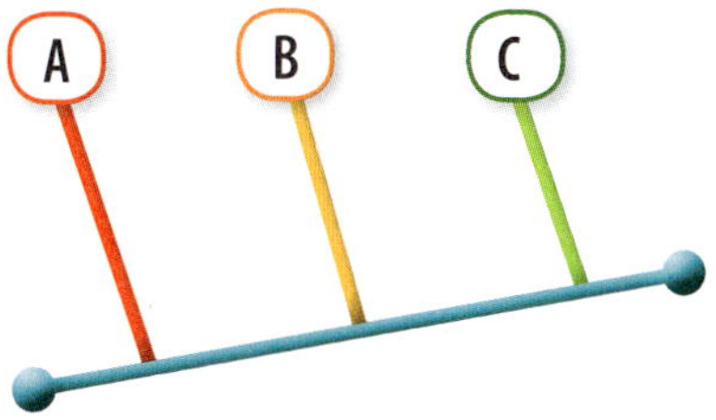

Data from still other traits might suggest that C is more closely related to A than to B.

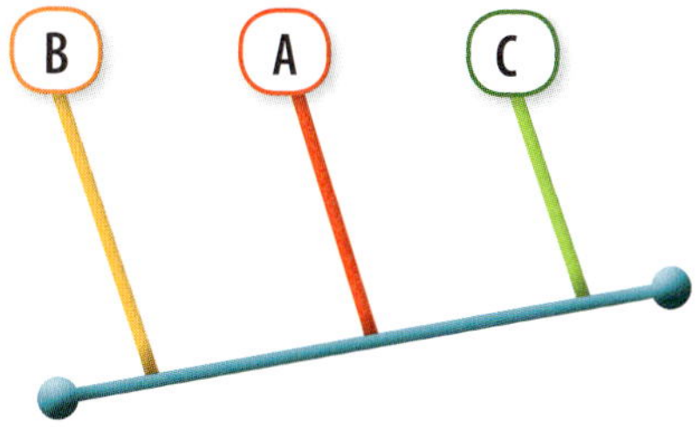

Scientists actually do encounter such conflicting data when they conduct a phylogenetic analysis. In fact, every phylogenetic tree contains some data that appears to be out of order. Scientists then have to decide which tree to choose. They decide based on something called **parsimony**, which means that they select the simplest option, or the one that requires the fewest evolutionary changes. What do you think might account for differences in data?

Scripture Spotlight

In **Matthew 12:46–50**, Jesus describes how people can be as closely related to Him as His own mother and brothers were. What characteristic makes us like family?

Look at this example of a cladogram. Groups of animals that share similar characteristics are called *clades*. Notice how the clades are nested within larger clades.

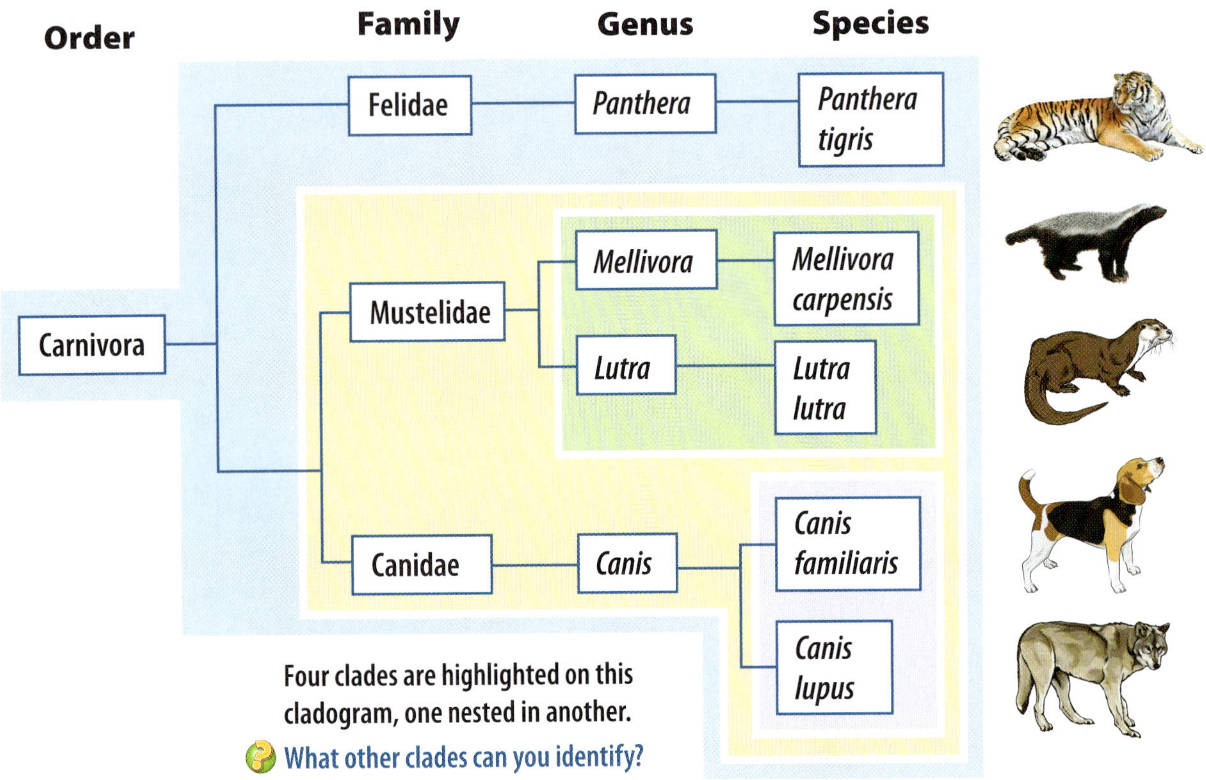

Four clades are highlighted on this cladogram, one nested in another.

What other clades can you identify?

By doing phylogenetic analyses on many different clades, scientists create trees or bushes that they think best approximate evolutionary history. As you might guess, not all scientists agree with each other about how the bush should look, and changes continue to be made as they gather more data.

Explore-a-Lab

Structured Inquiry

How many ways can the same organisms be grouped into different cladograms?

Your teacher will provide your group with an assortment of silverware and utensils. In your group, discuss the characteristics of the items and sort the items into two groups. Draw the beginnings of a cladogram on a large sheet of paper and record these two groups. Then sort the two groups into two more groups each. Record these new groups on the cladogram. Observe other groups' cladograms. Compare the different ways the items were grouped. How might a scientist interpret your cladogram?

In spite of the fact that phylogenetic analysis is based on the assumption that evolution has occurred, it can still be a useful tool for creationists as well as evolutionists. A cladogram is a valid way to represent the similarities and differences between different organisms, but it is important to remember that there are limits to what questions it can answer. Consider these two questions:

1. Which animals are most closely related to each other?
2. Did organisms evolve?

Cladistics, which is another name for the process of phylogenetic analysis, can suggest answers to Question 1, but it cannot answer Question 2. It assumes that evolution happened and tries to figure out the evolutionary history of organisms based on similar characteristics.

We examined biological evidence related to Question 2 in the last lesson. We looked at observable data and compared different interpretations of that data. Let us do the same thing with cladistics. The information gathered about organisms and entered into the computer is data. The suggestions of evolutionary relationships generated by the computer and selected by the scientists are interpretations of that data based on evolutionary assumptions. Scientists who believe in the biblical story of Creation might use the same data to draw conclusions about the similarities of various organisms created by God.

Concept Check Assess/Reflect

Summary: What is phylogenetics? Organisms are classified based on body processes, obvious physical characteristics, cellular structure, and DNA. Organisms can also be classified based on assumptions that they evolved from common ancestors. This process of grouping is called phylogenetic analysis, or cladistics. The results of the analysis are displayed using phylogenetic trees, or cladograms. Often the cladograms show conflicting data, and scientists will choose the simplest option.

1. How have advances in technology changed the way organisms are classified?
2. Why do scientists use parsimony to select a phylogenetic tree?
3. What are some limits in the use of phylogenetic analysis?

People in Science

Extend

Get to Know
Leonard Brand

Leonard Brand is a professor of biology and paleontology at Loma Linda University in California. He was educated at Cornell University, earning his PhD in ecology and evolutionary biology. As a biologist, one of Brand's areas of expertise is in taphonomy. Taphonomy is the study of organisms after they die. The events that take place after an organism dies, such as decomposition, burial, and fossilization, are all part of taphonomy.

As a paleontologist, Brand applies his understanding of taphonomy to the fossilized remains of organisms that lived long ago. He researches the processes that cause different types of fossils to form. His research often includes an analysis of the geological processes involved in fossil formation, such as weathering and erosion.

Brand has done extensive research into the taphonomy of vertebrates. He has studied whale fossils in Peru, which are preserved in the sedimentary rock of the Pisco Formation. He has studied fossilized turtles in Wyoming, where deposition in the Bridger Formation has preserved many turtle remains in sedimentary rock. Throughout his research, Brand aims to show that science and religion can work together.

Called to Serve

Leonard Brand has written extensively on ways that science and religion can work together. He believes that scientific theories are important "stepping-stones in our search for truth."

Concept Check

1. How can studying geological processes help us understand fossil formation?
2. Why is it important for Brand to think about ways that science and religion can work together?

The Geoscience Research Institute

Does science give us the answers to philosophical questions about our origin and destiny and about our purpose for living? A significant conflict of science and religion is between the theory of evolution with its billions of years for the progressive development of life and the biblical account of the Creation of life by God in six literal days a few thousand years ago.

The **Geoscience Research Institute (GRI)** is an official research institute of the Seventh-day Adventist Church. The GRI was founded in 1958 to advise the Seventh-day Adventist Church on issues relating science to faith. The Institute uses both science and revelation to study the question of origins because it considers that both sources of information can contribute to our understanding of Earth's history. The Institute serves the Seventh-day Adventist Church in two major areas, research and education.

Raúl Esperante, PhD, is a scientist engaged in paleontology research for the GRI in Peru and Spain. He was part of a team that studied a 370-acre area in Peru containing 346 fossilized whales. The nearly perfect preservation of these whales indicates that the whales had to have been buried very quickly—not over millions of years.

What caused so many whales to die in one place? What caused the rapid burial? Esperante and his colleagues theorize about what may have happened. One wonders if this may be more easily explained by a global flood that once covered Earth, and the rapid geological processes after that flood. Could these unusual events possibly explain how 346 whales were buried so rapidly and completely in Peru?

Science and technology are successful methods of applying knowledge to improve our understanding of how we do things and how things work. They have improved our standard of living — from better shelters, time-saving appliances, life-saving medical advances, and new types of recreation to faster and safer methods of communication and transportation. However, our knowledge that God was active in Creation means that questions of origins should be studied in the context of the information revealed in the Bible. The GRI provides the resources to respond to questions about science and faith.

Concept Check

1. Why was the GRI established?
2. What does the GRI do to try to settle the perceived conflict between religion and science?

Chapter 1 Review Assess/Reflect

Study Guide

Lesson 1

1. Origins of life attempt to describe where the world came from and how life began. Two basic ideas about the origins of life are Creation by God and life arising through the evolutionary process. Intelligent Design is the theory that the Universe and all life are the result of a designer. Evolution is the theory that all species evolved as a result of natural selection.
2. The Miller-Urey experiment attempted to simulate the conditions of early Earth and the process by which the first life formed.
3. Darwin theorized that new species evolved as a result of natural selection.
4. Microevolution refers to small changes in a species over time. Macroevolution refers to large-scale evolutionary changes over time. While microevolution is supported by considerable evidence, macroevolution has not been shown to have taken place.

Lesson 2

1. Evidence for different theories of origins can be found by studying embryology, vestigial structures, homology and analogy, genetic similarities, design, and fossil records. Interpretation of this evidence is used to support theories of the origins of life.
2. The creationist and naturalistic interpretations of the observable data concerning the origins of life are quite thorough and extensive. Creationists interpret the fossil record as representative of the sequence of species, burial during the Flood and later geologic activity. Evolutionists interpret the fossil record as the evolution of life forms that all arose from common ancestors.
3. Vestigial structures may be the result of the loss of genetic material, or they may not even be vestigial at all. It may be that scientists have not yet discovered the function of these structures.
4. Homologous structures have similar construction but differ in function. Analogous structures have different construction but the same function.

Lesson 3

1. Scientists use the process of systematics to classify organisms. Systematics uses physical characteristics, body processes, and molecular data to classify organisms.
2. Phylogenetics classifies organisms based on shared, common characteristics that were not shared by the presumed ancestors.
3. A phylogenetic tree graphically displays the grouping of organisms based on multiple shared characteristics that are not shared with organisms thought to have been their ancestors.

Vocabulary Check

Fill in the blanks with the correct vocabulary word.

1. The idea that life is actually widespread throughout the Universe is known as __________.
2. The idea that living organisms never change is called __________.
3. Microevolutionary changes in a species can be introduced through genetic __________.
4. When Redi found no maggots forming on meat inside a closed container, his evidence refuted the idea of __________.
5. Body parts that are thought to have no function are called __________.
6. A human hand and a whale's flipper have similar construction but different functions, so they are __________ structures.

Multiple Choice

Choose the best answer.

7. Phylogenetics is used by evolutionary scientists to support which statement?
   A. Species with analogous structures are distantly related.
   B. Species with homologous structures are closely related.
   C. Species with similar genes are closely related.
   D. Species with similar genes are distantly related.
8. Whose experiments and studies involved combining electricity and nonliving compounds?
   A. Aristotle
   B. Miller and Urey
   C. Linnaeus and Redi
   D. Jean-Baptiste van Helmont
9. A paleontologist discovered a new species of snail that emerged rapidly in prehistoric times. Which evidence is most likely being studied?
   A. genetic similarity
   B. embryology
   C. fossil records
   D. homologous structures
10. What does natural selection select?
    A. genes
    B. ancestors
    C. analogous structures
    D. homologous structures

Check Point

Answer the following questions.

11. Francesco Redi had to **control variables** in his experiment regarding spontaneous generation. Describe how he controlled the variables.
12. How did Darwin **classify** the Galápagos finches?
13. There is an increasing number of wild Asian elephants being born tuskless. Although poaching elephants for their ivory has decreased, it still exists. Based on this information, what can you **infer** about the connection between poaching and the changing elephant species?
14. A scientist is studying the structure of lizard legs and crab legs. Are these homologous or analogous structures? Explain.

Chapter 2

Types of Plants

Scripture Spotlight

Although plant species are rich in diversity, there are similarities in design. Learning how God intertwined these similarities in such a diverse kingdom can help you understand Him better. You will read the following passages in this chapter.

Genesis 1:11–12 (p. 55)
Jeremiah 17:7–8 (p. 76)
Revelation 22:2 (p. 80)

The Big Idea

As in all of Creation, God created unique flora. Flora, or plants, are diverse in their form and distribution, but they can be grouped according to certain shared characteristics.

What similarities and differences do you see in the plants pictured here?

Animals share Earth with between 300,000 and 400,000 species of plants.

Inquiry Kick-Off **Engage**

Have you ever eaten a cherry and had to spit out the pit (seed)? Have you eaten sesame seeds? How would you compare the two types of seeds? In your ***Science Journal***, you will observe four different types of seeds and compare them with the plants they produce.

Science Journal

Lesson 1

Objectives

- Describe how plants are classified.
- Compare and contrast vascular and nonvascular plants.
- Compare seed-bearing with non-seed-bearing plants
- Explain alternation of generations in nonvascular and vascular plants.

Vocabulary

vascular system
xylem
phloem
nonvascular plant
gymnosperm
angiosperm
spore
sporophyte
gametophyte
gamete
zygote
alternation of generations
sori
frond
sporangia

Essential Question

How Are Plants Classified?

There are many different species of plants in the world. Some scientists' estimates put the number near 400,000! When you take a walk through your neighborhood, what plants do you see? You might see maple trees. You might also see moss. These two plants look very different from one another, and yet they are both plants in the kingdom Plantae. What characteristics do maple trees and moss share? How are these plants different from one another?

You may recall Plantae is one kingdom in the domain Eukarya. Other organisms in this domain are protists (kingdom Protista), fungi (kingdom Fungi), and animals (kingdom Animalia). What makes plants different from the other organisms?

The Kingdom Plantae includes a diverse variety of organisms, such as moss and a maple tree.

Classifying Plants Explain

As you learned earlier in your study of science, the Kingdom Plantae is made up of organisms that are multicellular and produce their own food through photosynthesis. What other characteristics of plants do you remember? Recall that Carolus Linnaeus devised an orderly system for scientists to use to study God's Creation. This system groups plants together based on their similarities, but it also gives scientists a way to single out specific plants. Why is it important to be able to classify different plants?

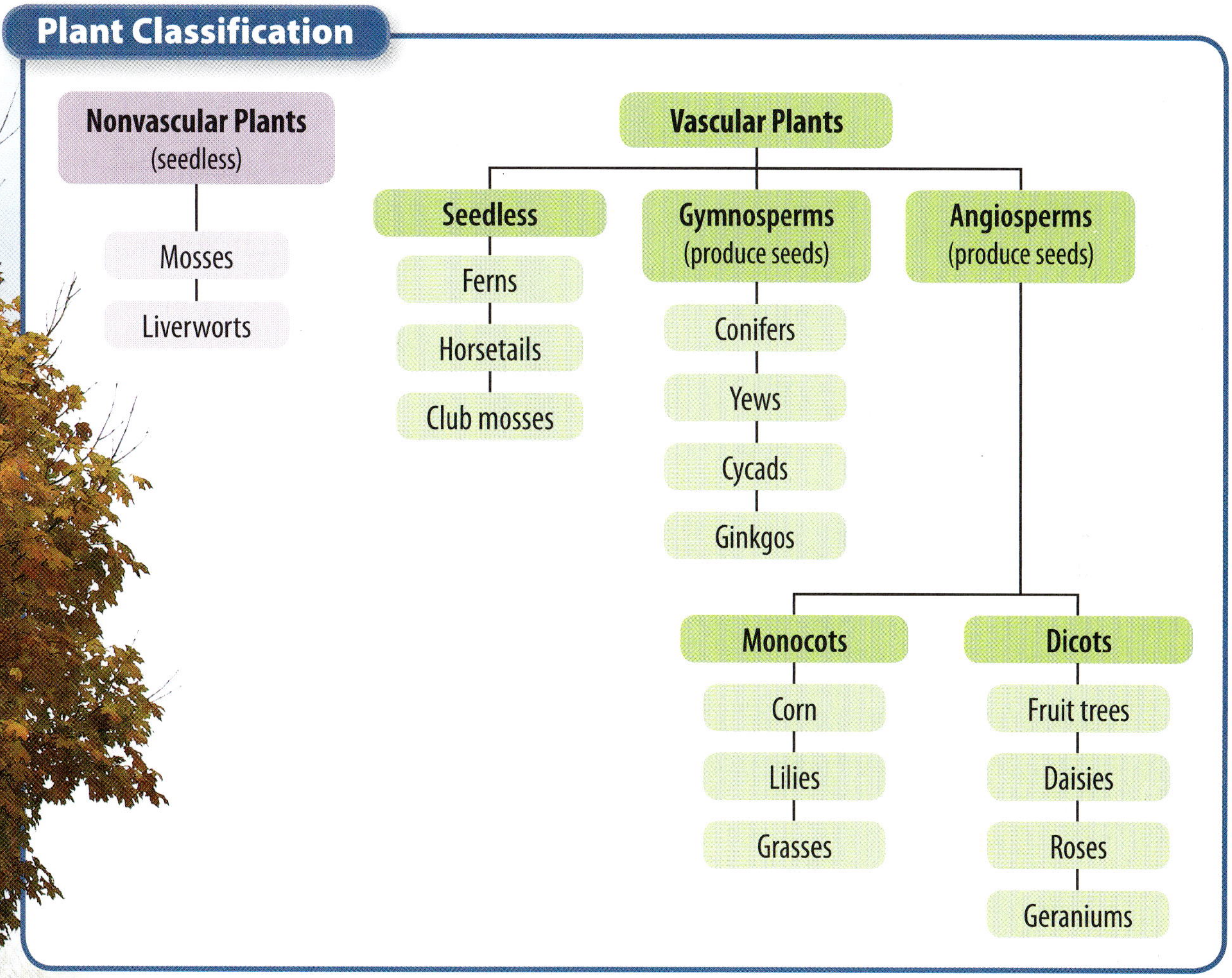

Plants are often classified based on two main characteristics. Botanists, scientists who study plants, group plants based on whether they have vascular tissue. You may remember from earlier classes that vascular tissue allows for the transport of materials within the plant. Plants are also classified by whether they have seeds. Why do you think these characteristics are used to classify plants? What other characteristics would you use to classify plants?

Structured Inquiry

Discover

Record your work for this inquiry. Your teacher may also assign the related Guided Inquiry.

Science Journal

Water in a Vascular Plant

How is the movement of water in a vascular plant affected by salt?

SAFETY: Always use caution while handling a scalpel.

Materials

- 4 100-mL beakers
- 25-mL graduated cylinder
- 3 g salt
- stirring rod
- masking tape
- 4 celery stalks (as close to the same size as possible) with leaves
- red food coloring
- distilled water
- scalpel
- medicine dropper
- paper towel
- metric ruler

Procedure

1. **Measure** 25 mL of distilled water in each of four beakers. Add 3 g of salt to two of the beakers of water. Stir to dissolve the salt.
2. Add four drops of red food coloring to each beaker. Stir to distribute color evenly.
3. Cut the bottom ends off the stalks of celery. Use the metric ruler to make sure the stalks are the same length.
4. Use a balance to find the mass of each stalk. **Record** the measurements in your ***Science Journal***.
5. Place a celery stalk, cut end down, into each beaker. Allow the celery stalks to sit overnight.
6. After 24 hours, **observe** the celery stalks. Measure the mass of each stalk and record these measurements in your ***Science Journal***.

Analyze Results

Create a bar **graph** comparing the mass of the stalks before and after the experiment. Is celery a vascular plant? What observations support your conclusion?

Create Explanations

1. How is the movement of water in a vascular plant affected by salt?
2. How does osmosis explain the differences you observed?
3. What effect would less salt have on the celery stalks? Give a reason to justify your answer.

Vascular versus Nonvascular

Like all living organisms, plants need nutrients to live. God created plants with two distinct ways of transporting the nutrients needed to sustain life. The vascular plants have a **vascular system** that moves water and nutrients from one part of the plant to another by the processes of diffusion and osmosis. What kind of plants do you think might have the largest vascular system? Why might some plants need a larger vascular system than others?

A vascular plant moves nutrients through a series of tubes called xylem and phloem. **Xylem** transports water and inorganic nutrients, such as nitrogen and minerals, from the roots to the rest of the plant. **Phloem** transports the food the plant makes to all the cells in the plant, including its roots. Do you think the xylem or the phloem does more work in the plant? Why?

These vascular tissues have functions other than transport. They help to support the plant. Why is this especially important for land plants? Because these plants have a vascular system, they can grow to great heights—for example, there are North American redwoods that are more than 100 m tall!

Other plants, called **nonvascular plants**, lack vascular tissue. Both vascular and nonvascular plants rely on diffusion and osmosis to absorb water and nutrients. Vascular plant use phloem and xylem to move materials from roots to stems to leaves and back. Nonvascular plants are unable to do this. How does the lack of vascular tissue affect these plants? Where would you expect these plants to grow most commonly? Why?

Can you find the xylem and phloem in the cross-section of the red cherry tree stem?

Explore-a-Lab

Structured Inquiry

How does water move through a nonvascular plant?

Obtain a 4-cm piece of dry sphagnum moss, 20 mL of colored tap water, a sheet of colored construction paper, a sheet of white drawing paper, a large plastic spoon, a stopwatch, and a hand lens. Observe the dry moss with the hand lens. How would you describe it? Draw the moss and write a description on a sheet of drawing paper. Place the spoon on the construction paper. Fill the spoon half full with water. Be careful not to splash water onto the paper. Set the moss so that half of it is in the water and the other half hangs above the colored construction paper. Start the stopwatch. Observe the moss with the hand lens. Watch for signs that the water is moving through the plant. When you see water drip from the moss onto the construction paper, stop the stopwatch. Record the time it took for the water to travel through the moss. Compare this time with the time from other groups in your class. Were they similar? What is the process of moving water through a nonvascular plant called?

These nonvascular plants are called liverworts. These tiny plants, usually less than 2 cm tall, are found in seepage areas in cool, shady areas.

Why do you think liverworts do not grow taller?

The processes of diffusion and osmosis move water and nutrients differently than xylem and phloem do. You learned that *diffusion* is the natural tendency of molecules to move from an area of greater concentration to an area of lesser concentration. If water is the substance being diffused, and it is being diffused across a semipermeable membrane such as a plant cell, the process is called *osmosis*. In osmosis, water can move across the membrane, but other substances cannot. Water moves both ways across the membrane until the concentration of the other substance or substances (salt, other minerals, etc.) is about the same on both sides of the membrane. Diffusion is the method by which gases are exchanged during photosynthesis. Osmosis is the process at work when roots absorb water and when the plant moves water upward to the rest of its structures.

Nonvascular plants tend to live in damp environments. How would living in a damp environment help a nonvascular plant?

Nonvascular plants do not have phloem or xylem and are unable to transport water and nutrients great distances like the vascular plants can. Because nonvascular plants lack these transport tubes, they also lack a strong structure to hold the plant upright and allow it to grow tall. Examples of nonvascular plants include mosses and liverworts, which are only 2 to 3 cm tall. You most likely will find nonvascular plants growing in groups. How do you think these plants benefit from growing in groups?

Seed-Bearing versus Non-Seed-Bearing

As you have seen, plants are classified by whether they are vascular or nonvascular. They are also classified according to their method of reproduction. **Gymnosperms** are vascular plants—most of which are trees—that produce seeds that are not enclosed in fleshy fruits. Instead, the seeds of most (but not all) gymnosperms are between the scales of woody cones. Structures that, at pollination, become eggs in these female cones are fertilized by the sperm contained in the pollen from male cones on other branches of the same tree or on other trees. After fertilization, the female cones fall to the ground so the seeds can be deposited in the soil. Gymnosperms include pine and cedar trees, as well as shrubs, such as the pygmy pine and the Mormon tea bush.

Female cone growing on the Lebombo cycad (*Encephalartos setaceous*)

Where would the eggs of this female cone be located?

Angiosperms are vascular plants that produce seeds in the ovaries of flowers. Some seed-bearing plants have flowers that contain both male and female parts. These plants are able to fertilize their own gametes. Other seed-bearing plants have flowers that have either male parts or female parts but not both. In many cases, these male and female flowers grow on completely different plants. Bees, butterflies, and other pollinators carry pollen from the male flowers to the female flowers. Do you think it would be better for a plant to be able to fertilize itself or to be fertilized by another plant? Why? Once a gamete is fertilized by the sperm contained in the pollen, the plant develops a fruit to protect the growing seed, which is safely tucked away inside. Animals and people spread the seeds when they eat the fruit. Marigolds, oranges, and water lilies are examples of angiosperms. What gymnosperms and angiosperms do you see growing at your school? At your home? In your town?

Other plants do not make seeds. Instead, they use spores for reproduction. A **spore** is a reproductive cell that can develop into a new individual without being fertilized, or fused with another cell, in ideal conditions. Spore-reproducing plants include vascular plants such as ferns and nonvascular plants such as mosses. Besides producing spores, what other similarities do these plants have?

Because the single-celled spore is small, it can be blown in the wind or carried by an animal to a new location. This single cell can survive poor conditions and will not begin dividing into a multicellular organism until conditions are favorable. Conditions are often unfavorable, so a single plant will produce thousands of spores. Most likely, only a few will survive and keep the species going.

Scripture Spotlight

When God put plants on Earth on the third day of Creation, did he plant seeds or were the first plants fully grown by His command? Read **Genesis 1:11–12** to find out.

Alternation of Generations

All plants go through a two-part cycle in their reproduction. During one part of the cycle, the plant produces spores. This form of the plant is called the **sporophyte** generation. Each spore can develop into a new plant. This new plant that develops from the spore represents the second part of the cycle. This form of the plant is called the **gametophyte** generation, because it is the form of the plant that produces **gametes** (egg and sperm). Recall that sperm and egg are gametes and that when they unite, a **zygote** forms. The zygote develops into an embryo. This embryo will develop into the sporophyte plant and the cycle continues—sporophyte to gametophyte to sporophyte to gametophyte, and so on. This amazing two-phased mode of reproduction is called **alternation of generations**. While this cycle occurs in all plants, these two generations—sporophyte and gametophyte—look different for nonvascular and vascular plants.

Nonvascular Plants

The gametophyte stage is the dominant stage of the life cycle of nonvascular plants such as mosses. Follow the list and diagram below to see the gametophyte and sporophyte stages.

1. The mature gametophyte produces gametes—egg and sperm.
2. The sperm must swim to the egg, so water must be present. The egg and sperm combine to form a zygote that develops into a sporophyte plant.
3. The much smaller sporophyte plant grows out of the gametophyte plant.
4. In mosses, the sporophyte produces a tiny spore-producing container at the end of a thin stalk. This spore-producing container is called a *capsule*. When the spores are mature, they are released.
5. The spores fall to the ground, and if conditions are right, germinate to form a structure called a *protonema*. The protonema develops into a new gametophyte plant, and the cycle continues.

Alternation of Generations in Mosses

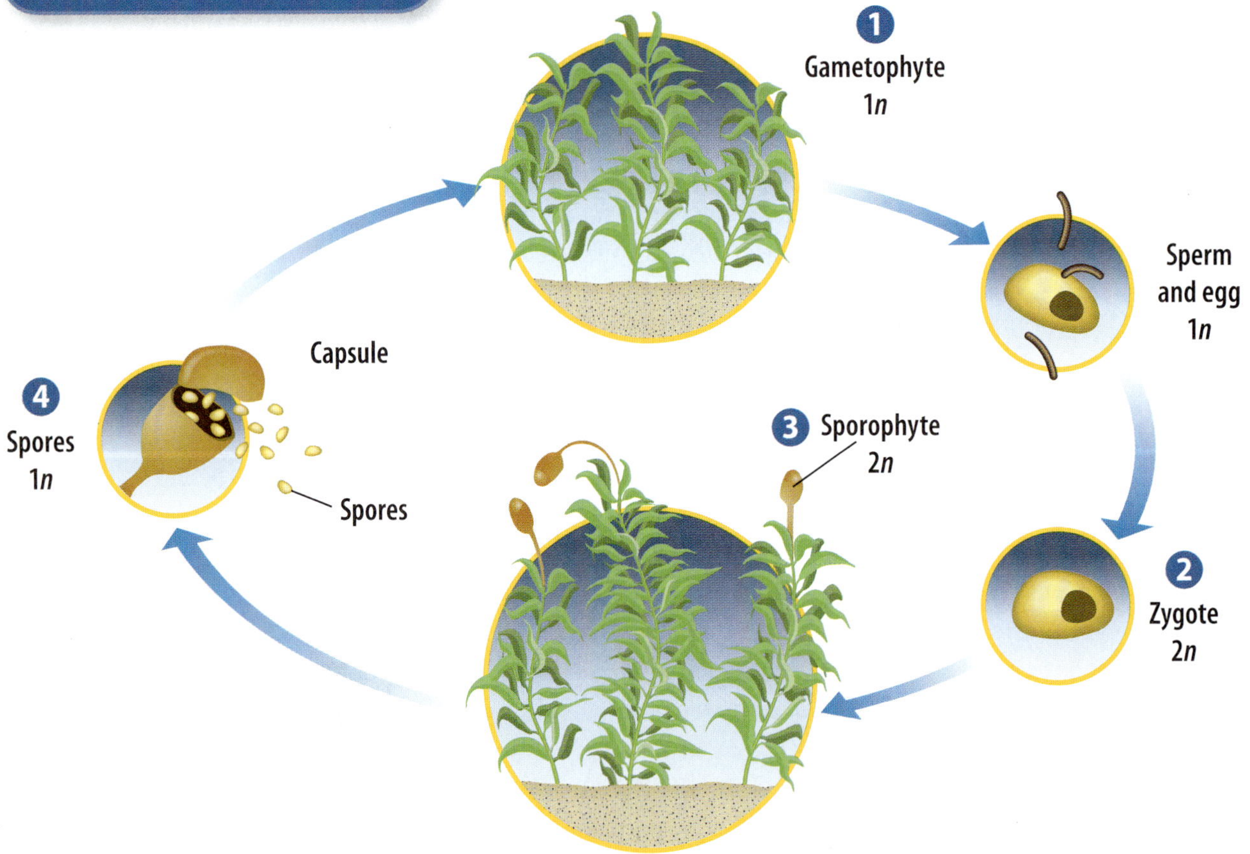

Why are both sporophytes and gametophytes essential to moss reproduction?

Vascular Plants

The sporophyte stage is the dominant stage of the life cycle of all vascular plants such as ferns.

1. In ferns, the reproductive structures of the sporophyte are called **sori** (singular: sorus). These structures, often located on the underside of the fern **frond**, or leaf, contain **sporangia** (singular: sporangium) that contain spores that fall to the ground.
2. When conditions are right, the spores germinate to form a very small gametophyte plant called a *prothallus*.
3. This gametophyte plant produces female and male reproductive structures that produce eggs and sperm. As in mosses, the sperm of the fern must have water to swim to the egg. Then fertilization takes place, and a zygote is formed. The zygote becomes the sporophyte plant.
4. The tiny sporophyte grows and develops fronds and a rhizome, or underground stem. When the fronds are mature, the sori release spores, and the cycle continues.

Alternation of Generations in a Fern

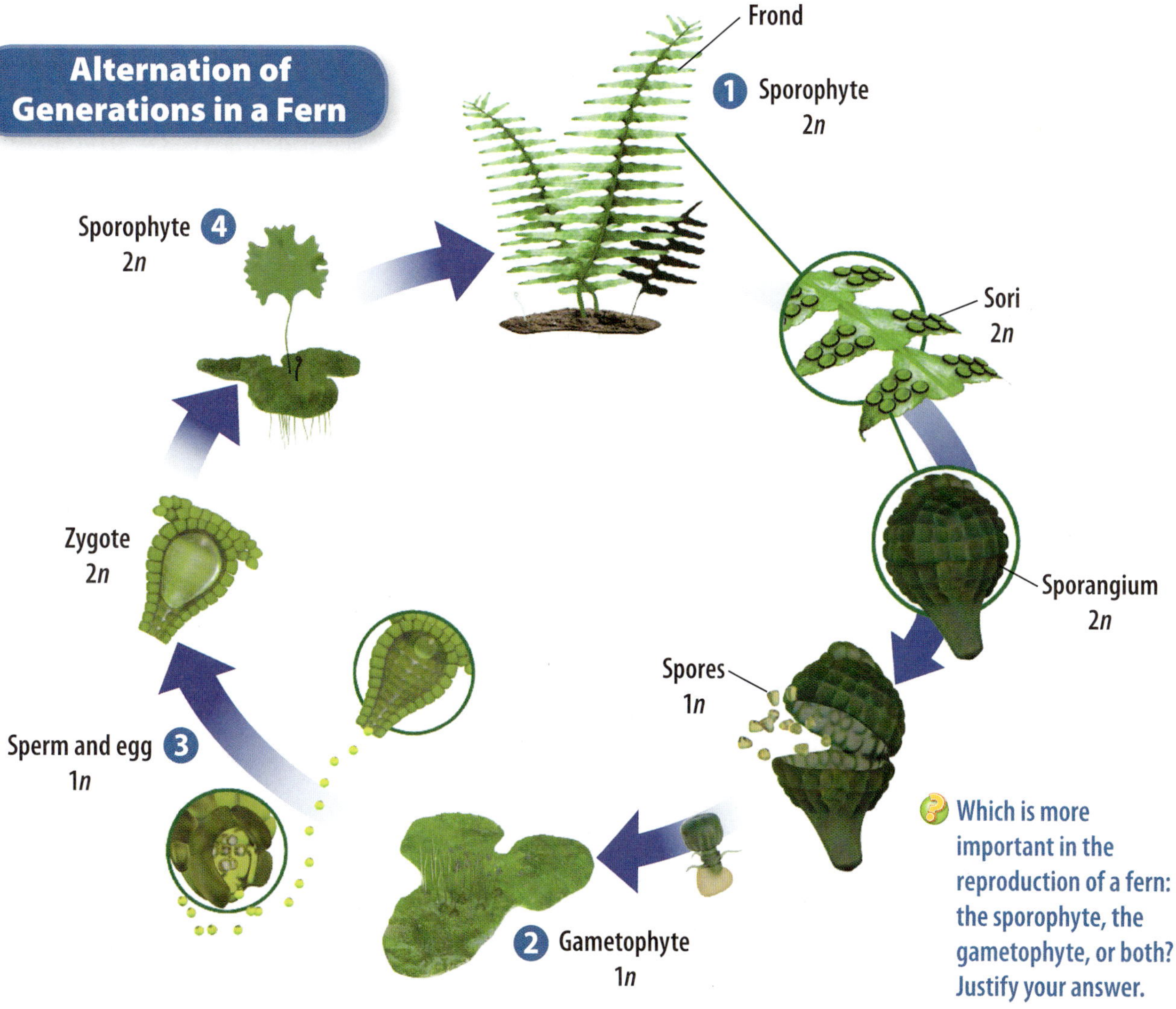

Which is more important in the reproduction of a fern: the sporophyte, the gametophyte, or both? Justify your answer.

Math in Science

You may recall that gametes have half the number of chromosomes as a regular cell. When an egg and a sperm form a zygote, the organism again has the full number of chromosomes. The netted adderstongue fern, *Ophioglossum reticulatum*, has 1 260 chromosomes. How many chromosomes would you find in a spore from the plant? How many chromosomes are in the gametophyte phase of the plant? How many chromosomes exist in the sporophyte phase? Why do you think a fern would need so many chromosomes?

Concept Check Assess/Reflect

Summary: How are plants classified? Plants are classified by whether they are vascular or nonvascular and by whether they form seeds or spores. Mosses are examples of nonvascular plants. They obtain nutrients and water through diffusion and osmosis. Vascular plants, such as the many species of trees, contain xylem and phloem to transport water and nutrients. Vascular plants also use diffusion and osmosis. The difference is that in vascular plants there is specialized tissue that allows for transport over a greater distance. Some plants, such as mosses and ferns, reproduce with spores, while others, such as marigolds and water lilies, reproduce with seeds. Of the seed-bearing plants, some (such as oranges) protect their seeds within fruits, while others (such as pine trees) have exposed seeds in cones. All plants have a two-part life cycle called alternation of generations.

1. What are the two main ways that plants are classified?
2. Compare and contrast diffusion with osmosis.
3. How do vascular plants compare with nonvascular plants?
4. How do sporophytes compare with gametophytes?

Lesson 2

Essential Question

What Are Spore-Producing Plants?

If you've ever hiked through a shady forest, you've probably seen and touched the soft mosses and feathery ferns that grow best in these damp, cool areas. Mosses and ferns are plants that reproduce using spores rather than seeds. What advantages do plants that reproduce via spores enjoy? What challenges do these plants face?

As you have learned, kingdoms are split into phyla (singular: phylum) to make groups of organisms easier for scientists to study. In the kingdom Plantae, the word *divisions* is used instead of *phyla*. Plants that use spores to reproduce are grouped into 6 of the 12 divisions in Kingdom Plantae: Bryophyta, Hepaticophyta, Anthocerophyta, Pteridophyta, Lycophyta, and Equisetophyta.

Objectives

- Compare and contrast the groups of nonvascular plants.
- Describe the groups of vascular plants that produce spores.
- Identify the structure of a fern.

Vocabulary

rhizoid

rhizome

Division Bryophyta Explain

How many kinds of mosses have you seen? There are more than 14,000 species of mosses, which belong to the division Bryophyta. Most species live on land. Throughout the world, they can be found growing soft and thick on a variety of surfaces, including shaded forest floors, tree trunks, fences, and rocky outcroppings. Why are mosses and similar plants the first to grow in an area of primary succession?

Mosses are nonvascular plants. Because they lack xylem and phloem, they are short and cannot stand upright on their own. For this reason, many mosses grow in groups. Some mosses drape over their surroundings. The smallest mosses are only a few millimeters in length, while the largest can be up to 1 m long. However, most are only about 2.5 cm tall. How tall are the mosses you see in your neighborhood?

Mosses do not have roots. Instead, they have rootlike structures, called **rhizoids**, which anchor them to a surface. The rhizoids of mosses absorb water and nutrients from the environment, and the mosses use diffusion and osmosis to transport the materials throughout their tissues. Do you think distributing nutrients throughout the plant through diffusion and osmosis without having a structured transport system makes it difficult for a moss plant to thrive? Explain your reasoning.

Mosses often grow in barren areas where nothing else can grow.

How might mosses change an environment so that it can support other kinds of plant life?

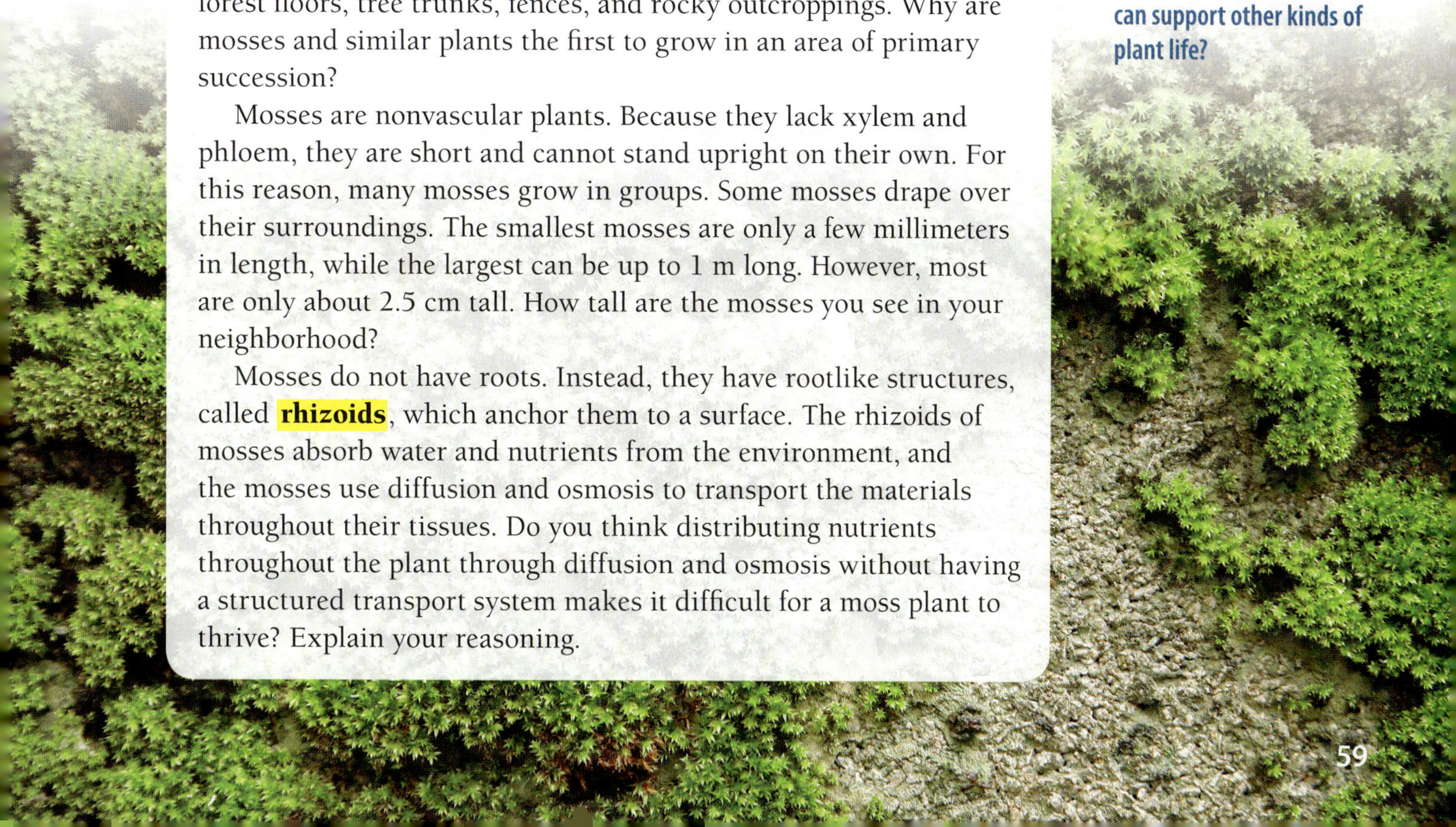

A rhizoid is the base of a moss plant. A stem grows from the rhizoid, and leaflike structures sprout on the stem. Most moss leaves are quite thin, with just one layer of cells that conduct photosynthesis. How do you think moss plants are affected by having fewer cells conducting photosynthesis than other plants, such as oak trees, have?

Mosses can also reproduce through fragmentation, a type of asexual reproduction. This happens when a small piece breaks off of the larger plant and establishes a new plant. What might cause a piece of the moss to break off?

Rhizoids use diffusion and osmosis to absorb and move water and nutrients to all parts of the moss.

What places in your neighborhood would be ideal moss habitats? Look to see whether you can locate any mosses.

Related Divisions

At one time, scientists grouped mosses in the Bryophyta division with liverworts and hornworts. After advances in technology detected differences among the plants, scientists decided liverworts should be in their own division called Hepaticophyta, which contains more than 8000 species. Scientists placed hornworts in the division Anthocerophyta, which contains about 100 species. What information do you think scientists learned that caused them to change the classification of liverworts and hornworts?

Hepaticophyta and Anthocerophyta share some traits with Bryophyta. Hornworts and liverworts are nonvascular plants with structures similar to those of mosses. The plants in all three of these divisions reproduce with spores and grow best in moist, shaded areas. The differences are in the leaves and rhizoids.

Explore-a-Lab

Structured Inquiry

What variations can I observe in the division Bryophyta?

During this inquiry, you and your partner will travel to different stations. At each station is a living bryophyte. Use a hand lens to examine each bryophyte. Sketch each specimen, using colored pencils, on a sheet of drawing paper. Label the parts that are present: gametophyte, sporophyte, spore capsule, and rhizoids. Write the name of the specimen next to each sketch. Compare your sketches of each specimen. How are they alike? How are they different?

Division Pteridophyta Explain

Ferns belong to the Kingdom Plantae, division Pteridophyta. Scientists estimate that this division contains somewhere between 9 000 and 15,000 species of vascular plants that live in a variety of habitats and climates. Most live on land, while some are tiny, floating water plants. Some live on other plants.

The majority of ferns prefer moist, shady locations, so they thrive in the tropics and in shady areas. Some fern species have adapted to harsher conditions. Ferns that are found in cooler, drier climates are typically smaller, mosslike varieties tucked under large rocks or in streambeds, where they are less exposed. Some ferns are small, soft-tissued plants with leaves as small as 2 cm in height. Other ferns grow sturdy support structures. Tree ferns are the largest members of this division. They can grow 25 m tall, and their leaves can be as long as 5 m. Why is there such a difference in the size of ferns and mosses?

Unlike the plants in divisions Bryophyta, Hepaticophyta, and Anthocerophyta, ferns have vascular tissue and thus have true roots. The roots are branched, thin, and irregular in appearance. In most ferns, the roots grow laterally from underground stems called **rhizomes**. Like the stems of flowering plants, rhizomes can be thin and flexible, or large and tough. They contain vascular tissue that transports water, minerals, and food. How do the roots and rhizomes work together to help the plant? How do the roots and rhizomes help the ecosystem? Other ferns have aboveground stems, some of which are thick and woody, similar to tree trunks.

Rows of sori are clearly visible on this fern. Each sorus is made of a cluster of sporangia. Each sporangium contains 50 to 2 000 spores, depending on the species of fern.

Why must ferns produce and release so many spores?

The leaves of ferns are called fronds, and they can sprout anywhere along the rhizome. Fronds can be simple and uniform or highly divided, having an almost lacy appearance. They come in a variety of shapes, including triangular, round, and kidney shaped. In cooler climates, fern fronds die back in winter as the plant becomes dormant. In early spring, the new fronds of nearly all ferns emerge from the soil as tightly coiled fiddleheads that gradually unfurl. What purpose do you think fiddleheads serve?

Like the Bryophytes, ferns do not produce seeds. Recall that they reproduce with spores that form in sori. The sori are found on the underside of the pinnae in rows of brown, green, yellow, or black structures. Why do you think the small size of the spores is important to the success of fern reproduction?

These fiddleheads are emerging from the ground alongside a brook.

Why do you think ferns emerge from the soil in the fiddlehead shape?

Structured Inquiry

Discover

Record your work for this inquiry. Your teacher may also assign the related Guided Inquiry.

Science Journal

Counting Spores

How many spores does one fern plant produce?

Procedure

1. Use the hand lens to **observe** the sori (spore cases) on the underside of the pinnae (leaflets) of the fern frond. They will look like light green, brown, black, or yellow spots.
2. Count the number of sori on one pinna (or leaflet) and **record** your answer.
3. Repeat Step 2 with five other pinnae. **Calculate** the average number of sori per pinna. Record your answer.
4. Count all the pinnae on the frond that contain sori. Then calculate the number of sori on the frond. Record your answers.
5. Use your hand lens to count the number of sporangia per sorus. You will need to look closely to see this. Count the number of sporangia in at least five sori. Record your data in your ***Science Journal***. Assume that each sporangium contains 64 spores. Calculate the number of spores per frond. Record your answer.
6. Assume that each fern has approximately 20 fronds. Calculate the total number of spores produced by a fern plant.

Materials
- fern frond with sori
- hand lens

Analyze Results

Describe the spores you observed. Share your calculation of the number of spores per plant and explain how you obtained that number.

Create Explanations

1. How many spores does one fern plant produce?
2. Why do ferns produce so many spores?
3. Why don't all pinnae contain sori?
4. Many of the numbers you calculated in this activity are estimates. How could you increase the accuracy of these estimates?

Related Divisions

Pteridophyta is by far the largest division of vascular plants that do not bear seeds, but it is not the only one. Lycophyta is also a division of Kingdom Plantae that reproduces with spores and has vascular structures. There are about 1200 species in the Lycophyta division, including club mosses and spike mosses, which are small evergreen plants with needlelike or cone-like leaves. Some look like tiny pine trees, although they are not in the same division as pine trees. The main difference between ferns and the plants in division Lycophyta is the structure of the leaves. Fern fronds have many veins that take nutrients to the edge of each part of the frond. The plants in division Lycophyta have a single vein in each leaf. How do you think having just one vein affects the plants in division Lycophyta?

Equisetophyta is another division of vascular plants that do not bear seeds. This division contains the plants commonly known as horsetails, which grow in moist areas all over the world. Horsetails are made up of hollow stems jointed into many sections. A ring of tiny leaves grows at each joint. Some stems also have branches that grow between the leaves, making the plants look fuzzy and giving them the name *horsetails*. Horsetails have cone-like structures in which they produce spores.

Horsetail species range in height from 30 cm to 10 m. In addition to the stems above the ground, horsetails have underground rhizomes. If the horsetail stem above ground dies, new plant stems can grow from the rhizome. The cells that make up the outer covering of horsetails contain the mineral silica, which gives the stems and branches a rough texture. What advantages might this rough texture give to horsetails? Early American settlers discovered that horsetails were useful in cleaning pots and pans and called them *scouring rush*.

Ancient Romans used the horsetail plant to make tea, and used the plant for food. Native Americans also used Equisetum as medication for sexually transmitted diseases and urinary problems. The stems are rich in silica and are used for polishing metal or as a fine sandpaper.

What other uses are there for this plant?

Explore-a-Lab

Guided Inquiry

What effect does the water have on the spores?

Equisetum, commonly known as horsetail, is the only living genus in a family of vascular plants that produce by spores rather than seeds. Place some Equisetum spores in a Petri dish. Observe the spores with the magnifying lens. Record your observations. Then add 1 drop of water to the spores. Use the magnifying lens to observe the spores again. Record your observations. How does this effect help the Equisetum survive in its environment?

Concept Check Assess/Reflect

Summary: What are spore-producing plants? Plants that use spores to reproduce are grouped into 6 of the 12 divisions in Kingdom Plantae. Mosses are in the Bryophyta division, liverworts are in the Hepaticophyta division, and hornworts are in the Anthocerophyta division. These three divisions contain nonvascular plants. Vascular spore-producing plants are found in the Pteridophyta (ferns), Lycophyta (club and spike mosses), and Equisetophyta (horsetails) divisions. Fern spores develop in clusters called sporangia on the underside of fronds. Groups of sporangia are called sori. Lycophyta and Equisetophyta have spores that develop in cone-like structures.

1. Describe three differences between a moss and a fern.
2. What is the function of a rhizome?
3. What is the difference between sori, a sporangium, and a spore?
4. How are the sporophytes and gametophytes of a fern different?
5. What are the characteristics of the plants in divisions Hepaticophyta and Anthocerophyta?
6. How do the plants in divisions Lycophyta and Equisetophyta differ from the plants in division Pteridophyta?

Lesson 3

Essential Question
What Are Gymnosperms?

Objectives

- Describe the characteristics of gymnosperms.
- Compare and contrast the common groups of gymnosperms.
- Describe the alternation of generations cycle in gymnosperms.

Vocabulary

evergreen

deciduous

conifer

What do you think of when you hear the names *bald cypress*, *Virginia pine*, *sago palm*, and *giant sequoia*? What traits do you think they share? Why? In what type of environment do you imagine each would grow? Research these plants and find a picture of each. You will notice that although they look very different, they have similar characteristics that scientists use to group them together. What similar characteristics can you see in the pictures you found? These plants do not produce spores, like mosses and ferns do. Instead, they produce seeds. But, unlike the flowering plants, the seeds of these plants are not enclosed or protected in a fruit. Their seeds are exposed. This group of vascular plants with exposed seeds is known as the gymnosperms.

Gymnosperm Divisions Explain

Besides having naked seeds on the cone scales, many gymnosperms have needle-like leaves and are evergreen. Today there are about 950 known species of gymnosperms. Based on the plants' unique characteristics, scientists have grouped these species into four divisions of Kingdom Plantae. The table on the next page describes the gymnosperm divisions. As you study the table, can you think of times when you have used the plants as a resource?

This yellow-rumped warbler is perching in a juniper bush. The berrylike structures on the juniper are its cones.

How do these cones compare to pinecones? How can these berrylike structures be considered cones?

Gymnosperms			
Division	**Description**	**Examples**	**Natural History**
Cycadophyta (100 species)	Leaves like palm fronds or ferns; thick stem like a trunk	Sago palms	Wildlife eat the coverings off the cones. How does this affect the cycad species?
Ginkgophyta (1 species)	Fan-shaped leaves; leaves that drop in the fall; fleshy cones with seeds	*Ginkgo biloba* trees	The leaves of this tree are used for medicinal purposes. Most of the species in this division are extinct. Why do you think scientists created a division for a single living species instead of grouping *Ginkgo biloba* with another division?
Gnetophyta (about 71 species)	Nearly leafless; slender branches; flowers with fleshy, red cones resembling berries.	*Ephedra*	Some species of *Ephedra* have traditionally been used by indigenous people for a variety of medicinal purposes, including treatment of hay fever and the common cold. Why is knowing about the medicinal use of plants by indigenous people important today?
Coniferophyta (550–575 species)	Cone-bearing plants with needlelike or scalelike leaves; some species with berrylike cones	Pines, firs, yews, redwoods, spruces, junipers, cypresses	How are these plants important economically? How are they important to the environment?

The plants in divisions Cycadophyta, Gnetophyta, and Coniferophyta are mainly **evergreens**, plants that retain their leaves all year long. However, a few species—such as the larch tree in division Coniferophyta—are **deciduous**. Plants that are deciduous drop their leaves in the fall. The one living plant species in Ginkgophyta, *Ginkgo biloba*, is also deciduous.

Cycadophyta

Cycads are evergreen plants with stems that resemble thick trunks. The leaves of some cycads resemble ferns, while the leaves of others resemble the stiff leaves of palm trees. Cycads range in height from 2 m to 20 m. They grow only in the tropics and subtropics. The bases of the leaves are arranged in a tightly packed spiral. Some cycad plants produce male cones, and other cycad plants produce female cones. Cycads look a little like palm trees, but they are not palms. Palms are flowering plants, while cycads are cone bearers. You will learn about flowering plants in the next lesson.

When fully mature, the female king sago will produce a basketball-sized structure that will house the developing seeds.

Structured Inquiry

Discover

Record your work for this inquiry. Your teacher may also assign the related Guided Inquiry.

Science Journal

Methods of Classification

What characteristics are used to classify pine trees?

Procedure

1. Study the illustrations.
2. Start with the pine limb labeled A. Begin at Step 1a of the identification key in your ***Science Journal*** and proceed through the key to identify pine A. **Record** your results.
3. Repeat Step 2 with samples B–G.

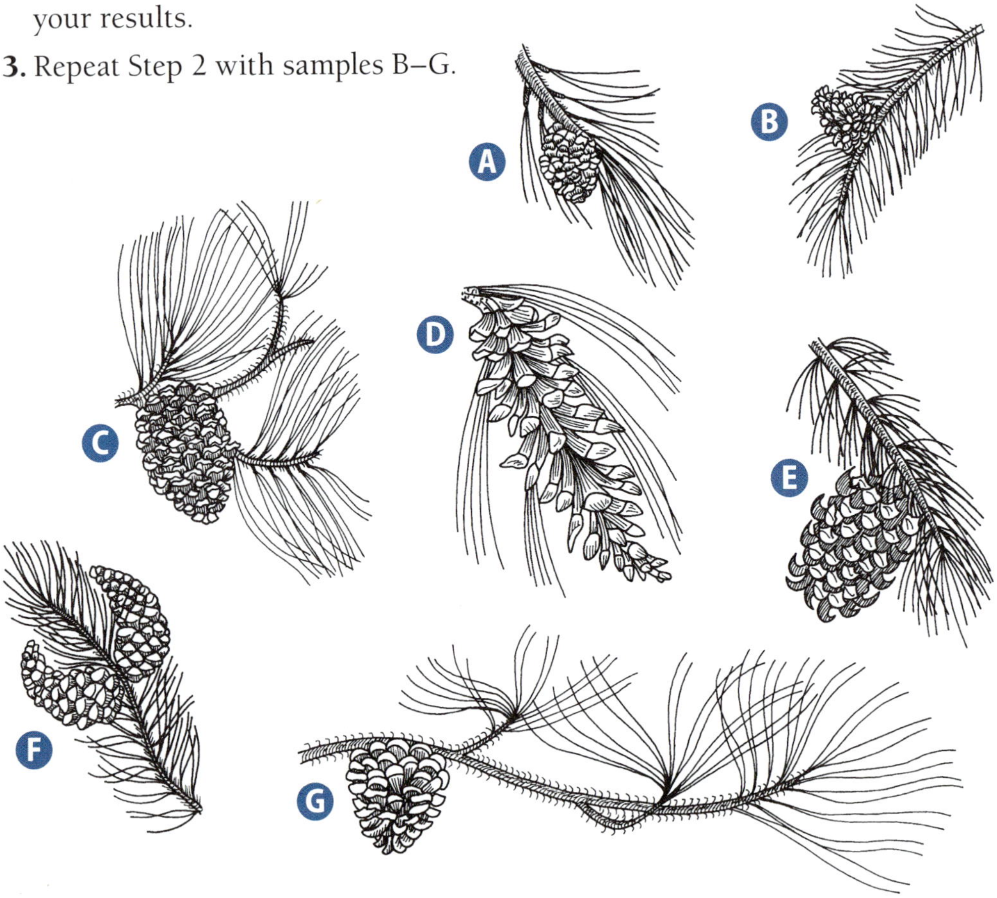

Analyze Results

Compare and discuss your data table with those made by your classmates.

Create Explanations

1. What characteristics are used to classify pine trees?
2. How is the *Pinus contorta* different from the *Pinus banksiana*?
3. Why can't you start at Step 2a in the key?

Ginkgophyta

Ginkgophyta is an unusual division with only one living species—*Ginkgo biloba* trees. These trees are either male or female. The male tree bears cones that contain pollen. The female tree produces soft, fleshy cones that look like apricots. The female gingko is infamous for the smell of its seeds. Their odor is slightly rotten and can sometimes be detected over long distances. How do you think this odor helps the tree complete its life cycle? The delicate, fan-shaped leaves of the ginkgo give it its nickname—the maidenhair tree.

Ginkgo trees are deciduous. In the fall, their leaves turn a golden yellow before falling off.

Most other gymnosperms are evergreen. Why do you think the ginkgo is deciduous?

Gnetophyta

The division Gnetophyta includes only three living genera: *Ephedra*, *Gnetum*, and *Welwitschia*. The 65 species of *Ephedra* are mainly shrubs and vines that grow in dry areas all over the world. This genus includes ma huang, a Chinese plant that is the source of the drug ephedrine. This drug is used in decongestants. The more than 30 species of *Gnetum* include woody vines, shrubs, and small trees that grow in tropical and desert climates. *Welwitschia* has only one species, a plant also named Welwitschia. This unusual plant grows only in desert areas of southwestern Africa. It has an extremely large storage root and two long, permanent leaves that, torn by the wind, appear to be many leaves. Like all gymnosperms, the plants of division Gnetophyta reproduce by means of spores and cones. Gnetophytes form male and female cones on separate plants, like the plants of divisions Cycadophyta and Ginkgophyta.

The plants in division Gnetophyta grow a tube that allows the eggs easy transport to the pollen tubes during fertilization.

How is this method of fertilization different from that of other gymnosperms?

The giant sequoia tree, found in California, can reach heights of 95 m.

Coniferophyta

The species of plants in the division Coniferophyta are mainly evergreen trees and shrubs that grow in a variety of regions all over the globe. Conifers dominate the enormous forests of the northern regions of Europe, Asia, and North America. They range in size from the 8 cm of the pygmy pine of New Zealand to the 110 m of the redwoods of California. Like the plants in divisions Cycadophyta, Ginkgophyta, and Gnetophyta, conifers have male and female cones. However, unlike the plants in those three divisions, the male and female cones of conifers develop on different branches of the same shrub or tree. What advantage might this arrangement give to conifers?

Gymnosperm Life Cycle **Explain**

Unlike mosses, ferns, and other plants that reproduce with spores, the gymnosperms reproduce with seeds. As you learned in Lesson 1, the sporophyte generation is the dominant stage of the life cycle of all vascular plants, including gymnosperms. In **conifers** and other gymnosperms, the gametophyte phase occurs entirely within the cones. A person looking at many species of gymnosperms would not be able to tell that the gametophyte phase is happening. How does this differ from the gametophyte phase in mosses?

Have you ever pulled apart the outer pieces of a pinecone? How did they feel? The cones are made of many individual plates, called scales. The reproductive structures of the conifer lie on the scales within the cones. Conifers and other gymnosperms produce both female and male cones. The female cones are woody, large, and easy to see. They produce eggs (ovules). The male cones are much smaller. They produce pollen. In some conifer species, female cones are found lower on the plant than the male cones. How do you think this helps to ensure fertilization?

Follow the list and the diagram on the next page to see how a conifer reproduces.

1. Pollen is formed on the scales of male cones. The pollen is part of the gametophyte stage of the plant. It has the chromosome number 1*n*, or half the number of chromosomes present in the parent plant.
2. Ovules are formed at the base of the scales of female cones. This is also part of the gametophyte stage of the plant. The ovules mature, forming egg cells. The eggs have half the genetic material present in the parent plant—1*n* chromosomes.
3. In the spring, male cones release pollen in great, dusty clouds. Carried by the wind, some pollen grains fall on female cones. At the time of pollination, a pollen grain sprouts a pollen tube. The pollen tube reaches the female gametophyte and releases two sperm cells. One will unite with the egg. The resulting zygote has the chromosome number 2*n*, the full number of chromosomes characteristic of the species.
4. A sporophyte zygote develops into an embryo and then into a seed within the female cone. The seed's chromosome number is 2*n*. It is protected by a thick seed coat. When conditions are right, the seeds are released.
5. The seeds fall to the ground. They sprout, and new gymnosperm sporophytes grow.

Alternation of Generations in Conifers

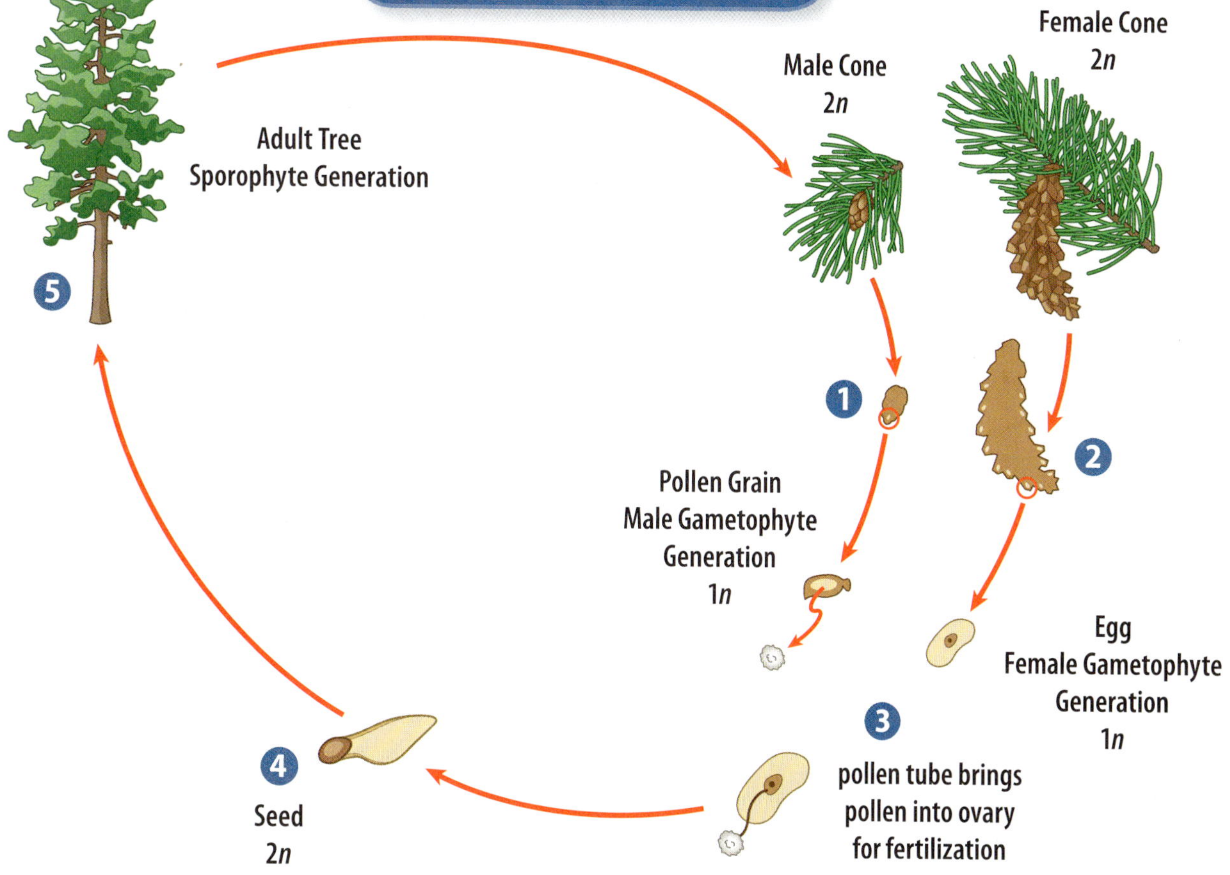

In conifers and many other gymnosperms, fertilization does not require water, as is the case for mosses and ferns. Instead, the pollen grains of the male cones are transported by the wind. The sperm they produce can move within the relatively moist environment of the female cone's scales. As a seed is forming, the embryo develops within the protected, moist environment of the female cone. A supply of food and water is packaged, along with the embryo, inside the hard, outer seed coat. The food supply will sustain the embryo when it first begins to grow into a new gymnosperm sporophyte. Can you think of other organisms that God created with a similar design for their offspring?

Concept Check Assess/Reflect

Summary: What are gymnosperms? *Gymnosperm* is the name for vascular plants that produce exposed seeds; that is, their seeds are not enclosed or protected inside a fruit. Kingdom Plantae has four divisions of gymnosperms: Cycadophyta, Ginkgophyta, Gnetophyta, and Coniferophyta. The four divisions together contain about 950 known species. Cycads are evergreen plants with leaves that look like ferns or palm trees and with thick, trunk-like stems. *Ginkgo biloba* trees are the only living species in Ginkgophyta and are one of the few deciduous gymnosperms. The more than 95 species of Gnetophyta include Ephedra, which was used for medical purposes by indigenous people. Conifers are mainly evergreen trees and shrubs that dominate the enormous forests of the northern regions of Europe, Asia, and North America. Like other plants, alternation of generations is part of the reproductive process of gymnosperms. To reproduce, male cones release pollen. Pollen spreads to female cones, where the egg is fertilized and seeds are produced. The seeds grow into the large sporophyte tree or shrub.

1. What two characteristics define the gymnosperms?
2. Why might a person observing a conifer have difficulty identifying the gametophyte state?
3. How do evergreen and deciduous plants differ?
4. How is the gametophyte stage in gymnosperms different from the gametophyte stage of mosses?
5. Explain the major difference between the conifers and the other three types of gymnosperms.

Lesson 4

Essential Question

What Are Angiosperms?

Angiosperms, the most numerous and widespread type of plant, are flowering vascular plants that encase their seeds in a fruit that grows from the female part of a flower. The fruits and many vegetables we eat come from plants that are angiosperms. Think about your favorite fruits. Where are their seeds located? Make a list of fruits that have seeds inside them. Some foods that we call vegetables, such as tomatoes, are actually fruits because they contain the seeds of the plant.

Division Magnoliophyta Explain

You most likely can name many angiosperms. These plants belong to the division Magnoliophyta. There are more than 300,000 known species of plants in this division, between 75 and 80 percent of the total number of species in Kingdom Plantae. Angiosperms thrive from the tropics to the tundra and from the seashore to the highest mountaintops. They form the base of most land food chains and provide the bulk of the food supply available to animals. How do you and your family use angiosperms?

There is no mystery about why angiosperms are so prevalent. Many have tough leaves that are resistant to drought and cold. Like ferns and gymnosperms, angiosperms have a vascular system that moves water and nutrients well. Like gymnosperms, many angiosperms contain chemicals that help them resist diseases and insects. Their fruits protect seeds as they mature. The fruits are attractive to animals and people, who eat them and spread the seeds found inside. The tough coats of the mature seeds protect the seeds until they are ready to germinate. Angiosperms are efficient at pollination and seed dispersal. Which of the characteristics just listed explains why angiosperms are so widespread? Which animals and environmental factors assist in the pollination and seed-dispersal processes?

Animals represent an important step in the fertilization stage of plants in division Magnoliophyta.

Objectives

- Identify the characteristics of angiosperms.
- Describe the alternation of generations cycle in angiosperms.
- Compare and contrast monocots and dicots.
- Explain the common types of fruit.

Vocabulary

cotyledon
stomata
stamen
anther
pistil
ovary
style
stigma
monoecious
dioecious
polygamous

Two Kinds of Angiosperms

The division Angiospermae is divided into two classes: Liliopsida (also known as Monocotyledonae, or *monocots*) and Magnoliopsida (also known as Dicotyledonae, or *dicots*). Several characteristics are used to separate the angiosperms into monocots and dicots. Have you ever opened a bean seed and looked inside? If you have, you might have observed two large food storage areas. These are the **cotyledons**, which function as the first "seed leaves" of a plant embryo. The number of cotyledons in a plant embryo is one characteristic that distinguishes monocots from dicots. Monocots, such as corn, have one cotyledon; dicots, such as beans, have two cotyledons.

Scientists today are revising aspects of the monocot/dicot classification, but the characteristics on which these modifications are being made occur on the molecular level. The reclassification of plants is an issue that is not resolved, and, for classification purposes, it is still useful to note the differences between monocots and dicots. Explore some characteristics of these two classes in the table below. How many differences do you notice between the two classes? Which do you encounter more frequently? How do you use monocots and dicots as resources?

	Angiosperms	
	Monocots	**Dicots**
Full name	Monocotyledon	Dicotyledon
Veins (part of the vascular system)	Parallel in leaves; vascular bundles scattered uniformly in the stem 	Form a netted pattern in leaves; vascular bundles arranged in a ring in the stem 
Petals	Usually occur in multiples of three 	Usually occur in multiples of four or five 
Seed	One cotyledon 	Two cotyledons 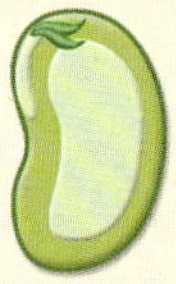
Roots	Develop from lower end of embryo (radicle) 	Develop from nodes in the stem 
Examples	Lily, iris, corn, bamboo, onion, tulip, rice, oat, and banana	Bean, carnation, lettuce, rose, cactus, and most nonconiferous forest trees

Parts of an Angiosperm

Roots

Have you ever tried to pull a plant out of the ground? Depending on the plant, pulling it out might have been a difficult job because of structures called roots. Roots anchor the plant into the ground.

Roots are the structures many plants use to take in the water and dissolved nutrients, mainly through osmosis, that they need to make food through photosynthesis. Some roots have root hairs that help in the intake process. The root hairs protrude from the *epidermis,* or outer layer of cells that covers the roots (and the other parts of the plant). Plants store the food they make in their roots, keeping it there for later use. Root tips are covered with a structure called a root cap, which helps protect the root as it pushes through soil and rock. As the root grows, cells in the root cap are rubbed off and replaced with new cells. Why is this necessary? Water and dissolved minerals move into the root, where the xylem and phloem are located. The xylem carries water and dissolved nutrients to the rest of the plant, while the phloem carries food made by photosynthesis in the leaves.

Many plants have numerous long roots of the same size, which are called fibrous roots. Grass is an example of a plant with fibrous roots. Other plants have one main root, called a taproot, and a few smaller roots. Carrots and beets are examples of taproots.

Stems

Unlike nonvascular plants, all vascular plants have stems. Besides transporting materials up and down between the roots and the leaves, what other jobs do stems carry out for the plant? How do we use plant stems? Leaves grow out of the stem at areas called nodes, and the spaces between nodes are called internodes. In *annual* plants, the entire plant dies at the end of the growing season, but in *biennial* plants, the lower part of the stem survives to sprout buds during a second growing season. In *perennial* plants, part of the plant's stem survives, year after year, to sprout buds at each new growing season.

The stems of many angiosperms contain chloroplasts, so they can perform photosynthesis, just as leaves do. This type of stem is called herbaceous. Asparagus, grasses, and cacti are plants in which photosynthesis takes place mainly in the stems. Plants such as shrubs and trees have woody stems. Some stems wrap around objects as they grow so that they can better support the plant. What plants with climbing and twisting stems can you name?

Some plants have stems that perform specialized functions. *Stolons* (also called runners) are stems that grow along the ground instead of standing upright. Where they touch the ground, they often take root and form new plants. Strawberries are an example of a plant with stolons. *Tubers* are underground stems. New plants can grow from the "eyes," or buds, of a tuber. Potatoes are an example of a plant with a tuber stem. Other plants such as irises and some ferns have horizontal underground stems called *rhizomes*. New plants can sprout from these stems. Another type of underground stem, found in crocuses and gladiolas, is called a *corm*. It, too, can produce new plants. Corms make up the majority of the plant and are sometimes confused with *bulbs*, which look similar. Bulbs, found in tulips or onions, are very short stems with fleshy leaves.

Scripture Spotlight

In **Jeremiah 17: 7–8**, humans are compared to trees. How is your faith in God like a tree, even during bad times?

Twig Structure

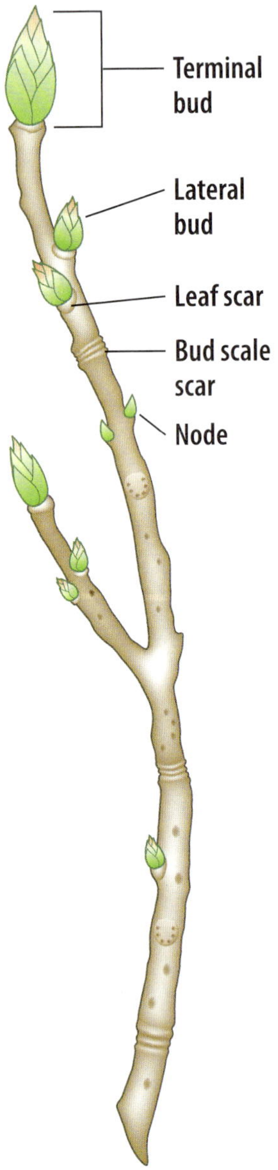

Explore-a-Lab

Structured Inquiry

What are the parts of a twig?

Get a sample of one of the twigs provided. Observe the twig closely and use the diagram below to find the labeled structures on your twig. Draw a detailed diagram of your twig and label the structures you can find. Repeat this process with two more of the stems provided.

Leaves

You have learned that leaves are the structures that many plants use to make food. Carbon dioxide enters a plant through tiny holes, called **stomata**. These structures in the epidermis of leaves have two *guard cells* that open and close to allow oxygen and carbon dioxide to move in and out of the leaf. Using special structures called chloroplasts, leaves capture light energy from the Sun and use it to convert carbon dioxide, water, and nutrients into chemical energy in the form of glucose, a simple sugar. Oxygen is also produced during photosynthesis. It leaves the plant via the stomata. Water also diffuses into and out of the plant through the stomata. Given what we know about plants, why are they essential to all life on Earth?

Leaves usually have a flat area called a *blade*. In some plant species, a stalk called a *petiole* attaches the leaves to the stem. Leaves can be categorized by many different characteristics. Leaves that are whole are *simple*, while leaves that are divided into smaller parts called leaflets are *compound*. Leaves are also categorized by their venation pattern (the pattern of their veins). *Pinnate* leaves have one main vein, while *parallel* leaves have several veins running next to one another, and *palmate* leaves have veins that radiate out from the base of the leaf.

The way leaves are arranged on a stem is another way that they are categorized. Some plants have leaves that extend from the same spot on a stem, on opposite sides. Others have leaves that alternate—one on one side, and then another on the other side, a little higher up. Still others are *whorled*, which means that two or more leaves extend from the same spot on a stem.

Parts of a Flower

Have you received flowers from someone—perhaps for a birthday or other occasion? In what other ways are flowers part of your life?

A flower is connected to its stalk, called a peduncle, at a fleshy area called the receptacle. Right above the receptacle are the *sepals*, which protect the flower while it is still a bud. All of the sepals together are called the *calyx*. Just above the sepals are the petals, which are most likely the first thing you notice about a flower, especially if they are colorful or if they have interesting patterns, dots, or lines. All of the petals together are called the *corolla*.

What does your sense of smell tell you about the flower? Think about why the flower looks and smells this way. If you look more closely, you will see the reproductive parts of the flower that lie inside the petals. The male part of a flower is the **stamen**. It is made of a stalk-like *filament* that supports a bulb-shaped structure. The bulb is the **anther**, which contains structures that produce the spores that develop into pollen grains. The female part of the flower is the **pistil**.

The female part of the flower is made of three structures:

- The **ovary** lies at the enlarged base of the pistil. Within it lie the ovules, where eggs form. After fertilization, ovules develop into seeds, and the ovary develops into a fruit.
- The stalk that rises up from the ovary is called the **style**.
- The top of the style is the **stigma**. Often, the stigma is covered with hairs or a sticky material that traps pollen grains.

Parts of a Flower

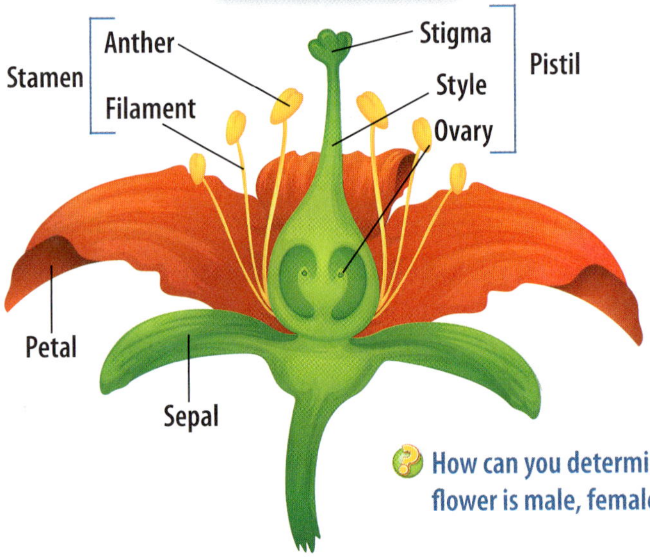

How can you determine whether a flower is male, female, or both?

Types of Flowers

Some flowers have sepals, petals, stamens, and one or more pistils. They are called *complete flowers*. Flowers that are missing one or more of these parts are called *incomplete flowers*. If a flower has both stamens and one or more pistils, it is a *perfect flower*. If it is missing stamens or pistils, it is an *imperfect flower*. A **monoecious** plant has perfect flowers or imperfect flowers of both sexes, while **dioecious** plants have the male (staminate) and female (pistillate) flowers on different plants. Plants that have both perfect flowers and imperfect flowers of both sexes are **polygamous**. What might be an advantage to having imperfect flowers on the same plant? On different plants?

Recall that organisms can be classified based on symmetry. Something is symmetrical if you can divide it in half, and each half is almost identical. Types of symmetry include radial (the body parts are equally spaced around a central point) and bilateral (the drawing of a line across the organism in only one location will divide the organism into two almost identical halves). Flowers that have radial symmetry, such as roses, are called *regular*, while flowers that have lateral symmetry, such as snapdragons, are called *irregular*.

Structured Inquiry

Discover

Record your work for this inquiry. Your teacher may also assign the related Guided Inquiry.

Science Journal

Dissect Flowers

How do the parts of different flowers compare?

SAFETY: Always use caution while handling a scalpel.

Materials

- 4 different flowers
- dissecting pan or glass pan
- dissecting needle
- dissecting scissors
- tweezers
- scalpel
- hand lens or dissecting microscope
- metric ruler

Procedure

1. Make a detailed drawing of the flower you are given to dissect. **Observe** the petals and sepals and label them on your drawing. Look for lines or dots on the petals. Add these to your drawing. Using scissors and tweezers, carefully remove the sepals and petals. Measure them with the metric ruler. **Record** the data.
2. Locate and observe one stamen in the flower. Observe the filament and the anther. Label these three parts on your drawing. Remove one stamen from the flower and observe it with a hand lens or dissecting microscope. **Measure** the stamen. Record the data.
3. Gently dissect an anther. Observe the pollen inside with a hand lens or dissecting microscope. Make a detailed drawing of what you see.
4. Locate and observe the pistil. Observe the stigma, style, and ovary. Record a description of the stigma's texture and appearance. Observe a pistil under magnification. Make a detailed drawing of what you see. Measure the pistil. Record the data.
5. Follow the tube-shaped style down from the stigma until you reach the ovary. Carefully open the ovary with a scalpel. Place the ovary under magnification. Look for seeds or unfertilized eggs. Make a detailed drawing of what you see.
6. Look back at the characteristics of monocots and dicots. Observe the flower for these characteristics and determine whether it is a monocot or a dicot. Record the data. Repeat Steps 1–6 for 3 more different flowers.

Analyze Results

How are the flowers you dissected alike? How are they different?

Create Explanations

1. How do the parts of different flowers compare?
2. If lines or dots are present on flower petals or sepals, infer their function.
3. Consider the characteristics of the flowers you examined. Which are monocots? Which are dicots? What characteristics did you use to make this classification?

The Formation of Fruits Explain

Scripture Spotlight

Revelation 22:2 discusses the tree of life that bore fruit each month and with leaves that were for healing the nations. What kind of healing is meant by this scripture? How do we use plants to heal in the modern world?

What do you think of when you hear the word *fruit*? Do you think of apples, oranges, and bananas? These are very familiar fruits. What about cucumbers and tomatoes? Are those fruits? Botanically speaking, a fruit is the ovary of a plant. It contains the plant's seeds, and its job is to protect and disseminate the seeds. Fruits include sugar snap peas, green beans, corn grains, wheat grains, sunflower "seeds," maple "seeds," coconuts, cucumbers, berries, pineapples, and figs. How many more can you name? Recall that the exposed seeds of gymnosperms lie at the base of the scales of the female cone. There is no fruit. In angiosperms, seeds develop within the ovary of a flower. As the seeds develop, the ovary enlarges to protect them and to attract animals and humans, who will eat the fruit and spread its seeds.

Classification of Fruits

Type	Description	Example
Simple	Develops from a single pistil (with simple or compound ovary) in one flower	Pear, apple, oak acorn, wheat, pea, cherry, milkweed, avocado
Accessory	Develops from simple or compound ovaries and other tissues	
Aggregate	Develops from several pistils in one flower	Strawberry, blackberry, tulip trees, magnolia
Multiple	Develops from several flowers growing together in a particular arrangement on the main stalk	Pineapple, fig, sweet gum, mulberry, sycamore, breadfruit

Angiosperm Life Cycle Explain

In Lesson 1, you learned about the alternation of generations in plants. Angiosperms are vascular plants, so the 2*n* sporophyte is their dominant generation. This is the plant you see, whether it is a bean plant or a maple tree. Which image in the diagram do you think is the sporophyte generation? Why?

The gametophyte generation of an angiosperm begins when the anthers and ovaries of a flower produce spores. Those spores contain the *n* number of chromosomes. They develop into the male and female gametophytes, which are so small that you almost can't see them. Locate the gametophytes in the diagram. From this perspective, you can see that the pistil and stamen are not the true male and female reproductive organs of the plant. They are spore producers. The tiny gametophytes—the pollen grain and the embryo sac—actually form the sperm and egg.

The Formation of a Seed

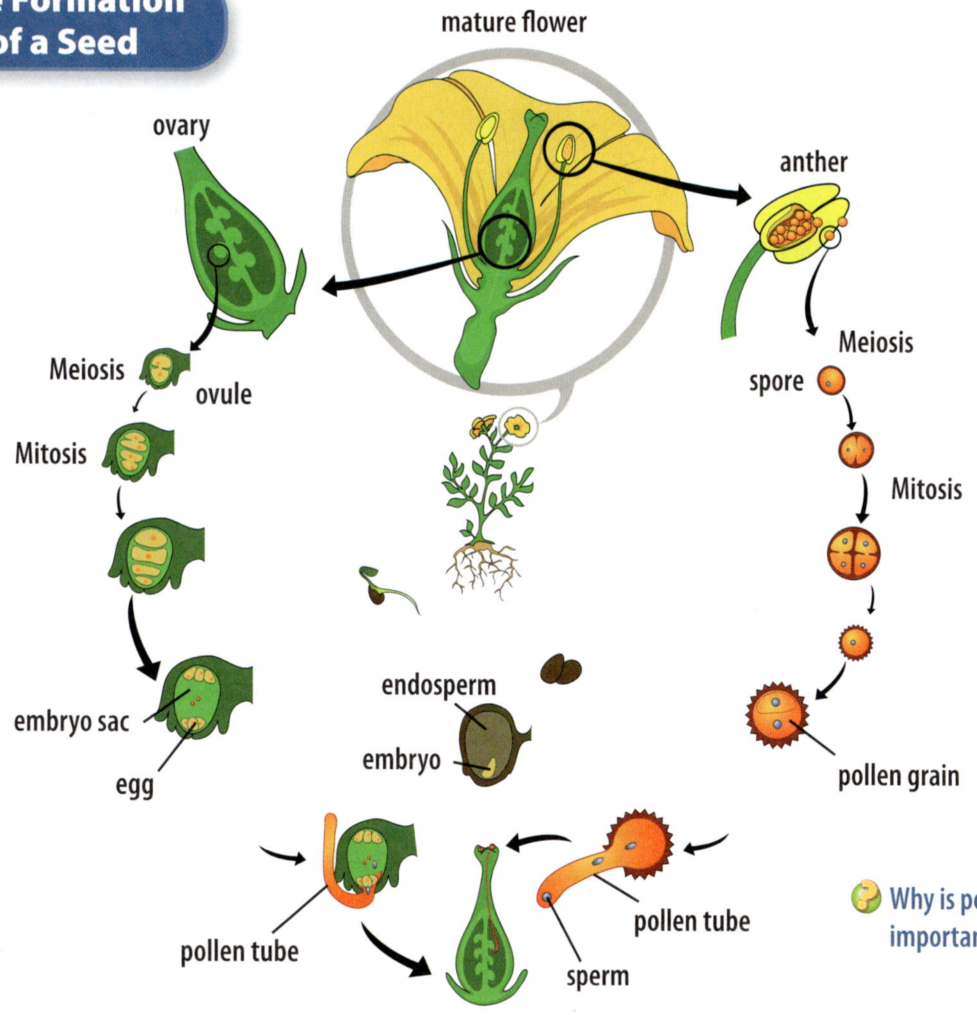

Why is pollination so important in this process?

Check out your ***Science Journal*** for a Guided Inquiry that explores the structure and function of cotyledons.

Extend

Pollination can occur in several ways. Pollen can be carried from the anther of a flower to the stigma of the same flower (if the flower is self-pollinating) or to another flower by the wind. Pollen can also be carried by animals. Flowers use fragrances, color, and nectar to attract pollinators, such as birds, butterflies, and bees. Why do you think fragrances and colors attract pollinators? As the organisms feed on the flower's nectar, they bump against the anthers, and pollen is transferred onto them. When they visit another flower, the pollen rubs off them and onto the stigma and grows a pollen tube toward the ovary. In the process, two haploid sperm form.

When the pollen tube reaches the ovary, one sperm will fertilize the egg to form a diploid zygote. The zygote develops into an embryo.

What happens to the second sperm? Recall that during meiosis, more than one haploid cell is produced. In angiosperms, one of the haploid cells becomes the egg that is fertilized by one of the sperm. The other sperm fuses with two of these haploid cells to produce a triploid ($3n$) nucleus that develops into the endosperm. The endosperm provides nourishment for the embryo as it grows.

Formation of a Seed	
Male Cell Division	• Meiosis occurs in anther • Male reproductive cells are formed, contained in pollen
Female Cell Division	• Mitotic and meiotic cell division occur in ovary • Results in single egg cell and two secondary nuclei • All are contained in embryo sac, the female gametophyte
Pollination	• Pollen germinates when it lands on stigma • Pollen tube grows toward ovary; two sperm cells produced • Pollen tube enters embryo sac
Fertilization	• Double fertilization occurs • Sperm fuses with egg to form a zygote • Sperm fuses with secondary nuclei to form endosperm (food source for growing embryo)
Seed and Fruit Growth	• Zygote grows into an embryo • Plumule (two tiny leaves) form in embryo, attached to hypocotyl, an embryonic stem • Cotyledons form in embryo • Ovary develops into a fruit • Embryo develops into a seed • Seed sprouts, seed coat breaks open, and cotyledons emerge • Plumule develop into first true leaves • Hypocotyl gives rise to stem below the cotyledons

Germination and Growth of a Seed

The parts of the flower that you identified in the Structured Inquiry are essential to reproduction in angiosperms. In those flower parts, both the reproductive cells and the gametophytes form. Use the pictures below as you learn about the formation of a seed.

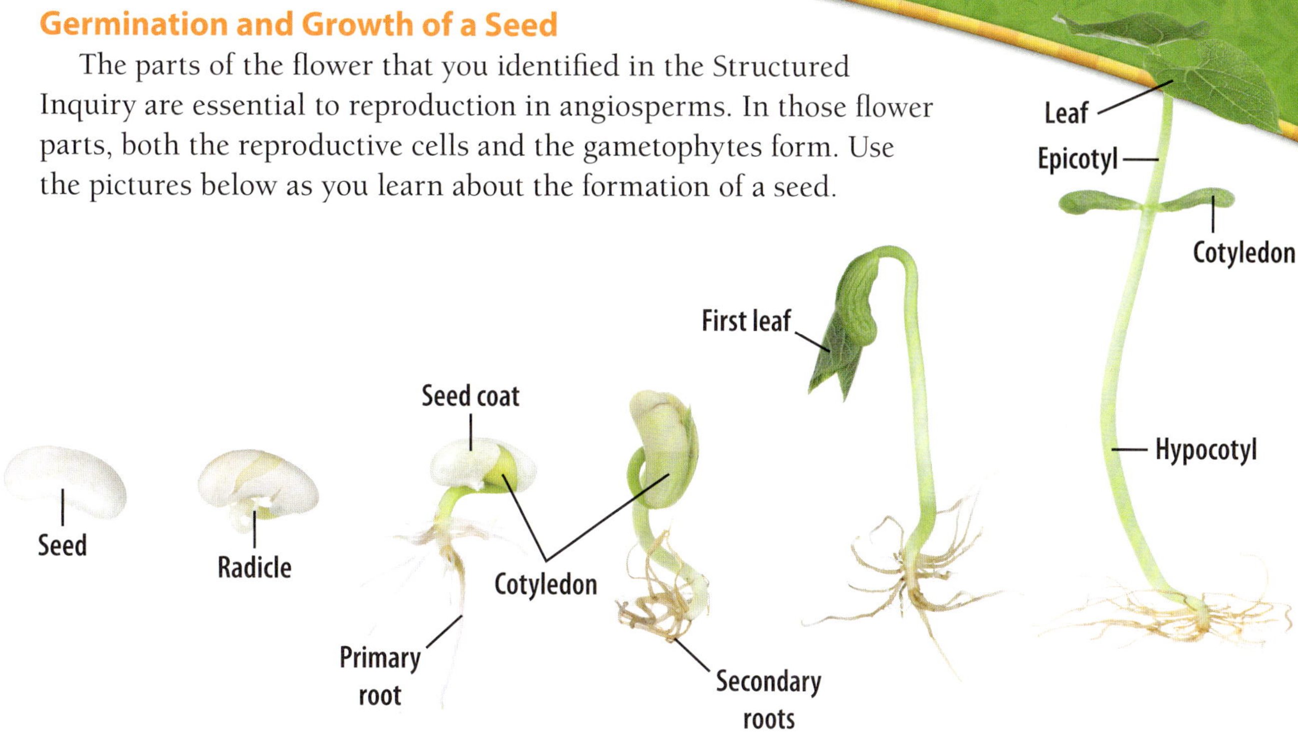

Concept Check Assess/Reflect

Summary: What are angiosperms? Angiosperms, the most numerous and widespread of all plants, are the flowering vascular plants that encase their seeds in a fruit that grows from the female part of a flower. They are efficient pollinators and masters of seed dispersal. They may be monocots (one cotyledon) or dicots (two cotyledons). The parts of an angiosperm include roots, stems, leaves, flowers, and fruits. The parts of a flower include the peduncle, receptacle, sepals (collectively called the calyx), petals (collectively called the corolla), stamen (including the filament, anther, and pollen), and pistil (including the stigma, style, ovary, and ovules). Flowers may be classified as complete or incomplete, perfect or imperfect, and regular or irregular. Fruit may be classified as simple, accessory, aggregate, or multiple. There are many subtypes of simple fruits. Because angiosperms are vascular plants, their dominant phase in the alternation of generations is the sporophyte. After pollen is carried from the anther to the stigma (pollination), the eggs are fertilized, and seeds develop.

1. Name and describe the structures that make up the male part of a flower.
2. Name and describe the structures that make up the female part of a flower.
3. Compare and contrast the two main groups of angiosperms.
4. In what ways is the alternation of generations different in gymnosperms and angiosperms?
5. What is double fertilization? How does the product of double fertilization aid a seed as it grows?

People in Science

Extend

Get to Know
Elizabeth Knight Britton

Have you ever visited a botanical garden? Many botanical gardens have plants from all over the world. They often have greenhouses where flowering plants bloom all year long. New York City (NYC) is home to a wonderful botanical garden thanks to an amazing woman named Elizabeth Knight Britton. In 1891 she helped establish the New York Botanical Garden. This was not an easy feat for a young woman living during the latter part of the Victorian Era.

Elizabeth Knight Britton graduated from what is now Hunter College in NYC in 1875. She became an expert on bryophyta. During this time, most young women were expected to keep house, entertain guests, plan household menus, embroider, paint, and play the piano. But not Elizabeth Britton—she was out collecting and cataloging plant specimens! She loved fieldwork and travelled to West Virginia, the Adirondack and Appalachian Mountains, Puerto Rico, Cuba, and Jamaica to collect and study mosses.

Today, many people are concerned about endangered plants and animals. But Elizabeth Britton was way ahead of her time. As early as 1902, she was concerned about wildflowers becoming extinct. She helped establish the Wildflower Preservation Society to teach the public the importance of these flowers.

Elizabeth Knight Britton left a huge legacy of scientific papers and specimens. She is remembered as a world-class botanist, for her role in establishing the New York Botanical Garden, and for creating a worldwide understanding of the need to preserve endangered plants and flowers.

Concept Check

1. What features might Elizabeth Britton have looked for while studying and classifying mosses?
2. In the late 1800s, it was not easy for women to freely choose their careers. What personal characteristics do you think Elizabeth Britton had that helped her to achieve her success?

Careers in Science

Extend

Forensic Botanist

Forensic science is the use of physical evidence to help solve crimes. Most often, the evidence we think of includes fingerprints, footprints, and, most recently, DNA evidence. But forensic botanists are a completely different group of scientists; they use plant remains to solve crimes.

Plants are everywhere. They produce large, easily seen objects such as seeds, fruit, leaves, and twigs. They also produce microscopic particles such as spores and pollen. Plants grow in specific environments, they grow in specific climates, and they produce different parts during different times of the year. All of these traits can be used by a forensic botanist to determine where a crime occurred, at what time of year it occurred, how long a person has been dead, and if a body has been moved before discovery by the police.

Pollen and spores decay very slowly. They have even been found in ancient Egyptian tombs more than 4000 years old!

Suppose a forensic botanist identifies a certain type of pollen stuck in the ear wax of a murder suspect. And this pollen just happens to also be present on the clothing of the victim. Now assume that this pollen grows only in a certain place. The police can reasonably conclude that both the suspect and victim were in the same place at the same time. A case could be built around this evidence to link the suspect to the crime. Some forensic botanists analyze stomach and intestine contents to determine what a victim last ate. They know how long it takes certain plant materials to decompose. This can give the police an idea of how long the person has been dead.

If you like the idea of solving crimes, a career as a forensic botanist could be for you. Forensic botanists attend a university and take courses in botany, chemistry, and mathematics. They are able to provide both written and oral reports on their findings. They also learn to analyze crime scenes critically and logically using the evidence provided. They work in places such as police departments, crime laboratories, and coroners' offices.

Concept Check

1. Where do you think forensic botanists work most often, in the field or in the lab? Explain your answer.
2. Suppose a crime suspect's shoes were covered with tiny thorns. How could a forensic botanist use this evidence to help solve the crime?

Chapter 2 Review Assess/Reflect

Study Guide

Lesson 1

1. Plants can be classified by whether they are vascular or nonvascular. Nonvascular plants include mosses and liverworts. Vascular plants are divided into seedless plants, gymnosperms, and angiosperms. Seedless vascular plants include ferns and horsetails. Gymnosperms are divided into several groups, including conifers, yews, cycads, and ginkgos. Angiosperms are divided into monocots and dicots.
2. Nonvascular plants lack vascular tissue, xylem and phloem, and cannot transport water and nutrients from one part of the plant to another. Vascular plants have xylem and phloem, which allows them to move water and nutrients from one part of the plant to another.
3. Plants can be classified by how they reproduce. Most gymnosperms produce seeds in woody cones. Angiosperms produce seeds in the ovary of a flower. Other plants reproduce using spores.
4. Plants go through a reproductive process characterized by the alternation of generations, in which the plant alternates between a spore-producing plant, the sporophyte, and a gamete-producing plant, the gametophyte. In nonvascular plants the gametophyte is the dominant plant (the largest and most visible). In vascular plants, the sporophyte is the dominant plant.

Lesson 2

1. The nonvascular plants are small and have small rootlike structures called rhizoids that enable them to absorb water and nutrients directly into their cells. Groups of nonvascular plants include mosses, liverworts, and hornworts. All of these plants grow in damp, shady places.
2. Spore-producing vascular plants include Pterophyta, the ferns. The division Lycophyta includes the club mosses. The horsetails are included in the division Equisetophyta.
3. Ferns have horizontal stems called rhizomes. The leaf of a fern, which comes in a variety of shapes, is called the frond. The frond is made up of many smaller leaflets. Newly emerging fronds have a coiled shape and are called fiddleheads. Sori contain several spore-producing structures called sporangia and are found on the underside of the leaflets.

Lesson 3

1. Gymnosperms are seed-bearing, vascular plants. The seeds of gymnosperms are not encased in fruit but are exposed. Many gymnosperms, such as cycads and conifers, are evergreens, but some are deciduous. Most have needlelike leaves.
2. Groups of the gymnosperms include the cycads, the gingkos, gnetophytes, and conifers. See the chart on page 67 for the descriptions of these groups.
3. The alternation of generations cycle in gymnosperms includes the large dominant tree, which is the sporophyte stage. The male cones produce the pollen, which is part of the male gametophyte. The female cone produces the ovules, which are part of the female gametophyte. The sperm from the pollen combines with the egg to form a zygote. The zygote develops into the seed, which drops to the ground to become the sporophyte tree.

Lesson 4

1. Angiosperms are flowering, seed-bearing, vascular plants. Their seeds are encased in a fruit.

2. Flowers are the reproductive structures of angiosperms and are specialized for the production of spores that produce the tiny gametophytes (sperm and eggs). When the sperm contained in the pollen makes contact with the stigma of a flower, a tube is formed through which the sperm travels to the ovary and fertilizes an egg.

3. Angiosperms are classified as either monocot or dicot. Monocots have seeds with one cotyledon, flower parts in patterns of 3s, and leaves with parallel veins. Dicots have seeds with two cotyledons, flower parts in 4s and 5s, and leaves with netted veins.

4. Fruits may be classified into four categories: simple, accessory, aggregate, or multiple. Simple fruits develop from a single pistil in one flower. Aggregate fruits develop from several pistils in one flower. Multiple fruits develop from several flowers that grow together.

Vocabulary Check

Fill in the blanks with the correct terms.

1. The small structures on mosses that consist of a stalk and a capsule are the __________ generation.

2. An angiosperm embryo that has only one cotyledon is a(n) __________.

3. The system that moves water and food through a series of tubes in some plants is the __________ system.

4. The tiny structures in the sorus of a fern that produce spores are called __________.

5. Trees that lose their leaves annually are referred to as __________.

6. A conifer that retains its needles year round is a(n) __________.

7. The fruit of an angiosperm forms from the __________ of a pistil.

Multiple Choice

Choose the best answer.

8. Which is absent in angiosperms?

A. seeds **B.** cones **C.** flowers **D.** fruits

9. Which is not needed for a plant leaf to perform photosynthesis?

A. sunlight **B.** carbon dioxide **C.** glucose **D.** water

10. Which is a true statement about the gametophyte generation compared with the sporophyte generation?

A. It is dominant in ferns.
B. It is dominant in gymnosperms.
C. It is dominant in angiosperms.
D. It is dominant in mosses.

11. What quantity represents the chromosome number of the gametophyte generation in mosses?

A. $\frac{1}{2}n$ **B.** n **C.** $2n$ **D.** $3n$

Check Point

Answer the following questions.

12. If the chromosome number in the sporophyte generation of a plant is 32, what is its number in the gametophyte generation? Justify your answer.

13. In terms of chromosome numbers, how might meiosis and fertilization be thought of as opposite processes?

Chapter 3

How Plants Survive

Scripture Spotlight

Jesus used parables about plants to teach spiritual lessons. As you learn about how plants make food and adapt to their environments, be alert for applications that can strengthen your faith. You will read the following passages in this chapter:

Micah 3:8 (p. 91)
Acts 1:8 (p. 91)
Genesis 1:26 (p. 110)
Revelation 11:18 (p. 110)
John 3:19–21 (p. 103)
Psalm 144:12 (p. 98)
John 12:32 (p. 103)

This tree sits at the top of a hill and must withstand a lot of wind. Notice the direction in which the branches grow.

The Big Idea

God designed plants with the ability to make their own food and adapt to their surroundings. As caretakers of God's Creation, humans should work to protect plant species.

How might humans change environments and affect plant survival?

Inquiry Kick-Off

Engage

Plants must produce seeds to help species survive. What are some ways that seeds travel? How does this help a species survive? Plants have adaptations that allow their seeds to travel in a variety of ways. In your *Science Journal*, you will explore adaptations seeds have for travel.

Lesson 1

Essential Question

How Do Plants Capture Energy?

Objectives

- Summarize the process of photosynthesis.
- Explain the difference between the light and dark reactions of photosynthesis.
- Distinguish between the structure and function of xylem and phloem.
- Explain the process of transpiration in plants.
- Describe common plant hormones and their functions.

Vocabulary

simple sugar
vascular tissue
vascular cambium
transpiration
meristem
cork cambium

Is energy important? Where do humans get energy to move, live, and reproduce? Where do plants get their energy? What do you know about the energy cycle of our planet? Where does all of this energy come from and where does it go? In this lesson you get to explore the very basics of life!

How Plants Make Food Explain

What are some characteristics that set plants apart from other living organisms? One characteristic you may have listed is that plants make their own food. Their structure allows them to transport that food throughout their roots, stems, and leaves.

As you recall, the majority of plants use the process of photosynthesis to get the energy they need. Photosynthesis can be summarized as the following chemical equation:

$$6CO_2 + 6H_2O + \text{light energy} \rightarrow C_6H_{12}O_6 + 6O_2$$

Leaf Antatomy

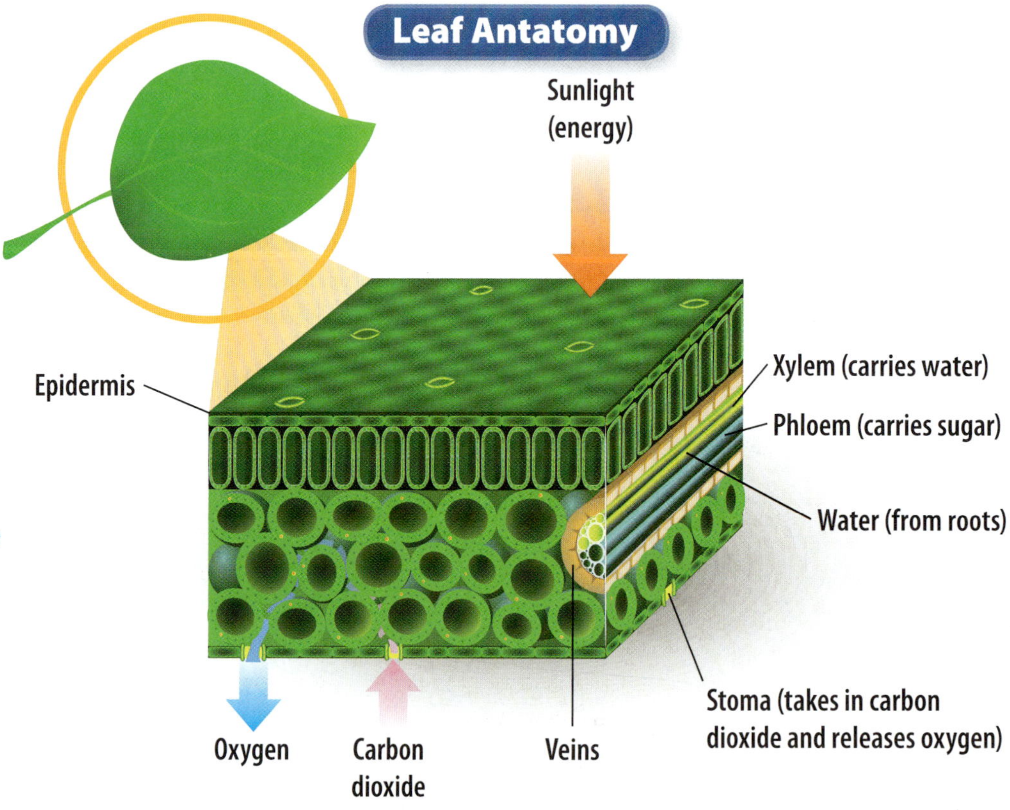

How does the leaf's structure enable the plant to make its own food and utilize the energy from that food?

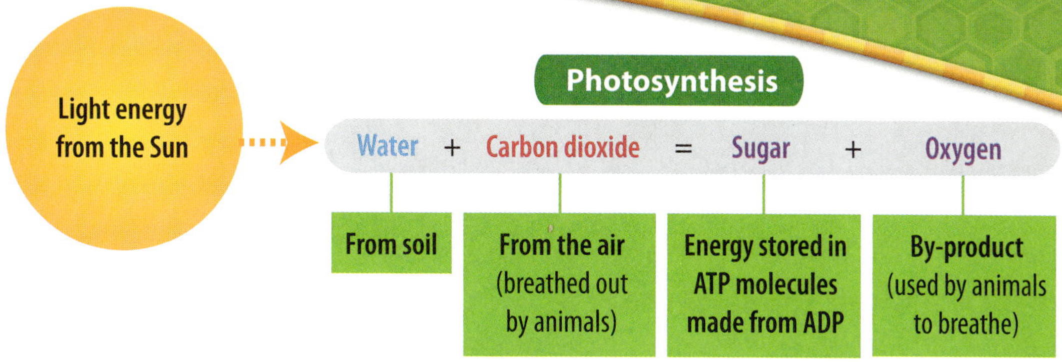

Recall that plants take carbon dioxide (CO_2) from the air and, using energy from sunlight, rearrange the atoms and combine them with atoms from water (H_2O). This rearrangement of atoms occurs in *chloroplasts*, which are organelles in plant cells that contain the green pigment *chlorophyll*. Two products are formed:

- *Glucose* is a **simple sugar** that cannot be broken down into smaller carbohydrate units. Each molecule of glucose has 6 carbon atoms, 12 hydrogen atoms, and 6 oxygen atoms: $C_6H_{12}O_6$.
- *Molecular oxygen* is a molecule that contains 2 oxygen atoms.

The chemistry of photosynthesis looks simple in its summary equation, but many steps are involved that use and produce important compounds. These compounds include:

- $NADP^+$—nicotinamide adenine dinucleotide phosphate
- NADPH—nicotinamide adenine dinucleotide phosphate plus an electron from hydrogen
- ADP—adenosine diphosphate
- ATP—adenosine triphosphate
- PGAL—glyceraldehyde 3-phosphate

The steps of photosynthesis are easier to understand if you consider them in two parts—the light reactions and the dark reactions.

Scripture Spotlight

Plants get their energy from the Sun. Where do Christians get the power for their lives? Compare **Micah 3:8** and **Acts 1:8**.

Explore-a-Lab

Structured Inquiry

Under what conditions do plants take in carbon dioxide?

Label three test tubes: *Control, Dark,* and *Light*. Pour 75 mL of spring water into a 250-mL beaker. Add 2 mL of bromothymol blue to the water. Using a straw, slowly blow into the solution, causing it to gently bubble for one minute. What gas are you adding to the solution? What color did the solution change to? Fill each of the test tubes about three-fourths full of the solution. Seal the *Control* tube with a stopper. Place a 5-cm length of elodea into each of the other two tubes. Gently push the plants into the solution with the straw. Seal with stoppers. Place the *Control* and *Light* tubes in light for 24 hours. Place the *Dark* tube in a dark location for 24 hours. Then, observe the solutions in each test tube. Did the color change? Why or why not?

Structured Inquiry
Discover

Record your work for this inquiry. Your teacher may also assign the related Guided Inquiry.

Science Journal

Transpiration in Plants

How can we observe the amount of water plants release into the air through their leaves?

Materials
- 200 mL of water
- graduated cylinder
- 4 clear plastic cups
- plastic wrap
- elastic band
- scissors
- plant shoot or leaves with their stalks
- petroleum jelly
- masking tape
- windowsill with full sunlight or bright light

Procedure

1. Pour 100 mL of water into each of two cups.
2. Cover the tops of the cups tightly with plastic wrap. Secure the plastic wrap around the cup with an elastic band or masking tape.
3. Use the point of a pencil to punch a very small hole in the center of the plastic wrap of each cup. The hole should be large enough for only the stalk of the leaf/plant to pass through.
4. Push the plant shoot or leaf stalk through the hole in one cup. Be careful not to damage the leaf or plant. Make sure the leaf or plant is in the water.
5. Use the petroleum jelly to seal around the holes in the plastic wrap for both set-ups. One of the cups will not have a plant in it.
6. Place a plastic cup upside down over the leaf/plant stalk. Repeat this with the other set-up. One cup will enclose the plant or leaves; the other cup will enclose the empty set-up. The empty set-up will serve as the control. Seal the seam where the two cups meet with masking tape.
7. Place your cup assemblies in sunlight. **Observe** your cups every 10 minutes for an hour. What differences do you see?
8. Leave the assemblies in place for a day. Remove the tops, plastic wrap, and leaves. **Measure and record** the amounts of water that remain in the cups.

Analyze Results

What differences did you observe between the two cups?

Create Explanations

1. How can we measure the amount of water plants release into the air through their leaves?
2. What was the independent variable in this experiment?
3. Was water transpired by the leaves? Explain.
4. Suggest an experiment you could perform to test variables that might affect transpiration.

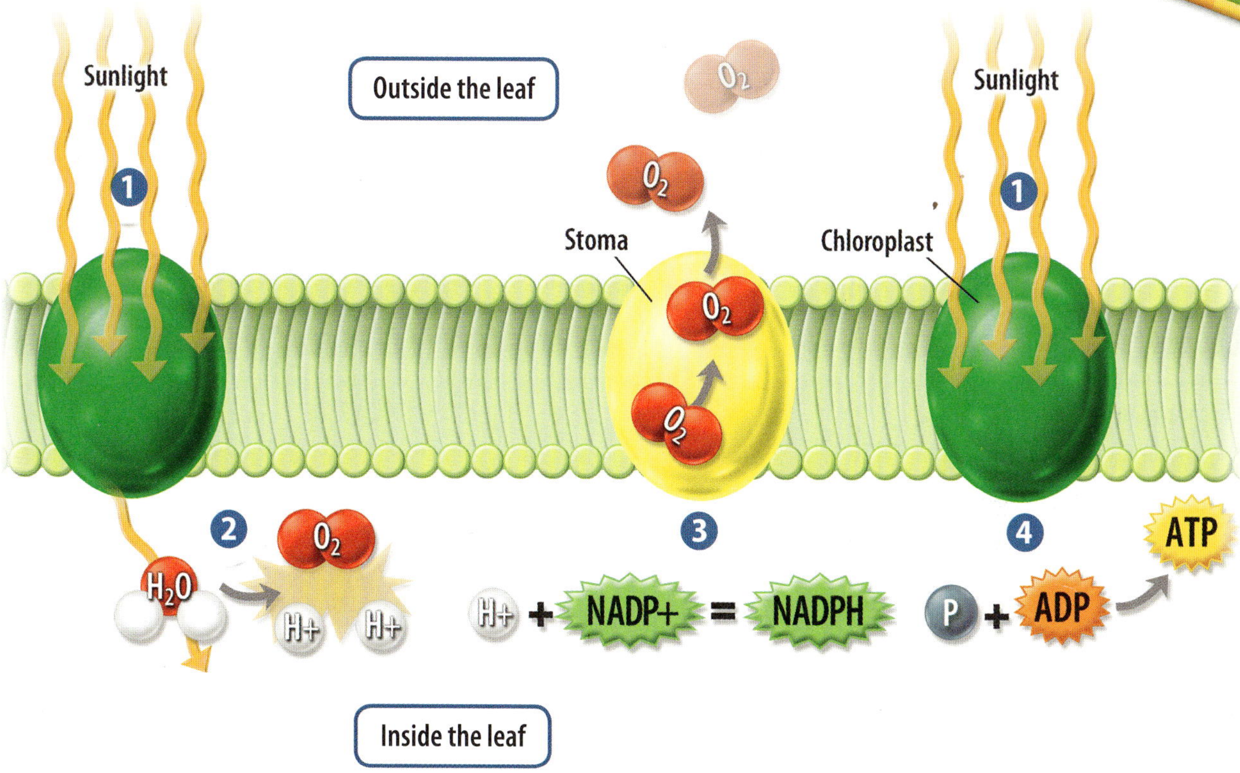

The Light Reactions

When light strikes the leaves, it is absorbed by the chlorophyll in the chloroplasts. The light energy splits water into oxygen and hydrogen. Oxygen is released during this process. Because these steps take place only in the presence of light, this part of photosynthesis is called the light reactions. Refer to the diagram of the light reactions as you study the steps of this process in more detail.

1. Light energy is absorbed by the chlorophyll in the leaf. This light energy is used in two ways.
2. The light energy is used to split the water molecule into hydrogen and oxygen.
3. The hydrogen produced is added to NADP$^+$, which already exists in the cells of the leaf, to make NADPH. The oxygen is not needed and is released by the leaf.
4. The light energy absorbed is also used to add a phosphate group to ADP, which also already exists in the cells of the leaf, to make ATP.

The products of the light reactions are NADPH, ATP, and oxygen. The NADPH and ATP are energy-rich compounds and are needed for the dark reactions in photosynthesis.

Check out your *Science Journal* for a Structured Inquiry that explores testing whether photosynthesis is occurring.

Extend

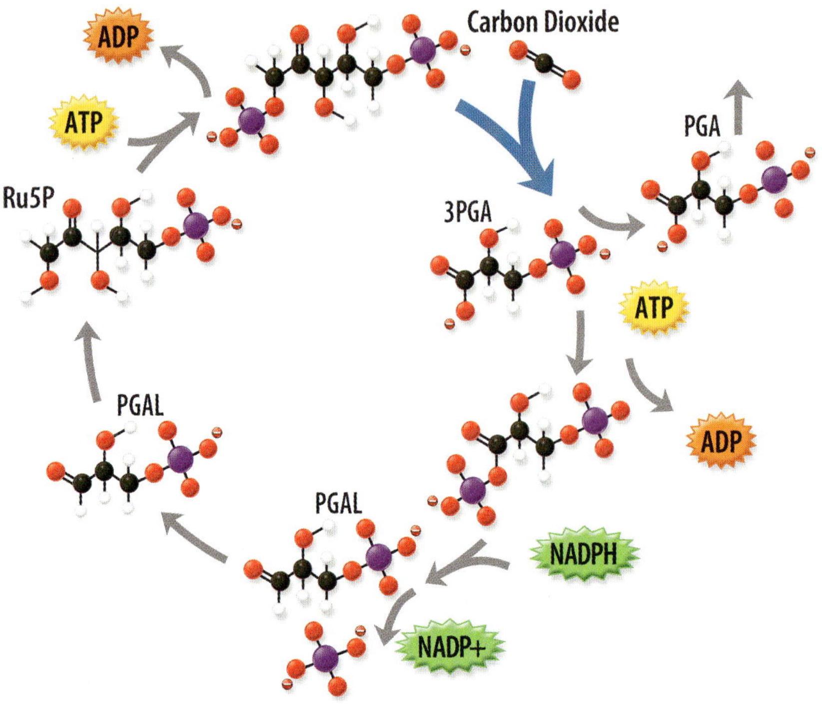

In the chloroplast, many chemical reactions trap energy and produce food.

Why is oxygen sometimes called a "waste product" of photosynthesis?

The Dark Reactions (The Calvin Cycle)

In another part of the chloroplast, the Calvin cycle occurs. The Calvin cycle is sometimes called the "dark reaction" pathway because it does not need light energy. Refer to the diagram of the dark reactions as you study the steps of this process.

1. Carbon dioxide (CO_2) enters the plant from the atmosphere.
2. Carbon fixation occurs. Carbon dioxide molecules are incorporated into three-carbon molecules in the chloroplast.
3. Chemical reactions cause ATP and NADPH to donate electrons, resulting in ADP and $NADP^+$. The electrons combine with three-carbon molecules to form PGAL molecules. This is the reduction phase.
4. Some PGAL molecules are reused to help keep the Calvin cycle in motion.
5. Some PGAL molecules leave the cycle and are used to build the organic molecules that make up the cell. *Starch* is one larger molecule that can be formed. It is made from many glucose molecules bound together. The fibrous material in plants, *cellulose*, can also be formed.

The light reactions and the Calvin cycle depend on each other. ATP and NADPH from the light reaction drive the Calvin cycle. In turn, $NADP^+$ and ADP from the Calvin cycle are needed to accept the energy trapped during the light reactions.

How Plants Use Food Explain

Why is the glucose produced during photosynthesis important to plants? Recall that cellular respiration is a process in which cells obtain the energy they need to survive. This is not the respiration that you might think of as breathing. In cellular respiration, cells use oxygen to break down glucose and store its energy in molecules of ATP. About 40 percent of the energy in glucose is transferred to ATP. The rest is used to power the life processes of the plant and is eventually released as heat. Where do animals get the oxygen required for cellular respiration? Where do plants get their carbon dioxide? Cellular respiration uses the oxygen produced by photosynthesis, and photosynthesis needs the carbon dioxide produced by cellular respiration.

Cellular respiration is critical for the survival of most organisms because cells use ATP to power virtually all of their activities. What cellular activities are powered by cellular respiration? Plants use the chemical reactions of cellular respiration to harness the energy stored in the glucose made during photosynthesis. In cellular respiration, glucose combines with oxygen. It releases water, carbon dioxide, and the energy the cell uses to power its activities. Is there ever a time when cellular respiration might stop?

The process of cellular respiration can be summarized as below. How does this equation relate to the one for photosynthesis?

$$C_6H_{12}O_6 + 6O_2 \rightarrow 6CO_2 + 6H_2O + \text{energy}$$

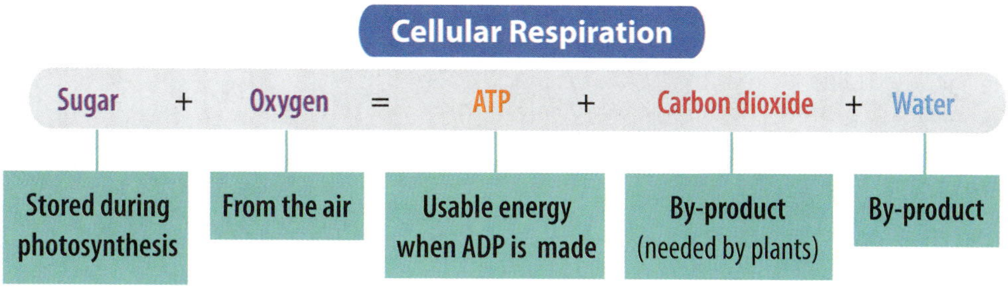

Explore-a-Lab

Guided Inquiry

How do photosynthesis and cellular respiration work together?

The processes of photosynthesis and cellular respiration occur in cycles. Make a diagram detailing the cycles of photosynthesis and cellular respiration. Make sure your diagram is labeled clearly and shows what occurs at the chemical level. How are the "waste products" of one process the "raw materials" of the other? Then use the materials supplied by your teacher to create 3-D models of the processes.

How Do Plants Transport Materials?

Most photosynthesis in plants takes place in the leaves. An important raw material for photosynthesis is water. The water needs to travel to the leaves. The food produced in the leaves needs to reach the rest of the plant to provide energy for cellular activities. You may wonder how all this happens. The transport of water and food in a plant is accomplished through its vascular system.

Two types of **vascular tissue** are *phloem*, which carries food, and *xylem*, which carries water and minerals.

Xylem

You may have seen a tree trunk that has been cut to expose a cross-section of the stem. If you look closely, you will see that the center of the trunk is usually darker than the areas surrounding it. The darker center is the heartwood, which is made up of old, nonfunctioning xylem cells. What function might these dead cells have in the tree?

A cross-section of an elderberry stem

Which layer of this trunk produces the xylem and the phloem?

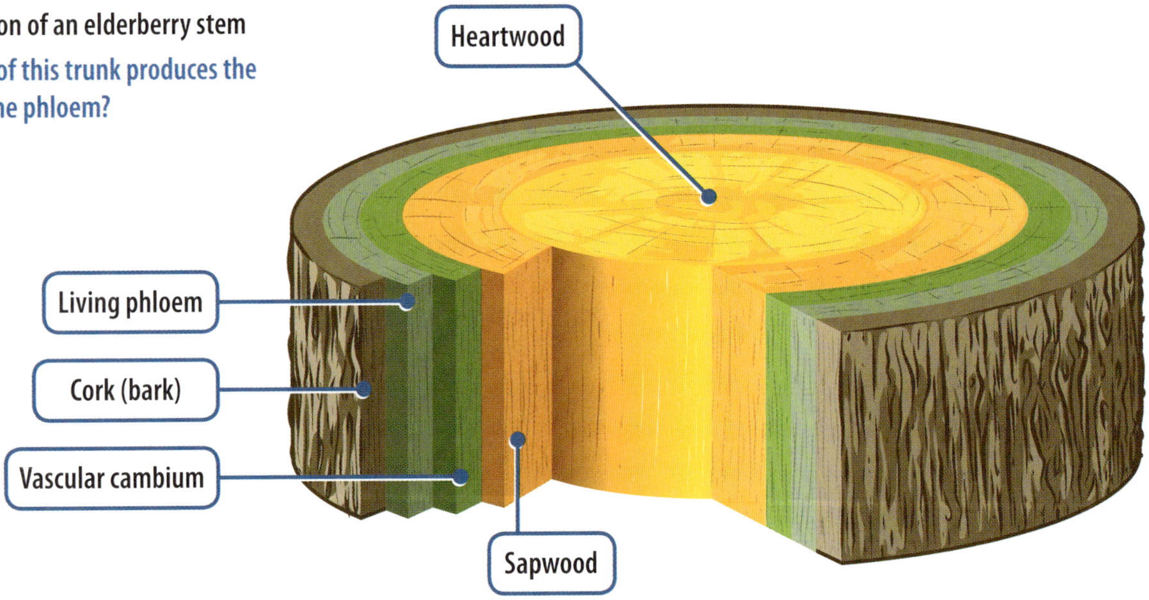

Surrounding the heartwood are "rings" of active xylem that transport water and minerals up the stem of the plant. These layers of active xylem are called the *sapwood* of the tree. Sapwood moves the water and minerals without using energy. A plant uses the water transported by the xylem in the process of photosynthesis. Where other than the stem of the plant would you find xylem? Why?

Next to the sapwood is the **vascular cambium**. This layer produces new xylem and phloem cells. The other layer of the tree trunk is the bark. It protects the living cells inside the tree trunk.

Phloem

Vascular tissues also move the glucose produced during photosynthesis to all parts of the plant. The vascular tissue that transports glucose is the phloem. Although xylem can move water and minerals in only one direction—from the roots to the leaves—phloem can move glucose both up and down the plant. Why is movement of materials in both directions important to the plant? The tree trunk also has dead phloem cells, which make up the cork. Why is the cork important to a tree? Do you think a tree can survive without its cork?

In the roots, energy is used to remove sugar from the phloem. The sugar can then be used for energy or converted to a storage form, such as starch. The increased concentration of sugar or starch changes the shape of the root. What are some plants that have this feature?

Transpiration

Water is pulled up the xylem during a process called **transpiration**, which is water loss from leaves as it escapes through their *stomata*. Each *stoma* is an opening in the surface of the leaf that allows gases to enter and exit during photosynthesis. What are these gases? In most plants, when light is present, the stomata open so the plant can photosynthesize. This also increases the rate of transpiration.

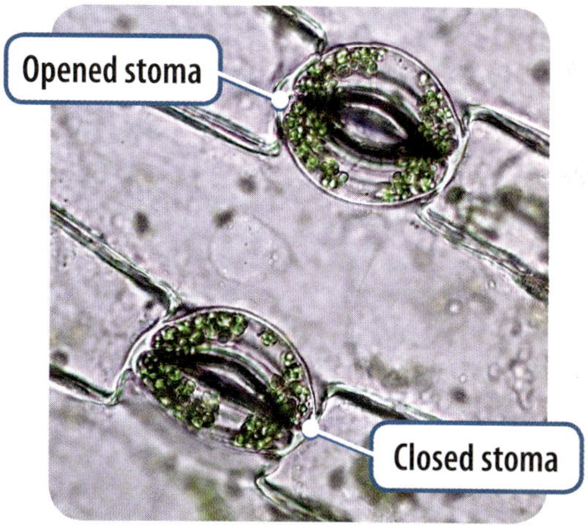

Two kidney-shaped guard cells border a stoma.

As transpiration occurs, the concentration of water molecules in leaf cells decreases. Through *osmosis*, water molecules move from the xylem in the leaf's veins into the leaf's other cells. This process causes the concentration of water molecules at the top of the xylem's water column to decrease. Thus, transpiration serves as a molecular "tug" that keeps water moving up from the roots into the leaves of even the tallest trees.

Explore-a-Lab

Guided Inquiry

How are stomata and the process of photosynthesis related?

Tear off a piece of an outer leaf of iceberg lettuce. Bend the leaf in half. Use tweezers to peel off a small piece of the epidermis (the transparent skin) from the underside of the thicker part of the leaf. Prepare a wet mount of this tissue. Examine the specimen under the microscope. Count the total number of stomata in your field of view. Record. Then count the number of stomata that are open. Record. Make a second slide of the leaf epidermis. Place a few drops of salt solution on the leaf when preparing the wet mount. Count and record the total number of stomata and the number of open stomata. Calculate the percentage of open stomata on each slide and record. How did the saltwater affect the stomata? Why? How are stomata and photosynthesis related?

Plant Growth **Explain**

Some of the largest of all living things are plants. The world's largest tree (by volume) is the General Sherman, a giant sequoia located in Sequoia National Park in California. It stands 84 m tall. Its circumference at its base is 31 m. How can plants grow to be so large?

Plant growth happens as the tips of roots and shoots grow longer. This type of growth at the tips of plants is called *primary growth*. As stems and branches grow longer, they need more support. They get this support as the stems grow thicker. Growth in plants in which stems grow thicker is called *secondary growth*. Both primary and secondary growth originate in growth regions of the plants called **meristems**. These growth regions are located in different parts of a plant. Cells in a meristem divide continually, creating new tissues for the plant.

Scripture Spotlight

How does David describe the growth of boys in **Psalm 144:12**? As you grow up, which characteristics might be like the primary growth of plants and which might be like the secondary growth?

Process of Plant Growth

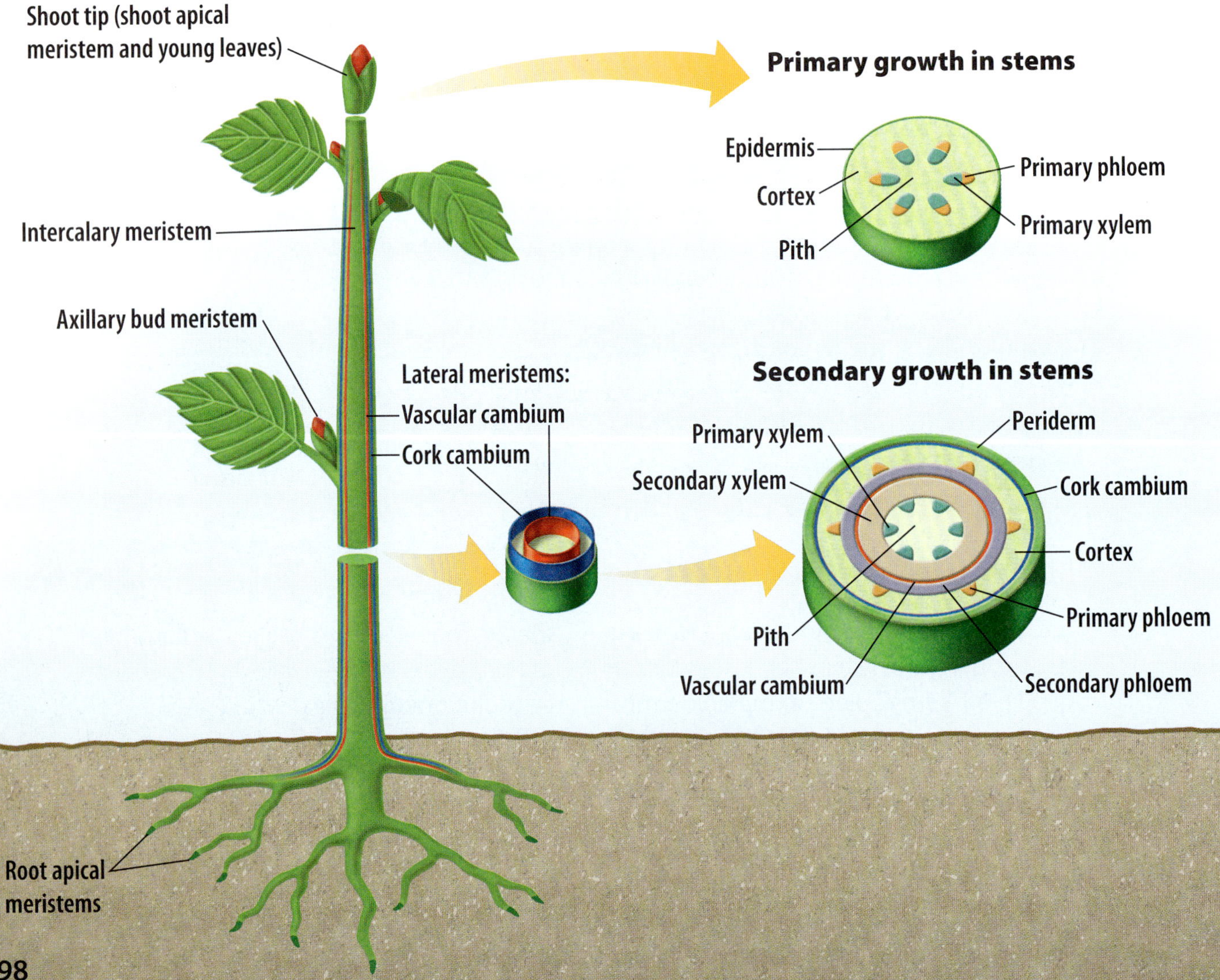

Apical meristems at the ends of these branches make the branches grow longer.

The meristems responsible for primary growth (growth in height) include:

- *Apical meristems* are found at the tips of stems and roots. They produce an increase in length at the farthest growth point, for example, at the top of a tree or at the tip of a root.
- *Intercalary meristems* are found between mature tissues in stems near the base of stems and leaf blades. Cell division in this part of the plant helps to elongate parts of the plant. For example, intercalary meristems are common in grasses. They appear at the base of grass blades and allow damaged or cut grasses to rapidly regrow.

The meristems responsible for secondary growth (growth in diameter) include:

- The *vascular cambium*, which is located between the xylem and phloem, produces an increase in the plant's vascular tissues, xylem and phloem. As the xylem and phloem increase, the stems grow wider.
- The **cork cambium**, which is located outside the phloem, produces cork tissue. Cork is made of cells that secrete a waterproof substance to protect the plant against water loss and then die. Cork, cork cambium, and phloem make up the outer layer in woody stems. Together, they are called *bark*.

The General Sherman tree in Sequoia National Park is estimated to be 2300–2700 years old.

How is tree bark important to this plant?

Math in Science

The General Sherman tree has a base circumference of 31 m. Calculate the diameter of the tree at its base.

Plant Hormones Explain

Plants produce *hormones*. The hormones send signals to specific cells, signaling the cells to "switch on" certain genes. The active genes direct the synthesis of enzymes and proteins that bring about a change in the cell's structure or function. How are plant hormones similar to hormones in animals?

Types of Plant Hormones

Hormone	What It Regulates	What Do You Know?
Auxins	Increase the length of most plant cells; contribute to growth and elongation of the plant	When might there be an abundance of auxins in a plant?
Abscisic acid	Suspends cell growth and division when the plant is ready for a state of dormancy; acts as a coping mechanism for stress during the growing season	How might a plant become stressed during its growing season?
Ethylene	Encourages ripening and dropping of leaves and fruit; slight pressure permits leaves and fruit to break free from stem	How does ethylene affect fruit such as apples, bananas, or avocados?
Gibberellins	Growth-regulating; stimulate plant growth via rapid stem elongation	Why might gardeners or farmers apply gibberellins and cytokinins to their plants? What would be some advantages and disadvantages of adding these hormones to crops?
Cytokinins	Induce plant cells to undergo mitosis; encourage increased growth in the roots and stems of plants	

Plants have unique processes and specialized tissues to make food and move materials back and forth through their structures. They grow to be some of the largest organisms on Earth. How is it that these organisms, anchored to one place, can be successful and survive in their environments?

Farmers are scientists who need knowledge of plants, the environment, and engineering.

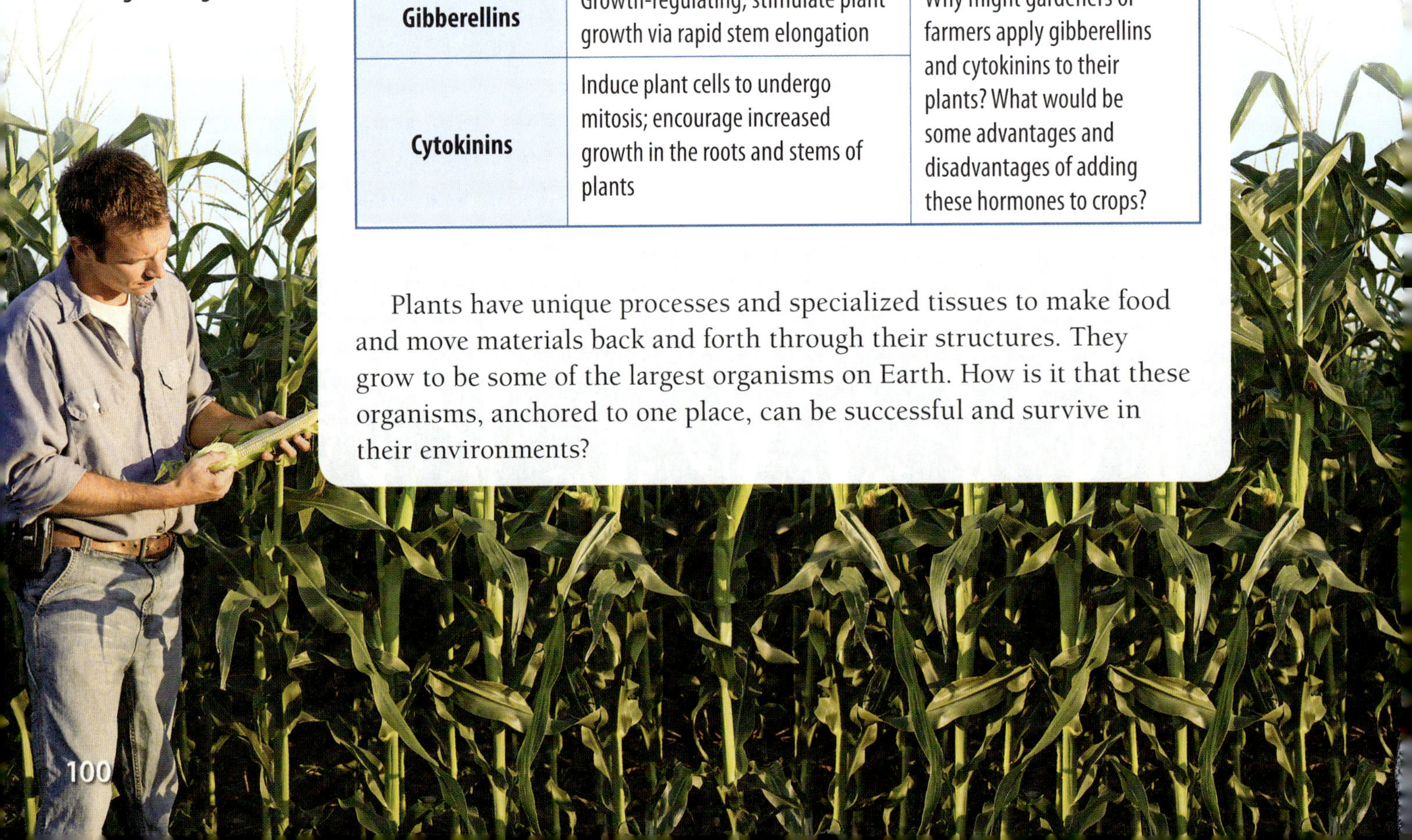

Explore-a-Lab

Structured Inquiry

How can we demonstrate the effect of a plant hormone?

How do you know whether a pear is ripe? How do you know when it is too ripe? Decide, as a group, what you will look for to determine the ripeness of a pear. Write your ideas on a sheet of paper. Then your teacher will provide your group with pears and apples. Label the pears *1, 2,* and *3,* using masking tape and a marker. Observe the pears and note their ripeness. Place one pear in a paper bag. In a second paper bag, place one pear and one apple. Place one pear and two apples in a third paper bag. Seal all the bags shut. Every 24 hours, open the bags and check the pears for ripeness. Record your observations. Determine the time it takes for the pears to ripen in all three configurations. Make a graph of the average ripening time versus the number of apples in the bag. Which pear ripened fastest? What variable caused the differences in ripening time? Why do you think that happened? If you did this activity without closing the bags, would the outcome change? If so, how?

Concept Check **Assess/Reflect**

Summary: How do plants capture energy? There are two parts to photosynthesis: the light reactions (require light) and the dark reactions (do not require light). In the light reactions, atoms are rearranged to form simple sugars, which plants use as food. Oxygen is a waste product of this process. The dark reactions use ATP and NADPH to capture released energy in PGAL, which is used to build starch and cellulose.

Food and water are transported through vascular tissue, phloem and xylem. In the transpiration process, plant leaves lose water through tiny holes called stomata. Water is pulled up the xylem during transpiration and exits the plant through the stomata. Meristem cells divide continually, creating new tissues for the plant as it grows. Plant hormones send messages to cells to "switch on" certain genes, which direct growth.

1. How are the light reactions and the dark reactions different?
2. What is the difference between transport and transpiration in plants?
3. Carbon dioxide is an important part of photosynthesis. Predict the effect on plant growth of increased amounts of carbon dioxide in the atmosphere due to climate change. Explain the reason for your prediction.
4. Apples, melons, and avocados are known to release the hormone ethylene, even at low temperatures. How does this knowledge affect how you would store these fruits with other fruits and vegetables?

Lesson 2

Essential Question
What Are Plant Adaptations?

Objectives
- Describe plant tropisms and give examples.
- Cite examples of how plants are adapted to their environments.

Vocabulary
tropism
phototropism
heliotropism
geotropism
thigmotropism

The Sonoran Desert, covering territory in California, Arizona, and Mexico, is home to plants perfectly adapted to the environment, such as the barrel cactus and the chain fruit cholla. How are they adapted to withstand high temperatures during the day and low annual precipitation? How are those adaptations different from those of plants living in the arctic?

Unlike animals, plants are rooted to one place, so they are forced to respond to stimuli and to changes in environmental conditions in place because they are unable to relocate themselves. What would happen if you moved a plant from a very different environment to your environment?

Tropisms Explain

What does a plant do in response to sunlight? What are some other stimuli that cause plants to respond? Plants will respond with movement. Some responses are relatively slow; they are so subtle, you may not notice them. Other responses are rapid. These movements are adaptations that help a plant survive in its environment. Any movement in a plant that occurs in response to a stimulus is called a **tropism**. Movement toward a stimulus is a *positive tropism*. Movement away from a stimulus is a *negative tropism*. Plants are very sensitive to their environment and have various forms of tropisms to ensure their survival.

Immature sunflower buds following the Sun.

In the sunflowers' bud phases, where would you expect the auxin concentration to be the greatest?

Explore-a-Lab

Structured Inquiry

How can we demonstrate phototropism?

Your teacher will provide your group with a small potted plant. Using a piece of tape or a sticker, mark one side of the pot. Place the plant in a sunny window, with the marked side toward the window. Sketch the plant or take digital photos of the plant. Keep the soil moist during the course of the activity. Over the next three days, observe the plant's growth. Each time you observe the plant, make another sketch or take another photo. After three days, turn the plant so that the marked side of the pot faces away from the window. Observe the plant for three more days, sketching or photographing with each observation. How could you use this information if you grew your own plant at home?

Response to Light

The directional response of plants to light is called **phototropism**. It requires the absorption of light by proteins known as phototropins. Phototropins change to a new shape when they absorb certain wavelengths of light. The new shape causes phototropins to act as "on" switches for a gene that regulates the activity of auxins. Recall that auxins are plant hormones. Auxins can cause the positive phototropism of stems. What does it mean to have a positive phototropism? Light on the bright side of the stem causes auxin to move to the shaded side of the stem. The cells with auxin will grow faster and cause the stem to curve toward the light. Most plants are phototropic. How would this be different in a plant that has negative phototropism? Do you think there are any plants that lack phototropism? Why or why not?

Some plants have a very rapid response to sunlight, in which the plant will follow the Sun throughout the day. This response is called **heliotropism**, or solar tracking. Some examples of plants that exhibit heliotropism are the sunflower you see on the previous page, as well as alfalfa, cotton, and soybean. The cells of these plants have very elastic cell walls. What is the advantage of a plant being heliotropic? What are the disadvantages?

Immature sunflower plants, for example, exhibit heliotropism as soon as the sprouts emerge from the soil. As the Sun rises, seedlings, buds, and leaves face the east, absorbing warmth and energy. As the day progresses, the plants track the movement of the Sun across the sky, turning to the west at dusk. During the night, the plants return to an eastward-facing position, ready for the morning sunrise. About one day before sunflower buds blossom into fully developed flowers, the heliotropic behavior ends and the flower heads remain in an eastward-facing position. Why do you think this happens?

Scripture Spotlight

Imagine that God is the stimulus and that people are the plants. Compare **John 12:32** with **John 3:19–21** and identify examples of positive and negative tropism. Which best describes you?

Response to Gravity

Geotropism is the response of a plant to the effects of gravity. The roots of plants are positively geotropic. They grow toward the area of greatest gravitational attraction. Which direction would that be on Earth? Plant stems are negatively geotropic. They grow up, away from Earth's gravitational pull. Geotropic growth will continue even as a plant is turned sideways or upside down. In other words, no matter what you do to a plant within Earth's atmosphere, it will still grow roots down, stem up. Do you think plants have the same response to gravity while orbiting Earth in the International Space Station?

In many plants, geotropic responses seem to be regulated by auxin. Suppose a seedling begins to grow with its roots and stem horizontal to the ground. Auxin might accumulate along the lower side of the roots and stem. Root tissue growth is inhibited by auxin, so cells on the underside would slow their growth. Greater growth on the upside would turn the root down. Stem tissue cells are stimulated by auxin. Their growth would speed up on the underside of the stem, causing the stem to bend upward, thus exposing the plant to more sunlight. Have you seen a tree that has slid with dirt down a hill? What happened to the shape of the tree?

Math in Science

Some *Rhododendron* species exhibit thermotropism, which is movement in a plant in response to temperature changes. If a *Rhododendron* leaf is hanging straight down at 8:00 a.m. at −4° C and moves to a horizontal position by 8:20 a.m. at 0° C, what is the rate of change in degrees Celsius per minute?

This Banyan tree is growing on the ruins of a temple.

Where does this tree exhibit negative geotropism? Where does it exhibit positive geotropism?

Structured Inquiry

Discover

Record your work for this inquiry. Your teacher may also assign the related Guided Inquiry.

Science Journal

Investigating a Tropism

How can we identify plant tropisms?

Procedure

1. Soak four radish seeds in a Petri dish with water overnight.
2. In a clean Petri dish, place the four radish seeds equal distances apart.
3. Cut a coffee filter to fit inside the Petri dish. Place the coffee filter on top of the seeds.
4. **Measure** and pour 15 mL of water on the coffee filter.
5. Cut a circle of cotton quilt batting to fit inside the Petri dish. Place it over the filter paper. Replace the lid of the Petri dish.
6. Hold the covered Petri dish at a right angle to the table. Look at the bottom and check the position of the seeds. They should not move. If they do, reposition them, add another layer of batting to hold them, and check again.
7. When you are sure the seeds will not move, tape the Petri dish closed.
8. Tape the Petri dish to the inside door of a cupboard or closet, with the bottom of the dish (showing the seeds) facing out. Close the door.
9. **Observe** the seeds daily for five days. **Record** your observations with drawings of what you see.

Materials

- 4 radish seeds
- 2 Petri dishes
- scissors
- coffee filter
- 15 mL of water
- graduated cylinder
- cotton quilt batting
- pencil
- strong tape
- cupboard or closet

Analyze Results

What happened to the stems of the emerging radish seedlings? What happened to the roots? Why was closing the cupboard or closet door so important in this experiment?

Create Explanations

1. How can we identify plant tropisms?
2. What factor might account for the plant responses you observed? How do stems and roots respond to this factor?
3. How might the responses you observed help a plant survive in its environment?
4. What would happen if you rotated the Petri dish 90 degrees once the stems and roots began to grow? Why?

Response to Touch

Thigmotropism is a plant's movement or growth response to touch. Some plants are actually more sensitive to touch than humans. Thigmotropism is very important to climbing plants and vines that rely on solid objects for their support. When tendrils and stems touch a surface, they grow in a curl or spiral that helps them hold on. This is positive thigmotropism. Morning glories, climbing beans, and vine cucumbers all use their thigmotropic response to attach to whatever support structure is available. How does their climbing movement offer them an important advantage for survival?

Roots also depend on touch sensitivity to navigate their way through the soil. The general touch response in roots is negative. This means that when a root "feels" an object, the root grows away from the object. This is how roots are able to grow around rocks and other obstacles in the soil, allowing roots to follow the path of least resistance.

Some plants respond to touch not with growth, but with rapid movements that occur independently of the direction of the stimulus. The sensitive plant, for example, folds its leaves together rapidly and droops when touched. The response occurs because of chemical changes in the plant that cause water to leave certain cells very rapidly. The movement is an adaptation that may discourage herbivores from eating the drooping leaves, may prevent water loss as the folded leaves transpire less water vapor, and may dislodge unwelcome insects. The Venus flytrap is another example of a plant that is sensitive to touch. It reacts to the touch of an insect by closing its leaves around it. Why does the flytrap trap the insect? What is the relationship between the flytrap and the fly?

This Venus flytrap has just trapped a fly in response to touch.

What needs have been met by this adaptation?

Adaptations in Plant Populations Explain

Adapting to changing environments is important for species survival. How do plants respond to the environment around them? They can't leave a harsh environment, so how do plants protect themselves from cold, heat, drought, and pests? For species survival, the species must be able to adapt. In God's perfect design, the information needed to adapt is coded into the plants' DNA. God created a wide variety of plant species with various adaptive capabilities.

Seed Dispersal

Plants, like any living organism, need to reproduce to survive. Often, it is critical that the seeds be moved far from parent plants. What would be some of the advantages of seedlings developing at a distance from mature plants? How might mature plants hinder the growth of seedlings? One example of adaptation is the many different methods of seed dispersal in plants. Seed dispersal is the spreading of seeds at a distance from the parent plant. Study the different types of dispersal in the table below. Are any of the plants listed in the table native to your neighborhood?

This milkweed pod is is releasing its seeds.

What characteristics of the seeds allow for wind dispersal?

Seed Dispersal		
Dispersal Method	**Description**	**Examples**
Wind	• Lightweight fruits with wispy, feathery, or winglike "arms" • Easily blown by the wind	Dandelion, milkweed, maple
Water	• Fruits that float to new growth sites • Can travel in rivulets, streams, rivers, and oceans	Coconut, cattail, lotus
Animals	• Gummy or barbed seeds or fruit • Stick to animals that pass by; seeds hitch a ride • Seeds are eaten; pass through in animals' feces	Burdock, blackberry, cherry
Bursting	• Seeds ejected by force • Seed pods dry out in the sun; pod contracts and pops open, spreading the seeds	Geranium, witch hazel, violet

Like all cacti, the saguaro cacti in the Sonoran Desert open their stomata at night.

How might opening stomata at night help a desert plant survive?

Desert Adaptations

In a desert ecosystem, temperatures are extreme and water is scarce. Short grasses, sagebrush, creosote bushes, and cacti thrive. These plants exhibit characteristics that promote their survival and reproduction in the desert's harsh conditions.

The major factor affecting plant survival in the desert is the availability of water. Desert plants exhibit several adaptations to get and store water. Examples include combination root systems (long taproots and shallow root systems), specially shaped leaves to collect water, stem tissue that can expand quick to store storm water, and the ability to remain dormant until rains come. Desert plant adaptations also include small or spinelike leaves, a waxy outer cuticle, vertical leaves, leaves covered with fine hairs, and a rapid reproduction cycle that takes place within just a few days after a rain.

Concept Check Assess/Reflect

Summary: What are plant adaptations? Plants respond to stimuli and changes in environmental conditions in a variety of ways. Tropism is plant movement in response to the direction of a stimulus. Movement toward a stimulus is a positive tropism, while movement away from a stimulus is a negative tropism. Plants' directional response to light is phototropism. Geotropism is the response of a plant to the effects of gravity. Thigmotropism is a plant's movement or growth response to touch. Other plant adaptations include seed-dispersal and structural adaptations for ecosystems.

1. Give an example of a tropism and explain how it might be positive or negative.
2. Make a list of plant adaptations that allow desert plants to conserve water.
3. What plant adaptations would you expect to see in the poor soil conditions of a bog ecosystem?

Lesson 3

Essential Question

How Do Humans Affect the Survival of Plant Species?

What are some places in your community where plants have not survived? What native plants in your area or region have gone extinct or are endangered or threatened? Plants that are well adapted to their environments survive, thrive, and reproduce. What about plants that are not adapted to a changing environment? What becomes of them? What human activities might cause a decline in the number of plant species? What activities could help the survival of a plant species?

Extinction Explain

During catastrophic environmental events in Earth's history, such as floods, climatic changes, and tectonic events, species of plants, as well as animals and other organisms, have disappeared in large numbers. The dramatic changes to their ecosystems challenged many species beyond their ability to adapt. When all members of a species have died off, the species becomes **extinct**.

Most of what we know about extinction is based on the fossils left behind. Environmental disruption, such as extreme climatic changes and meteorological events, affects the numbers and ranges of plants, and many do not survive. The result is that thousands of plant species that lived long ago are now extinct.

Humans Contribute to Extinction

How do people depend on plants? What would happen if many of the plants on Earth disappeared? Because we see so many plants around us almost anywhere we go, the disappearance of entire species from Earth may seem like an exaggeration. However, species do disappear, and some scientists estimate that 99 percent of all species that ever lived have become extinct. In addition, the rate of extinctions has been increased by human activities as they change the environment so that species can no longer survive. Why are many scientists and others concerned that some plant species may disappear completely?

Objectives

- Explain how humans accelerate the rate of extinction.
- Explain the primary factors that cause plant species extinction.
- Discuss how humans work to protect threatened and endangered species.

Vocabulary

extinct
native species
exotic species
invasive species
threatened species

Scripture Spotlight

How do **Genesis 1:26** and **Revelation 11:18** relate to how well we care for Earth?

***Brighamia insignis*, or alula plant, is a critically endangered plant, native to the cliffs of Kauai, Hawaii. The plant's only pollinator, a sphingid moth, is believed to be extinct. Human volunteers pollinate the remaining alula by hand to prevent extinction.**

Seeds are collected and grown in botanical gardens. How might this help the conservation of the species?

Scientists estimate that about 300,000 plant species exist on Earth today. Of those, the Center for Biological Diversity has evaluated about 12,900 of them. Its studies indicate that 68 percent of the evaluated plants are currently threatened with extinction.

Like all living organisms, each plant species is adapted to live in a certain range of environmental conditions. Unlike most animals, plants are unable to move as environmental conditions change.

Habitat loss is the primary cause of species loss. Urban development, fire, agriculture, and logging are some of the causes of habitat loss. Habitat loss can be harmful to more than one species at a time. It can destroy entire natural communities. Habitat loss is also a problem because it fragments a habitat. When a habitat is fragmented, it is divided into smaller areas. The "edges" around an area of land will have a set of conditions that is different from the conditions in the interior of the area. As land is fragmented, the amount of border areas increases. How might this affect plant species living in the fragmented area?

How does global warming affect habitats? Global warming has caused, and may continue to cause, dramatic changes to the distribution of plants around the world. In some places, plant species can survive these changes as their seeds spread to more favorable conditions. How do plants spread their seeds? What might happen to a species of plants if its current habitat was fragmented and separated by large areas of developed land?

Another factor that can contribute to plant species loss is pollution. What are some pollutants that commonly affect the air, water, and soil? What is the source of these pollutants? How might these pollutants be reduced or eliminated?

There are many reasons why we should reduce our impact on the extinction rate. One is a moral reason. Humans have a responsibility to be stewards and caretakers of all Creation. When one species is lost, many other species in the same habitat can be affected. Extinction disrupts the balance of nature. There are also economic as well as environmental impacts. For example, the extinction of plant species affects our ability to continue to discover new medicines. The Svalbard Global Seed Vault stores seeds so that we have the ability to revive a struggling plant species. If we continue our current level of impact on the extinction rate, what are some changes we can expect to see in our environment, even in the near future? What are some ways we might change the way we get food, shelter, and medicines?

Exotic Species **Explain**

Plants that live naturally in an environment are called **native species**. Native species are well adapted to their environments. They interact with other organisms in predictable, stable ways. One of the most widespread native species in the United States is the goldenrod plant, which is found in every state except Hawaii. What benefits do native plants bring to other organisms in the local environment?

The stability of ecosystems is disrupted with the introduction of **exotic species**, which are non-native species that may be introduced by natural processes or by human activity. Humans may introduce exotic species either intentionally or accidentally. When an exotic species harms the environment, it is called an **invasive species**. An invasive species spreads rapidly and competes with native species. Why are invasive species often more successful than native plant species? What things are they competing for? What are some of the ways an invasive species may change the ecosystem? Many invasive species cause economic damage. What human industries might invasive species affect?

The nodding thistle, *Carduus nutans*, is an invasive species. It is native to Eurasia and was most likely brought to North America as a garden plant. Now it invades meadows, prairies, and pastureland across the United States and Canada. It grows in large clumps that crowd out native plants and hinder the movement of livestock. Each *C. nutans* flower can produce as many as 1200 seeds. The seeds spread in the wind and attach to the fur of animals and the clothes of hikers. They can lie dormant in the soil for 16 years and still germinate.

Controlling an invasive species is difficult. The table below lists some control methods.

The nodding thistle costs Americans and Canadians billions of dollars annually.

Controlling Invasive Species

Type of Control	Description of Control	What Do You Know
Mechanical	Plants are pulled from the ground by hand or machine.	How cost effective is this type of control on a large farm?
Chemical	Plant-killing chemicals, called *herbicides*, are used.	How might herbicides affect native plants?
Biological	A natural enemy from the native range is introduced to control the invasive species	How might this affect the native species of plants and animals?
Ecosystem management	Habitats can resist invaders better if they are managed properly.	What can you do to properly manage your yard at home or your schoolyard?

Structured Inquiry

Discover

Record your work for this inquiry. Your teacher may also assign the related Guided Inquiry.

Science Journal

Model an Invasive Species

How can you model the effects of an invasive species?

Procedure

Materials
- green, tan, and red card stock
- scissors
- 1 sheet of construction paper

1. Imagine you are a rancher raising cattle for organic beef on natural pastureland. The nodding thistle plant invades your land. To begin, **measure** and cut 2 cm × 2 cm squares from card stock. You will need 30 green, 5 tan, and 4 red squares.
   - Green squares represent pasture grass plants. Label each with a value of $10.
   - Tan squares represent cattle. Label each with a value of $100.
   - Red squares represent invading thistles. They have no economic value.
2. **Model** your pasture by placing 30 green squares on the construction paper. Model what happens during seven years using these rules:
   - Place five cows on the land each year.
   - Each cow eats three grass plants before it is sold (remove grass and cows from the pasture). A cow that does not eat enough grass dies (remove from the pasture).
   - At the end of a year (after the cattle are gone), each remaining grass plant reproduces another grass plant.
3. For the first three years, assume that no thistles invade. For years 4–7:
   - A single new thistle invades each year. Cattle cannot eat thistle.
   - It kills two grass plants in the pasture (remove them from the pasture) before the next year's cattle arrive.
   - Thistles remaining from a previous year have no added effect.
4. Organize your data in a **graph**. Then use numbers to **calculate** the value of your business each year *after* you have sold your cattle and *after* your grass plants have reproduced. Use the data table in your ***Science Journal*** to record your data.

Analyze Results

What does your **model** show? What conclusions can you draw from it?

Create Explanations

1. How can you model the effects of an invasive species?
2. What advice would you give farmers about nodding thistles?

Threatened and Endangered Species Explain

Shrinking habitats and invasive species pose a risk to the survival of native species. Recognizing that risk, in 1973, the U.S. Congress passed the *Endangered Species Act* to protect species from extinction, which makes it illegal to carry out activities that threaten the continued existence of listed species or their habitats. The act classifies selected species as either endangered or threatened. An endangered species is "in danger of extinction throughout all or a significant portion of its range." A **threatened species** is "likely to become an endangered species within the foreseeable future."

The Endangered Species Act has helped many species come back from the brink of extinction. Humboldt Bay, California, is home to two endangered flowering plants: the beach layia and the Humboldt Bay wallflower. The coastline environment in which these plants live is a fragile, but productive, ecosystem. These endangered plants are adapted to poor soil and harsh ocean storms. But invasive plant species have forced them to near extinction. Since 1995, the U.S. Bureau of Land Management and the California Conservation Corps have been working to restore the Humboldt Bay habitat. Workers and volunteers have pulled up invasive European beach grass and other exotics and replanted native dune grasses. Their efforts have paid off. The habitat has been restored, and the population size of both endangered plants has increased.

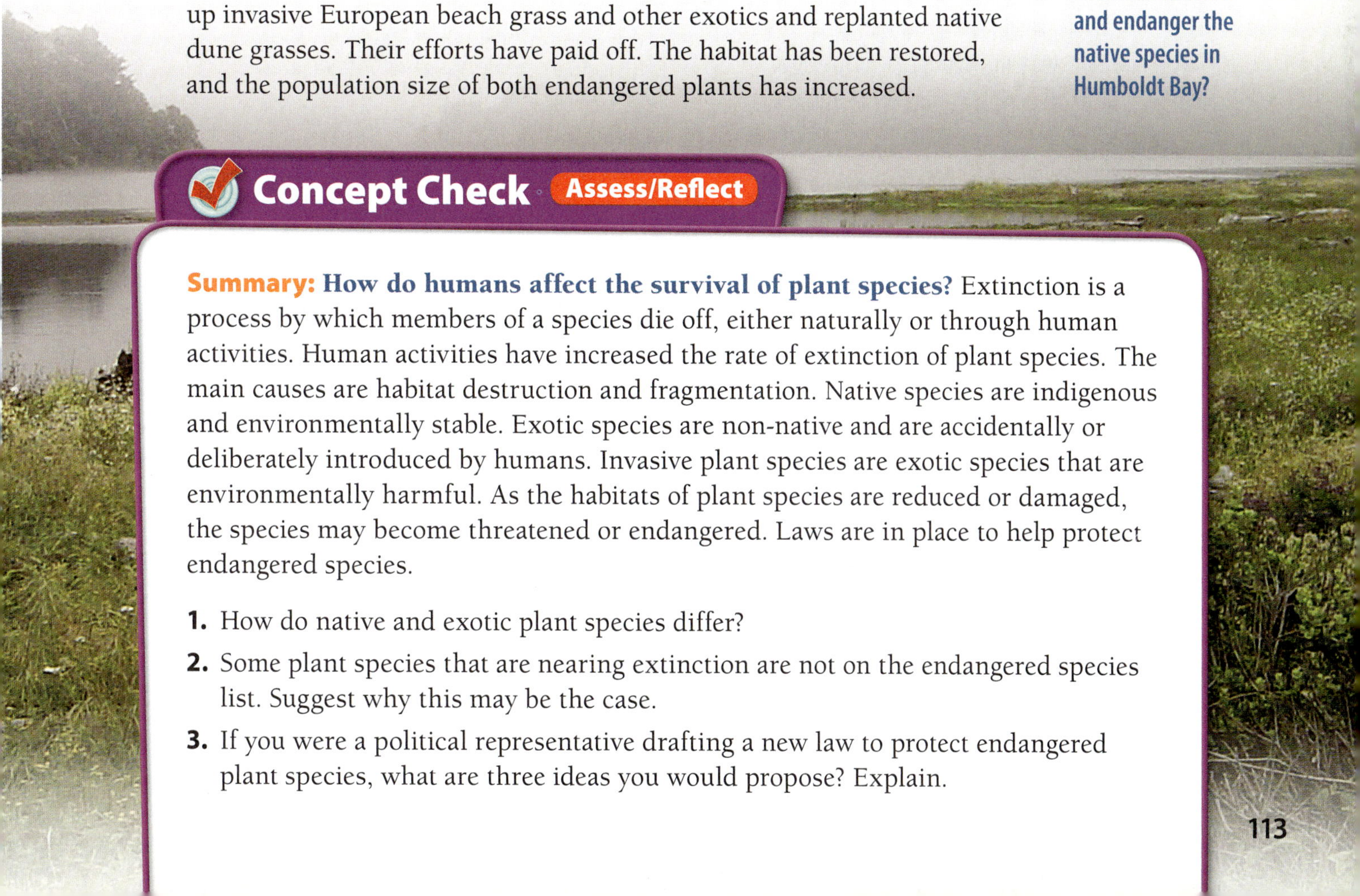

Humboldt Bay, California

How might invasive species change the environment and endanger the native species in Humboldt Bay?

Concept Check Assess/Reflect

Summary: How do humans affect the survival of plant species? Extinction is a process by which members of a species die off, either naturally or through human activities. Human activities have increased the rate of extinction of plant species. The main causes are habitat destruction and fragmentation. Native species are indigenous and environmentally stable. Exotic species are non-native and are accidentally or deliberately introduced by humans. Invasive plant species are exotic species that are environmentally harmful. As the habitats of plant species are reduced or damaged, the species may become threatened or endangered. Laws are in place to help protect endangered species.

1. How do native and exotic plant species differ?
2. Some plant species that are nearing extinction are not on the endangered species list. Suggest why this may be the case.
3. If you were a political representative drafting a new law to protect endangered plant species, what are three ideas you would propose? Explain.

Careers in Science

Extend

Urban Ecologist

An ecologist studies the relationships between organisms and the human-dominated ecosystems in which they live. The urban ecologist hopes to find solutions to urban environmental problems. Training to become an urban ecologist includes college course work that integrates ecology with engineering, architecture, anthropology, economics, and law.

So how do ecologists account for the role of humans in an environment when humans change the natural rules that most other organisms live by? Now that human activity has grown into a global ecological force, some ecologists are finding that traditional ecological beliefs no longer provides ways of understanding systems dominated by human activity.

With over half of the human population living in urban settings, many in cities of epic size, the science of urban ecology is a hot topic. The rapid urbanization that is taking place has caused an instant increase in urban poverty and urban food insecurity. By 2020, the developing countries of Africa, Asia, and Latin America will be home to some 75% of all urban residents. For many developing countries it is difficult to cope with this development, and they struggle to create sufficient employment for their citizens. Developing countries also have increasing problems with the disposal of urban wastes and wastewater while maintaining air and river water quality.

Many organizations that conduct urban ecology research do so with the goal of building healthier, safer, and more beautiful cities. Some issues that urban ecologists face and some solutions they have created are:

- Food security and nutrition—Through the use of urban agriculture, the efficiency of national food systems is increased by supplying products that rural agriculture cannot easily supply (perishable products, products needing a quick delivery upon harvest).
- Economic impact—Growing your own food saves household expenditures on food. Poor people in poor countries generally spend a large part of their income on food.
- Social impact—Participants in urban agricultural projects may feel enriched by the possibility of working constructively, building to improve their community, and working together to produce food and other products for consumption and for sale.
- Contributions to urban ecology—Programs influence the greening and cleaning of the city by turning run-down open spaces into green zones and maintaining buffer and reserve zones free of housing, which affects the microclimate (shade, temperature, CO_2 levels).

The importance of urban agriculture is increasingly accepted by international organizations such as UNCED (Agenda 21), UNCHS (Habitat), FAO (World Food and Agriculture Organization), and CGIAR (international agricultural research centers).

Concept Check

1. Describe the different types of work an urban ecologist and an ecologist do. Why are both their jobs so important?
2. List some of the problems an urban ecologist faces when working with the rapid increase in population in developing countries.

Science and **Technology**

Extend

Genetic Conservation in the National Seed Laboratory

We need to eat to survive. The National Seed Laboratory (NSL) wants to make sure that we will continue to have a variety of foods to choose from, so it collects and stores seeds. The millions of seeds inside the laboratory will be relied upon to feed future generations for years to come. NSL's brick building on the grounds of Colorado State University in Fort Collins, Colorado, holds a unique collection of almost 250,000 seed samples collected from around the world. The seeds represent about 3000 different plant species, with each sample containing between 1500 to 3500 seeds.

Why is it so important to gather and store these seeds? Without concentrated conservation efforts, many native plant species will decrease or disappear within our lifetime. As these species decrease, there is an effect on the plant communities and the native animals that depend on them. Seed banks can provide vital insurance against species extinction by conserving plant seeds for research and restoration. Stored seed resources can protect native populations against natural catastrophes such as wildfires and flooding. The plant world is constantly changing; new disease strains evolve, new insect challenges develop, and invasive plant species move in. Seed banks protect us from the destruction of genetic diversity by land development and habitat fragmentation. These events have led to a genetic conservation program preserving wild sources of plant *germplasm*—the genetic material from which new plants can be grown.

Genetic conservation is the practice of protecting plant genes and seeds from loss and extinction so that the genetic material for plant species will be available for the use of future generations. Seed banking is the most efficient and cost-effective method of genetic conservation available. Drying seeds and placing them in cold storage allows large amounts of germplasm to be preserved in a small space. The dried seed samples are small in size, are easily handled, and require little maintenance. The seeds usually remain able to germinate, for a long period, and in many instances indefinitely. However, seed banks can be expensive to run, both because of the need to maintain low temperatures and the necessity for germination tests, growth trials, and regeneration. Plant germplasm must be preserved. No one knows what kinds of diseases or future events will happen in the years to come, so it is imperative that we maintain as much genetic material as possible.

Concept Check

1. What is plant germplasm?
2. What are some of the threats to the future of today's plant germplasm?

Chapter 3

Review Assess/Reflect

Study Guide

Lesson 1

1. Through the process of photosynthesis, plants convert light energy, carbon dioxide, and water into sugar and oxygen. Photosynthesis takes place in two steps: the light reactions and the dark reactions.
2. The steps of photosynthesis that take place only in the presence of light are called the light reactions. The dark reactions, also called the Calvin cycle, can take place without light energy. The light reactions and the Calvin cycle depend on each other.
3. Xylem is made of dead cells, and it moves water in one direction throughout a plant. Phloem transports glucose and other nutrients to be stored and used by the plant. Phloem can move materials throughout the plant in both directions.
4. Transport refers to the movement of food and water through a plant. Transpiration is the transport of water. When water evaporates from the stomata of a plant's leaves, water is pulled up through the xylem of the plant. The process of transpiration does not use energy.
5. Plant hormones send signals to specific cells, signaling the cells to "switch on" certain genes. The active genes direct the synthesis of enzymes and proteins that bring about a change in the cell's structure or function.

Lesson 2

1. A tropism is any movement in a plant that occurs in response to a stimulus. Movement toward a stimulus is a positive tropism. Movement away from a stimulus is a negative tropism. Phototropism is a response to light. Geotropism is a response to gravity. Thigmotropism is a response to touch.
2. Plants are adapted to disperse seeds in different ways. Desert plants have combination root systems, leaves to collect water, stems that store water, and the ability to remain dormant.

Lesson 3

1. Humans accelerate the rate of extinctions by changing the environment so that species can no longer survive.
2. The primary cause of plant species extinction is habitat loss, which results from urban development, fire, agriculture, and logging. Habitat loss results in fragmentation, which causes an increase in border area and thus a change in environmental conditions. Pollution and the introduction of exotic species are other human causes that increase the rate of extinction.
3. In 1973, the U.S. Congress passed the *Endangered Species Act* with a goal to protect species from extinction. The act has helped many species come back from the brink of extinction. People can also help to restore habitats.

Vocabulary Check

Fill in the blank with the correct vocabulary word.

1. The loss of water through the stomata of a leaf is __________.
2. Glucose is a(n) __________ that cannot be broken into smaller carbohydrate units.
3. Both primary and secondary growth of plants begin in regions called __________.
4. A species that lives in an area naturally is a(n) __________.
5. An exotic species that poses a risk of environmental or economic harm is a(n) __________.
6. A plant response to sunlight is called __________.

Multiple Choice

Choose the best answer.

7. Which statement is true of the dark reactions?
   A. They make ATP.
   B. They break down glucose.
   C. They require the Sun's energy.
   D. They require carbon dioxide.
8. How does fragmentation affect a plant species?
   A. It makes it easier for plants to get exposure to the Sun.
   B. It allows for seeds to be dispersed via the wind.
   C. It causes native plant species to overpopulate.
   D. It limits the amount of habitat a plant has available.
9. Which of these are found in vascular bundles?
   A. stomata
   B. meristems
   C. xylem and phloem
   D. chloroplasts and chlorophyll
10. When do introduced species become invasive?
    A. when they displace native plants and disrupt natural systems
    B. when they are planted in large numbers
    C. when they are listed as threatened or endangered
    D. when they become native to an environment
11. Which of these is not a type of invasive species control?
    A. chemical
    B. mechanical
    C. biological
    D. fragmental

Check Point

Answer the following questions.

12. **Compare** the roles of ATP and NADPH in the dark reactions of photosynthesis.
13. Explain why the light reactions do not require carbon dioxide.
14. Explain why plants are able to continue the dark reactions when they are not able to use the light reactions.
15. Suggest a method you could use to **gather and record data** to determine whether a species exhibits sonotropism, which is movement in response to sound. Explain why your method would work.
16. The paperbark tree lives in coastal wetlands and freshwater swamps. It is an invasive species in Florida but a threatened species in its native Australia. **Develop a hypothesis** to explain how this might have occurred.

Chapter 4

How Animals Survive

Scripture Spotlight

As you read about the intricate dynamics of ecosystems and the fascinating adaptations of the creatures that live there, think about the wisdom and power of your Creator. You will read the following passages in this chapter.

Jeremiah 17:8 (p. 121)
James 1:11 (p. 121)
Psalm 119:11 (p. 134)
Proverbs 10:14 (p. 134)
Hebrews 10:22 (p. 145)

The Big Idea

God designed living organisms to interact with one another and their environment. He gave them the ability to adapt as their environments change.

What are some factors that can affect the woodland caribou's habitat?

The woodland caribou, *Rangifer tarandus*, inhabit Canada's northern forests. These animals are unique because they are able to digest lichens due to special enzymes present in their digestive system.

Inquiry Kick-Off **Engage**

How are animals' claws adapted to help them survive? God designed animals to be able to thrive in their environments. In your ***Science Journal,*** you will do an inquiry where you will discover which types of food are best for different claw adaptations.

Lesson 1

Essential Question

What Affects Ecosystem Dynamics?

Objectives

- Describe ways in which species compete.
- Compare and contrast the biotic and abiotic factors of ecosystems.
- Describe the flow of energy in an ecosystem.
- Describe the natural cycles that characterize ecosystems.
- Describe factors that can affect the balance in an ecosystem.

Vocabulary

competition
trophic level
nitrogen fixation
denitrification

How would you describe an ecosystem? Now think about where you live and name one ecosystem found in your area. What living things and nonliving things are parts of that ecosystem? How do the living things and the nonliving things in the ecosystem interact with one another? Think about the ecosystem you have described. How does it change? Which of these changes are natural, and which are caused by humans? Which are positive, and which are negative? As you study this lesson, think about how changes in an ecosystem affect humans as well as other organisms in the environment.

Populations and Competition Explain

Imagine hiking through a ponderosa pine forest in Colorado. The ponderosa pines are a population of trees adapted to the area. Recall that a *population* is a group of individuals of the same species living in the same place at the same time.

As you continue hiking, you notice a blue grouse, a mother black bear and her cub, and small brightly colored green hairstreak butterflies. Each of these species represents a population in the ecosystem. What other animal and plant populations might you expect to find here? How might the populations of animals and plants interact with one another?

Organisms need resources, such as food and water, to survive and reproduce. If resources are limited, organisms will compete with one another for these resources. **Competition** is the struggle among organisms to obtain the necessities for survival and reproduction.

Northern pike are primarily fish eaters.

What organisms do you think the northern pike compete with for resources?

Sometimes, organisms in the same population must compete with one another. This is known as *intraspecific competition*. For example, when two maple trees are planted next to one another, they compete for sunlight, water, and soil nutrients. By sharing these resources, it is possible that neither tree will flourish or that only one will live. When populations of different species compete with one another, it is known as *interspecific competition*. Now suppose the maple tree is planted next to a giant redwood tree and competing for the same nutrients. Which tree do you think is more likely to win this competition? Whether the competition is intraspecific or interspecific, the outcome is the same. Organisms that outcompete others are more likely to survive, reproduce, and pass on their characteristics to future generations. What factors might affect the level of competition in an area?

Avoiding competition is an effective strategy for species survival as it allows organisms to spend more energy on other activities. For example, suppose that individuals of five species of warblers need to nest in the same tree. If they nest in different parts of the tree, they can spend more energy on foraging and reproduction rather than spending limited energy on competition for the same space. This is an example of sharing a space. In what other ways can organisms avoid competition?

Characteristics of Ecosystems

If you were asked to identify factors that characterize an ecosystem, what would you say? How do those factors interact? How do ecosystems differ? Do the factors that affect ecosystems change or are they static?

As you know, some parts that make up an ecosystem are *biotic factors*, which are the effects of a living organism on other organisms or on its nonliving environment. A mountain lion preying on mule deer is an example of a biotic factor.

Also recall that temperature, precipitation, the amount of sunlight, oxygen concentration, and soil quality are some of the *abiotic factors*, or nonliving factors, in an ecosystem. Abiotic factors are important in determining what organisms live in an ecosystem. That is because only species that are adapted to the conditions of the ecosystem will survive. What are some of the abiotic factors in a temperate forest ecosystem? How do those factors vary from the abiotic factors of a tundra ecosystem? How do the abiotic factors at the base of a tree differ from those at its top?

Abiotic and biotic factors do not act independently. A change in one factor will often result in a change to other factors. Think about a pond that receives full sunshine throughout the day. The populations of fish and other species that live in the pond can survive in the daily changes that take place in the abiotic factors of the pond, such as the lack of sunlight and the cooling of the temperature at night.

Scripture Spotlight

Read **Jeremiah 17:8** or **James 1:11**. What interaction of biotic and abiotic factors is described there?

Now suppose that a large building is constructed next to the pond. The building will shade much of the pond most of the day. How will the abiotic factors of the pond change? How will this change affect the biotic factors? Could this change in biotic factors then have an effect on the new abiotic factors? Why do you think God created a world with so many types of interactions?

Energy Flow through Ecosystems Explain

In addition to biotic and abiotic factors, the flow of energy is one of the most important factors that determines what kinds of and how many organisms live in an ecosystem. Energy flows through an ecosystem in one direction—from the Sun or inorganic molecules to producers and then to consumers. When organisms eat other organisms, energy is passed along.

Food Chains and Food Webs

As you know, you can show the feeding relationships among producers and consumers in a *food chain*. Consider the relationships among organisms in a stream ecosystem such as the one below. What are some food chains that might be part of a stream? Would all the organisms in a stream food chain live in the water?

Although energy in an ecosystem flows from the Sun through producers to consumers, the energy does not follow a straight path. The feeding relationships in an ecosystem are more complex than can be shown in a food chain. The feeding relationships in an ecosystem form a complex web of individual food chains, called a *food web*. A food web links all the food chains in an ecosystem. What interconnecting food chains can you find in the stream food web?

The blue and green arrows in this food web show food chains. The red arrow shows that metamorphosis takes place.

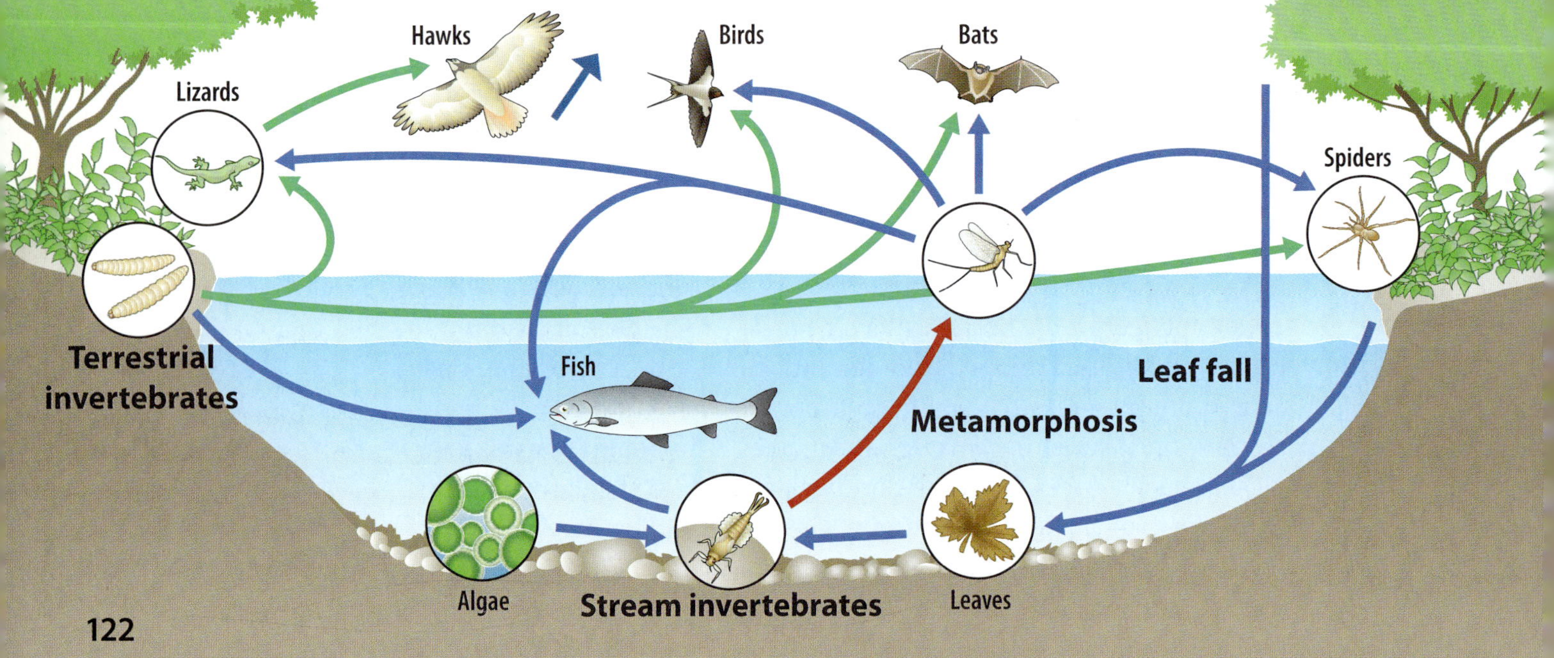

Trophic Levels

What is the path of energy from the Sun to the bass in the stream food web? How is that different from the path for the snail? Scientists study how energy moves through a food web by assigning trophic levels. A **trophic level** is based on an organism's source of energy.

Trophic Levels	Description	Examples
First Trophic Level	The lowest level of any ecosystem is occupied by the *producers*. Organisms produce energy by *photosynthesis*, or *chemosynthesis*.	Grasses, trees, phytoplankton What organisms in the stream ecosystem would be producers?
Second Trophic Level	Made up of *primary consumers and herbivores*, organisms that eat plants	Snails, cows, deer, squirrels, grasshoppers How many herbivores can you find in the stream?
Third Trophic Level	The highest level of any ecosystem is occupied by the *secondary consumers*, organisms that eat other organisms. Includes *carnivores*, organisms that eat meat, and *omnivores*, organisms that eat both plants and animals	Bats, snakes, trout, bears, wolves What are the carnivores and omnivores that are part of the stream ecosystem?
Decomposers	*Decomposers* break down the wastes and remains of all living things in an ecosystem. Decomposers are essential for recycling materials.	Bacteria, protists, fungi What decomposers might be in the stream food web?

Explore-a-Lab

Structured Inquiry

How would the removal of trees along a stream affect the stream ecosystem?

Look carefully at the picture on page 122. With a partner, write down all the living organisms you can see. Next to each organism, identify the trophic level that the organism represents. Construct a chart to display your listing. Now draw a food web that shows the relationships between the organisms you have listed. Do this by drawing an arrow from an organism to the organism that eats it, for example, crayfish → raccoon, not raccoon → crayfish. What are other first, second, and third trophic-level organisms that would be part of the stream ecosystem but that are not shown in the illustration?

Structured Inquiry

Discover

Record your work for this inquiry. Your teacher may also assign the related Guided Inquiry.

Science Journal

Lawn Food Web

What organisms make up the food web in a lawn?

Procedure

1. Find an area in a lawn to study.

2. Tie the two ends of the string together and stretch out the string to form a circle on the lawn.

3. Look carefully at the patch of lawn inside your circle and try to collect at least one example of each animal you find. Wear disposable gloves and use the tools provided to catch the animals you find. Place the animals in the plastic cups. Be careful not to injure or kill any of the animals.

4. Use the ruler to **measure** three 10 cm × 10 cm squares that are well spaced in your circle.

5. Use the garden trowel to dig up the first 2–3 cm of soil in each square to find and collect the animals living there.

6. Once you have made your collection and **recorded** your findings, use the lawn guide provided to identify the animals you have found and list them in the data section. For animals you cannot identify, describe them as best you can.

7. Record whether each animal is a primary, secondary, or tertiary consumer.

8. Construct a diagram of a food web that includes all the animals you identified by drawing a line from each animal to what you think it might eat. Be sure to include plants in your food web.

Materials

- bug box magnifier
- disposable gloves
- garden trowel
- lawn guide
- magnifying lens
- plastic cups (clear) with slotted lids
- plastic spoon
- ruler
- string 2 m in length
- tweezers

Analyze Results

How did your food web **compare** with those of your classmates?

Create Explanations

1. What organisms make up the food web in a lawn?

2. Were the food webs of each group exactly the same? Why might they differ?

3. Do any of the organisms that you collected occupy more than one level in the food web?

4. Compare the lawn with the forest floor. Where do you think you would find a greater diversity of life? Explain.

Energy Transfer and Loss

You know that organisms use energy in the food they eat to power life processes, such as movement or growth. During these processes, energy is lost as heat. What effect will this have on the amount of energy that an organism receives from the trophic level beneath it? How will this loss affect the number of consumers in an ecosystem?

Scientists use a pyramid as a tool to show the feeding relationships and energy loss among organisms in an ecosystem. The relative areas of the pyramid reflect the relative amounts of energy available at each trophic level. Notice in the energy pyramid that the largest group of organisms at the bottom (first trophic level) of the pyramid is the producers. As you move up the pyramid, and thus up the trophic levels, the areas get smaller. What does this indicate about the amount of energy available as you move up the pyramid? If you wanted to get the most energy from the food you eat, from which level of the pyramid should you choose your food? What foods did you eat from each level today?

Remember that very little (only about 10%) of the energy is passed on between the levels of a food chain.

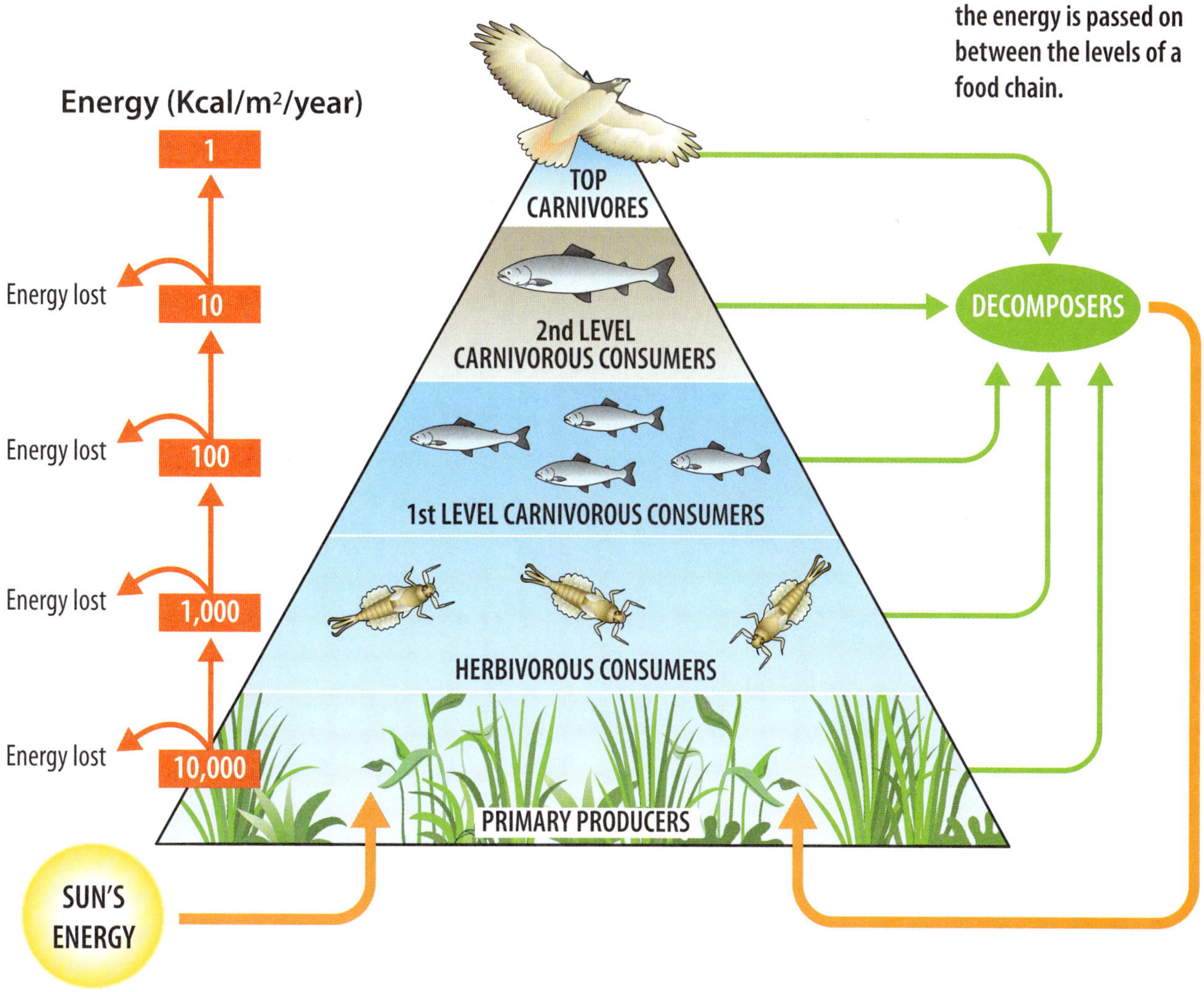

Balance in Ecosystems

You have just seen how energy flows through ecosystems, but all organisms need more than energy to survive. What materials do organisms need to stay alive and healthy? Where do these materials come from? Are these materials available in unlimited supply? How do ecosystems maintain a balance of the nutrients that living organisms need?

The Cycling of Nutrients

Consider one important nutrient found in all living things—carbon. It occurs in a variety of forms and has a variety of functions. Where do organism get this carbon? What happens to the carbon when organisms die? Let's look at the paths that carbon can take in any ecosystem. Follow the diagram as you read about each step.

What do you suppose the carbon cycle was like in the Garden of Eden?

Carbon Cycle

1. Plants take in carbon dioxide, which contains carbon, from the air and use it to make carbon-containing sugars during photosynthesis.

2. Some carbon is released into the atmosphere as carbon dioxide that is produced when plant cells break down sugar during cellular respiration. Why is cellular respiration necessary in plant cells?

3. The carbon in the sugars may also be stored in the plant's tissues. If the plant is eaten, the carbon is transferred to the animal that eats it. If the plant is not eaten, what happens to the carbon in the plant? Is it ever used? If so, how?

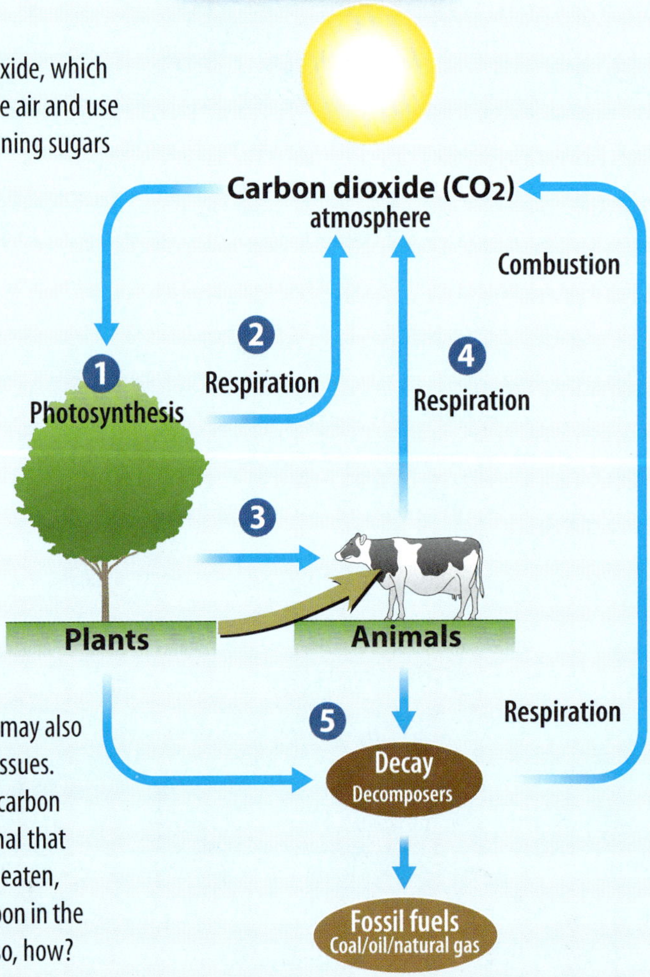

4. Carbon is then released into the atmosphere as carbon dioxide that is produced during cellular respiration in animal cells.

5. When plants and animals die, decomposers break down their tissues. The decomposers use the carbon-containing tissues as food. What happens to the carbon dioxide that is produced in this process?

Why would this cycling of carbon be important to all living things? What would happen if carbon didn't cycle in this way?

Carbon is only one of the materials that cycles in ecosystems. Another important nutrient is nitrogen, which is found in the bodies of organisms and in wastes from animals. The diagram shows how nitrogen cycles through an ecosystem. Notice two especially important processes, nitrogen fixation and denitrification.

- Most organisms can only use nitrogen once it has been combined with other elements. **Nitrogen fixation** is the process of changing nitrogen gas from the air into nitrate compounds that plants can use. Nitrates are compounds that have oxygen and other elements in them. Nitrates are the only form of nitrogen that most plants can use.
- During **denitrification**, bacteria convert nitrates back into nitrogen gas by removing the oxygen atoms. The resulting nitrogen gas then makes its way back into the atmosphere.

Faith Connection

Are you a producer or a consumer in your home, school, or church? Can you be both? If so, how balanced are you?

What processes can you identify in the diagram of the nitrogen cycle?

Nitrogen Cycle

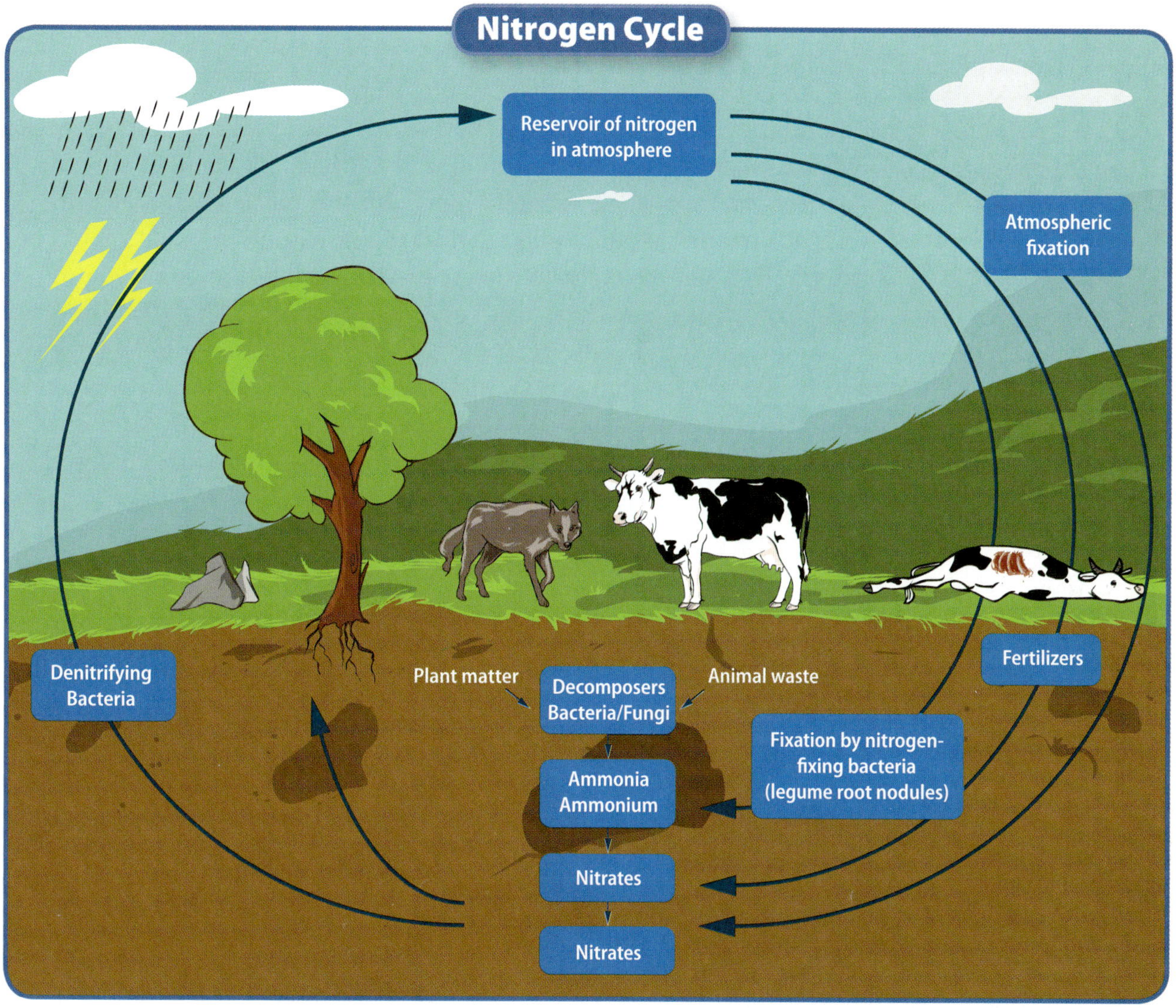

Ecosystems Out of Balance

Have you ever seen pictures of the area where you live from 50 or more years ago? What changes did you notice? What caused these changes? Were the changes positive or negative? Ecosystems are constantly changing. The changes can be small or large. They can happen suddenly or slowly. They can be caused by living things or nonliving things.

Perhaps you have lived somewhere that experienced flooding, fires, or mudslides. How did these events affect the ecosystem you are part of? These events happen and cannot be controlled by humans. But some factors that cause changes in ecosystems are caused and/or controlled by humans. Some of these human-caused factors include pollution, introduced species, habitat alteration, overhunting, and overfishing. What are some examples for each of these factors? What can humans do to help restore an out-of-balance ecosystem? How can humans prevent ecosystem damage before it happens?

Population Changes

As you learned, any changes to one factor in an ecosystem can affect other factors in the ecosystem—both biotic and abiotic. A change might cause a specific species to die out because it is not adapted to survive the new conditions. What would happen if the species that died out were a consumer population? The loss of a consumer population might cause only a small change to the ecosystem, or the change could destroy the entire ecosystem. There is no one answer. The result depends on the kinds and number of species within an ecosystem and on what and how large the change is. Which do you think would have a greater impact on the ecosystem—the loss of a producer species or the loss of a secondary consumer? How would the changes differ with the loss of each type of species?

Lesson Activity

Work with a group to choose one of these essentials materials:

- oxygen
- sulfur
- phosphorus
- water

As a group, research the material to find out how it cycles in ecosystems. Then choose how to present the information to the rest of the class. For example, you might use computer presentation software, a skit with props, or a poem.

How are these natural forms of recycling evidence of a wise and loving Creator?

Think about the stream ecosystem that was discussed earlier in this lesson. Suppose a darter fish that eats small invertebrates, such as microscopic organisms, as well as larger insects and worms died out. The producers—the algae, the phytoplankton, and the terrestrial plants—would continue to produce. The other consumers that depended on the darter fish would still have their other sources for food. How would the amount of zooplankton available to the other consumers change? How would the elimination of the darters affect the competition for zooplankton and aquatic insect larvae? Which consumer populations might grow larger? Which might grow smaller? Why?

Lesson Activity

Research the issues of wolf restoration in the United States. Find out the arguments for and against wolf reintroduction. Then choose a side to represent in a class debate. Meet with other members of the class who also agree with your opinion to prepare for the debate. Did the information presented in the debate influence your opinion? If so, how?

Do you think wolf reintroduction has a positive or negative effect on the environment? Why?

Concept Check Assess/Reflect

Summary: What affects ecosystem dynamics? Members of the same population or of different species may compete for water, habitat, food, or mates. Biotic factors in an ecosystem are the effects of a living organism on other organisms and on its nonliving environment. Abiotic factors in an ecosystem are the nonliving things, such as water, soil, and temperature. Biotic and abiotic factors in an ecosystem interact in a number of different ways. Energy flows through an ecosystem as plants produce food using energy from the Sun, which then continues through the food chain. A food chain is a single set of feeding interactions. A food web consists of a series of interconnected food chains in an ecosystem. Nutrients, such as carbon and nitrogen, are kept in balance in an ecosystem by being recycled. The balance of ecosystems can be upset by natural means such as flooding, fires, or mudslides. It can also be upset by human activity such as pollution, the introduction of new species, habitat alterations, and irresponsible hunting practices.

1. What are several types of organisms that are essential for the recycling of materials in an ecosystem?
2. Describe the path of the flow of energy in an aquatic ecosystem.
3. Which is a greater threat to a mountain ecosystem: a beetle that kills coniferous trees or a wolf that hunts and eats deer? Explain your response.
4. Can two organisms live in the same ecosystem and not compete with each other? Explain.

Lesson 2

Essential Question

What Are Animal Adaptations?

Objectives

- Describe physical adaptations.
- Compare and contrast innate and learned behavioral adaptations in animals.

Vocabulary

cryptic coloration
disruptive coloration
countershading
Batesian mimicry
Müllerian mimicry
taxis
estivation
habituation
classical conditioning
operant conditioning
imprinting

Look at the photograph on this page. What do you see? Look closely. Can you see anything else besides the tree bark? The picture includes an animal. Have you found it yet? All animals have a variety of adaptations that help them survive in the living conditions of their specific ecosystem. The animal in the picture is adapted so that it looks like the tree bark. How does this adaptation help the individual animal? What animals do you think might be around you on the playground that you can't easily see?

Physical Adaptations Explain

What does the foot of the common housefly have in common with your tongue? They both have sensory structures that can taste! Taste is just one of the many kinds of physical adaptations you will find among animals. A *physical adaptation*, you will recall, is a body part or shape that helps an animal survive in its environment. Examples of physical adaptations include the webbed feet of a duck, the sharp teeth and claws of a carnivore, feathers on a bird, scales on a fish, and the beak of an owl. What are some physical adaptations that you have? How do they help you live in your environment?

An animal in this picture is using a physical adaptation to hide.

What animal is hiding?

Camouflage

The spider in the picture was difficult to see because it displays a physical adaptation called *camouflage*. Recall that camouflage is a trait that enables an animal to blend in with its environment or to hide in plain sight. Many animals, including the spider, have a body color, shape, or pattern that matches their surroundings. How does this kind of adaptation help an animal? The table below shows a few different forms of camouflage. How do you think camouflage is evidence of God's careful Design plan?

Different Forms of Camouflage			
Body Shape	**Cryptic Coloration**	**Disruptive Coloration**	**Countershading**

The body shape of this katydid helps it blend in with its background. What other trait also helps this insect blend in?	**Cryptic coloration** is the use of color to blend into the background. The Smith's dwarf chameleon is one of the few chameleons that can change its color to match its background. What advantage does this strategy have over a fixed color pattern?	The spotted coat of a leopard is an example of **disruptive coloration**, which helps it blend in among grasses and the leaf litter on the ground. This trait also helps the leopard successfully stalk its prey.	Many fish and whales blend into their backgrounds by using **countershading**, which means their dorsal surfaces are darker, and their ventral surfaces are lighter.

Warning Coloration

Warning coloration is the opposite of camouflage. Animals that use warning coloration may have distinctive patterns, contrasting colors, or bright colors. These animals may also be venomous, have sharp spines and stingers, or have foul odors. All of these adaptations help to reinforce the connection between the warning colors and danger or unpleasantness. Predators that attack such brightly colored animals usually learn to avoid others of the same species in the future. What are some brightly colored organisms that you avoid?

Organisms can have one or more traits that are part of their camouflage.

Why might camouflage also be important for organisms that are awake at night? What kinds of camouflage might they display?

Explore-a-Lab

Structured Inquiry

How does coloration provide protection from predation?

While your partner is turned away from you, spread out 40 jelly beans (10 each of four different shades of green) in a random pattern on a green poster board. Then have your partner turn around and pick up, one at a time, as many jelly beans as possible in five seconds. Record the number and the shades of the jelly beans that were "caught." Repeat this process at least three times. Do you see any patterns in your results? How would you explain your results? How might the same relationship between coloration and predation apply in the wild?

Mimicry		
Type	**Definition**	**Example**
Batesian mimicry	Refers to two (or more) animals that appear similar, but only one of them is dangerous to predators	The scarlet king snake is not poisonous, but its red, yellow, and black stripes resemble those of the poisonous coral snake. Predators that have had bad encounters with coral snakes will also avoid the king snake.
Müllerian mimicry	Occurs when two (or more) animals look alike, and both have strong defenses against predators. After a bad experience with either species, predators will avoid both.	The yellow jacket wasp and the bumblebee both have bright yellow and black colors and use powerful stings.

The blue-ringed octopus can be hard to spot when it is calm and hard to miss when it is provoked. It preys on crabs, shrimp, and small fish.

How is the blue-ringed octopus able to match the color and texture of its background when it wants to hide?

Innate Behavioral Adaptations **Explain**

Have you ever quickly pulled your finger from a hot object without thinking about it? Why did that happen? Did you have to learn to do that? How does this response help you survive in your environment?

In addition to the ability to develop physical adaptations, God also gave animals the ability to develop behavioral adaptations. Some of these behavioral adaptations are inherited while others are learned. Inherited behavioral adaptations are called innate behaviors. An *innate behavior* is a set of inborn actions in response to a stimulus. An animal does not have to learn the behavior; it inherits these behaviors from its parents. For example, a newly hatched sea turtle will automatically move toward the ocean. Innate behaviors may help an animal locate a food source, find a mate, escape a predator, or catch prey. What innate behaviors were you born with?

Reflex

The simplest kind of innate behavior is a reflex. A reflex is an automatic response to a stimulus. A reflex may involve as few as two nerve cells—a sensory neuron and a motor neuron. The sensory neuron detects the stimulus and fires. The impulse triggers a response in the motor neuron. The motor neuron may cause a muscle to contract or a gland to secrete a chemical. Most insects have simple startle-and-escape reflexes that work in this way. What types of reflexes do humans exhibit?

Taxis

Another innate behavior is **taxis**. It is a movement toward or away from a stimulus. Birds, seals, and even the dung beetle navigate by starlight. A diving beetle uses a taxis toward light to find its way down and back up again in water.

Migration

As you know, migration is the movement of animals—usually in large numbers. Studies confirm that migration is also an innate behavioral adaptation. In one study, scientists switched eggs between the nests of herring gulls, which do not migrate, and lesser black-backed gulls, which do migrate. Despite being reared by nonmigrating parents, the lesser black-backed gulls migrated, leaving their foster parents behind.

The breeding habitat of the golden plover is the Arctic tundra of northern Canada and Alaska. They migrate south to South America. It is one of the longest known migratory routes (more than 40,000 km).

What environmental factors might influence when and where golden plovers migrate?

Check for Understanding

Do humans migrate? Explain.

In other experiments, migrating species of birds were kept indoors. Light and temperature in the birds' cages were kept the same all year. Yet, somehow the birds "knew" when the seasons changed outside. They gained weight before their normal time of migration. They became restless at night when migration time drew near. For many migrating bird species, day length and changing temperature are cues for migration. How do you think climate change is affecting the timing of bird migration?

North-south migrations are familiar to us, but there are other types of migration as well. Pacific salmon migrate from the streams they begin life in to the open ocean and then back to their home stream every 3–4 years. Some mammals and birds migrate between higher and lower elevations in mountain ranges. For example, elk spend summers grazing in high alpine meadows. In winter, they move down to valleys and forests to feed. Some migrations occur twice a day. In the oceans, tiny organisms, called zooplankton, and some fish can descend to 1000 m during the day, where there is no sunlight. They will rise to the sea surface at night to feed on the tiny, photosynthetic organisms called phytoplankton. What are the advantages and disadvantages to this adaptation?

Hibernation

Many other animals do not migrate. How do these animals find enough food to survive? You will recall that *hibernation* is a state of inactivity similar to sleep. What happens to processes in the animal's body during hibernation? How does hibernation solve the problem of not being able to find enough food?

In northern temperate forests, some animals increase their body weight by as much as half in late summer, when food is more readily available. Animals can rely on their stores of body fat to survive long, cold winters during which food is typically scarce. If they undergo a dormant or semidormant state in winter, animals are able to lower their energy needs significantly. What is one way that they might reduce their energy need? Why is the body temperature of a dormant animal lower than normal?

Bats and small hummingbirds may go into a semidormant state for a few hours every day. Some mammals, including mice and squirrels, can stay dormant for many weeks. This longer period of dormancy is called hibernation. During hibernation, animals reduce their rate of energy use by as much as 95%. Hibernation is not a total shutdown, however. Periods of arousal can elevate the body temperature of a hibernating animal. What conditions might trigger these periods of arousal? Why might arousal be important? How might this arousal be dangerous?

Scripture Spotlight

Some animals store food or water for later use. What do **Psalm 119:11** and **Proverbs 10:14** suggest storing?

Estivation

For some animals, the problem is not caused by cold weather, but by hot weather. **Estivation** is a dormant state that helps animals survive a lack of resources in hot, dry weather. When rivers dry up in summer, mudskippers retreat into underground burrows. They remain inactive there until the water rises and brings food with it.

Most estivating species can also protect themselves against water loss during dormancy. Frogs and toads that estivate have stores of water in the bladder that they use to keep their cells alive. Some frog species have physical barriers that slow water loss from their bodies. They estivate in cocoons made of dried mucus or multiple layers of shed skin.

The mudskipper uses its fins to move on land. It digs a deep burrow to survive long, hot periods that dry up the riverbed it lives in.

How do you think it gets air while in estivation?

Explore-a-Lab

Structured Inquiry

Which covering is the best insulator against the cold?

You will need a thermometer, bowl of ice water, cotton balls, fake fur, felt, feathers, plastic wrap, and zip-top sandwich bags. Use the thermometer to find the temperature of the bowl of ice water. Fill the plastic bag with cotton balls, and insert the thermometer into the middle of the bag so that it is covered with cotton balls. Repeat this step for the other coverings, making sure the thermometer is always surrounded by the material. What animals have covering similar to each of the materials you tested? Where do those animals live? What covering had the coldest temperature?

Other Innate Behaviors

Although animals are born with innate behaviors, they are not necessarily simple. Some animals have elaborate courtship rituals that they use to find and win a mate. The complex patterns of signs and signals that honey bees use to find food and communicate are also innate behaviors. Hunting for food, food gathering, nest-building activities, and defenses of colonies or territories are all innate behaviors. Why do you think some behaviors are innate while others are not? How is the success of social insects related to their innate behaviors?

Male grouse fan their tail feathers, puff up their chests, and make noises to attract females during mating season courtship displays.

Is the female's response to the male grouse's display innate? Why or why not?

Learned Behavioral Adaptations **Explain**

Learned behaviors are acquired during the lifetime of the individual. How do animals learn behaviors? Are learned behaviors easily obtained?

Habituation

Habituation is one kind of learned behavior in which an animal learns to stop responding to a particular stimulus over time. Think about what it would be like to work in a restaurant that makes your favorite foods. At first, all of the smells are exciting and make you hungry quickly. However, after a while, you stop noticing the smells. Why? What are other examples of this kind of learned behavior? Why is habituation an important behavioral adaptation?

Classical Conditioning

Classical conditioning is a kind of learned behavior. It is learning to associate one stimulus with another, unrelated one. Honey bees, for example, can be trained using food bowls of different colors. If a blue dish with pure water sits next to a yellow dish with sugar water, worker bees quickly learn that yellow is food and blue is not. When the solutions are switched (sugar to blue and water to yellow), the bees continue to feed at the yellow dish for a while. Then, by trial and error, they learn that the blue dish contains the food.

Science Journal

Check out your *Science Journal* for a Guided Inquiry that explores the concept of classical conditioning. **Extend**

Operant Conditioning

Operant conditioning occurs when an animal remembers the outcome of past events and modifies future behavior accordingly. Rewards reinforce behavior. A rat learning to run a maze to find a food reward is an example of operant conditioning. How do horse trainers use operant conditioning? What other animals can be trained in this way?

This dog owner is using operant conditioning to train her dog to "shake hands."

What behaviors would you want a pet to learn? How would you use operant conditioning in training a pet to do that behavior?

Structured Inquiry

Discover

Record your work for this inquiry. Your teacher may also assign the related Guided Inquiry.

Science Journal

Finger Maze

How can you learn to complete a maze without seeing it?

Procedure

Materials
- 9 × 12 card stock
- drinking straws
- masking tape
- scissors

1. Draw a simple maze similar to the one shown below. Make sure the lines are about 2.5 cm apart.

2. Tape straws on the lines of your maze.

3. Blindfold a classmate and have him or her trace the maze with one finger. Tell the classmate not to lift the finger until he or she has found the way out of the maze.

4. Time each of five different trials. **Record** your results in the table in your ***Science Journal***.

5. Rotate the maze 180° and check the time that it takes the classmate to find his or her way through the maze. Record the time.

Create Explanations

1. How can you learn to complete a maze without seeing it?

2. Do you think that your classmate learned the maze's pathway? Explain.

3. What do you think would have happened if you had let the classmate study the maze before being blindfolded? Explain.

Imprinting

Imprinting is a kind of learned behavior. It was described by Konrad Lorenz in 1935. Lorenz found that very young Greylag geese learn to follow the first moving object they see between 13 and 16 hours after hatching. Imprinting can occur only during that "critical period." If it does not occur then, it never will.

Physical and behavioral adaptations help animals survive. But can these adaptations help animals survive in ecosystems changed by human activity? Why do you think imprinting is an important adaptation?

Check for Understanding

Horses will come to the sound of feed being poured into a trough. What type of learned behavior is this?

Lesson Activity

In groups of three, choose an animal that you are interested in learning more about. Then research the animal to find out about its behavioral adaptations. Create a poster about your animal to present to the class. Make sure that you include a picture of the animal you chose and highlight how the behavioral adaptations make it successful in finding food and living in its habitat.

Discuss how these behavioral adaptations are similar to or different from those of the animals chosen by other groups.

What other behavioral adaptation would make your animal even more successful? What would be the cost of that adaptation?

Concept Check Assess/Reflect

Summary: What are animal adaptations? Animals have adaptations that improve their chances of survival and reproduction. Physical adaptations include camouflage, warning coloration, and mimicry. Innate behavioral adaptations are those that an organism is born knowing how to do or an organism's ability to respond in a specific way without having to learn how. Innate behavioral adaptations include reflexes, taxis, migration, hibernation, estivation, and other behaviors. Learned behavioral adaptations are acquired during an individual animal's lifetime. Learned behaviors include habituation, classical conditioning, operant conditioning, and imprinting.

1. How does hibernation help animals cope with winter?
2. Suggest a reason why imprinting during a single, critical period is adaptive.
3. Can you assume that any coloration you see in an animal is cryptic? Why or why not?
4. Explain the type of behavior that happens when a bell rings at the end of a class.

Lesson 3

Essential Question

What Changes Occur in Ecosystems?

Objectives

- Describe how humans affect Earth's land.
- Describe how to preserve ecosystems.

Vocabulary

deforestation

sustainable agriculture

fracking

How much impact do you think an individual can have on an ecosystem? What actions have you taken in the past week that would affect ecosystems? Did your actions have positive or negative effects on ecosystems? What were the effects? How did your actions affect the animals in ecosystems? Some human influence on ecosystems is obvious, such as the building of roads or the damming of rivers. Other actions do not have such obvious impacts. What are some examples of less obvious human impacts? How do you think animals in these ecosystems are able to adapt to the changes? How might the animals react if they can't adapt?

Changes to Earth's Land Explain

What materials do we take from the Earth? You may recall that these natural materials gathered from Earth are called *natural resources*. The land we live on is our most basic and vital resource. Every time we change it, we trigger a series of effects that spread like ripples in a pond. Which of these "ripples" might be good? Which might be bad?

Deforestation

Consider the widespread removal or clear-cutting of trees from an area, or **deforestation**. What happens when a forest is cut up into many smaller forests to build roads or provide land for other kinds of development? Does this change affect only the local area? How might small changes lead to widespread consequences? How might the mud and silt that gets washed off the areas of road construction affect the surrounding ecosystem? What is the impact of road building in your community? What can be done to lessen the negative impact of road building?

This work crew is building a road through the Amazon rain forest.

Could this road building in the Amazon affect people in the United States and Canada? Explain your reasoning.

Nutrient Depletion

In some areas, forests are burned to create farmland or ranches on which to raise cattle. However, when the land is planted and the crops are harvested, the nutrients that have been taken in from the soil by the plants are lost to the soil, as is the amount of organic matter. As a result, the soil is less fertile and less able to hold moisture, and the land is unable to support crops or enough grass on which to raise cattle. So people continue to clear more land to create more farmland and grassland. What do you think happens to the land areas that can no longer support crops? What affect do you think nutrient depletion has on the various trophic levels in the ecosystem's food web?

Sustainable Farming

The problem of nutrient and organic matter depletion in soil due to farming can be reduced by using sustainable agricultural practices. **Sustainable agriculture** refers to farming methods that help keep the land productive. Sustainable farmers often plant cover crops, such as clover and rye, during times that the land would normally lie bare. Cover crops help keep the land from washing away, and they help the soil absorb moisture. They provide a habitat for beneficial organisms that help slow the growth of weeds. What other benefits might these cover crops offer?

Explore-a-Lab

Structured Inquiry

How does sedimentation affect aquatic ecosystems?

One of the effects of removing trees and plants from an area is that the topsoil is no longer protected and can be washed into streams and rivers by rain. Form a group of four students. Carefully cut the top off a clear plastic 2-L bottle. Then fill it with water until it is half full. Next, have each person in the group place a flex straw into the bottle and blow bubbles. Record the level of difficulty using a scale of 1–5, where 1 = easy to blow bubbles and 5 = impossible to blow bubbles.

Next, place the straws in the bottle and pour pea gravel into the bottle until it forms a layer that is 2 cm deep. One end of each straw should be under the surface of the layer of pea gravel. Blow bubbles once again and record how easy or difficult it is to do so. Then pour in a cup of coarse sand so that it forms a layer that is 2 cm deep and repeat the process. Finally, pour in a cup of silt so that it forms a layer that is 2 cm deep and repeat. Discuss your findings. How do the results of this experiment inform us about how aquatic organisms are affected by sedimentation? What other changes will result? What nutrients does the soil lose when the topsoil washes away? What effect does this have on plants growing there?

Sustainable farmers who raise cattle can rotate their livestock from one grazing area to another. This practice allows the plants a longer growing period, and plants in the grazing areas become denser. How will denser plant growth help the land and the cattle? What other practices can you think of that would help keep farmland profitable?

Air Changes and Ecosystems **Explain**

You have learned that some of the most valuable natural resources are fossil fuels—coal, oil, and natural gas. Why are these resources so important? What are some ways you used these resources today? Why are these resources limited? How does acquiring, refining, and using these resources lead to pollution? How can pollution affect you?

Global Warming

Burning fossil fuels releases carbon dioxide into the atmosphere. Carbon dioxide is one of many greenhouse gases in the atmosphere. Greenhouse gases trap the heat that is produced from the radiant energy of the Sun. This greenhouse effect is necessary for all life on Earth. However, many scientists believe that human actions are increasing the amount of greenhouse gases in the atmosphere. They believe that rising global temperatures are related to this increase in greenhouse gases. How might you be affected by these predicted changes? Are there places on Earth that will be less affected by global warming than others? Explain. Global climate change is important because climate determines the distribution of organisms and the character of ecosystems. Global warming can result in changes to ecosystems all over the world. What can you do to help reduce global warming? What can state and local governments do?

Effects of Global Warming

Climate	Processes	Species	Ecosystems
• Warmer climates • Earlier springs • Shorter winters • Altered precipitation patterns	• Melting glacial ice • Lower lake levels and altered streamflow • Increased coastal erosion • Rising sea levels • Increased wildfires and floods	• Decline in populations of some species • Changes in ranges and migration patterns of some species • Changes in breeding times	• Shrinking sea-ice ecosystems • Increased shrubs in tundra • Changes in forests at transition zones • Changes in food webs • Changes in wetlands

Directed Inquiry

Discover

Record your work for this inquiry. Your teacher may also assign the related Guided Inquiry.

Science Journal

Model the Greenhouse Effect

How does increasing carbon dioxide in the air affect temperature?

Materials

- aluminum foil
- 2 100-mL beakers
- 2 thermometers
- plastic wrap
- water
- 50-mL graduated cylinder
- lamp or sunlight
- 2 effervescent tablets
- stopwatch

Procedure

1. Wrap the outside of the two beakers with aluminum foil. **Measure** and pour 50 mL of water into each beaker.
2. Stretch plastic wrap over the top of both beakers. Poke the thermometers through the plastic wrap in each beaker.
3. Place the two beakers side by side in strong sunlight or 40 cm from a lighted desk lamp. **Record** a starting temperature in your ***Science Journal***.
4. Add two effervescent tablets to the water in one beaker and quickly reseal the plastic wrap. **Observe** what happens.
5. Record the temperatures in the beakers every minute for 30 minutes.

Analyze Results

How is the air in the two jars different? **Use numbers** to **graph** your results.

Create Explanations

1. How does increasing carbon dioxide in the air affect temperature?
2. How does this model demonstrate the greenhouse effect on Earth?
3. Why should we care whether greenhouse gases change?
4. In what way might increasing temperature affect animal populations?

Air Pollution

As you know, Earth's atmosphere provides living organisms with many of the essential materials they need to stay alive. What atmospheric gases do organisms use? Can they get these materials anywhere else?

Sometimes, solid particles and gases that can harm ecosystems get into Earth's air. As you know, these harmful substances are called *pollutants*.

Some air pollution is naturally occurring, such as gases from volcanoes, dust, pollen, mold spores, and methane emissions from termite mounds. Methane is a potent greenhouse gas. Another major contributor of the total methane released into the atmosphere comes from the large number of cattle raised for human consumption. How does air pollution affect humans and animals?

Once in the air, oxides of sulfur and nitrogen (SO_2 and NO_3) that are released by burning fossil fuels react with water in clouds to form *acid rain* (HNO_3 and H_2SO_4). Acid rain dissolves calcium in the soil and destroys the waxy outer layers of leaves, which help reduce water loss from the plant. Deprived of the calcium from the soil that they need for growth and the covering on their leaves, trees grow more slowly and can eventually die. Acid rain can also increase the acidity of lakes and streams, which can affect fish populations, especially the young of most species. Below a certain level of acid, fish eggs cannot hatch. How would these changes in fish populations affect the aquatic ecosystem? How do you think the increased acidity affects plant populations?

Acid rain does occur naturally. So why are human-made sources a problem?

Nitrogen oxides and organic compounds in air react to form ground-level ozone. Ozone, a gas, is a major part of smog in cities. Although Earth needs an ozone layer for protection from the Sun's harmful ultraviolet rays, ozone can be dangerous when it forms near the ground. Ground-level ozone can affect ozone-sensitive plants and the ecosystems they are part of. It can interfere with the ability of plants to carry on photosynthesis and make plants more susceptible to disease, insect damage, harsh weather, and the effects of pollution. These effects, in turn, can affect entire ecosystems by altering habits, changing nutrient cycles, and reducing species diversity. How might a change in a nutrient cycle affect a third-level consumer? Draw a diagram to show the cause-and-effect steps of how this might happen. How does a decrease in species diversity make an ecosystem more susceptible to ecological damage?

Explore-a-Lab

Structured Inquiry

How can acidic groundwater affect plants?

Label four small potted plants of the same kind *Water* and label another four of the same kind *Acid*. Mix 25 mL of vinegar into 575 mL of water. Dip pH indicator paper into the vinegar water and the untreated water. Compare the color of the paper with the pH chart on the package. Record the pH of the water. Add 30 mL of untreated water to each "Water" plant and 30 mL of vinegar-water to each "Acid" plant every day for five days. Take photos of the plants every day and note your observations. Discuss your findings with the class. What conclusions can you make? How might your experiment differ from the effects of acid rain in the natural environment?

Water Changes and Ecosystems

Life on Earth depends on the constant recycling of clean water. How do organisms use water to survive? What are some adaptations of plants and animals for getting and keeping water? What are some sources of water? What processes of the water cycle help make sure that water is recycled? How do you use water for recreational activities? How would your life be different if water were in very short supply?

Every pollutant added to water poses a threat to animals and other organisms. When water from precipitation runs off buildings, rivers, and other objects, it can pick up toxic materials, such as chemicals and disease-causing organisms. Water pollution can also result from other sources, such as factories, construction, and fracking. **Fracking**, or hydraulic fracturing, is the process of drilling and injecting fluid into the ground at high pressure to break apart shale rocks and release natural gas resources. How might this process pollute water? What other sources of water pollution can you identify? How can you help limit water pollution?

How can the pesticide applied to this lawn pollute water?

You might be surprised to learn that farms and a homeowner's lawn can be a source of water pollution. Runoff from fertilizer used on farmers' fields and homeowners' lawns increases the nitrogen content of water. Too much nitrogen stimulates algae growth. When the algae die and decay, the oxygen is removed from the water. How would this affect water habitat food webs? Could the effects spread to land food webs?

Many forms of water pollution are difficult to see, but some pollution is difficult to overlook. The Great Pacific Garbage Patch is estimated to be twice the size of Texas and swirls in the Pacific Ocean roughly between the coast of California and Hawaii. Currently, an estimated 11 million tons (and growing) of floating plastic covers an area of nearly 13 million km^2 in the Pacific Ocean, about 1125 km northeast of the Hawaiian Island chain and 1610 km from the coast of California.

Explore-a-Lab

Guided Inquiry

How can you clean polluted water?

Pour some saltwater into a beaker. Record how the water in the beaker looks. Add coffee, oil, and flour to the water. Stir the water in a counterclockwise direction. Can you see through the water? Record your observations. Add packing peanuts, fishing line, food coloring, and sawdust to the water. Stir the water again, and record your observations. What effect do you think this water would have on animals living in a marine ecosystem? Do you think the animals could survive? Why or why not? Use various materials and tools to clean up a polluted water supply as much as possible. There are constraints:

- Only provided materials can be used to clean up the water supply.
- At least 90% of the original volume of water must remain in the water supply.
- All team members must be involved in the clean-up effort.

In this model, what did the coffee, oil, and flour represent? What did the packing peanuts, fishing line, food coloring, and sawdust represent? What tool was most useful in removing the solid pollution? What tool was most useful in removing the liquid pollution? Do you think the water you cleaned would support animal life? Why or why not? Why is it important to keep Earth's oceans free of pollution?

Scripture Spotlight

What kind of water is described in **Hebrews 10:22**? What kind of spiritual pollution is mentioned? How do we get rid of it?

Oil spills on land and at sea can cause ecosystem damage.

How do you think animals could survive a disaster like this?

Preserving Ecosystems Explain

One important way to help preserve ecosystems is by learning to make informed decisions about environmental issues. A few basic steps are the "backbone" of any decision-making process.

Making Environmental Decisions

1. Identify the decision that needs to be made.
2. Gather information about the topic, including the range of possible choices. Where might you look for reliable information about environmental issues?
3. Identify the values that are relevant to the issue. Values might include economic, ethical, scientific, or social. What other values might be related to an environmental issue?
4. Assess the alternatives against the values you have identified.
5. Identify the positive and negative consequences of each alternative.
6. Use all the information to make an informed decision.

Lesson Activity

With a group, identify a particular environmental issue that involves a particular animal species in your region. What effect does this issue have on the animals in the local habitat? Identify what decisions need to be made about the issue. Use the decision-making steps to form a group response to the issue. Present your decision to the rest of the class. Be sure to back up your decision with facts, and use visuals to support your facts. After your presentation, take a class vote to see whether students agree with your decision.

Was your presentation effective in convincing a majority of students? What action could you take to support your opinion?

Concept Check Assess/Reflect

Summary: What changes occur in ecosystems? Human activities can cause severe changes to ecosystems. Earth's land is affected by deforestation, nutrient depletion, and sustainable agriculture. The air is affected by pollution, which can lead to an increase in greenhouse gases and global warming. Air pollution also leads to acid rain. Earth's water is affected by human activities that produce toxic chemicals, such as factories, construction, fracking, fertilizer use, and oil spills. There are many ways in which we can reduce our impact on our environment.

1. Explain how the greenhouse effect influences Earth's climate.
2. In what ways do sustainable agriculture practices help farmland?
3. Describe one step you can take to reduce your negative impact on Earth's ecosystems. Explain how that step will help preserve the goods and services that Earth's ecosystems provide.

Lesson 4

Essential Question

How Do Organisms Respond to Environmental Changes?

At Creation, God created living organisms with the ability to adjust to many, but not all, changes in their environment. Today, organisms continue to respond to changes. Even with the adaptive ability that God gave them, there are limits beyond which living organisms can adjust. Nevertheless, many organisms demonstrate an amazing ability to adjust to subtle and dramatic changes, ensuring a diversity of life on Earth. What types of changes can an ecosystem adjust to or recover from naturally? What organisms would be able to survive a fire? What strategies might these organisms use to survive? Explain why humans should or should not intervene to help more plants and animals survive natural disasters.

Objectives

- Describe how the Chesapeake Bay ecosystem changed over time.
- Explain the limiting factor that affected the Chesapeake Bay estuary.
- Describe the process of ecological succession.

Vocabulary

estuary

dead zone

This blue crab has just been washed ashore.

What changes to the crab's habitat would likely make it impossible for this crab to survive? What role do humans play in keeping the habitat inhabitable?

The Chesapeake Bay: A Case Study Explain

The Chesapeake Bay is located along the northeastern coastline of the United States. It is the largest estuary in North America and the second-largest estuary in the world. An **estuary** is the area where the freshwater of a river meets the saltwater of the ocean. The Chesapeake Bay ecosystem is made up of the Bay, its rivers and streams, and all their living organisms. The water composition in the Bay ecosystem varies as you move down the Bay closer to the ocean water. How do you think the salinity of the water changes as you move down the Bay? The water in all parts of the ecosystem contains many dissolved materials, including bacteria and trace elements that are important to many biological processes. Dissolved salts are important to the life cycles of many species. For example, some fish hatch in freshwater or brackish water and then move to saltier water as they mature. Dissolved oxygen and carbon dioxide are critical to the organisms of the Chesapeake Bay. How do organisms use these two gases?

The Chesapeake Bay is known for its beauty and its bountiful plants and animals. Common plants include submerged grasses, which add oxygen to the estuary water. Animals include more than 300 species of fish and numerous shellfish and crab species, such as the blue crab on the previous page. Its water conditions for many years were ideal for oyster fisheries. What are some conditions that the species living in the Bay would need to survive?

However, by the 1970s, parts of the Chesapeake Bay were identified as **dead zones**, areas of water with very low levels of dissolved oxygen. When a dead zone develops, most marine species either die or they leave the area if they can move. Why would low oxygen levels cause the death of organisms? One cause of the dead zones was pollution that contributed to algae blooms. When the algae are alive, they prevent sunlight from reaching the bottom of the bay; when the algae die and decay, they use up the oxygen in the water. What might be the sources of pollutants that contaminate the waters of the Bay? How would blocked sunlight affect organisms that live in the Bay's waters?

In 2010, the Environmental Protection Agency (EPA) took action. It established limits for the amount of pollutants that can flow into the Bay each year. Restoration projects that restored habitats have helped some, and stewardship programs have helped to foster and expand action by citizens and communities. Today, the Bay's health has slowly improved, but many parts of the ecosystem remain in poor condition. The water in parts of the Bay remains polluted, and the populations of many species have not rebounded. Bay grasses have not returned to the levels that are needed. How might the lower populations of grasses affect the Bay food web? In what ways would this affect humans?

The Bay, however, is still very productive. In fact, it supports more than 27,000 species and produces about 227 million kg of seafood a year. What must people do to save the Bay from continued decline?

Math in Science

Over the past 60 years, the area of shallow wetland habitats along the Chesapeake Bay has decreased. The table below shows average numbers per year of some species of water birds observed there. Calculate the percentage change in each population between 1950 and 1999. Using the "major food source" key, develop a hypothesis to explain what your percentages show.

Average Numbers of Water Birds Observed in the Chesapeake Bay Wetland

Species	1950–1959	1985–1999	Percentage
Redhead*	76,000	2000	
Northern pintail*	40,000	3000	
Common goldeneye**	22,000	6000	
Canada goose***	178,000	386,000	
Brant***	13,000	22,000	

Major food source for the species:
* = submerged aquatic vegetation; ** = invertebrates; *** = agricultural fields

The blue heron is an important species in the Chesapeake Bay ecosystem.

Where does the blue heron fit in the Bay food web?

Directed Inquiry

Discover

Record your work for this inquiry. Your teacher may also assign the related Guided Inquiry.

Science Journal

What Are the Limits?

What abiotic factors have the greatest effect on the rate of seed germination?

Materials

- 6 six-pack planters
- 36 bean seeds
- 10-mL graduated cylinder
- aluminum foil
- graph paper
- light source
- plant labels
- potting soil
- refrigerator
- trowel
- water (spring)

Procedure

1. Write the name of the factors listed below on the plant labels.
   - Light/room temp/5 mL water
   - Dark/room temp/5 mL water
   - temp 7°C/dark/5 mL water
   - temp 21°C/dark/5 mL water
   - 2 mL water/light/room temp
   - 10 mL water/light/room temp
2. Fill each cell of each six-pack planter with potting soil.
3. Use your index finger to make a hole about 1 cm deep in the potting soil of each cell.
4. Fill each of the holes you made in Step 3 with one bean seed. Gently push the soil over the seed and press down lightly to make sure the soil surrounds the seed in each cell.
5. Place one of the plant labels you made in each of the six-packs.
6. Set each six-pack in the required area and monitor the seeds over the next two weeks. Be sure the conditions called for on the label are met.
7. **Observe** the seeds daily and **record** the total number of seedlings that germinate each day.

Analyze Results

Record your data in the table provided in your ***Science Journal***. Construct a **graph** of the data.

Create Explanations

1. What abiotic factors have the greatest effect on the rate of seed germination?
2. In this activity, what was the independent variable, and what was the dependent variable?
3. What are other abiotic limiting factors that would most likely affect germination?
4. Which of the factors you investigated seemed to have the most negative effect on seed germination?

Limiting Factors

As you just read, the Chesapeake Bay was once an ideal place for oyster fisheries. Its carrying capacity for populations of oysters was very large. Recall that *carrying capacity* is the largest number of individuals of one species that an ecosystem can support without harmful effects. However, today the Bay's carrying capacity for oysters has shrunk to less than 1% of what it once was. You know that pollution contributed to this reduction in population size. But how?

Pollution not only affects the biotic factors in an ecosystem, but it can also affect its abiotic factors—either directly or indirectly. In the case of the Chesapeake Bay, pollution indirectly caused the amount of dissolved oxygen in the water to decrease. What was the cause-and-effect chain of events that led from the water pollution to the decrease in dissolved oxygen?

The long period of low dissolved oxygen in the Bay killed most of the oysters. Oxygen is a limiting factor in the Bay's ecosystem. A *limiting factor*, you will recall, is something that keeps the size of a population from getting too large. What are other limiting factors in the Bay ecosystem? Which of these factors also affect land ecosystems? What changes in limiting factors would be needed to restore the Chesapeake Bay to its original condition?

Explore-a-Lab

Structured Inquiry

How do fertilizers affect an aquatic ecosystem?

Label four quart jars *A* (no fertilizer), *B* (0.5 g of fertilizer), *C* (1 g of fertilizer), and *D* (2 g of fertilizer). Fill each jar with 750 mL of pond water. Add the specified amount of plant fertilizer to each jar. Place the jars in the same light conditions. Make and record your observations about the changes in each jar every day for two weeks. Discuss your findings as a group. What do your results suggest about the effect of adding fertilizer to the pond water? What does the response to fertilizer tell you about how quickly organisms can respond to changes in limiting factors? Where might you find this occurring in the real world? What might happen to a pond that experienced an influx of fertilizers?

Ecological Succession

As you have seen, human activities can change ecosystems. Natural events also change ecosystems. Sometimes the changes happen very quickly. For example, the winds of a hurricane can change an ecosystem almost immediately. What changes might occur in an ecosystem as a result of a hurricane? What other natural events might cause sudden changes?

Even without natural disasters and changes caused by humans, ecosystems change. In fact, all ecosystems are constantly changing. As you know, *ecological succession* is the gradual replacement of one type of biological community by another through a series of changes.

Primary Succession

Recall that in *primary succession*, communities develop in an area where soil or organisms do not exist. You can see this process in the picture below. What kinds of events might create areas that do not have soil or living organisms?

Primary Succession

Primary succession begins with newly exposed rocks and progresses over many years to a climax community.

Would the organisms in a climax community be the same in all parts of North America? Why or why not?

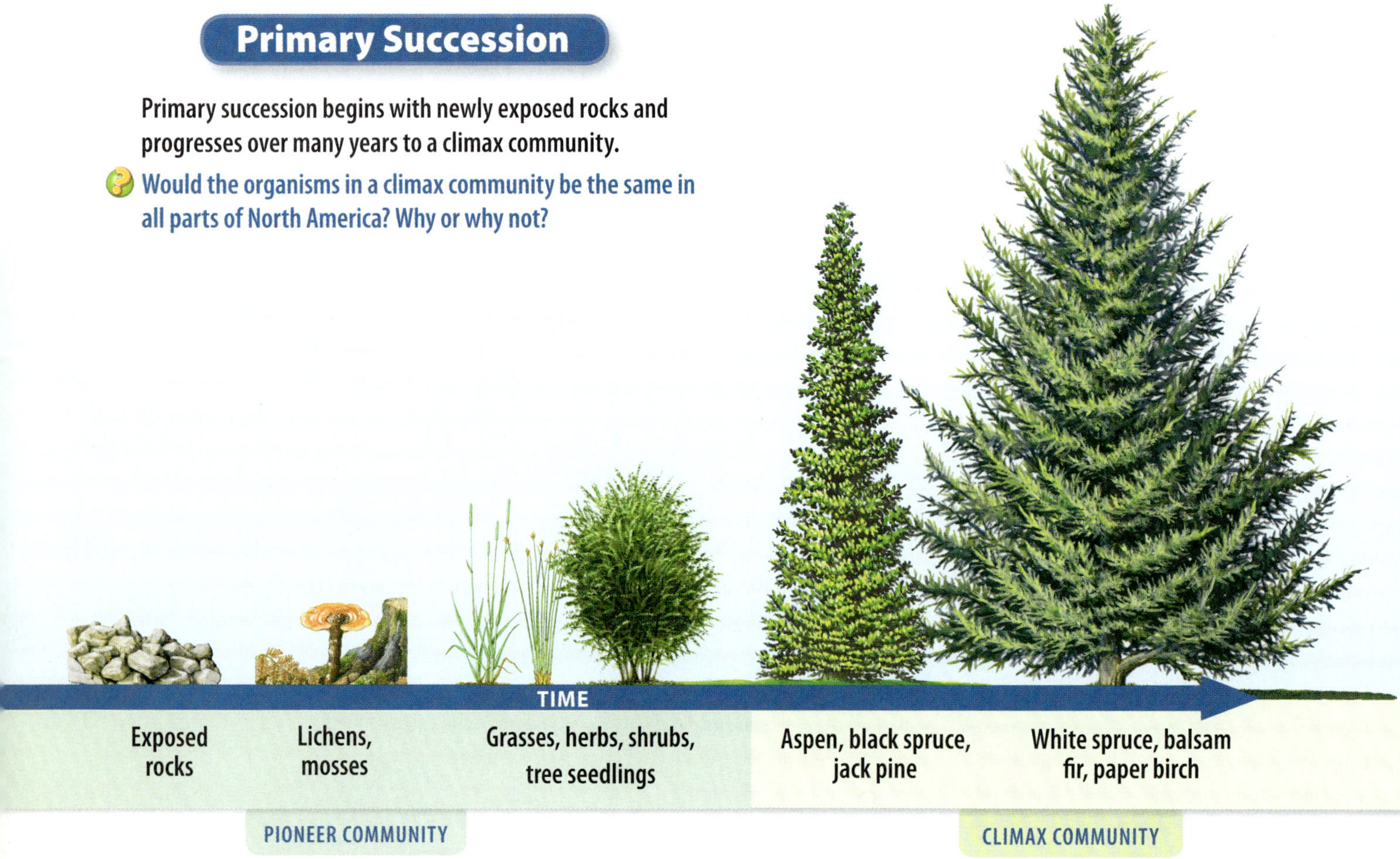

What characteristics must a pioneer species have to survive in the harsh conditions of a pioneer community? Where might you find some pioneer species in your community? What changes would have to happen in the area for plants and animals to survive? Where does the soil come from that the grass needs to grow?

Secondary succession

As you know, *secondary succession* occurs when an ecosystem has been disturbed but soil and organisms remain. The picture shows the steps of secondary succession. What events might disturb an ecosystem enough so that secondary succession takes place? Where in the secondary succession process do you think animals first inhabit the area? Why? Why do you think that no low-growing grasses appear in the climax community? What do you think will happen to the climax community over time? Which do you think results in a climax community more quickly—primary succession or secondary succession?

Secondary Succession

A deforested area in Idaho changes from a meadowland to a coniferous forest in about 150 years.

Why is the coniferous forest considered the climax, or end stage, for this succession?

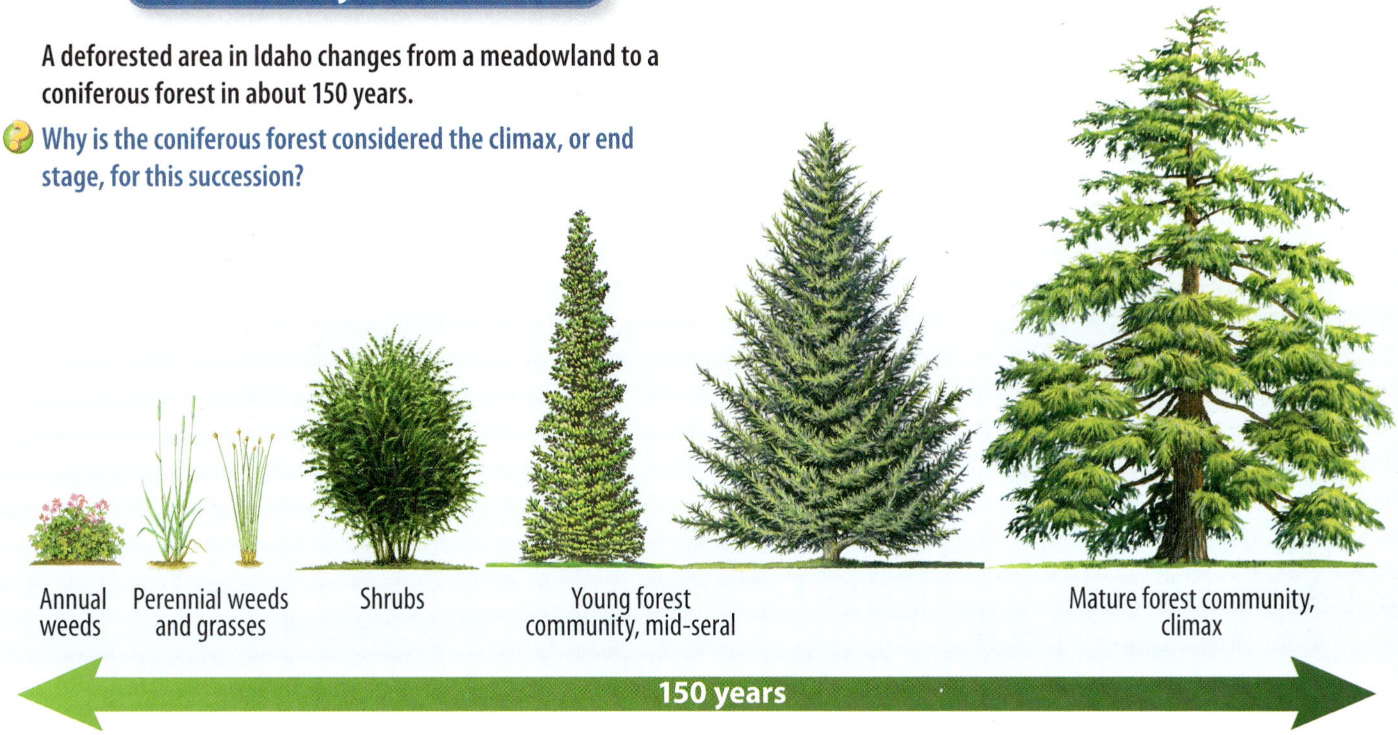

Succession in a Pond

A natural event such as a glacier or a sink hole sometimes results in a different kind of ecosystem—a pond. Over hundreds or thousands of years, the pond can be transformed into a land community.

1 In deep freshwater, rooted plants do not grow because sunlight does not reach the deep water. What kinds of organisms would live in deep freshwater?

2 Sediments are carried into the pond, making the pond shallower. Birds and other organisms carry seeds to the pond. Rooted and floating plants begin to grow. What animals might inhabit the pond at this stage?

3 As sediments continue to build up, reeds and grasses begin to grow at shallow edges of the pond, forming a marsh community. After some time, the pond floor is close enough to the surface that plants can grow all the way across the floor. When this happens, the pond becomes a marsh. What causes sediments to build up? How will the animals at this stage differ from those at Stage 2?

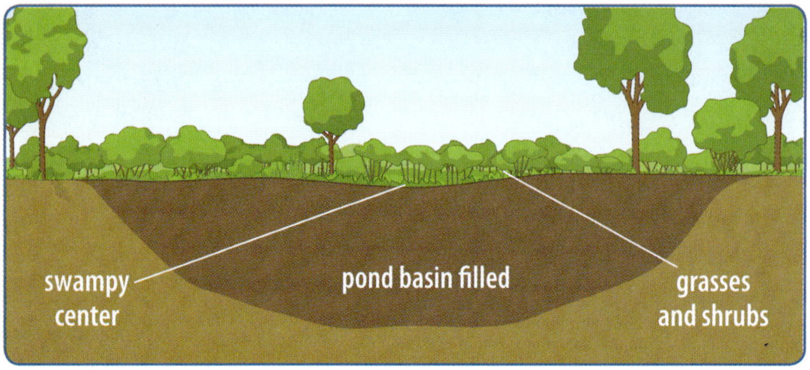

4 Over time, soil fills in the entire pond, and a land community forms with a swampy center. As soils become drier, tree species become established. What will happen to this land community as time passes?

Explore-a-Lab

Structured Inquiry

What can you observe in succession in a pond?

Place a test tube in test tube rack to hold it upright. Add 1 cooked wheat grain, several small pieces of cut hay, and a small amount of soil (do not use commercially packaged soil). Fill the test tube about halfway full with spring water. Add 1–2 droppers full of pond water. Make sure to get some of the organic material in the dropper. Set the test tube in bright but indirect natural light. Observe the test tube every couple of days over the next 2 weeks and observe any changes you see. Record your observations. When you make your observations, include an observation of a sample of your pond water under the microscope. Draw sketches of the organisms you observe. Use books and Internet resources to identify the organisms you see. How do your observations compare with the observations of your classmates? What organisms did you observe first? Over time, how did the populations change? What are some ways in which people influence succession in ecosystems?

Concept Check **Assess/Reflect**

Summary: How do organisms respond to environmental changes? The Chesapeake Bay estuary has experienced many environmental changes that resulted in dead zones. Pollution led to the formation of algae blooms. EPA intervention and restoration programs have helped to improve the Bay's ecosystem conditions. Oxygen is one example of a limiting factor. A limiting factor is something that keeps a population from becoming too large. A natural event that causes change within an ecosystem is ecological succession. Ecological succession is the gradual replacement of one type of biological community by another and can be primary or secondary. Over hundreds or thousands of years, ecosystems change from one type to another through the process of succession.

1. What effect did pollution have on the populations in the Chesapeake Bay estuary?
2. How does dissolved oxygen act as a limiting factor in the Chesapeake Bay estuary?.
3. Compare and contrast primary succession with secondary succession.
4. Why are short-lived organisms more likely to adapt to a rapidly changing environment than long-lived organisms?

People in Science

Extend

Get to Know William Derham

William Derham was a minister with a great curiosity about the natural world. He was born in Stoughton, Worcestershire, England, in 1657 to a poor family. His family's poverty made it necessary for him to enroll as a *poor boy* at the University of Oxford.

Derham received a Bachelor of Arts in 1679 and a Master of Arts in 1683 from Trinity College, Oxford. As a clergyman, Derham served in many different capacities. He was Chaplain to Lady Grey and later was appointed Chaplain to the Prince of Wales.

In addition to his ministry duties, he practiced medicine and pursued his interest in science and natural history. He wrote *The Artificial Clockmaker*, which discusses calculating numbers to movement, the motion of pendulums, and how to make chimes and set them to musical notes. Many of his works, such as *Physico-Theology: or, a Demonstration of the Being and Attributes of God*, focused on how the natural world proved the existence of God.

As a member of the Royal Society, a prestigious group of scientists, Derham wrote many scientific articles. His articles covered many areas of science, including animal behavior. In his writings, Derham discussed the migration of birds. Birds have an instinct to migrate when temperatures change to places where food is more abundant. They may fly thousands of miles to reach a place that meets their needs. Derham noted that some birds have long wings to aid them in their flight. Other birds, such as quails, Derham stated, have especially strong muscles that aid them in their flight. These observations showed Derham the wisdom and mercy of God in giving animals the instincts and structures needed to survive.

Derham also wrote about his observations of the deathwatch beetle. He placed a male and female beetle in a box and observed them for several weeks. These beetles make a tapping noise, which Derham deduced was used to attract a mate.

Called to Serve

Derham wrote several books reasoning that God's existence can be evidenced from the natural world. He supported his belief through his studies and writings of scientific observations.

Concept Check

1. How did Derham's belief in God affect his scientific pursuits?
2. What did Derham do to demonstrate his beliefs?

Smart Collars for Wildlife Conservation

Remember life before smartphones? The same way that the smartphone changed how we communicate, the *smart collar* for animals changes how we manage and understand wildlife populations. The collars use a combination of GPS technology and an accelerometer. An accelerometer detects and measures the vibration and the speed of an animal wearing the collar. It helps us understand the animal's surroundings better; it can tell us whether the animal is moving uphill or lying down, and when it takes its next step. The device can transform our knowledge of wildlife by delivering a 24-hour data stream of an animal's activities. This information helps us better understand how animals respond and adapt to their environment.

Biologists can more accurately study the eating habits, metabolic rates, sleeping patterns, times spent resting, and times spent running, along with other essential information from wildlife. Being able to measure and record data can change how we work to conserve wildlife.

The collars are used on animals that encounter and live among humans, such as mountain lions, wolves, and bears. The data stream from a smart-collared animal shows, for example, when the animal has last eaten and how likely it is to go beyond its normal range in its search for food. Biologists can also "see" that the animal caught prey and can identify what type of prey it is.

Scientists can learn much more about animal adaptations and survival in general through the use of these collars. With data constantly streaming in about both predators and prey, national park and wilderness managers can predict interactions within the ecosystems they oversee: who might eat or be eaten and where both predator and prey might go, driven by their survival instinct or hunger.

Concept Check

1. The smart collar uses a combination of two technologies; what are they?
2. How do smart collars help scientists study animal behavior?

Chapter 4 Review Assess/Reflect

Study Guide

Lesson 1

1. Organisms will compete with each other for water, food, habitat, and mates. Intraspecific competition occurs between organisms in the same population. Interspecific competition occurs between populations of different species.
2. Biotic factors are the effects of a living organism on other organisms or on its nonliving environment. Abiotic factors are the nonliving factors in the environment.
3. Energy flows through the environment from the Sun to producers and then to consumers. Food chains may have many levels of consumers. A food web describes the independent food changes in an ecosystem.
4. The carbon cycle recycles carbon through an ecosystem. Producers take carbon dioxide from the air and convert it to sugars. The carbon is transferred to the consumers that eat the producers, and they release it to the air as carbon dioxide through respiration. When the consumers die, decomposers release the carbon dioxide back into the ecosystem. Nitrogen is recycled through an ecosystem as bacteria take in nitrogen from the air and fixate it into compounds that can be used by other organisms. Different bacteria release nitrogen back to the air during denitrification.
5. Natural factors, such as floods, wildfires and mudslides, can affect the balance in ecosystems. Human causes of changes in ecosystems include pollution, introduced species, habitat alteration, overhunting, and overfishing.

Lesson 2

1. Physical adaptations are traits that help an animal survive and thrive. Some physical adaptations that protect prey from predator attacks are camouflage, warning coloration, and mimicry.
2. Both innate and learned behavioral adaptations are actions in response to a stimulus. Innate behavioral adaptations are inborn, whereas learned behavioral adaptations are acquired during the lifetime of the organism.

Lesson 3

1. Human actions that affect land are deforestation, nutrient depletion, and sustainable agriculture. Air pollution can cause global warming, acid rain, and increased ozone levels.
2. The steps on p. 146 can help with making environmental decisions that help to preserve ecosystems.

Lesson 4

1. In the Chesapeake Bay estuary, pollution led to the formation of algae blooms. The algae blocked sunlight from reaching the bottom of the Bay, and as the algae died off, they used up dissolved oxygen in the water. Dead zones formed as a result.
2. The amount of dissolved oxygen in the water is a limiting factor in the Chesapeake Bay estuary. As oxygen levels decreased, organisms either died or moved to other habitats.
3. Ecological succession is the gradual replacement of one type of biological community by another. Primary succession occurs where soil or organisms do not exist. Secondary succession occurs when an ecosystem has been disturbed, but soil and some organisms remain.

Multiple Choice

Choose the best answer.

1. Which organisms break down dead organisms?
   A. producers
   B. decomposers
   C. herbivores
   D. consumers
2. Which adaptation do skunks, hornets, and coral snakes have in common?
   A. disruptive coloration
   B. countershading
   C. cryptic coloration
   D. warning coloration
3. Which of the following is based on an organism's source of energy?
   A. taxis
   B. trophic level
   C. estivation
   D. fracking

Check Point

Answer the following questions. For items 4–6, use the diagram.

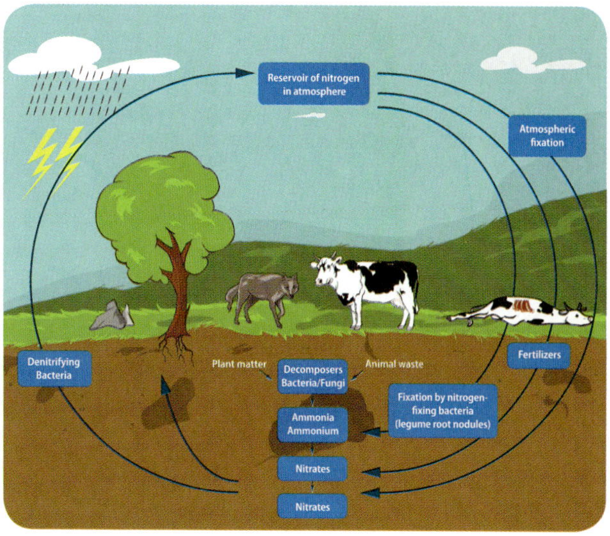

4. Identify which organisms are necessary for the maintenance of the cycle in the diagram and explain your choices.
5. Explain what this diagram shows.
6. How might the ecosystem shown be affected if the grass species died out?
7. Why is far less energy available to top predators than to first level consumers in a food web?
8. **Compare** and contrast innate behaviors with learned behaviors.
9. What is the difference between Batesian mimicry and Müllerian mimicry?
10. How does the use of cover crops help keep land productive?
11. Describe one way that a healthy lake could develop into a dead zone.
12. Why can small perennial weeds and grasses not serve as pioneer species during primary succession?

Unit 2

The Human Body

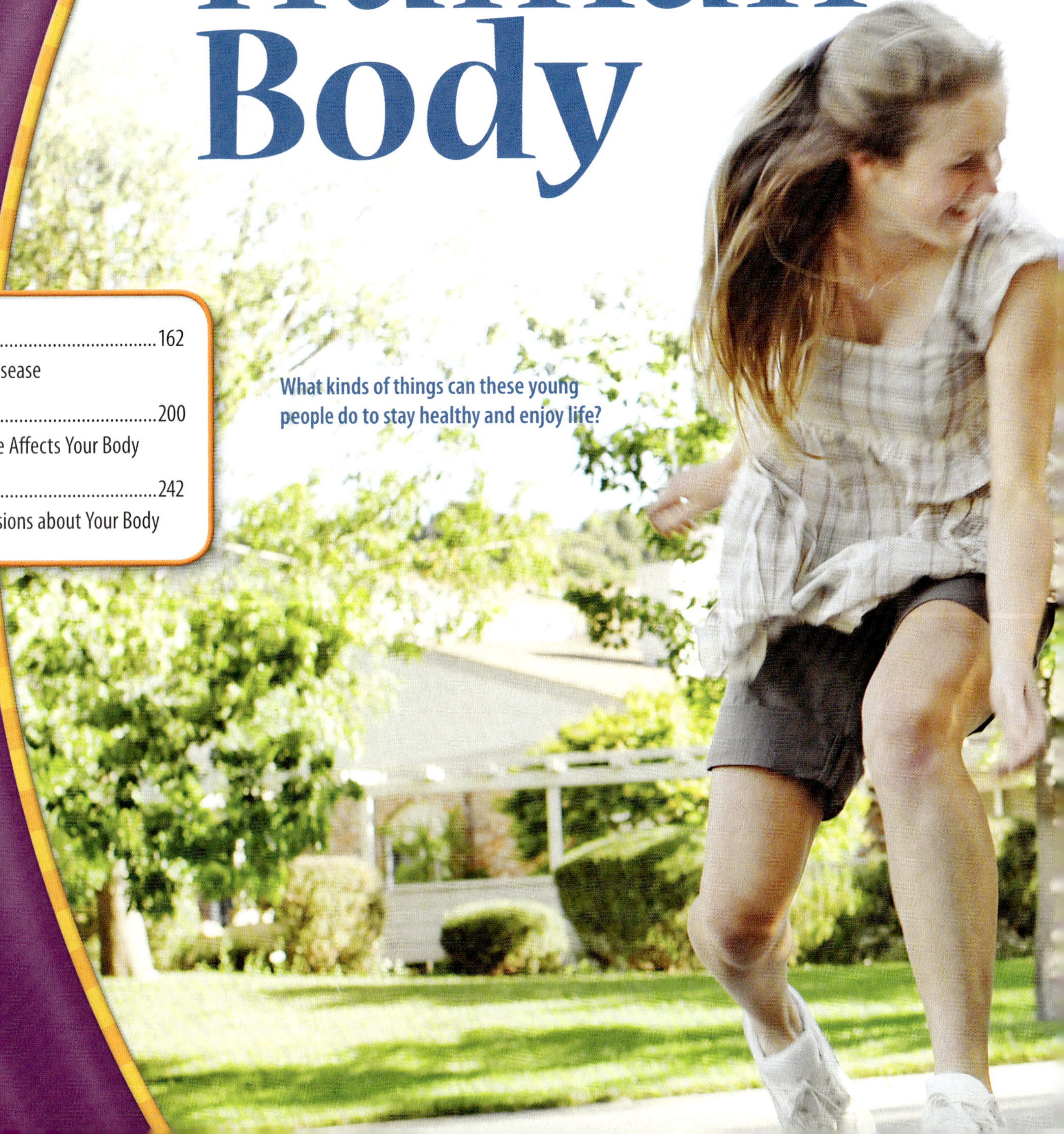

What kinds of things can these young people do to stay healthy and enjoy life?

Unit Overview

How do you feel today? In Chapter 5, you will learn what science has discovered about factors that affect your health. Those factors include the genes you inherited, the environment around you, and the personal choices you make to listen to the evidence.

Chapter 6 focuses on those personal choices. They include whether to smoke tobacco, use alcohol and other drugs, or exercise and eat well to stay fit. After looking at the evidence and considering the risks and benefits of the possible choices, you will make a plan for your body. That plan will help assist you in staying healthy!

Chapter 7 looks at how to make wise decisions when it comes to your health. It can sometimes be difficult to make some decisions, especially when friends pressure you to do some things while commercials and ads pressure you to do others. What is best for your health? How can you base your decisions on the evidence and your own values?

Your teacher may assign an Open Inquiry lab and a Lifestyle Challenge activity. Use your *Science Journal* to record your work.

Chapter 5

Genes and Disease

Scripture Spotlight

Sickness and disease entered our world because of sin. When Jesus was on Earth, He healed people of their diseases. When Jesus comes again, sickness and disease will be gone forever. You will read the following passages in this chapter.

3 John 1:2 (p. 166)
Proverbs 3:1–2 (p. 176)
1 Corinthians 6:19–20 (p. 186)
Matthew 12:34–35 (p. 190)
John 13:35 (p. 190)

A family tree can be a helpful tool for showing how genes are passed from generation to generation.

The Big Idea

Your health can be affected by genes passed on to you from your parents, certain environmental factors, and your own daily choices.

How is your family tree connected to your health?

Inquiry Kick-Off **Engage**

What controls your health? In your ***Science Journal,*** you will model how much of your health is influenced by the genes you inherited, how much is influenced by the environment where you live, and how much is influenced by the choices you make.

Science Journal

Lesson 1

Essential Question

What Are Genetic Disorders?

Objectives

- Explain how genetic disorders occur.
- Describe several common genetic disorders.
- Identify common genetic tests.
- Explain how genetic counseling can help parents.

Vocabulary

genetic disorder
mutagen
sickle-cell anemia
cystic fibrosis
color blindness
Down syndrome

How well can you see at night? Most animals have better night vision than humans. How are their eyes different from ours? Some people have autosomal-dominant, congenital, stationary night blindness. This is a genetic disorder that causes the retina, which contains cells that detect light and color, to function improperly. This condition makes it difficult for a person to see and distinguish objects in low light. What would happen if a genetic scientist discovered how to "turn on" genes that improved your vision at night? Would you consider having this procedure done if you had night blindness? Why or why not? Your life from the moment of conception is influenced by genetics. How do genes affect your ability to do things, such as see in the dark?

Inheriting a Disorder Explain

Recall that mitosis is a type of cell division where each daughter cell contains the same amount of genetic material as the parent cell. As you read this page, mitosis is occurring in your body. What would happen if mitosis stopped happening? What would happen if there were an error during mitotic division? Fortunately, cell divisions occur with relatively few errors so that cells contain a normal number of chromosomes and an intact set of genetic information. Even if an error is made, cells have mechanisms to correct it. However, some errors may not get repaired. In this case, errors that result in an abnormal number of chromosomes or in changes in the DNA of cells can lead to a genetic disorder. What might cause errors in cell division to occur?

The size and shape of an owl's eyes aid in its excellent night vision but keep the owl from being able to move its eyes. The owl's extremely flexible neck muscles help make up for this, allowing the owl to move its head 270° side to side.

What is different about this owl's eyes and your own? How would vision like an owl's help you function? What would be the drawbacks?

A **genetic disorder** is a health problem caused partly or totally by a defect in one or more genes or chromosomes. Genetic disorders may not affect a person's ability to function normally. For example, some men have a genetic disorder known as male pattern baldness. This genetic disorder results in a loss of hair. Unlike male pattern baldness, some genetic disorders are much more serious and even life threatening.

Male pattern baldness is a genetic disorder that does not become apparent until adulthood. However, some genetic disorders are apparent at birth. One example is cleft palate, a split that starts at a baby's lip and may extend through the soft palate. A cleft palate, which is one of the most common birth defects, can occur in varying degrees. A cleft palate forms early during pregnancy and is caused by genetic and environmental factors. How can doctors treat this condition?

Other genetic disorders may never be identified, having no noticeable effect on the person's life. Because some people have genetic disorders without knowing it, it is difficult to determine how frequently they occur. However, about three in every one hundred newborns in North America probably have a genetic disorder. Scientists have identified thousands of genetic disorders, but most of them are very rare. These numbers may seem staggering, but in reality only one person in several thousands or even millions may have a genetic disorder of concern.

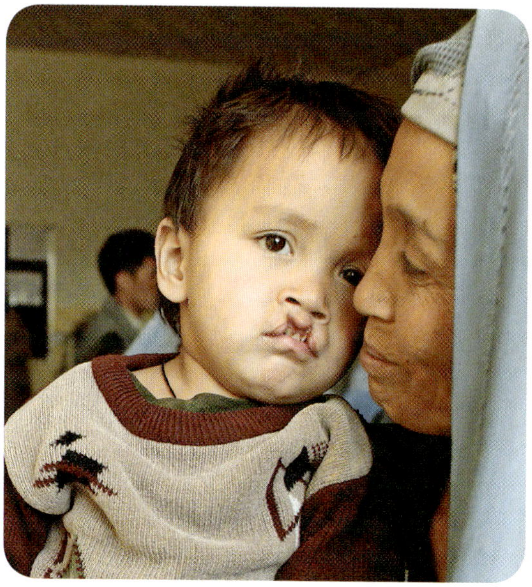

A cleft palate is easily corrected and rarely seen in developed countries. Many groups raise money to send doctors to developing nations to perform corrective surgeries on children with this disorder.

How do you think a cleft palate affects the physical development of a child?

Common Genetic Disorders Explain

What causes a change in the number of chromosomes or in the DNA so that a genetic disorder develops? One cause can be a mutation, or change, in just one gene or in a group of genes. A mutation can be the result of an environmental factor that causes a change in the DNA of an organism. Such an environmental factor is called a **mutagen**. Mutagens include radiation, such as X-rays or nuclear radiation. Other mutagens include chemicals, such as dioxins, asbestos, those in tobacco, and others. How might you become exposed to harmful levels of radiation or chemicals? What can you do to reduce the risk of exposure to mutagens?

A single-gene disorder results from one mutated gene. Scientists can attribute as many as 4000 diseases to single-gene disorders. The mutated gene is typically recessive. Recall that a recessive gene is not expressed when it is paired with a dominant gene. What is necessary for a recessive gene to be expressed by a person? If both parents have a dominant and recessive allele for the disease, what is the probability of a child inheriting the disease? How does the probability change if one parent contains two recessive alleles for the disease? Single-gene disorders include sickle-cell anemia, cystic fibrosis, color blindness, and hemophilia. What other diseases are caused by single genes?

Scripture Spotlight

What does John wish for his friend Gaius in **3 John 1:2**? It is God's desire for you, too.

Sickle-Cell Anemia

Sickle-cell anemia is a genetic disorder that results when a person inherits two recessive alleles for the genetic disorder. Sickle-cell anemia describes a condition where the red blood cells, which contain the oxygen-carrying protein hemoglobin, form a sickle, or crescent-moon shape. The sickle shape does not move smoothly through blood vessels. Sickle cells are also sticky and tend to clump together. The clumps of sickle cells block blood flow to the limbs and organs. In addition, sickle cells do not live as long as healthy red blood cells. As the sickle cells die, the person ends up with fewer red blood cells than normal, leading to anemia. How might an anemic person feel? Other symptoms of this disorder include severe joint pain and kidney problems. Why would a genetic blood disorder cause joint pain and kidney problems?

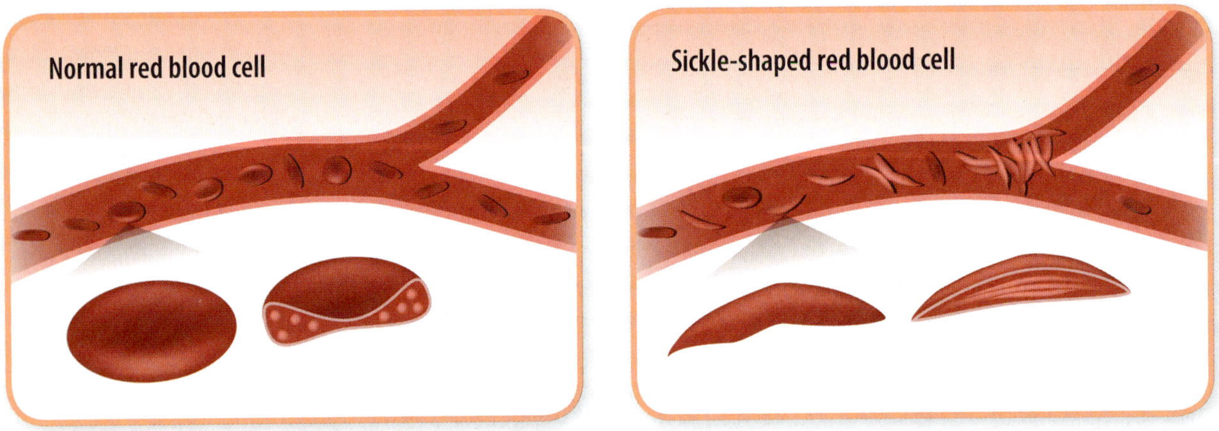

Normal red blood cells are circular, making it easy for them to slide through blood vessels. Can you pick out the sickle cell?

How could a blocked blood vessel affect the surrounding tissues?

Structured Inquiry

Discover

Record your work for this inquiry. Your teacher may also assign the related Guided Inquiry.

Science Journal

Sickle-Cell Anemia

What is the relationship between sickle-cell anemia and malaria?

Materials
- marker
- 2 paper cups
- 75 red beans
- 25 white beans
- coin

Procedure

1. Use the marker to label one of the cups *Gene Pool* and the other cup *Discard*.

2. Place the 75 red beans and the 25 white beans in the paper cup labeled *Gene Pool*. The red bean is a dominant allele for a normal, healthy blood cell. The white bean is a recessive allele for sickle-cell anemia.

Possible Genotypes
- **AA** – healthy red blood cells
- **SS** – sickle-shaped red-blood cells
- **SA** – one gene for healthy red blood cells and one gene for sickle-shaped red blood cells

3. Select two beans from the *Gene Pool* cup at random to represent a person. Record the genotype represented in the table provided in your ***Science Journal***, as well as the outcome for the patient.
- If the person has the SS genotype, he or she will have sickle-cell anemia. He or she will have resistance to malaria, but will eventually die due to sickle-cell anemia. Place the two red beans in the cup labeled *Discard*.
- If the person has the AA genotype, he or she will not be affected by sickle-shaped cells, but will not have malaria resistance and may die from malaria. Flip the coin. If it lands "heads" up the person survives a malaria infection (return the beans to the *Gene Pool* cup). If the coin lands "tails" up the person does not survive a malaria infection (add the beans to the *Discard* cup).
- If the person has the SA genotype, he or she will not be affected by sickle-shaped cells, will have resistance to a malaria infection, and will survive. Place the two beans in the *Gene Pool* cup.

4. Repeat Step 3 until all 50 individuals have been chosen and the data has been recorded.

Analyze Results

Graph the the data you collected. During how many rounds was malaria contracted? What percentage of the rounds contracted malaria? **Compare** your results with those of other groups.

People who have only one parent with the disorder inherit only one recessive allele, resulting in a milder form of the disease called sickle-cell trait. They rarely develop any symptoms, but they can pass the recessive allele on to their children. Today, all states require newborns to be tested for this lifelong disease, so treatment can be started right away. Why do you think this genetic test was adopted by all 50 states and parts of Canada?

Although the sickle-cell trait is a debilitating disorder, it provides a benefit to people who live in malaria-prone areas. Scientists have discovered that the sickle-shaped blood cells impair malaria parasites' growth and development. The sickle trait does not prevent a person from developing malaria, but it provides the person with a greater ability to survive the acute symptoms associated with malaria.

Cystic Fibrosis

A person diagnosed with **cystic fibrosis** has a mutated gene that causes the body to produce very thick, sticky mucus. The mucus builds up in the breathing passages, causing serious lung infections. This mucus also builds up in the pancreas, making it harder for the body to break down and absorb food. What is the function of the pancreas? Why might this gene also affect it?

The cystic-fibrosis gene can mutate in a thousand ways and cause a range of symptoms, including delayed growth in infants, severe digestive problems, lung congestion, pneumonia, and fatigue. How might the symptoms affect a person's daily life? Why might a genetic disorder that produces thick, sticky mucus cause fatigue?

Because this disorder is recessive, both parents must pass on the mutated gene to their offspring. About 1 in every 29 Caucasians has this gene. That makes it easier for two parents to pass the gene to their offspring. In fact, cystic fibrosis is the most common—and most deadly—genetic disorder affecting Caucasians in North America. Why might Caucasians be more susceptible to this gene mutation than other groups?

Lesson Activity

Singer Alex Pangman is known as Canada's sweetheart of swing. She is also one of the strongest supporters of organ donation. This is because she underwent a double lung transplant. Alex was born with cystic fibrosis. Use the Internet or other resources to learn more about life with cystic fibrosis. Use the Internet to research treatments, food recommendations, and other facts about the disease. Share your results with the class in a format of your choice.

What is daily life with cystic fibrosis like?

Treatment for cystic fibrosis includes antibiotics to fight infection and a special diet to help with digestion. What other lifestyle behaviors might benefit someone with this genetic disorder? What behaviors would have a negative impact?

Improved knowledge of the cystic fibrosis disorder and better medical care and medications allow people to live longer and healthier lives with the disorder. As afflicted people age, however, their condition may get worse, and medication and a healthful diet may not be enough. Some people need a lung transplant to survive.

Color Blindness

Also called color vision deficiency, **color blindness** is caused by mutations to different genes. Color blindness is a recessive disorder. To have the disorder, the offspring must inherit a mutated gene from both parents.

- Red-green vision defects affect about 8 in 100 males and 1 in 200 females of Northern European ancestry. People with this disorder see red and green as the same color. What difficulties might red-green color blindness cause in daily life?
- Blue-yellow vision defects affect males and females equally. This disorder occurs in fewer than 1 in 10,000 people worldwide.
- Complete color vision deficiency is a very rare condition, affecting only 1 in 30,000 people. People with this deficiency see only black, white, and gray.

How do any of the forms of color blindness affect how a person interacts in society?

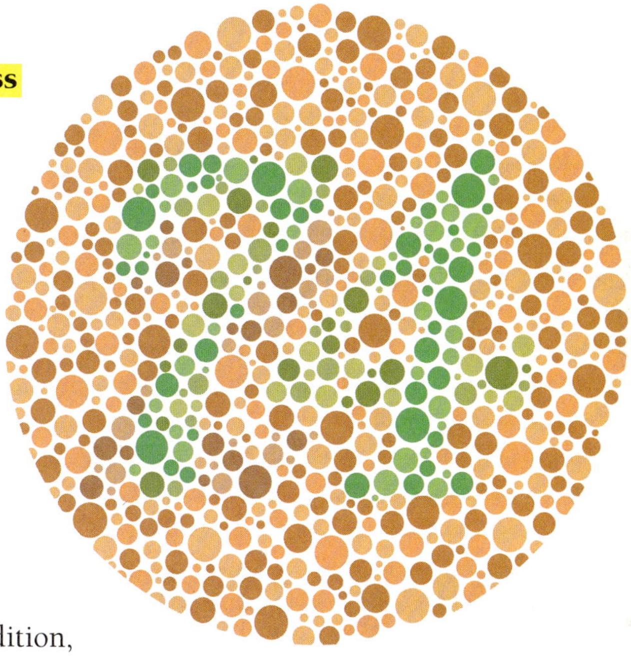

Can you see a number in this circle? What is it?

What kind of color blindness does this circle of dots test?

Explore-a-Lab

Guided Inquiry

What percentage of students had difficulty reading the cards?

Separate into teams of two. Obtain two series of note cards from your teacher. Using the first series of cards, one student should present the card and take notes. The other student should try to read what is on the card. After completing the first series of note cards, switch roles using the second series of note cards. After completing both series, gather as a class and discuss the results. One student should take notes based on the statistics gathered from the class.

Chromosomal Disorders Explain

Genetic disorders can also occur when all or part of a chromosome is missing or when there is an extra chromosome. Most humans have 23 pairs of chromosomes, or a total of 46 chromosomes. However, because of problems in cell division in the sperm, egg, or embryo, some fetuses end up with three copies of chromosome 21. This condition results in **Down syndrome**, a group of symptoms that may include delayed mental and social development. The syndrome is characterized by flattened facial features and upward-slanting eyes. These children are often small for their age.

Most forms of Down syndrome cannot be passed from parent to child. One rare form of Down syndrome involves extra genetic material attached to chromosome 21. This form can be inherited, but it affects only 4% of people with this syndrome.

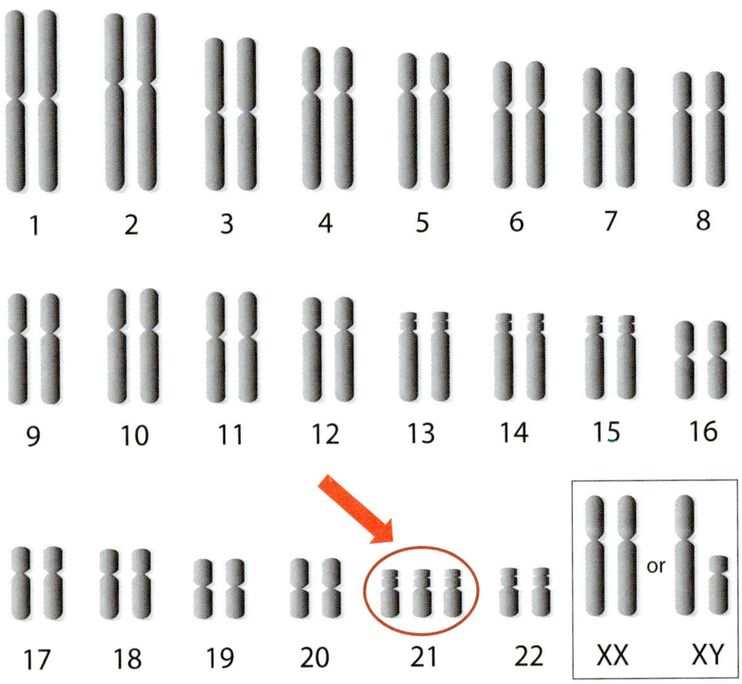

This is a chromosome map, or karyotype, from a person with Down syndrome. Notice that all but one of the chromosomes come in pairs. However, chromosome 21 consists of three and not two chromosomes.

Why are the chromosomes shown in pairs?

The risks of having a baby with Down syndrome increase as the age of the mother increases. When a mother is 35, the risk of having a baby with Down syndrome is 1 in 400. However, when a mother is 45, the risk rises to 1 in 30. Why do you think the age of a woman increases the risk of having a baby with Down syndrome? Do you think the father's age affects a child's chance of having Down syndrome? Why or why not?

Testing for Genetic Disorders Explain

Because every cell in your body, except for sex cells, has the same DNA, testing for genetic disorders is relatively easy. Doctors can analyze small samples of blood or body tissues, such as skin. Results can show whether that person carries gene disorders or mutated chromosomes. Tests can also indicate whether genes are missing or damaged. The two tests described below are used before birth. Both can determine whether a fetus has Down syndrome, sickle-cell disease, cystic fibrosis, or other genetic disorders. Why might it be important for doctors and parents to know about genetic disorders as soon as possible? How might parents feel if they elect to have one of the procedures performed?

Amniocentesis

Amniocentesis can be performed on pregnant women when the fetus is 16 to 18 weeks old. The doctor removes a small amount of the fluid (amniotic fluid) surrounding the fetus. Analysis of the chromosomes in fetal cells in the fluid shows the gender and age of the fetus. More important, this test can identify genetic problems. How might the doctor remove the fluid for testing?

Chorionic Villus Sampling (CVS)

This test can be done when the fetus is 10 to 12 weeks old. A small piece of the placenta is removed, and the genes in it are examined for genetic disorders.

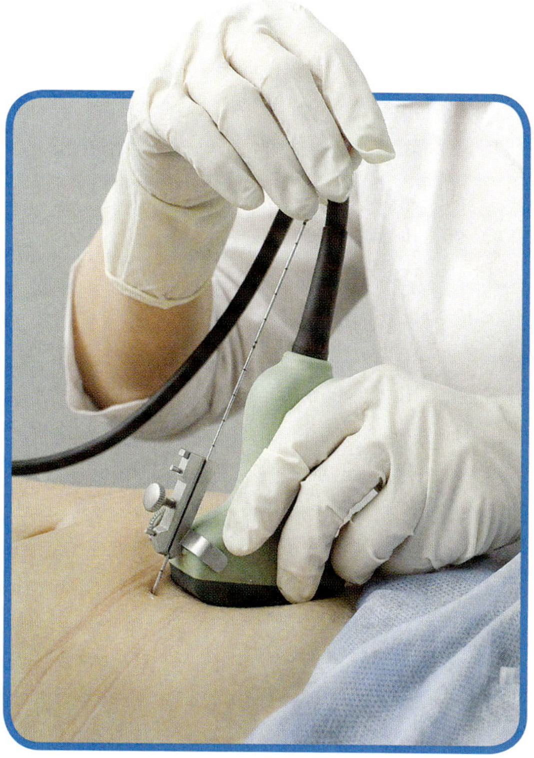

The fetal cells in amniotic fluid will be analyzed for genetic problems.

How is it possible to determine the presence of genetic disorders from such a small sample?

Lesson Activity

Separate into groups and select one of the following single-gene disorders: hemophilia, Huntington's disease, muscular dystrophy, thalassemia, or phenylketonuria. Use the Internet to learn more about the disorder, the symptoms, and the ethnic groups commonly associated with the disorder. Create a short presentation to report your findings. Members of each group should take turns explaining to the class what the group learned. After the presentations, gather as a class and discuss the information.

Which disorders are more common in certain ethnic groups?

Genetic Counseling **Explain**

Couples who are concerned about having a baby with a genetic disorder might work with a genetic counselor. This professional has a medical background, along with training in genetics. Often a counselor is part of a team that might include psychologists, social workers, and others. Do all parents have the opportunity to meet with a genetic counselor? Why or why not? How might the cost of genetic testing skew which parents receive genetic counseling? How could this affect a community?

A genetic counselor uses genetic material obtained from each parent and family medical histories to collect information. The counselor explains the risks of giving birth to children with certain genetic disorders. Then the couple must decide whether to have biological children. How might a close connection with God help during these difficult life decisions?

Faith Connection

Ellen White wrote about inherited tendencies before scientists were studying genetics. She wrote in *The Ministry of Healing* that "the physical conditions of the parents, their dispositions and appetites, their mental and moral tendencies, are, to a greater or less degree, reproduced in their children."

Parents might seek genetic testing in the following situations:

- A close relative has a genetic disease. One member of the couple might be carrying a recessive gene for this disease. If both parents have the gene, each of their children has a one-in-four chance of inheriting that disease.
- A couple already has a child with a serious genetic disease. The couple might be carrying recessive genes for that disease.
- The woman has had two or more miscarriages, or a baby who died before birth. Chromosomal damage can lead to a miscarriage.
- The woman is over age 34 and thus at higher risk for having a baby with Down syndrome.

Tests can also be done after a baby is born to determine whether he or she has a genetic disorder. The tests are performed to ensure the baby has the best start in life. For example, after babies are born, many states require them to be tested for phenylketonuria (PKU). This genetic disorder can cause mental delays. However, a restricted-protein diet can avoid any damage to the baby. What testing does your state or province require for newborns?

Lesson Activity

Two students will be parents, and a third will be a genetic counselor. The counselor will have a chart explaining the disorder detected during genetic testing of the parents. Conduct a role-play activity. Use empathy, scientific knowledge, and support to help the parents decide on the best course of action as they plan their family.

What decisions will you make?

Concept Check Assess/Reflect

Summary: What are genetic disorders? A genetic disorder is a health problem caused partly or totally by a defect in one or more genes or chromosomes. Some genetic disorders are the result of a mutation that affects a single gene. Other genetic disorders are the result of changes to part of a chromosome or to an entire chromosome. A mutation can result from an error during cell division. It can also be caused by an environmental factor, such as radiation or chemicals. Genetic disorders may be apparent at birth or may take years to present symptoms. Common genetic disorders include sickle-cell anemia, cystic fibrosis, color blindness, and Down syndrome. Parents can seek advice from genetic counselors to quantify the risk of a genetic disorder in their offspring. Many prenatal tests, such as amniocentesis and chorionic villus sampling, are routinely used to screen for genetic disorders in the fetus. In addition, many states now require genetic tests for disorders such as phenylketonuria to ensure the child gets the best start in life.

1. How is a mutation different from a mutagen?
2. Why are some diseases more common in certain groups of people, such as Caucasians or African Americans?
3. Why might it be important to find out whether you carry a recessive gene for a genetic disorder?
4. Why are most genetic disorders considered lifelong diseases?

Lesson 2

Essential Question

What Can Trigger a Genetic Disease?

Objectives

- Define a multifactorial disease.
- Explain how environmental factors can activate a genetic disease.
- Explain how a person's lifestyle can increase the risk of a genetic disease.

Vocabulary

multifactorial disease
cardiovascular disease
atherosclerosis
diabetes
obesity

Do you like eating pizza? Is pizza a healthful food choice? Does every person who eats pizza become obese? Why or why not? Obesity is one of many diseases that are caused by genetics and environmental factors.

A disease is an abnormal condition, or disorder, that affects the body. A disease can be caused by an infection. It also can be caused by a mutated gene, which can be passed from parent to offspring. What is an example of an infectious disease? How is the disease spread? What is an example of a genetic disease?

From your previous study of Lesson 1, you know that many genetic diseases, such as cystic fibrosis, are inherited. However, this is not the case with all genetic diseases. In fact, two people may have a gene for a particular genetic disease, but only one person develops the disease, while the other person remains healthy. Why?

Multifactorial Diseases Explain

The development of many genetic diseases is not always as straightforward as those diseases that are caused by single-gene disorders, such as cystic fibrosis. **Multifactorial diseases** result when several factors work together to cause the disease. Two main factors can trigger the development of a multifactorial disease—a genetic factor, or predisposition for disease, and an environmental factor.

Pizza does not cause obesity, but an unhealthful diet contributes to obesity.

How can pizza be part of a healthful diet?

Genetic Health Risks

The genetics of multifactorial diseases is complicated. These diseases are caused by more than one gene. Unlike disorders caused by single-gene mutations, multifactorial disease is not necessarily passed from the parent to the offspring. It can occur in isolation, where a child may exhibit the disease while the parents remain unaffected. However, the genetic component of the disease can cause the disease to occur in families. Although the diseases can cluster in families, there is no clear line of inheritance from parent to offspring. For these reasons, doctors have a difficult time determining a person's risk for developing a multifactorial disease.

Scientists classify the degree of relationship between family members to identify the genetic similarity between individuals.

A multifactorial disease may occur more frequently in one gender than in the other, but it is not linked by the sex chromosome. In addition, the genetic component of a multifactorial disease may occur more frequently in a specific ethnic group.

Relationships between Family Members

Degrees of Relationship	Percentage of Genes in Common	Example
First-degree relative	50%	Parents, children, siblings
Second-degree relative	25%	Aunts, uncles, nieces, nephews, grandparents
Third-degree relative	12.5%	First cousins

Environmental Health Risks

Genetics make up one component of multifactorial diseases. The environment also plays a major role in whether the disease will be expressed. The table below provides some hazards, sources, and health effects of common environmental contaminants.

Environmental Contaminants

Hazard	Sources	Health Effects
Air pollutants	Burning gasoline and other fossil fuels; industrial wastes	Asthma, bronchitis, cancer, other respiratory diseases
Arsenic	Groundwater; runoff from farms, industries, and mining	Skin problems, cancer, death
Lead	Paint in homes built before 1978; industrial waste; some imported products	Severe learning problems, slow growth
Mercury	Industrial waste; contaminated fish and shellfish; coal-fired power plants	Range of health problems, including cerebral palsy or learning disorders
Pesticides	Eating unwashed food; runoff from home and farm use of pesticides	Headache, dizziness, nausea, vomiting, cancer, Parkinson's disease

Laws limit the amount of harmful substances that industries can release into the air and water and the amount of hazardous chemicals that can be included in household and industrial products. Is there any amount of exposure that is safe for human health?

Lesson Activity

Separate into small groups. Each group should select one of the following common environmental contaminants: polychlorinated biphenyls (PCBs), pesticides, fungal toxins, phthalates, volatile organic compounds (VOCs), dioxins, asbestos, heavy metals, chloroform, or chlorine. Each group should use the Internet to identify the common sources for each environmental contaminant as well as the health effects. Each group should also identify how people can be exposed to the contaminate in their daily activities. Each group should prepare a poster presentation and an oral overview of the information for the class. After the presentations, regroup as a class and discuss the information.

How are you exposed to harmful pollutants every day?

Scripture Spotlight

In **Proverbs 3:1–2**, the Bible says that remembering God's teaching and keeping His commandments will add years to our lives. The Adventist Health Study, based in Loma Linda, California, has found that Adventist men in California live 7.3 years longer than non-Adventist men. The advantage for women is 4.4 years. What might be possible reasons for this?

Cardiovascular Disease Explain

Cardiovascular disease affects the heart and blood vessels and is the leading cause of disease in the United States. More commonly called heart disease, it involves a structural or functional abnormality of the heart or the blood vessels. Its causes include an inherited tendency to develop it and environmental factors that foster it.

Scientific studies have shown that very fine, airborne particles can contribute to heart disease. Common sources for these particles include car exhaust and factory smoke. These particles are also emitted during a forest fire.

Cardiovascular disease is an example of a case where an inherited trait can be activated or even made worse by environmental factors. In addition, a person's lifestyle can increase the risk of an inherited disease being expressed. For example, 1 in 500 people inherit a tendency to develop high levels of cholesterol. Cholesterol is a waxy substance in the blood. "Bad" forms of cholesterol can clog blood vessels, leading to a condition called **atherosclerosis**. What lifestyle choices can you make that might reduce your risk of developing cardiovascular disease?

Directed Inquiry

Discover

Record your work for this inquiry. Your teacher may also assign the related Guided Inquiry.

Science Journal

Survival Game

How do your actions affect your health?

Procedure

1. Working on your own, read the statements on diet, exercise, stress level, and smoking found in your ***Science Journal***. Decide which statement best describes you.
2. **Record** your answers on graph paper. Do not write your name on the paper. So be honest!
3. Total your points, paying attention to the pluses and minuses. Record your total score on your graph paper. Fold the paper in half.
4. Hand your paper in to your teacher, who will read the information to the class. Record the information from the class into the table in your ***Science Journal***.

Materials
- graph paper

Analyze Results

What information is common for the class? How does a low total of points affect lifespan?

Create Explanations

1. How do your actions affect your health?
2. Why do we sometimes make unhealthful lifestyle choices?
3. What is the problem with making predictions based on this activity?

Atherosclerosis happens when plaque builds up on the inside walls of the arteries. As the plaque forms, the arteries narrow, restricting or reducing blood flow, which can lead to heart attack or stroke. This condition can also lead to high blood pressure, a measure of how hard the blood pushes against the walls of the arteries. Healthy arteries stretch like elastic and help move blood through the body. When the heart pumps blood with more force, the arteries stretch to allow the blood to flow. This force over time can stretch the artery, making it more difficult to move blood effectively through the body. The tendency to develop high levels of cholesterol or high blood pressure is a genetic trait, but these diseases can be triggered by poor eating and exercise habits.

Science Journal

Check out your ***Science Journal*** for a Structured Inquiry that explores multifactorial disease.

Extend

Lesson Activity

Heart disease is the leading cause of death in the United States. How does this statistic compare with other countries with similar economics (first world or third world) around the world? Working in groups, use the Internet or other sources to research the five leading causes of death in the following countries:

Country	Top Five Causes of Death	Life Expectancy (men)	Life Expectancy (women)
United States			
Canada			
Japan			
United Kingdom			
France			
Cuba			
Bangladesh			
Afghanistan			
Haiti			
Ethiopia			

Prepare a table or graph that summarizes your findings. What do you notice about the results? Are the causes of death the same in all countries? As a group, consider why there may be differences. Share your thoughts with the class.

How do the leading causes of death differ among different countries?

Diabetes **Explain**

Diabetes results from problems in the production of the hormone insulin, which is produced by the pancreas and regulates the amount of sugar in the blood. Diabetes can lead to blindness, the loss of limbs, kidney failure, heart disease, and stroke. What are symptoms of diabetes that you know of?

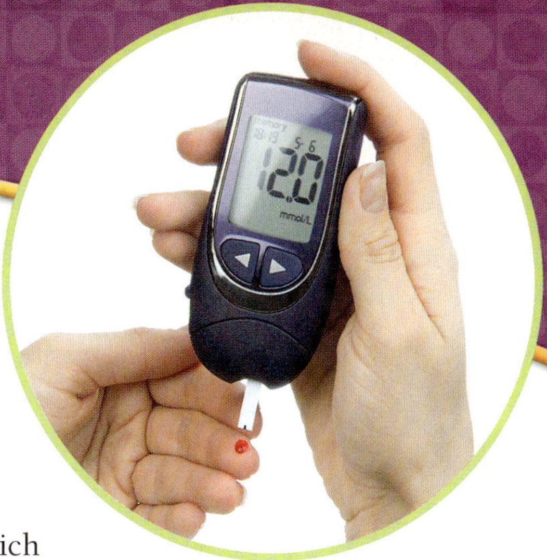

These test strips measure the amount of glucose in blood.

How does a person with type I or type II diabetes use the information on the test strips?

Some people inherit a tendency to develop diabetes, which scientists classify as type I and type II. Type I diabetes is associated with abnormal antibodies that attack certain cells in the pancreas. As a result, the person is unable to produce insulin and must receive daily doses of this hormone. How might a doctor diagnose type I diabetes in a young person? Type II diabetes is more common. A person with type II diabetes produces insulin, but the body cannot use it properly. As a result, the pancreas produces more insulin. Over time, the pancreas can no longer make insulin.

Some people can control type II diabetes by carefully choosing what they eat. What foods might a person with type II diabetes try to avoid? Others must receive doses of insulin. How would a person administer the insulin? How would he or she know how much to use? Most cases of diabetes are type II and begin in people who are 30 or older. However, type II is becoming much more common in younger people. Why do you think type II diabetes is becoming more common in children?

While diabetes has a genetic basis, the disease can be triggered by environmental factors. Researchers are trying to identify those factors, which might include viruses, the intestinal immune system, and environmental contaminants. Lifestyle choices can also worsen this disease. What lifestyle choices would worsen diabetes? Besides sugary snacks, what foods are naturally high in sugar and should be avoided by a diabetic person?

Obesity

Fat is an important part of a healthful diet. However, too much fat is unhealthy. **Obesity** is a condition of being overweight. An obese person has about 20% or more fat than is ideal for that person's height and body type. Some people might have inherited genes that help the body store food as fat, or they may have metabolic factors that make gaining weight easier. In addition, scientists have found that certain kinds of genes are more frequent in obese people. Why are more people at risk of obesity today than 30 years ago? Scientists are now linking obesity to changes in the bacteria that live in a person's gastrointestinal system. How can bacteria in your gut affect your weight?

Faith Connection

In 1902, Ellen White wrote, "Teach the people that it is better to know how to keep well than how to cure disease." (*Medical Ministry* section 12) What types of preventative measures could be used to help avoid some of the diseases mentioned in this lesson?

However, the main cause of obesity is lifestyle choices. Many of us eat more food than our bodies need, and we exercise less than we should. The extra energy is stored as fat. How does eating the diet that God provided to humans at Creation help reduce the risk of diabetes? Being obese puts people at greater risk of heart disease, diabetes, cancer, and a long list of other diseases. Obesity is now a national crisis, affecting our health and our health-care system.

Our health is affected by genetic factors, environmental hazards, and lifestyle choices. To stay healthy, we can be aware of genetic and environmental factors and practice making healthful lifestyle choices.

Explore-a-Lab

Structured Inquiry

Why do food choices influence the risk of developing diabetes?

Monitor the type and amount of foods that you eat at every meal for one week. Place this data into a table. Use the Internet to identify the amount of sugar in each food item and the types of foods and food combinations that are healthful.

Concept Check Assess/Reflect

Summary: What can trigger a genetic disease? A multifactorial disease has both a genetic and environmental basis. Common multifactorial diseases include cardiovascular disease, diabetes, and obesity. Each can be triggered by environmental factors and lifestyle choices. Cardiovascular disease is caused by a genetic tendency to develop the disorder, but environmental factors and lifestyle choices contribute to and exacerbate the disease. Type I diabetes seems to be triggered by environmental factors, while type II diabetes is worsened by lifestyle issues. Lifestyle choices can play a major role in triggering obesity, which is now a crisis in the United States, affecting both health and the health-care system.

1. Compare and contrast single-gene disease with multifactorial disease.
2. How does diet affect the risk of cardiovascular disease, obesity, or diabetes?
3. Compare and contrast the genetic factors with environmental factors associated with type I and type II diabetes.
4. Identify one common environmental contaminant, its source, and how it might affect the human body.

Lesson 3

Essential Question

What Is Cancer and How Is It Caused?

Cells are the building blocks of all living things. Recall that cells divide to create new cells with the same genetic content. How many cells make up the human body? How many cells have divided in the short time you have studied this page? You have learned earlier in your study of science that when cells no longer divide normally, they produce different types of cancer. Cancer is not a single disease but actually a group of diseases. In fact, there are more than 100 different types of cancer. What causes cancer? What are common forms of cancer you know about? What steps are scientists taking to find a cure for cancer? How can people reduce the risk of getting cancer?

Understanding Cancer Explain

Cancer can affect all body parts except the hair, nails, and teeth. *Lymphomas* are cancers of the immune system. The blood-forming organs can develop cancers called *leukemias*. Cancers of the connective tissues, including bones, are called *sarcomas*. *Melanomas* begin in cells that make the pigment melanin, including the skin, eye, or intestine. Cancers of other organs, including the skin and lungs, are called *carcinomas*. How might a doctor diagnose and treat each type of cancer?

Objectives

- Explain the relationship between cancer and cell division.
- Describe factors that contribute to the development of cancer.
- Identify the symptoms that might indicate cancer is present.
- Explain why cancers are difficult to treat.
- Identify ways to lower the risk of getting cancer.

Vocabulary

tumor
benign tumor
malignant tumor
metastasis
biopsy
oncogene
carcinogen
basal cell cancer
melanoma

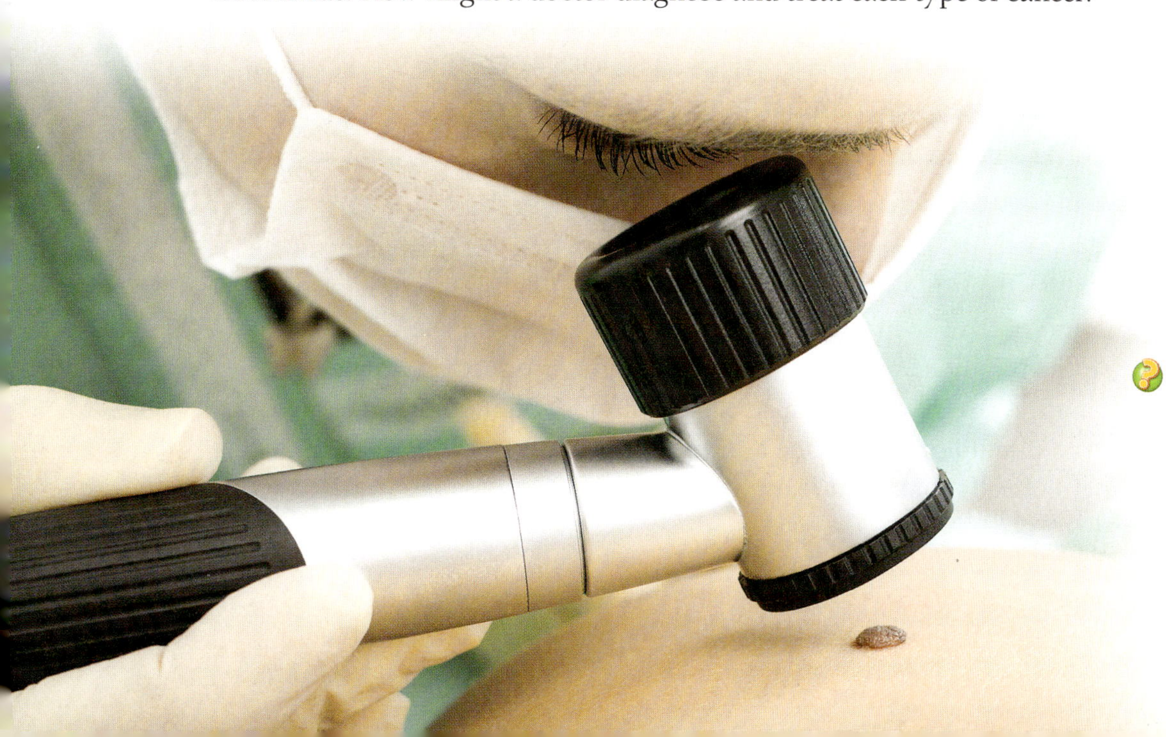

The doctor is examining a mole to determine whether it is a type of carcinoma.

How can you lower your risk of developing a carcinoma?

Under normal conditions, a cell follows a regular cycle of growth and division. Recall that during the cell cycle the cell will rest during *interphase* and divide during *mitosis*. The cell cycle plays a role in replacing damaged or dead cells. Mitosis replaces approximately 30,000 skin cells that die each minute.

The Cell Cycle

1 Prophase

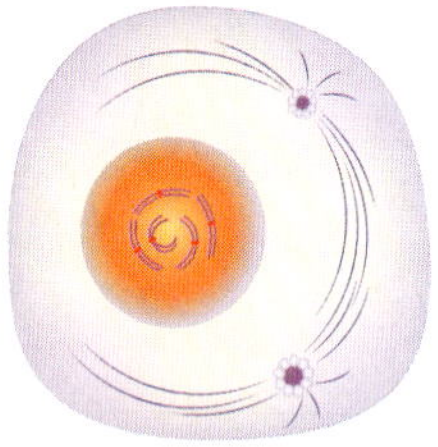

Chromosomes thicken

2 Prometaphase

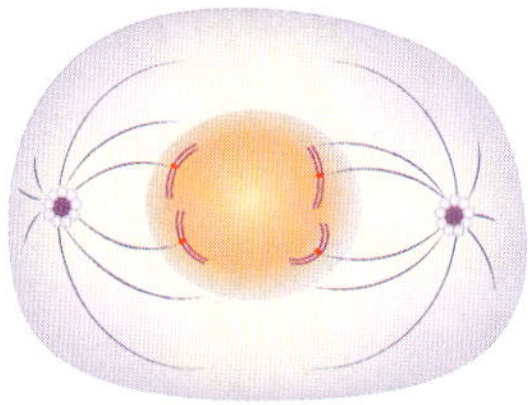

The nuclear membrane breaks down.

3 Metaphase

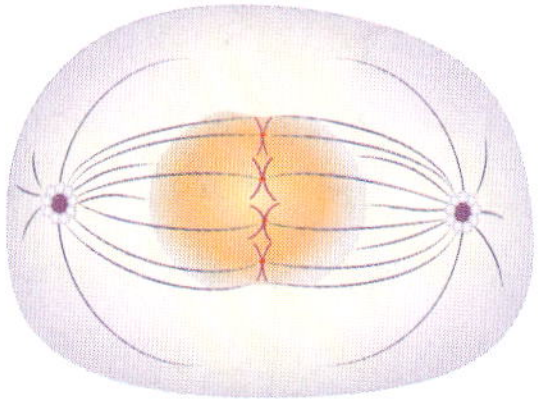

Chromosomes in the cell duplicate themselves. Each pair forms an X shape.

4 Anaphase

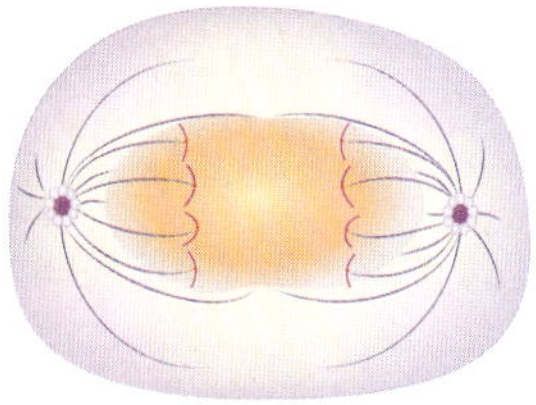

As mitosis begins, the doubled chromosomes line up along the equator of the cell.

5 Telophase

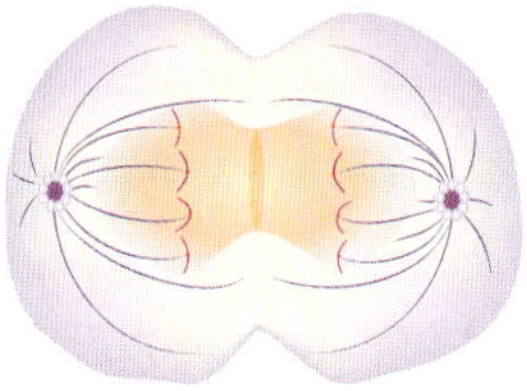

Next, each pair of chromosomes splits and moves to opposite ends of the cell.

6 Cytokinesis

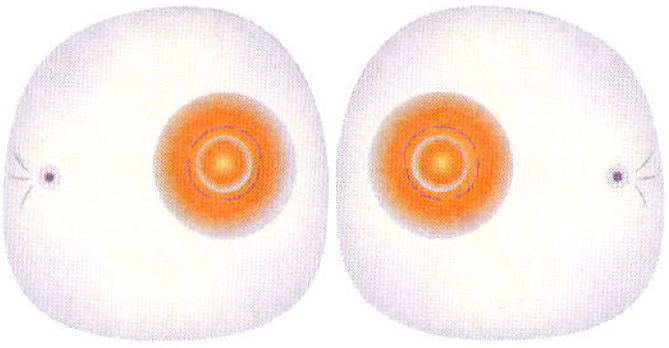

The cell divides in half, resulting in identical pairs of chromosomes in each daughter cell. The cycle continues as the two new cells begin to grow during interphase.

Notice that the cell cycle consists of six stages.

Which stage is the shortest? What happens to end division?

Genes control the production of enzymes that, in turn, control the cell cycle. Enzymes signal the cell when it is time to move from interphase to mitosis. They control all of the steps in the cycle, along with the length of the cycle. Problems in the production of enzymes during the cell cycle can cause cells to divide too quickly. The new cells start to pile up on top of one another. The result is a tumor. A **tumor** is a mass of cells that form from uncontrolled cell division.

Structured Inquiry

Discover

Record your work for this inquiry. Your teacher may also assign the related Guided Inquiry.

Science Journal

Taking a Biopsy

How can you take a biopsy of a tumor?

Materials

- clear gelatin
- hot water
- clear plastic cup
- ice cubes
- 2 to 3 pieces of fruit
- scalpel or long-blade knife
- syringe needle
- small plate
- glass slide
- microscope
- paper towels

Procedure

1. Receive gelatin and hot water from your teacher.
2. Prepare the gelatin in the plastic cup, following the instructions on the packet. Add several ice cubes so the gelatin sets faster.
3. Refrigerate for about 30 min.
4. Remove the cup from the refrigerator; the gelatin should be slightly firm. Add two to three pieces of fruit to the cup. Each group will use a different type of fruit. Return the cups to the refrigerator for another hour or longer so that the gelatin firms.
5. Remove the cup from the refrigerator. Decide the best procedure for taking a biopsy (small piece) of the fruit. Make as few cuts as possible.
6. Use the knife or scalpel and syringe needle to remove the biopsy and place it on a clean plate.
7. Select a portion of the biopsy sample and prepare it on a microscope slide.
8. **Observe** the fruit biopsy under the microscope.
9. **Record** your **observations** in your ***Science Journal***. Include a description of the "tumor," its size, and it's location (depth/orientation).

Analyze Results

Describe your procedure for taking the biopsy. If you could do something differently, what would you do? Describe what you see under the microscope. Draw your results.

Create Explanations

1. How can you take a biopsy of a tumor?
2. How did your fruit compare with other fruit samples that students biopsied?
3. How is this activity similar to an actual biopsy?

There are two kinds of tumors: benign tumors and malignant tumors. **Benign tumors** are noncancerous and grow more slowly than cancerous ones. Benign tumors are surrounded by membranes that prevent them from spreading.

Florida Hospital in Orlando, Florida, is operated by the Seventh-day Adventist Church. The hospital is internationally known for its commitment to treat cancer patients, giving attention to the whole person — spiritually, physically, and mentally.

Malignant tumors are made up of cancerous cells that are not encased in a membrane. As such, the cells can be carried to nearby tissues and organs by the bloodstream or lymph, where they infect healthy tissue. The spread of cancerous cells from their origin to other parts of the body is called **metastasis**.

As a tumor metastasizes, it can compete with healthy cells for nutrients. As tumors grow, they press against nearby tissues and organs, affecting their function. Tumors can also block circulation, interfering with blood flow to organs and tissues.

Knowing whether a tumor is benign or malignant can determine the course of treatment. To determine whether a tumor is benign or malignant, a doctor can perform a **biopsy**, taking a small sample of cells for testing.

The Causes of Cancer Explain

Any organism can get cancer. In humans, age, gender, occupation, lifestyle, and family history may increase the risk of cancer. Just as there are many types of cancer, there are many factors that can cause cancer. Below is information about the major causes of cancer:

Genetic

Some types of cancer, such as breast cancer, run in families. This suggests a genetic link to these types of cancer. In fact, scientists have identified genes called **oncogenes** that can cause cancer. What is the chance of developing cancer if you carry an oncogene?

Faith Connection

Many churches offer Breathe Free programs to help people stop smoking. If your church is planning a Breathe Free program, volunteer your time to help.

Tobacco Use

Smoking tobacco causes 85% of lung cancer in the United States, but the damage does not stop there. Chemicals in tobacco are **carcinogens**, substances that can cause cancer. Seventy of the 7000 chemicals in tobacco smoke are carcinogens.

The carcinogens in tobacco smoke go from the lungs into the bloodstream. The blood carries these carcinogens to every organ in the body, where they can interfere with the cell cycles in many body parts, not just the lungs, mouth, and throat.

In North America, smoking cigarettes leads to more than 440,000 premature deaths each year. How can you and your friends avoid smoking? Are you susceptible to smoke if you never light a cigarette? Secondhand smoke comes from lit cigarettes and carries carcinogens. About 3400 nonsmokers die from lung cancer every year from exposure to secondhand smoke. Do you live with or spend much time with any smokers? If so, your chance of developing lung cancer increases by 20 to 30 percent!

Diet and Physical Activity

Our lifestyle is our way of living. Eating poorly, not getting enough physical activity, and drinking alcohol can undermine health and may also increase the chance of developing cancer. In fact, about one-third of all cancers are linked to an unhealthful lifestyle. For example, being overweight can cause the body to produce too much of the hormones that encourage cancer growth. What do you think is the most common type of cancer in the United States and Canada? If you practice a healthful lifestyle, can you be certain that you will not develop cancer? Why or why not?

Viral Infection

Viruses are microscopic particles of RNA or DNA that reproduce by entering a living cell and manipulating the host cell to produce more viruses. Scientists estimate that 15 percent of all human cancers may have a viral origin. However, the progression from a viral infection to cancerous growth is a slow process. How can understanding viruses prevent cancer?

The Sun and UV Exposure

Most skin cancers are the direct result of exposure to the ultraviolet (UV) rays in sunlight. There are two major types of skin cancer—basal cell cancer and melanoma. **Basal cell cancer** is the most common type of skin cancer and can usually be treated by surgically removing the cancerous cells. **Melanoma** is much more serious and can be life threatening should it metastasize. Why is skin cancer more common in people who regularly visit tanning salons? What can you do to reduce your risk of getting skin cancer?

Melanoma

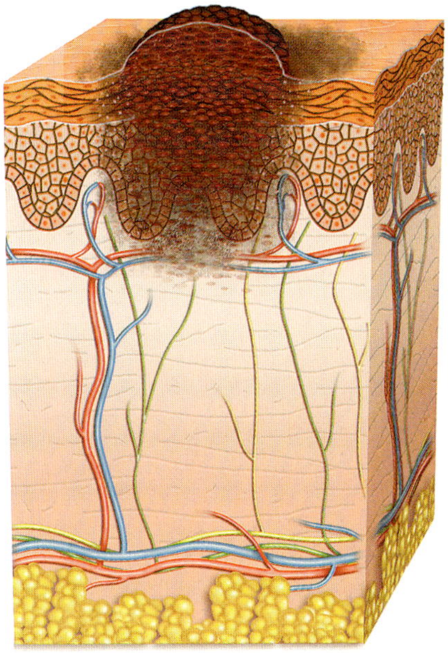

A mole or dark freckle can be like the tip of an iceberg. This diagram of melanoma shows what is happening under the surface of the skin.

What does this diagram tell you about how cancerous skin cells spread?

Radiation Exposure

Exposure to radiation can lead to a variety of serious cancers. Additionally, radiation exposure can affect your reproductive cells and increase the risk of certain cancers in your children.

X-rays can also cause mutations that lead to cancer. People are exposed to X-rays from three major sources. Natural radiation comes from the Sun and from radioactive elements, such as radon, that are found in certain rocks. A second source of radiation comes from various diagnostic tests such as X-rays and body scans. Why does a dentist place a lead apron on patients before taking X-rays of their teeth?

A third source of radiation comes from human activities, such as airport security scanners or certain industries that use radioactive materials. In what industries do you think workers have exposure to radiation? Do you think people who travel a lot should worry about the amount of radiation they are exposed to? Why or why not?

Cancer Symptoms Explain

Because cancer can take so many forms, the symptoms are varied. You can take an active role in your health by knowing the symptoms associated with cancer. The American Cancer Society has listed seven warning signs of cancer, which spell out the acronym *CAUTION*. Even if you have one or more of them, remember that you might just have a cold!

Cancer Warning Signs

Changes in bathroom habits (going more or less frequently or a change in "output")

A sore that does not heal

Unusual bleeding or discharge; for example, moles and freckles should not bleed

Thickening or a lump in the breast, mouth, or testicles

Indigestion or problems in swallowing

Obvious change in a wart or mole; a mole with an odd shape or an uneven color could be melanoma

Nagging cough or hoarseness

Other possible symptoms include an unexplained weight loss of 4.5 kg or more or fatigue that does not go away, even after you rest. If you or someone you know has one or more of these symptoms, consult a health professional immediately.

Scripture Spotlight

We have a moral obligation to care for our bodies. Read **1 Corinthians 6:19–20**.

Lesson Activity

Select one of the cancer warning tips provided earlier in the chapter. Create posters for the classroom or a school hallway to educate your schoolmates about the dangers, warning signs, and health tips. If time permits, prepare a public service announcement to be read during the morning announcements.

What other cancer information would be helpful to your classmates?

Cancer Diagnosis and Treatment

Identifying and treating cancer before it spreads greatly improves the survival rate. If a lump is detected, doctors use X-rays and other techniques to pinpoint its location and size.

After a diagnosis, the treatment depends on the type of cancer, whether it has spread, and the health of the patient. Possible treatment options include the following:

Cancer Treatment Options

- Surgery to remove some or all of the tumor
- Chemotherapy to destroy cancer cells with chemicals
- Radiation therapy to shrink tumors and kill cancer cells using radioactive substances
- Immunotherapy to use the patient's immune system to recognize certain cancers and destroy them
- Hormone therapy to use medication to slow or stop the production of hormones that encourage the growth of cancer cells

You don't have to be on a team to get physical activity. Just get together with a friend or two and do something active.

How is being physically active related to maintaining a healthful weight?

Lowering Your Risk of Cancer

You cannot change your heredity, and you cannot avoid all of the possible carcinogens in the environment. However, making healthful choices will greatly reduce your risk of getting cancer. In addition to the tips below, what other healthful lifestyle choices can you make?

Tips to Lower Cancer Risk

- Do not use tobacco. If you smoke, stop! Your body will have a chance to fight off carcinogens.
- Eat a healthful diet that includes fruits, vegetables, whole grains, and fiber. Avoid fried, fatty foods.
- Exercise at least 30 minutes a day. Being active helps you avoid disease.
- Do not drink alcohol.
- Protect your skin from the Sun. Use sunscreen with an SPF of at least 15 when outside. Wear a hat, sunglasses, and loose-fitting clothing.
- Get immunized for known viruses, such as hepatitis B and HPV.
- Avoid risky behaviors.
- Take advantage of tests that can help identify possible cancers during routine medical checkups.

Concept Check Assess/Reflect

Summary: What is cancer and how is it caused? Cancer is uncontrolled cell division that is caused by problems in the production of enzymes during the cell cycle. Cancers are named according to the body system affected—lymphomas affect the immune system, leukemia affects blood-forming organs, sarcomas affect connective tissues, and carcinomas affect the skin and lungs. The uncontrolled growth of cells produces tumors, which can be benign or malignant. A malignant tumor can spread, or metastasize, and produce cancer in other tissues throughout the body. It can be caused by environmental factors, such as radiation, UV exposure, poor diet, tobacco use, and viruses. Cancer can also be caused by genetics. Be aware of the warning signs of cancer by remembering the *CAUTION* acronym. Doctors treat cancer using a variety of treatments that might include surgery, chemotherapy, radiation, immunotherapy, and hormone therapy. Making healthful choices will help reduce your risk of cancer.

1. Describe one factor that could cause cancer. How can you lower your risk of developing cancer this way?
2. How does the cell cycle relate to cancer?
3. Why is cancer so difficult to treat?

Lesson 4

Essential Question

How Does Technology Keep People Healthy?

Objectives

- Describe the value of screening tests.
- Analyze the effects of technology on personal and family health.

Vocabulary

mammogram

CAT scan

MRI

PET scan

Do you think scientists will ever develop the "magic pill" that can cure all our ailments? Why or why not? Scientists conduct research to improve technologies and treatments to diagnose and treat cancer. Many advances were made using the cells from Henrietta Lacks, a woman who succumbed to ovarian cancer at the age of 31. The cells were harvested from Ms. Lacks in 1951 without her consent. Her unusually fast-growing cancer cells were used in more than 70,000 medical studies. Even though the cells continue to be used in medical research, it is now considered unethical to use cells without the consent of the donor. Why do you think this is so? If the cells were used in research to find a treatment for a type of cancer, is it acceptable that the donor did not consent to their use in medical research? What are other ways that scientists have developed technologies to keep people healthy?

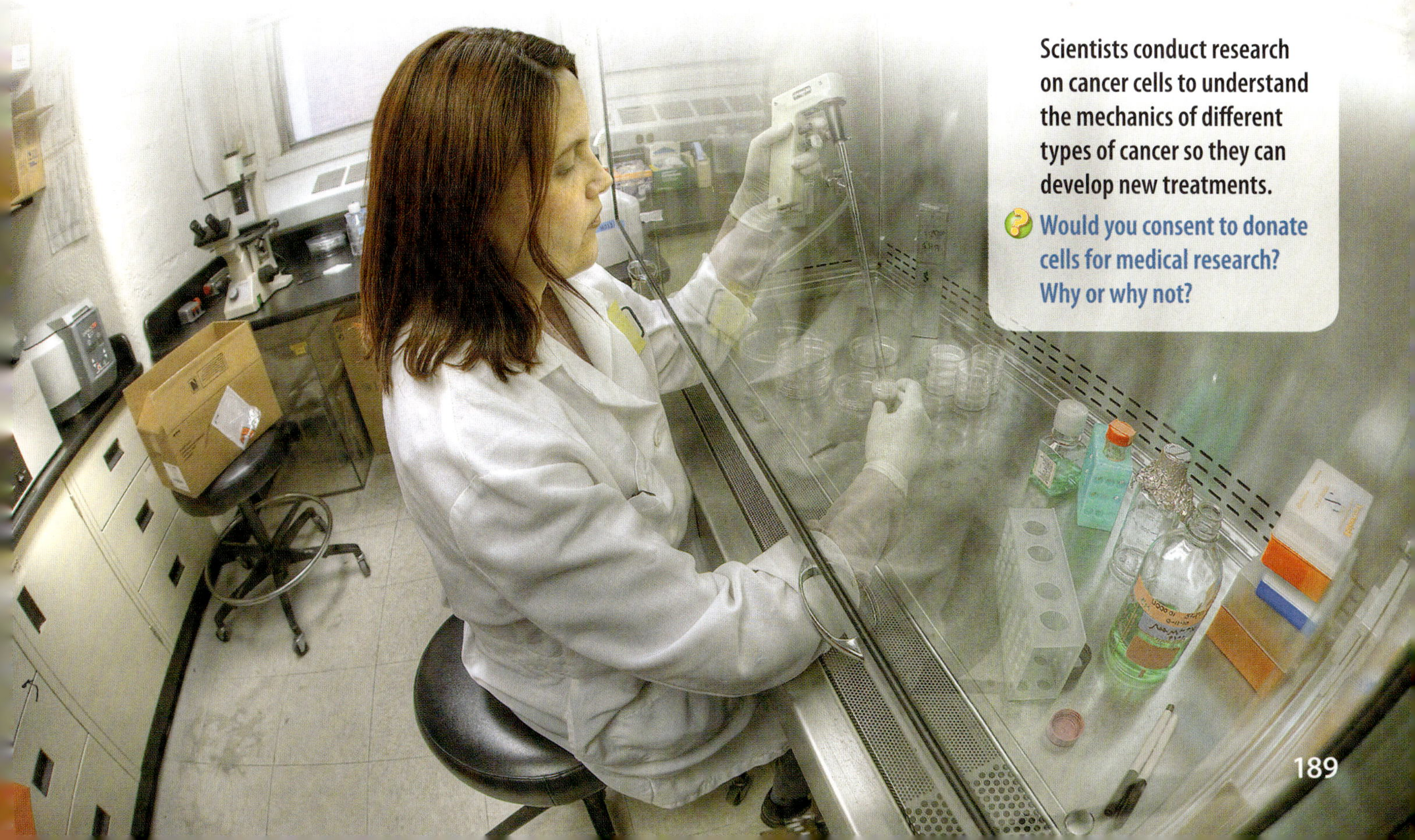

Scientists conduct research on cancer cells to understand the mechanics of different types of cancer so they can develop new treatments.

Would you consent to donate cells for medical research? Why or why not?

Knowledge Is Power Explain

For nearly every disease, the sooner it is detected, the better the chance of recovery. Why is early detection important? Diagnosing a disease may involve a simple screening test. Four simple screening tests are a blood analysis, a blood pressure test, a colonoscopy, and a bone density test. At what age do you think these tests are performed? How often are these tests performed?

Screen Tests

Test	Analyses
Blood analysis	• Check cholesterol levels • Measure insulin levels to check for diabetes • Monitor functioning of the liver and kidneys • Determine whether a male has prostate cancer • Screen for HIV or other sexually transmitted diseases (STDs)
Blood pressure	• Check for cardiovascular disease that can lead to heart attack or stroke
Colonoscopy	• Insert a camera into the colon to look for signs of cancer
Bone density	• Use an X-ray to measure the density of bone matter

Scripture Spotlight

Just like there are tests to check for physical diseases, the Bible describes tests that can be used to evaluate spiritual health. What tests are described in **Matthew 12:34–35** and **John 13:35**?

Specialized Tests Explain

Based on the results from screening tests, a doctor may order a more specialized procedure to identify the problem and offer the most beneficial diagnosis. These procedures require highly specialized equipment and technology. Does every hospital have access to this equipment? Why or why not? For example, a person with chest pain might undergo an *electrocardiogram* (ECG) in the emergency room. This test can help determine whether he or she is having a heart attack.

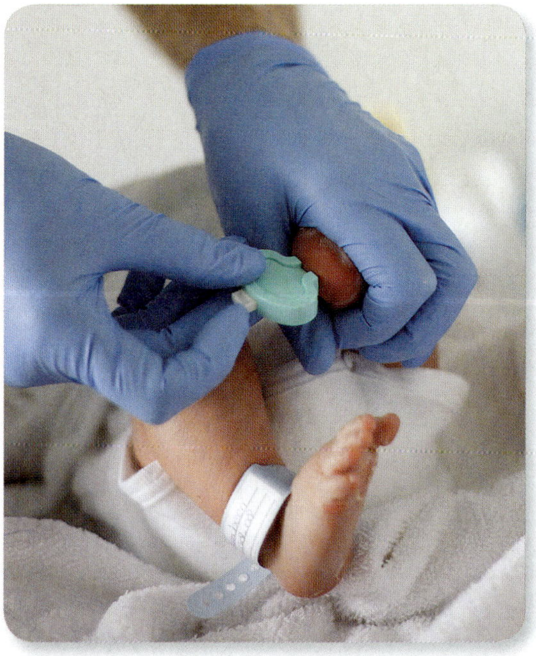

A tiny amount of blood is drawn from a newborn's heel to check for diseases.

Why do doctors test babies' blood when they are so little?

Mammograms

A **mammogram** is a special X-ray for soft tissue. It can find tiny tumors that cannot be felt. Guidelines for mammogram screenings vary among different health organizations, but all recommend that screenings start at age 40. The differences in guidelines have to do with the frequency of the screenings. Some organizations suggest yearly screening, while other organizations recommend a screening every two years. A woman should work with her doctor to decide what the best practice is for her. The risk of breast cancer for women younger than 40 is low, making periodic mammograms unnecessary for these women. However, at times a doctor will recommend a mammogram for a younger woman. What might be some reasons for this? Does the availability of a mammogram make the daily breast exam obsolete? Why or why not?

The doctor is using old technology (a magnifying lens) to take a closer look at an image produced by modern technology (a mammogram). The white tumor is easy to see in this mammogram, but it might have been impossible to find otherwise.

If this tumor is cancerous, what might happen if it is not removed?

Ultrasound Testing

Ultrasound testing uses high-frequency sound waves to create a picture of tissues inside the body. The technician moves a small unit back and forth over part of the body. As the sound waves bounce off tissue under the skin, a computer interprets the behavior of the sound waves to produce an image that can be analyzed. Ultrasound can be used to get a clearer picture of a cyst or tumor in the breast or in other parts of the body. A doctor might use ultrasound when a mammogram is not appropriate, such as to examine a lump in a young woman's breast. How else might doctors use ultrasound to examine the body?

CAT Scan

A **CAT scan**, or computerized axial tomography scan, is a specialized X-ray technique. The patient lies on a table while a doughnut-shaped X-ray machine moves in a circle around the patient's body. The machine produces many detailed cross-sections of the area of concern. The scan provides a clearer image than a regular X-ray. A dye may be injected into the patient to make blood vessels and other organs easier to see. Some patients may be asked to drink a liquid with dye in it to check the digestive tract.

Structured Inquiry

Discover

Record your work for this inquiry. Your teacher may also assign the related Guided Inquiry.

Science Journal

Creating Stethoscopes

What factors affect the quality of sound produced by a stethoscope?

Materials

- 3 large funnels
- 2 small funnels
- 40 cm of plastic tubing
- 40 cm of thin garden hose
- cardboard tube
- electrical tape
- stopwatch

Procedure

1. Construct a stethoscope by connecting one large funnel and one small funnel on either end of the plastic tubing. If necessary, secure the funnels with electrical tape.
2. Construct a second stethoscope by connecting one large funnel and one small funnel on either end of the garden hose. If necessary, secure the funnels with electrical tape.
3. Construct a third stethoscope by connecting one large funnel to the cardboard tube. Secure the funnel with electrical tape.
4. **Gather and record data** by placing the large funnel on the back and over the heart of your lab partner. Record the sound as loud, medium, or soft. Your partner will use the stopwatch to measure 30 seconds. During this time, listen and count the number of heartbeats.
5. Switch roles and repeat Step 4 for all of the stethoscopes.

Analyze Results

Compare the data results from the three stethoscope designs.

Create Explanations

1. What factors affect the quality of sound produced by a stethoscope?
2. Which stethoscope worked the best? Why?
3. What part of the best design do you think made it work best?

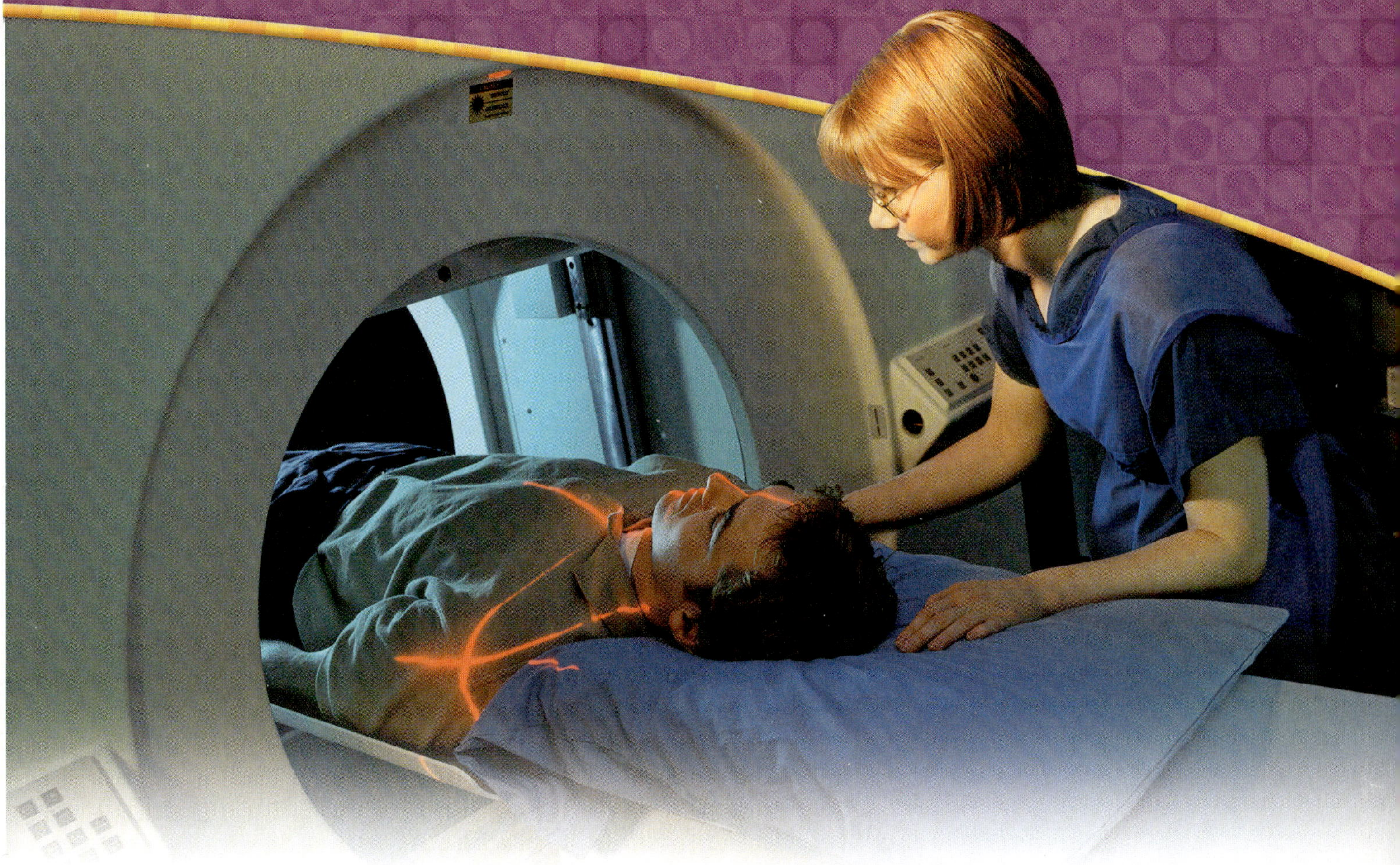

Only part of the patient's body may be inserted into a CAT scanner.

How would a technician prep a patient to perform a CAT scan to look for a brain tumor?

Technicians can use CAT scan images to study tumors, bone and muscle disorders, heart disease, and infections. The images also show blood clots, internal bleeding, injuries from accidents, and other conditions. The images can be used to plan biopsies and other surgeries. They help monitor internal healing and the results of surgery.

MRI

Magnetic resonance imaging (**MRI**) uses magnets, radio waves, and a computer to create images of internal tissues and organs. For this test, the patient lies on a narrow bed that slides into a tube. Around the tube is a giant magnet that produces a strong magnetic field. The magnetism and radio waves combine to produce a detailed image of internal structures. As with a CAT scan, dyes are sometimes injected into the patient to highlight certain tissues or organs.

An MRI operates differently from a CAT scan. Each is suited for different types of diagnoses. An MRI is suited for examining soft tissue, such as a ligament, tendon, or brain. In contrast, a CAT scan is better suited for bone injuries, lung and chest imaging, and detecting cancers.

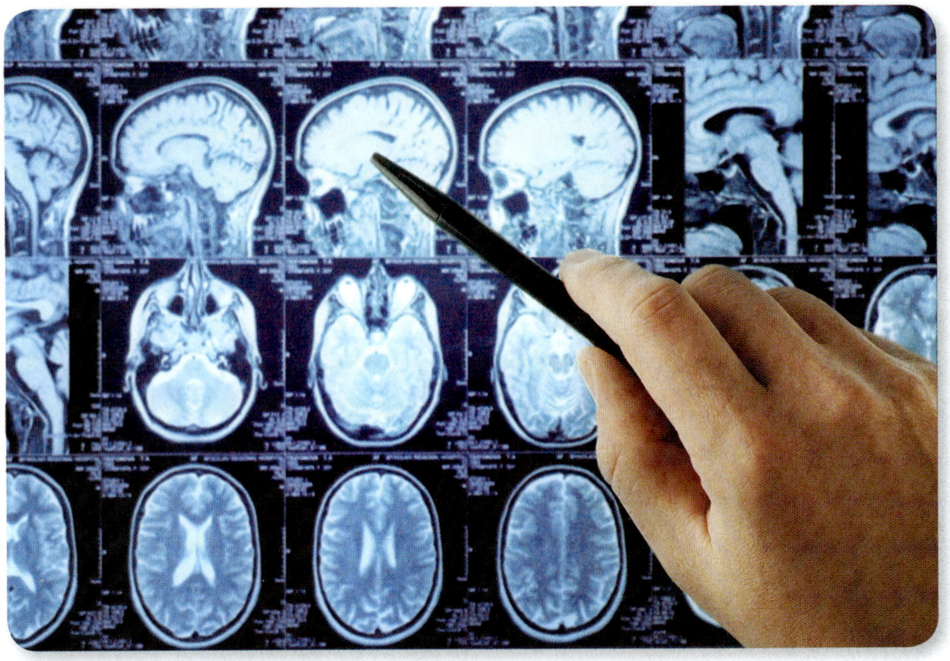

The MRI provides detailed images of slices of the body part of interest.

How do the image slices help the doctor?

CAT scans are widely used in emergency rooms because the procedure takes less than 5 minutes. An MRI, on the other hand, can take up to 30 minutes. One advantage of an MRI is that it does not use radiation, while CAT scans do. This radiation is harmful if there is repeated exposure.

PET Scan

A doctor may order a **PET scan** (positron emission tomography scan) to produce a three-dimensional, color image of a body part. Before performing the scan, a technician injects a small dose of a radioactive substance through a patient's vein. The blood transports the radioactive substance throughout the body. A PET scan then analyzes how the organs, such as the liver, break down the radioactive substance to analyze organ functions. The biggest advantage of a PET scan, compared with an MRI scan or an X-ray, is that it can reveal how a part of the patient's body is functioning, rather than just how it looks.

Math in Science

A company offered screening tests to its 50 employees at a total cost of $330,000. As a result, 60% of the employees made enough lifestyle and other changes to avoid the diseases for which they had been tested. The average cost of those diseases to the company through its insurance plan was estimated at $15,000 per person. How much money did the company save in insurance costs by funding these screening tests?

Smartphone Technology and Your Health Explain

CAT, MRI, and PET scans were developed specifically for scientific and medical purposes. However, technology has also resulted in consumer products that are now being used to maintain a person's health. Smartphones are just one example. Researchers have developed new medical applications (apps) for smartphone technology that allow doctors to conduct blood tests or other tests with just a tap on a screen. Results are now available in moments, rather than days or weeks, and can be done with only a tiny drop of blood, rather than vials. This kind of testing, in addition to being much faster, is also much less expensive. Smartphone apps and specialized attachments are sometimes as little as $10 compared with over $3000 for other portable devices. With this new technology, people in remote areas can now download an app, conduct their own testing, and wirelessly and instantly send the results to their doctor thousands of kilometers away.

Concept Check Assess/Reflect

Summary: How does technology keep people healthy? Diagnosing a disease at an early stage helps keep it from spreading and causing more damage. This approach can save lives. Some screening tests rule out disease. Some tests monitor people with a known disease to check whether the disease is progressing or whether a medication is working. Tests, such as a mammogram, an ultrasound, a CAT scan, an MRI, and a PET scan use complex equipment to make images of soft tissues inside the body. Doctors examine these images to identify irregularities and offer diagnoses. Computers take the data and create detailed and accurate pictures of internal organs without surgery on the patient. These specialized tests are only a few of the hundreds of new high-tech medical advances available to doctors and patients in the past few decades. Each new advance improves the output of technology that is often more cost effective.

1. Explain how technology has changed to improve the diagnosis of a person's health condition.
2. How can the same test be used for screening and for treatment?
3. Can we do too much health screening? Explain your answer.

People in Science

Extend

Get to Know

Dr. Frank Jobe

Dr. Frank Jobe is known for his pioneering work in the field of sports medicine and sports surgery. In 2013, at the age of 88, Dr. Jobe was honored by the National Baseball Hall of Fame for the impact he made on the game.

After serving his country in the armed forces, Dr. Jobe went on to earn his medical degree from Loma Linda University in California. There Dr. Jobe teamed up with Dr. Robert Kerlan to further the underdeveloped field of sports medicine. Their concept of sports medicine was a holistic approach that included understanding all patients' desired levels of activity and treating their injuries with a combination of physical therapy, surgery, and rehabilitation.

The duo cofounded what would become the Kerlan-Jobe Orthopedic Clinic in 1965, where they supervised medical treatment for the Los Angeles Dodgers; the Los Angeles Rams; the Los Angeles Lakers; the jockeys at Hollywood Park, Santa Anita, and Del Mar; and the California Angels, as well as a number of other professional and amateur athletes from across the country.

On September 25, 1974, Dr. Jobe made sports medicine history when he invented and performed an ulnar collateral ligament (UCL) reconstruction on Los Angeles Dodgers pitcher Tommy John. The UCL is a tendon located in the inner elbow, which is often injured by pitchers and other athletes who repetitively use similar throwing motions. Dr. Jobe's groundbreaking procedure, widely known as *Tommy John surgery*, has saved the careers of countless pitchers and other professional athletes. While the techniques used in this procedure may be considered routine by today's standards, they had never been used in such a way prior to 1974, and Dr. Jobe's success triggered a sports medicine revolution.

Dr. Jobe was recognized both nationally and internationally for his excellence in the practice and research of orthopedics and sports medicine. For 40 years, Dr. Jobe served as the team physician for the Los Angeles Dodgers. He was the orthopedic consultant for the PGA and Senior PGA Tours for 26 years and was named the emeritus physician for the PGA Tour. In Japan, Dr. Jobe's personal care for professional baseball players and the generous teaching and training he provided to Japanese physicians made him a national treasure to both baseball and sports medicine followers. Dr. Jobe died on March 16, 2014.

Concept Check

1. How did Dr. Frank Jobe make sports medicine history?
2. Why did the National Baseball Hall of Fame honor Dr. Jobe in 2013?

Science and Technology

Extend

High School Student Develops App to Detect Breast Cancer

Brittany Wenger was a Florida high school junior when she developed a revolutionary computer program to detect breast cancer. Using artificial intelligence, she created a detection tool to evaluate tissue samples for signs of breast cancer as her Google Science Fair project. Artificial intelligence is the creation of intelligent machines that work, think, and act like humans. The cloud-based computer program, which Brittany coded to think like a human, took grand prize at the 2012 Google Science Fair. The program is named "Global Neural Network Cloud Service for Breast Cancer."

Breast cancer affects one in eight women worldwide, and Brittany's family has been affected by it. Brittany says she was inspired by a cousin who struggles with breast cancer. She wanted to create a way for doctors to use a less invasive procedure, called Fine Needle Aspirate, in the process of having lumps examined. During this procedure, a section of the tissue is removed through a thin needle in order to be examined microscopically. Brittany developed an artificial neural network to help doctors take what used to be a safe but less reliable form of biopsy and turn it into a highly successful tool for detecting breast cancer.

Artificial neural networks are essentially computer programs coded to think like the brain. They are able to detect patterns that are too complex for the human brain. Brittany built her neural network using Java, and then deployed it to the cloud. She ran 7.6 million trials on it and found it to be 99.1% sensitive to cancer cells. The network will only improve as it processes more data. More data means the accuracy rate will go up and the inconclusive rate will drop. Networks like this one "learn" what to do by analyzing data they're given and will perform better as they get more data. As the accuracy rate of her program improves, Brittany's hope is to get the product into hospitals. According to Brittany, with some recoding and tweaking it would also be easy to adapt it to diagnose other types of cancer and medical problems.

Concept Check

1. What inspired Brittany Wenger to invent the Global Neural Network Cloud Service for Breast Cancer?
2. How will the artificial neural network program become more accurate?

Chapter 5 Review Assess/Reflect

Study Guide

Lesson 1

1. A genetic disorder is a health problem that is caused partly or totally by a defect in one or more genes or chromosomes. This defect can be caused during cell division or by a mutagen.
2. Sickle-cell anemia is a genetic disorder where red blood cells form a sickle shape. Cystic fibrosis is a genetic disorder in which the body produces very thick, sticky mucus that interferes with breathing and food absorption. Color blindness can make it difficult for a person to distinguish between red and green or blue and yellow, or see anything but black, white, and gray. Down syndrome occurs when a person has an extra copy of chromosome 21, and its symptoms are delayed mental and social development, flattened facial features, and slanting of the eyes.
3. During amniocentesis, a doctor removes amniotic fluid, which is analyzed for genetic disorders. Chorionic villus sampling involves removing a small piece of placenta to test for genetic disorders.
4. A genetic counselor works with parents to provide an analysis of the risk of giving birth to a child with a genetic disorder.

Lesson 2

1. A multifactorial disease results when several factors work together to cause the disease.
2. Environmental contaminants such as pollutants, chemicals, viruses, and bacteria can trigger the expression of genetic diseases.
3. Lifestyle factors such as poor diet and lack of exercise can trigger genetic diseases and make them worse.

Lesson 3

1. Problems in the production of enzymes during the cell cycle can cause cells to divide too quickly and form a tumor.
2. Cancer starts when a malignant tumor grows uncontrollably. It competes with healthy cells for nutrients, can interfere with their function, and can block circulation.
3. There are a number of symptoms that might indicate cancer. These are listed in the table on page 186.
4. Cancer can be difficult to treat because treatment is dependent on the type of cancer, whether it has spread, and the health of the patient.
5. There are a number of ways to reduce the risk of getting cancer. These are listed on page 188.

Lesson 4

1. Screening tests are very valuable tools because the sooner a disease is detected, the better is the chance of recovery from the disease.
2. Technology has several great effects on personal health. Specialized technology such as X-rays, ultrasounds, CAT scans, MRI scans, and PET scans give information about our bodies and help diagnose disease. Some consumer products such as smartphones can be used for medical testing and health monitoring as well.

Vocabulary Check

For each question, write a sentence that explains how each pair of terms is related.

1. cystic fibrosis—mutation
2. biopsy—metastasis
3. carcinogen—melanoma
4. multifactorial disorder—heart disease
5. MRI scan—CAT scan

Multiple Choice

Choose the best answer.

6. What is the most important factor in determining whether a child will have cystic fibrosis?
   A. where the child is born
   B. whether the child is the first born of these parents
   C. the genetic history of the parents
   D. whether an ultrasound test indicates the presence of cystic fibrosis
7. What could a biopsy show?
   A. which carcinogen caused a tumor
   B. whether a tumor was caused by environmental factors
   C. whether a lump is cancerous
   D. whether a lump is a mutagen
8. For which disease or condition would an MRI be the least helpful test?
   A. cancer **C.** cystic fibrosis
   B. heart disease **D.** obesity
9. Which of these is not a multifactorial disorder?
   A. sickle-cell disease **C.** skin cancer
   B. heart disease **D.** obesity

Check Point

Answer the following questions.

10. You know that you carry a genetic mutation, and you and your spouse are thinking about starting a family. What should you do?
11. A neighbor smokes two packs of cigarettes a day, but he is fairly healthy. His wife, who does not smoke, is very sick with bladder cancer. Infer what might have happened.
12. Explain what this chromosome map shows.

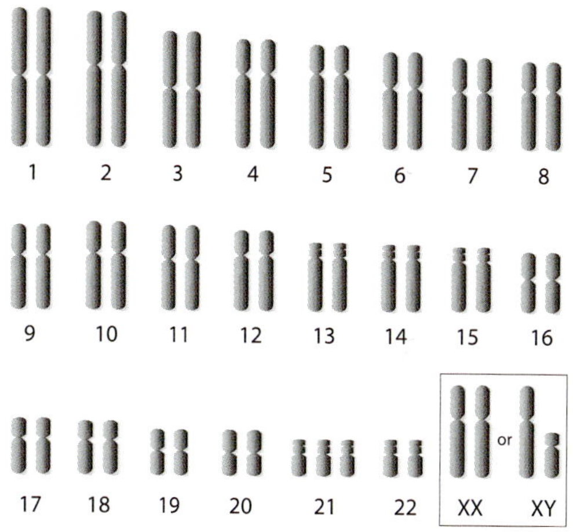

13. Why is tobacco called the most avoidable carcinogen?
14. Why haven't scientists identified all of the environmental carcinogens?
15. **Predict** whether lifestyle choices will contribute more to disease in 20 years than they do now. Explain your prediction.
16. Why is it difficult for scientists to **control variables** as they study multifactorial diseases?
17. What advice does **Proverbs 20:1** give us regarding alcohol consumption?

Chapter 6

Your Lifestyle Affects Your Body

Scripture Spotlight

God wants us to have good health. When Jesus was on Earth, He healed people of their diseases. You can strengthen your faith by making good choices and caring for the body God gave you. You will read the following passages in this chapter.

2 Samuel 22:7 (p. 209) Proverbs 31: 4–5 (p. 211)
Psalm 18:6 (p. 209) 1 Chronicles 11:9–19 (p. 230)

The Big Idea

Your health is affected by the actions and choices you make daily. God gave you the ability to make these choices. Making wise choices to live a healthy life leads to a healthy body and shows respect for God.

How do your actions and choices affect your body?

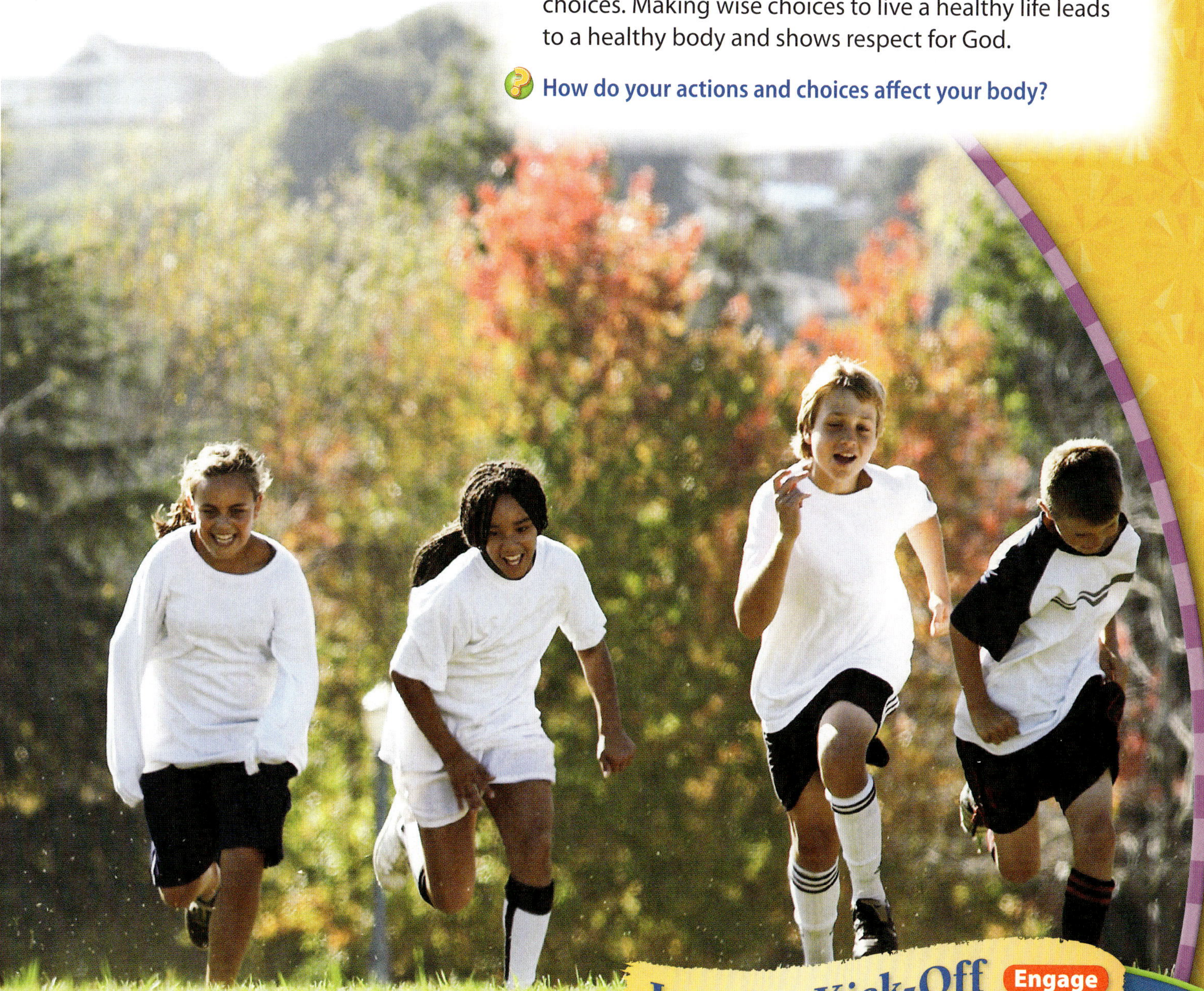

These teens are enjoying a race.

How can staying active keep you healthy in body and mind?

Inquiry Kick-Off Engage

What does your heart rate tell you about your overall health? How can doctors use this information to evaluate your health? In your ***Science Journal*** you will work with a partner to measure your heart rate.

Science Journal

Lesson 1

Essential Question

How Do Drugs Affect Your Body?

Objectives

- Explain how legal drugs can be harmful.
- Describe the major classes of drugs and their effects on the body.
- Explain why some people abuse drugs.
- Distinguish between physical and psychological addiction.
- Explain the addictive qualities of some drugs.

Vocabulary

drug
over-the-counter (OTC) medicine
prescription
addiction
physical dependence
withdrawal
psychological dependence

Millions of people take some type of medication every day. Many people take medications to stay healthy or to fight or prevent disease and infection. What do you think are the most common medications that people take every day? How can using medications benefit a person? People also use medications in physically harmful and illegal ways. Abusing medications can cause serious damage to individuals and the societies in which they live. Why do you think using medications only in the way they are prescribed by doctors is important? Science has provided many advances in the development and use of medicine. How has the availability of medication changed the world and treatment of diseases?

Medical Drugs Explain

Drugs can save your life, or they can kill you. A **drug** is any substance, other than food, that changes the way your body or mind works. If you take an aspirin for a headache, would this be considered using drugs? Explain why or why not. There are different groups of drugs that can do the following:

- Help prevent disease (Vaccines are one example.)
- Destroy or weaken pathogens (This group includes antibiotics, such as penicillin.)
- Ease pain (The medicines in this group range from ordinary aspirin to powerful painkillers.)
- Help keep your body healthy (Examples are allergy medications, heart medicines, cold medicines, and insulin.)

People can buy many kinds of medications at the grocery or drug store. This kind of drug is called **over-the-counter, or OTC, medicine**. Over-the-counter medicines can be purchased without a prescription. A **prescription** is an instruction that a physician or other authorized medical professional writes, specifying the type, dosage, and frequency of a medication for a patient. Examples of OTC medicines include pain relievers, such as aspirin, and cold/cough

Pharmacists have years of training to help customers use medications wisely.

Do you think a pharmacist can answer questions and give advice about medications better than a doctor? Explain your answer.

medication. Why do you suppose that some drugs are widely available without a prescription, while other drugs are regulated and available only by prescription?

Prescriptions help ensure that people get the medicine they need and take the amount that will help them. Many people need to take more than one kind of medication. Medications can interact with one another in many ways. For instance, a medication can cancel out the good effects of one drug while increasing the strength of another. Some interactions cause illness or death, so medical professionals watch for drug interactions. Why would it be important to mention any allergies (food or other) and other medicines you may be taking before a doctor prescribes medication?

People sometimes misuse medicines. Taking another person's prescription medicine can be dangerous because no doctor has reviewed the medical history of the other person to see whether the drug could be helpful or harmful. The medicine might also be too strong or too weak. How could this affect the person "borrowing" the medication?

Medicine misuse can take other forms. Some people think they will get better sooner if they take double doses of a medication. Others stop taking their medication when they start to feel better. How could these be unhealthful things to do?

Lesson Activity

Pass around the empty containers of several OTC drugs. Look carefully at the labels on these containers. Make a list of the active ingredients in the OTC drugs. What do you think the purpose of each medication is? **Compare** the labels examined by each group. Discuss how the active ingredients differ between the different medications. What else differs between the medications? Examine the dosage amounts of the medications. Discuss the varying dosages. Why would dosages be different for the same drug?

Structured Inquiry

Discover

Record your work for this inquiry. Your teacher may also assign the related Guided Inquiry.

Science Journal

Selecting the Best Aspirin

Which brand of aspirin is the best buy?

Materials

- several brands of aspirin samples
- copies of labels of aspirin samples
- balance

Procedure

1. **Calculate** the cost of each tablet of aspirin by dividing the price of the container by the number of tablets in the container. **Record** the price in the column of the data table in your ***Science Journal***.
2. Divide the price of each tablet by the number of milligrams of aspirin in each tablet. This will be listed as the *mg* of active ingredient and should be found on the label. Repeat for each brand of aspirin. Record your solutions in the data table.
3. Place 10 tablets of the first brand of aspirin on the balance and determine the mass in grams. Divide the mass by 10 to find the average mass of each tablet. Record the mass in the data table. Repeat for all of the brands of aspirin.
4. Determine the percentage of aspirin per tablet. Record your answer. Repeat for all of the brands.

Analyze Results

Compare the results of the various types of aspirin. Determine which data columns are the best indicators of which aspirin is the best buy. Select the aspirin brand that you believe is the best buy and explain why you believe this.

Create Explanations

1. Which brand of aspirin is the best buy?
2. What is the best way to determine which brand is the best buy?
3. Did a more expensive price per tablet always equal more aspirin per tablet?
4. Which aspirin brand would be the worst purchase to make? Why?
5. Do you think this lesson proves that there is a big difference between buying store-brand aspirin and brand-name aspirin? Explain.

Legal or Illegal? Explain

Legal drugs are medicines for use in avoiding or treating illnesses. An antibiotic that a doctor prescribes for you is an example of a legal drug. What else would be a legal drug? The use of legal drugs can help a person live a healthier life. What is the relationship between drugs and good health?

People, however, sometimes misuse legal drugs, and this can cause, not treat, illness. A person is not permitted to manufacture, possess, or sell *illegal drugs*. These drugs, which include cocaine and heroin, do not have a legal use or have only strictly limited legal uses. When someone other than a pharmacist sells or gives away prescription drugs, those drugs become illegal drugs. Suppose your father had some penicillin that he did not take when he was sick. Then you have an ear infection and decide to take a dose of the penicillin. Would you be using the penicillin correctly? Prescription drugs should be taken only by the person for whom they were prescribed. Often prescription painkillers and sedatives are abused or even sold illegally.

Alcohol and tobacco also are drugs. Are these legal or illegal drugs? It is against the law for you to buy either one of these products. The legal age to purchase alcohol in the United States is 21 years old. The legal age to purchase alcohol in Canada varies by province and territory but ranges from 18 to 19 years old. The legal age to purchase tobacco products is 18 years old for the majority of states in the United States. However, at least four states have raised the age to 19 years old, and there is currently legislation in some counties in the United States to raise the purchasing age to 21 years old. The legal age in Canada for purchasing tobacco ranges from 18 to 19 years old. These ages are different in other countries. Why do you think the ages vary across countries?

Smoking is legal; however, public and private businesses post "no smoking" signs throughout their facilities.

How are these efforts benefitting people of all ages?

Harmful Drug Effects

Why is it harmful for anyone to use illegal drugs or misuse legal drugs? Remember that all drugs, legal and illegal, change the way the body or mind functions. Remember that a drug's effect on an individual depends on the following:

- Amount used
- Weight of the user
- Age of the user
- Other drugs in the user's bloodstream
- Gender of the user
- Medical history of the user
- Mental state of the user

An overdose of most drugs can be damaging or fatal. Is the amount needed to overdose the same for all people?

Drug abuse can include the use of illegal drugs, illegally used prescription drugs, and misuse of OTC drugs. Regardless of how they are used, drugs change how the mind and body function. In drug abuse, the user might feel good for a while, but too often this comes with a steep price. Side effects of drugs can range from slurred speech, headache, loss of coordination, nausea, and memory problems, to unconsciousness and death. When illegal drugs are used or legal drugs are misused, the effects of the drug often lead to dangerous or unwanted circumstances. Users who are under the influence of a drug are less able to protect themselves from being robbed or sexually assaulted. Poor decision making is an effect of abusing drugs. What are some examples of users making poor decisions?

Aside from a fatal overdose, perhaps the greatest possible danger from some drugs is addiction. An **addiction** is a dependence on a drug. Not all drugs create addiction, but many do. A user with a **physical dependence** has a chemical need for the drug. If the person stops using the drug, he or she experiences **withdrawal**. The symptoms of withdrawal can include anxiety, nausea, headache, vomiting, chills, cramps, and the inability to sleep. In some cases, withdrawal can lead to death. As people become addicted to a drug, their dependence can cause them to act in extreme ways to continue obtaining their drug. Drug addiction can possibly lead to nonviolent and/or violent crime as the user resorts to whatever means necessary to acquire a high. This dependence and addiction typically bring on a downward spiral that increasingly damages the addicted person's health, relationships with family and friends, and career.

In addition to or instead of physical dependence, users of certain drugs can develop a **psychological dependence**. Psychologically dependent users believe they need a drug to feel good or even just to feel normal. Although these users might not have developed a physical dependence, the drug is still influencing how they are living their lives. How can using illegal drugs put your body and mind at risk?

People who abuse drugs spend more time by themselves or with other users, so friendships suffer or may end. Why would drug abusers spend more time alone? Drug abusers also take risks that can lead to accidents and injuries. Think about driving while using a drug or riding with a driver who is using a drug. In addition, drugs are expensive, so a drug abuser may do desperate things to get enough money to buy more drugs. Some resort to selling drugs to support their own addiction and thus spread the misery. Students who abuse drugs often do poorly in school.

If abusers of drugs are caught holding, using, or selling illegal drugs, they can be suspended from school or put on probation, put in a correctional facility for teens, or put in jail. Are the penalties the same for all drugs? Is this fair?

The consequences of abusing drugs are also great for adult users. If users are absent from the workplace or show up for work while high, there is a good chance that they will lose their job. This loss of income can cause users to become financially desperate and may lead to illegal activities. If adult drug users are caught with drugs, they will be arrested and could end up in jail or prison. How could these consequences affect people other than the drug abuser?

The surest way to avoid the negative health and social effects of drug abuse is to abstain completely. This is the only truly safe approach to drugs. It is important to live a life that you can be proud of, a life in which you maintain control of your mind and body. You can achieve this control by saying no to drugs and alcohol. How can drugs put your body at risk?

Lesson Activity

Every drug produces specific effects in its users. Use the chart on the next two pages to see whether you can guess the name of Drug X.

Drug X works on your body systems in the following ways:

- Makes it harder for you to control your body, causing you to look awkward and uncoordinated
- Slows your reaction time, impairing your performance on a sports team and making you a threat to yourself and others while driving a vehicle
- Interferes with your thinking and memory, making it hard for you to learn
- Impairs your judgment, influencing you to say and do things that might embarrass you later
- Makes you feel anxious and have panic attacks
- Increases your heart rate and blood pressure, which can lead to a heart attack as you get older

Can you determine which drug has been used?

Effects of Drugs

Categories of Drugs	Drug Effects	Legal or Illegal
Stimulants–Increase metabolism, mental altertness, and energy		
Nicotine	Damages heart and lungs; causes cancer	Legal (for users over the age of 18 years in most places)
Caffeine	Causes nausea, tremors, shaking, sweaty palms, rapid heartbeat, chest pain, anxiety, and seizures; can lead to dehydration	Legal; provides body with energy and increases alertness and concentration followed by tiredness unless more is taken.
Cocaine	Increases blood pressure and heart rate; can lead to stroke, heart attack; person may feel excited, anxious, panicked	Illegal
Amphetamine, methamphetamine, some prescription drugs	Increase blood pressure and heart rate; can lead to stroke, heart attack; person may feel excited, anxious, panicked; lead to dental problems	Amphetamine: legal and illegal (Often legal amphetamines are prescribed to help with behavioral disorders or narcolepsy but are used by others without a prescription.) Methamphetamine: illegal Prescription drugs: legal but only for the person for whom they are prescribed and only when used in the manner prescribed
Depressants–Decrease mental altertness, coordination, visual acuity, and memory		
Alcohol (beer, wine, liquor)	Dulls senses, impairs judgment; causes nausea and liver and heart disease with extended use	Legal (for users over the age of 21 in most areas)
Heroin, opium, some prescription drugs	Cause drowsiness, confusion, loss of coordination, nausea, slowed breathing	Heroin: illegal Opium: illegal, although some forms are legal as painkillers Prescription drugs: legal but only for the person for whom they are prescribed and only when used in the manner prescribed
Drugs with a mixed effect–Increase metabolism and decrease mental processes and coordination		
Marijuana, hashish	Causes slower reactions, loss of coordination, impaired learning and memory, increased heart rate, anxiety, panic attacks	Illegal (legal for medical purposes in some states to ease pain from long-term conditions, such as cancer)
Inhalants	Impairs senses; increase heart rate; cause anxiety	Illegal

Effects of Drugs *(continued)*		
Categories of Drugs	**Side Effects**	**Legal or Illegal**
Hallucinogens–Cause altered states of perception and hallucinations		
Ecstasy, "club drugs," "designer drugs"	Cause hallucinations (see and hear things that are not there), feelings of panic or anxiety, blurred vision, increased heart rate	Illegal
LSD	Increases body temperature and blood pressure; causes hallucinations, sleeplessness, body tremors, decreased appetite	Illegal
Steroids–Cause hormonal and other physiological effects		
Anabolic steroids	Lead to slow growth, anger, and aggression; cause feminine traits in males and masculine traits in females	Steroid prescription drugs: legal but only for the person for whom they are prescribed and only when used in the manner prescribed; can be used to reduce nausea, shrink tumors, and decrease swelling Performance-enhancing steroids: illegal

Scripture Spotlight

Look up the word *distress* in the dictionary. How is it related to why someone might use illegal drugs? Then compare **2 Samuel 22:7** with **Psalm 18:6**. How did David handle distress?

Concept Check Assess/Reflect

Summary: How do drugs affect your body? Drugs can be helpful if you are sick or in pain. They can help your body to heal. Medicines can be obtained with a prescription from a doctor or purchased over the counter (OTC). Examples of OTC medicines are items such as cold medicines and aspirin. Prescription drugs are meant to be used only by the person for whom they are prescribed and only by following the specified directions. Prescription drugs are often abused when used by someone other than for whom they were intended and when used in different amounts. Drugs that are not allowed to be manufactured, possessed, or sold, such as heroine and cocaine, are illegal drugs. These drugs are harmful, and you can become addicted to them. Drugs affect your body in different ways. There are stimulants that speed up body systems, depressants that slow down body systems, and hallucinogens that alter your mind.

1. How can legal drugs be harmful?
2. What is the difference between physical and psychological dependence?
3. Why is addiction difficult to overcome?
4. Provide an example of an illegal drug that is a depressant, one that is a stimulant, and one that is a hallucinogen. Explain the effects each of the drugs has on the body.

Lesson 2

Essential Question

How Do Alcohol and Tobacco Affect Your Body?

Objectives

- Describe how alcohol affects the body.
- Identify the factors that influence the effect of alcohol on the body.
- Describe how tobacco affects the body.
- Explain why smokeless tobacco is as dangerous as cigarettes.

Vocabulary

intoxication
cirrhosis
nicotine
emphysema
secondhand smoke

Some drugs are legal for adults to use, but that doesn't mean that you cannot become addicted to them. Nicotine and alcohol are examples. People who have become addicted to alcohol must continue to drink alcohol to avoid symptoms of withdrawal. Nearly every person who takes up smoking has a difficult time stopping. This is why it is so important never to start! Do you think these addictions are physical or psychological? Why? Other than addiction, what health issues can result from drinking alcohol or smoking?

Mental Effects of Drinking Alcohol Explain

In the United States, drinking alcohol in moderation is legal for adults over the age of 21. This does not mean that all adults drink alcohol; alcohol consumption is not an acceptable practice in the Adventist worldview. Although there is a large amount of risk associated with drinking alcohol, there is no positive benefit to this behavior. As a result, the choice to abstain is easy.

Choose to enjoy positive activities with good friends.

Do these friends need to drink beer to have fun together? Why or why not?

Historically, Seventh-day Adventists have been known for living a healthful lifestyle and abstaining from alcohol, illegal drugs, and tobacco. This is largely due to Ellen White's 1863 health vision. Although the church did not have information in the 1800s to scientifically back up what Mrs. White had to say, current research continues to confirm the health advice she gave. The first Adventist Health Study began in 1958. This study showed that California Adventists lived longer than other Californians and that fewer died of heart disease and smoking-related cancers. Do you think that abstaining from alcohol can reduce the risk of heart disease and certain cancers? Why or why not?

Some adults do consume alcoholic beverages on occasion. Alcohol consumption affects each person differently. How much alcohol makes a person intoxicated? That is not an easy question to answer. **Intoxication** means that a person's mental and physical abilities are impaired because of drug use. People often use the word *intoxication* to refer to alcohol intoxication, but it can refer to other drugs as well.

Scripture Spotlight

What does **Proverbs 31:4–5** say about wine? What reason does it give?

Explore-a-Lab

Structured Inquiry

How do alcohol and drugs impair a person's reactions and abilities?

Work with a partner and time how long it takes you to thread a nut all the way up a bolt. Record your time. Repeat this process two more times. Now put on a pair of mittens and try the task again. Record your time. Repeat the process two more times. Next, walk a straight line on a taped line on the floor. Record your time. Repeat this process two more times. Now put on a pair of cloudy sunglasses and have your partner turn you around three times. Try walking the taped line again. Record your time. Repeat this two more times.

Design a chart or table to collect and organize the data from this investigation. Compare the results of the activity. **Analyze** the differences in time from the unimpaired threading of the bolt and the threading while impaired. Discuss how wearing mittens or cloudy sunglasses relates to completing a task while impaired by drugs or alcohol. Discuss some of the physical abilities that alcohol and other drugs compromise.

Immediate Effects of Alcohol

An intoxicated person may have slurred speech or stumble while walking. He or she is more likely to take unreasonable risks, such as riding with a driver who is also intoxicated or performing a dangerous stunt. What other careless activities are associated with intoxication? Do all people who are intoxicated show these effects?

Can people drink the same amount of alcohol with different effects on their bodies? Why or why not? At your age, you are not old enough to drink legally. Alcohol is not legal until the age of 21 because of its dangerous effects. A teen's body is not yet fully developed. Alcohol would have a much greater intoxicating effect on you than on an older person, which can lead to harmful results. No matter who you are, drinking alcohol has no positive effects, so the best choice is to abstain completely. At any age, the amount of alcohol that can cause intoxication depends on several factors, including:

- **Food**—Having food in your digestive system slows the entry of alcohol into the bloodstream.
- **Body size**—A person with a smaller body feels alcohol's effects more quickly than a person who is larger.
- **Gender**—Alcohol enters the bloodstream more quickly in females than in males due to the lower volume of water in the female body.
- **Amount and rate of drinking**—Different types of alcoholic drinks contain different concentrations of alcohol. The more alcohol someone drinks, the more significant the effects on the body will be. Drinking alcohol faster than the liver can process it causes intoxication. The liver is able to process about one drink (350 mL of beer, 140 mL of wine, or 35 mL of 80 proof liquor) per hour.

Why do you think the legal age for adults to purchase or consume alcohol is 21? From 1920 to 1933 in the United States was a time of Prohibition. During this period, it was illegal to sell alcohol. This was considered by many to be a victory for public health and morals. Why was Prohibition ended?

Lesson Activity

As a class, list on the board every reason you can think of not to drink alcohol. Then the class will vote to see which reason is the most convincing. Do you think the class chose the most important reason? Why or why not?

What are some reasons to avoid alcohol?

You have learned that alcohol is a depressant. This means that it inhibits the function of the central nervous system and decreases the heart rate. These effects slow down the body's reaction time and cause the body's cells and organs to operate less efficiently. Alcohol affects the brain by slowing a person's perception and altering a person's senses, mood, and judgment. Why is it illegal to drink alcohol and drive a vehicle?

Long-Term Effects of Alcohol

Alcohol consumption has both immediate and long-term effects on your body. Over time, excessive drinking can damage liver cells to the point where scar tissue begins to accumulate, resulting in **cirrhosis**, a disease of the liver. As the scarring continues, the liver becomes less able to function as the scar tissue presses on blood vessels in the liver, limiting the blood flow, and eventually causing the death of liver cells. After a while, the liver will no longer be able to process alcohol or function normally. If the person stops drinking before the liver is too badly damaged, the liver may recover somewhat. Sometimes, even completely abstaining from alcohol cannot reverse the disease. About 25,000 Americans die of cirrhosis every year. Why do you think it is difficult for people to stop drinking to allow their livers to heal?

Cirrhosis is only one of the many harmful effects of drinking alcohol. Drinking alcohol causes other diseases, including cardiovascular disease, high blood pressure, and cancer. Alcohol also contributes to 40 percent of all traffic deaths in the United States. In addition, about one-third of those who commit suicide have been drinking. Alcohol is a major factor in domestic violence, child abuse, and workplace injuries. Alcohol consumption leads to so many unfortunate events because it alters a person's judgment.

It is important that you understand alcohol's effects. Not only are there negative health effects and social effects, but underage drinking is illegal. In addition to these effects, the use of alcohol and other drugs also interferes with your relationship with God. It is important to make a firm decision to abstain from alcohol. Making this decision in advance will make it easier to be unaffected by peer pressure and to refuse when someone offers you a drink. What are some things you can say if you are asked to drink alcohol?

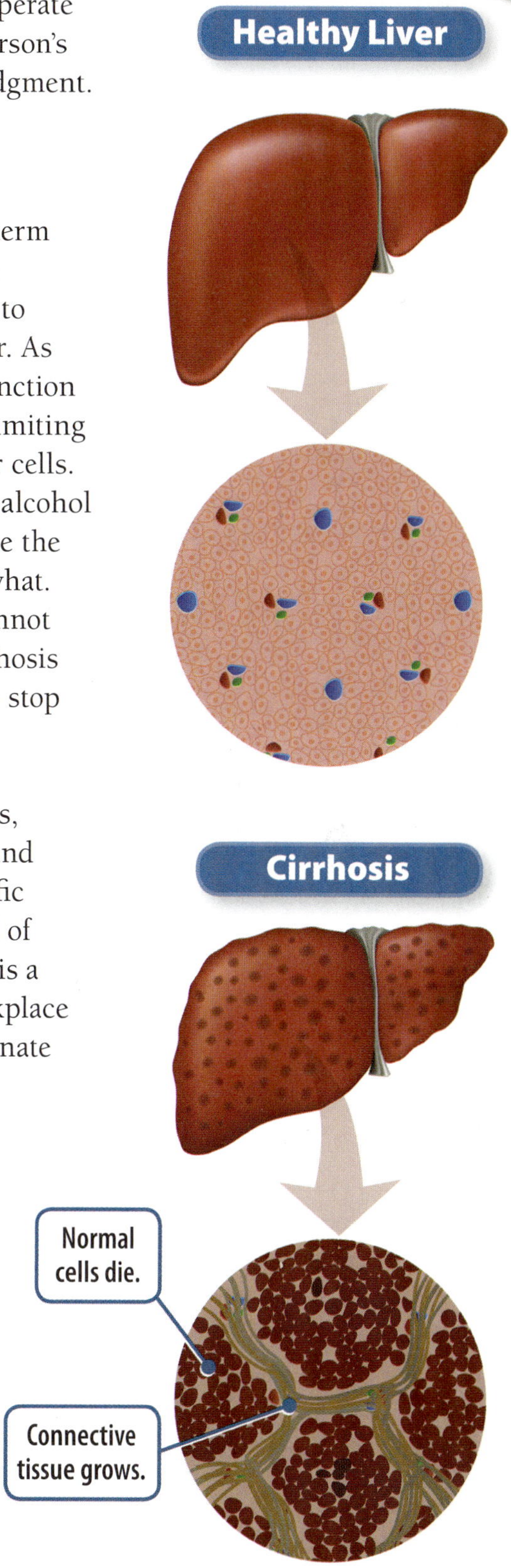

Tobacco and the Body Explain

At one time, people believed that smoking tobacco was healthy. The common belief at the time was that smoking would kill germs and prevent disease. Today we have many research studies that show the dangers of smoking tobacco. Ellen White was ahead of her time when she wrote in the 1800s, "Tobacco is a poison of the most deceitful and malignant kind, having an exciting, then a paralyzing effect upon the nerves of the body. It is all the more dangerous because its effects upon the system are so slow, and at first scarcely perceivable. Multitudes have fallen victim to its poisonous influence." (*Spiritual Gifts*, vol. 4, p. 128)

Tobacco use is an activity that can be socially encouraged. Why avoid it? Abstinence from tobacco will undoubtedly keep your body healthier. A chemical in tobacco called **nicotine** makes the central nervous system, heart, and other organs work faster. This might sound like a benefit, but it is not. As nicotine speeds up the heart rate and raises blood pressure, it increases the risk of a heart attack. Tobacco smoke also contains 7000 other chemicals and compounds, including many carcinogens, formaldehyde, carbon monoxide, and arsenic. Because of the dangerous and addictive effects of tobacco, the Surgeon General lists warnings for consumers on every tobacco product. Some of the statements include the following:

- Cigarettes are addictive.
- Cigarettes cause strokes and heart disease.
- Smoking can kill you.
- Smoking by pregnant women may result in fetal injury, premature birth, and low birth weight.

Smoking tobacco causes harmful, short-term effects from the very beginning; smoking can irritate the eyes and nose and cause a cough to develop. What physical signs give a person away as a smoker? Over time, the tar in tobacco smoke builds up in the lungs, damaging the tender tissues and inhibiting the gas exchange that normally occurs there. The tar often causes chronic coughing and shortness of breath. A smoker has a much greater likelihood of developing lung cancer and other lung diseases than does a nonsmoker.

Tobacco smoke also causes **emphysema**, a disease that damages the tiny sacs (alveoli) in the lungs where gas exchange occurs. The walls of the sacs harden, reducing their ability to absorb oxygen from the air and release carbon dioxide from the bloodstream. The alveoli lose their elasticity and cannot expand because of the hardened walls, and breathing becomes more difficult. Although emphysema has other causes, the most common one is irritation from the harmful chemicals in tobacco smoke. Is smoking now worth the risk of not being able to breathe later in life?

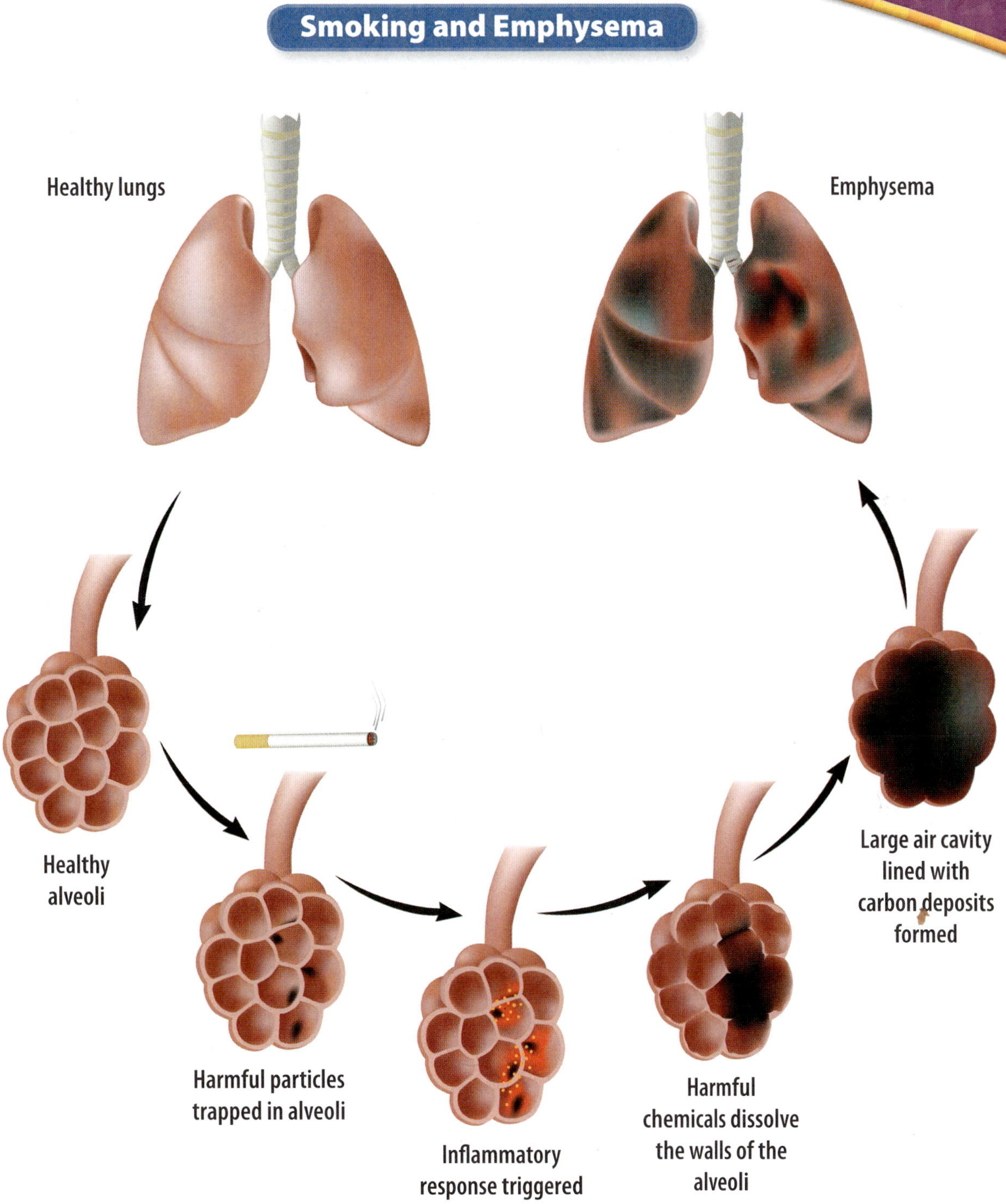

The pictures show a healthy set of lungs becoming damaged from the effects of emphysema.

What differences do you see in the images of the lungs in the beginning of the cycle and at the end?

Structured Inquiry

Discover

Record your work for this inquiry. Your teacher may also assign the related Guided Inquiry.

Science Journal

I Can't Breathe!

What does it feel like to not be able to breathe well?

Procedure

Materials
- drinking straw cut in 2 pieces: 2.5-cm section, 7.5-cm section
- balloons
- stopwatch

1. Blow up a balloon until it is about 30 cm long. Have a partner time how long it takes to blow up the balloon to this length. **Record** the time.
2. On a one-to-five scale, record how easy it was to breathe after blowing up the balloon, where five is the easiest breathing and one is the most difficult breathing.
3. Repeat Steps 1 and 2, but this time insert the 7.5-cm section of straw into the end of a balloon and blow through the straw to blow up the balloon to the same length as before. Use a new balloon each time.
4. Repeat Steps 1 and 2 again, this time using the 2.5-cm section of straw.
5. Switch roles with your partner and repeat Steps 1 through 4.
6. Determine and record your resting heart rate.
7. Run in place for one minute with unrestricted breathing. Determine and record your heart rate immediately after the minute of running.
8. Switch roles and repeat Steps 6 and 7.
9. Run in place for one minute with restricted breathing (with one end of an undamaged drinking straw in your mouth). Determine and record your heart rate. Rest until your heart rate returns to its resting rate.
10. Run in place for one minute with restricted breathing (with one end of a flattened drinking straw in your mouth). Determine and record your heart rate.
11. Switch roles and repeat Steps 9 and 10.

Analyze Results

Analyze the data. How did the restriction of breathing change the amount of work it took to blow up the balloon? How did that affect heart rate?

Create Explanations

1. What does it feel like to not be able to breathe well?
2. How can this activity relate to the harmful effects of smoking?
3. What else could you do to make breathing even more difficult?
4. Why was it more difficult to blow up the balloon using the straw?

Smoking and Cancer

Recall that carcinogens in tobacco can interfere with the growth cycle of lung cells by causing them to divide uncontrollably. The result is cancer. You also know that the carcinogens in tobacco smoke can spread throughout the body through the bloodstream, possibly causing cancer in the bladder, kidneys, pancreas, and stomach.

Even if you do not smoke tobacco by means of cigarettes or cigars, you may not be safe from its risks. If you are around smokers, you breathe in smoke from their burning cigarettes and from the smoke-polluted air they exhale. We call this airborne hazard **secondhand smoke**. The same carcinogens that affect smokers will enter and affect your body. Every year in the United States, secondhand smoke causes about 3000 deaths from lung cancer and 46,000 deaths from heart disease. How can you protect yourself from secondhand smoke?

If you chew tobacco, you will escape the damage that tobacco smoke can do to your lungs. But you are still at risk from the carcinogens and other harmful chemicals in tobacco. Regardless of how these chemicals enter your body, they will damage your cells in the same way as if you smoked the tobacco. This damage can result in cancers of the mouth and throat and increases your risk of heart disease and stroke. Like smoking tobacco, smokeless tobacco is addicting. Who do you think is more likely to develop mouth cancer, a smoker or a tobacco chewer?

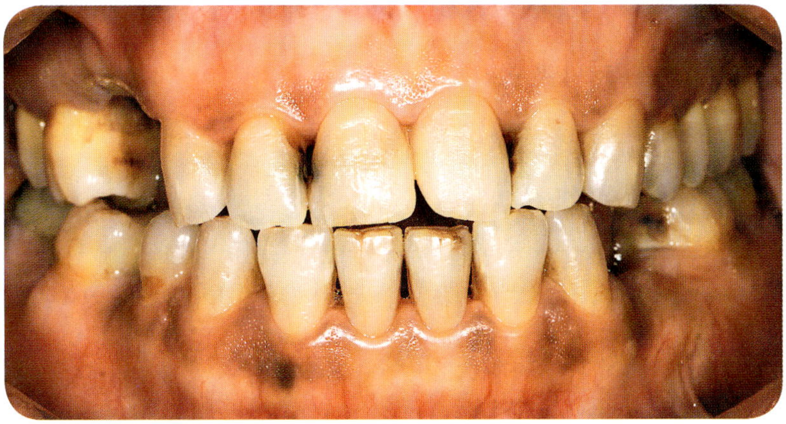

Smokeless tobacco is just as dangerous because it damages the mouth.

What warnings would you give to a person who uses or wants to try smokeless tobacco?

Math in Science

Smoking cigarettes is not only an unhealthful habit, but it can be expensive as well. The price of cigarettes varies. Find out how much a pack of cigarettes costs in your community. If a person smokes one pack of cigarettes per week, how much would this habit cost over a year (52 weeks)? How much would it cost if a person smoked one pack of cigarettes per day?

Check out your *Science Journal* for a Structured Inquiry examining how tobacco advertising attracts smokers. **Extend**

Alcohol, Tobacco, and Addiction **Explain**

Nicotine is as addictive as heroin and cocaine. The younger you start smoking, the more quickly you are likely to become addicted. Many smokers often voice their desire to quit smoking but struggle to quit due to the strong craving their bodies have for nicotine. Attempts to stop smoking typically cause physical discomfort similar to withdrawal symptoms. How does withdrawal from nicotine differ from withdrawal from other drugs?

Not everyone who drinks becomes addicted to alcohol. But those who do become alcoholics spend a lifetime struggling with the need to drink. However, even those who do not become alcoholics can suffer because of alcohol's influence on their lives. Getting drunk even one time can lead to a damaged reputation, injuries from taking a foolish risk, unplanned pregnancies, or death in a car crash. So why would someone drink? Many people choose to completely abstain from drinking alcohol, allowing them to remain in control of themselves and to make sound judgments. This policy of temperance is the one to live by; keep your body free of all things harmful and full of things to keep you healthy.

Concept Check **Assess/Reflect**

Summary: How do alcohol and tobacco affect your body? Alcohol and nicotine are two legal drugs that can be harmful to your body. Alcohol alters your mood and can put you at risk of making poor decisions. Excessive drinking can also destroy your liver and contributes to other serious diseases. Nicotine in cigarettes is highly addictive, and carcinogens in tobacco can damage your body by causing cancer or other diseases. Making smart choices as a teen can significantly improve the quality of the rest of your life. It is always most important to abstain from using any amount of alcohol or drugs; they have no positive effect on your body or mind and have many negative effects.

1. What are two short-term and two long-term consequences of smoking cigarettes?
2. What are two short-term and two long-term consequences of drinking alcohol?
3. How can abstaining from alcohol and/or tobacco lead to a healthier life?
4. Imagine that you have a friend who drinks. How can his or her behavior harm you? Imagine that another friend smokes cigarettes. How can his or her behavior hurt you?

Lesson 3

Essential Question

How Do Diet and Exercise Affect Your Body?

An anonymous author said, "If you don't take care of your body, where are you going to live?" Think about this quote. What are its implications? What are some ways you can take care of your body and practice healthful living? Physical activity is important to keep your body fit and strong, and this activity does not always take place in a gym. When you ride your bike, play volleyball, rake the leaves, or take the dog for a walk, you are doing physical activity. Diet is another important component for your health. What are some of your favorite snacks? Do you think these foods are good for your body? Before eating a food, you should think carefully about the effect it will have on your body. A balanced diet along with regular physical activity will help keep your body fit and healthy for as long as you live.

Objectives

- Identify the basic nutrients, their sources, and their importance.
- Identify important dietary guidelines.
- Explain how food labels help you make good choices.
- Explain how physical activity helps improve and maintain health.

Vocabulary

nutrient

endurance activity

aerobic exercise

anaerobic exercise

body composition

How, specifically, are these teens helping to maintain their health?

Diet and Your Health Explain

Do you like pizza? Would you want to eat it for every meal? Probably not. Eating the same food all the time would be boring, not to mention unhealthful. Eating a variety of foods is the best way to ensure you get all of the nutrients you need. **Nutrients** are the chemicals in food that your body uses for energy, growth, and cell repair. There are six basic nutrients that your body needs to flourish. These nutrients are carbohydrates, fats, protein, minerals, vitamins, and water.

Nutrient	Importance	Food Sources
Carbohydrates	• Supply energy • Digest easily	Bread, rice, potatoes, cereals, pasta, fruits, vegetables
Fats	• Supply energy • Store vitamins	Dairy products, oils, meat, margarine, nuts, peanut butter, salad dressings
Proteins	• Supply energy • Help with cell growth and repair	Dairy products, eggs, meat, fish, poultry, soy, legumes, nuts, meat analogs (meat substitutes)
Minerals	• Control body processes • Help build new cells • Carry oxygen in blood	Fruits, vegetables, dairy products, eggs, bread, cereals, meat, meat analogs (meat substitutes)
Vitamins	• Control body processes • Help build new cells • Fight disease	Fruits, vegetables, dairy products, eggs, bread, cereals, meat, meat analogs (meat substitutes)
Water	• Helps digestion • Removes waste • Main part of blood • Controls temperature	Drinking water, milk, juice, soup, fruits, vegetables

It is essential to be sure that the foods you eat provide your body the nutrients it needs. Also be sure that you are eating foods from each of the four major food groups. Food groups are made up of foods that contain the same types of nutrients. Eating the right amount of foods from each group provides you with a balanced diet. The four groups that provide you with a healthful diet are fruits and vegetables, grains, dairy foods, and protein. How often do you eat food from each group? Why is it important to eat a variety of foods from each group?

Food Group	Importance	Food Sources
Fruits and Vegetables	• Supply vitamins A and C, calcium, and iron • Help night vision • Fight disease • Provide fiber	Apples, grapes, plums, oranges, bananas, carrots, squash, peas, cabbage, broccoli
Grains	• Supply carbohydrates, iron, niacin, and thiamine • Provide energy	Bread, cereals, pasta, rice, corn
Dairy Foods	• Supply protein, calcium, and riboflavin • Strengthen bones and teeth • Help vision • Ensure healthy skin	Milk, cheese, yogurt, ice cream, cottage cheese
Protein	• Supply niacin and iron • Build muscle, bone, and blood cells • Ensure healthy skin and nerves	Legumes, nuts, meat, fish, poultry, eggs, peanut butter, meat analogs (meat substitutes)

Having a balanced diet means that the proper nutrients are obtained from the foods we eat and that all the food groups are represented within the diet. Eating balanced meals provides a person with more energy and overall better health. Follow seven dietary guidelines to have a more healthful diet.

Dietary Guidelines	
1. Eat a variety of foods.	**4.** Eat foods that contain fiber.
2. Be at your best weight, the weight at which you look and feel your best.	**5.** Eat less sugar.
3. Eat fewer fatty foods.	**6.** Eat less salt.
	7. Do not drink alcohol.

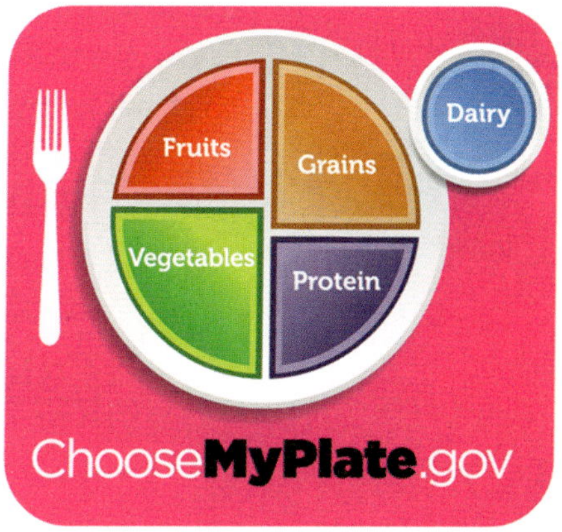

What do you think is the right balance between healthful and unhealthful foods in your diet? Explain your reasoning. Moderation and making wise choices are important with all things that we do, diet included.

One of the most rampant problems in our society is obsession with body image. It is important to feel good about yourself and your body. It is equally important to be fit and healthy. Being at your ideal weight is important for physical reasons; this is the weight at which you feel your best. Often we are misguided as to what the ideal weight is. Being over or under your desired weight is not healthy. Dieting to lose weight during the teenage years can be dangerous because the body is still growing and changing. Regular physical activity and a balanced diet can help you maintain your desired weight. What positive or negative impacts do you believe society has on a person's weight?

Becoming Informed about Your Diet

Knowing what your body needs is an important step in having a healthful diet. Ensuring that you are eating from the four food groups and following the dietary guidelines can lead to a successful diet. Understanding food labels can help with these tasks. Open your pantry when you are at home and look at some of the food labels on packages. Labels are required on all packaged foods to let consumers know what they are buying. Each label must identify the name of the product, the company that manufactured the product, the amount of the product, and the ingredients in the product.

Math in Science

Below is part of a food label for macaroni and cheese (Label A). The "% DV" column shows the percentage of fat, cholesterol, or another substance that this food contributes to a 2 000-calorie daily diet.

There is a newer product that manufacturers are promoting—veggie macaroni and cheese—as a more healthful option. The food label for this product is also shown (Label B).

About how many cups of each type of macaroni and cheese would provide the total amount of fat recommended for one day for the average person? Compare the labels. Which brand of macaroni and cheese would be the best choice if both were the same price? Why?

A

Nutrition Facts

3 servings per container

Serving size 1 cup (70g)

Amount per 1 cup

Calories 260

% DV*	
5%	**Total Fat** 3.5g
8%	Saturated Fat 1.5g
	Trans Fat 0g
5%	**Cholesterol** 15mg
24%	**Sodium** 580mg
16%	**Total Carbs** 47g
4%	Dietary Fiber 1g
	Sugars 6g
	Added Sugars 0g
11%	**Protein** 10g
0%	Vitamin A
0%	Vitamin C
10%	Calcium
10%	Iron

*Percent Daily Values are based on a 2,000 calorie diet. Your Daily Values may be higher or lower depending on your calorie needs.

	Calories:	2,000	2,500
Total Fat	Less than	65g	60g
Sat Fat	Less than	20g	25g
Cholesterol	Less than	300mg	300mg
Sodium	Less than	2,400mg	2,400mg
Total Carbohydrate		300g	375g
Dietary Fiber		25g	30g

B

Nutrition Facts

2 servings per container

Serving size 1 cup (70g)

Amount per 1 cup

Calories 260

% DV*	
5%	**Total Fat** 3.5g
8%	Saturated Fat 1.5g
	Trans Fat 0g
3%	**Cholesterol** 10mg
23%	**Sodium** 550mg
16%	**Total Carbs** 47g
12%	Dietary Fiber 3g
	Sugars 7g
	Added Sugars 0g
10%	**Protein** 10g
0%	Vitamin A
15%	Vitamin C
10%	Calcium
15%	Iron

*Percent Daily Values are based on a 2,000 calorie diet. Your Daily Values may be higher or lower depending on your calorie needs.

	Calories:	2,000	2,500
Total Fat	Less than	65g	60g
Sat Fat	Less than	20g	25g
Cholesterol	Less than	300mg	300mg
Sodium	Less than	2,400mg	2,400mg
Total Carbohydrate		300g	375g
Dietary Fiber		25g	30g

Learning how to read the labels on food can further help you make good decisions about what you put in your body. The ingredients in a product must be listed, beginning with the ingredient present in the largest amount. Nutrition labels will present the number of calories along with the amount of fat, carbohydrates, sugar, and sodium present in a serving of the food. A food that might seem healthful may actually be very high in sodium or sugar content.

Calories and Metabolic Rate

Your food supplies energy as calories. A *calorie* is a unit measure of the amount of energy in food. Some foods contain more calories than others. Good weight management relies on knowledge of energy input and output. Your energy input is the number of calories you take in as food and drink. At your age, you typically need between 2 000–3 000 calories per day. Energy output is the amount of calories you burn as energy each day. *Metabolism* is the conversion of fuel from the food and drink we consume into the energy used to power our bodies. Have you ever heard someone talk about having a fast or slow metabolism? The rate at which you use your energy is your metabolic rate. When you are sleeping, your metabolic rate is slower, and when you are active or exercising, your metabolic rate increases. The higher your metabolic rate, the more calories your body is using. How does your calorie usage change as your metabolic rate slows?

Physical Activity and Your Health Explain

How important is physical activity? Current studies show that every year, about 250,000 Americans die from causes related to a lack of physical activity. Did you know that only 60 minutes of physical activity a day is enough to keep young people in good health? For adults, only 30 minutes a day is enough.

How much physical activity do you get in a day? How much TV do you watch every day? A recent study of high school students showed that just over 12% were physically active for 20 minutes three or more times a week. The remaining 88% did not even get that much exercise. At the same time, 70% of these students watched TV for at least one hour on each school day. About 35% watched television for three or more hours on school days. What relationship do you think exists between the amount of TV watched and the amount of physical activity engaged in?

Faith Connection

Is exercise mentioned in the Bible? Why do you think this is?

Structured Inquiry

Discover

Record your work for this inquiry. Your teacher may also assign the related Guided Inquiry.

Burning Calories

How are calories measured?

SAFETY: Safety note: Students with peanut allergies should avoid contact with the peanuts.

Materials

- goggles
- 3 large paperclips
- ring stand
- wire ring
- large rubber stopper
- balance
- lighter
- water
- 100-mL graduated cylinder
- aluminum soda can (355 mL)
- scissors
- thermometer
- forceps
- shelled peanuts

Procedure

1. Construct the calorimeter as shown in the illustration in your ***Science Journal.***
2. Fill the graduated cylinder with 200 mL of water and add this to the soda can.
3. Place the thermometer into the mouth of the can. **Measure** the initial temperature of the water and **record** it in the data table.
4. Use the balance to find the mass of a single whole peanut. Record the mass in the data table.
5. Holding the cork in your hand, place the peanut on the bent end of the paperclip. Carefully use the lighter to light the peanut on fire. Place the burning peanut beneath the bottom of the soda can. Hold it there until it stops burning.
6. When the burning stops, take the temperature of the water and record the results. Allow the peanut to cool and use the forceps to carefully place the remains on the balance to find its mass. Record the mass in the data table.

Analyze Results

Calculate the change in mass of the peanut. Calculate the change in temperature. Calculate the amount of energy in the peanut (in kcal) by using the equations in the ***Science Journal***. Do you think a peanut is a high-calorie food based on these results?

Create Explanations

1. How are calories measured?
2. Are all foods that may be high in calories, such as peanuts, bad for you? Explain why or why not.
3. How would your observations be different when comparing full-fat chips and a reduced-fat version using the calorimeter?

Avoiding physical activity and eating too much food has unfortunate consequences. Recall that obesity occurs when a person is significantly above a normal, healthful weight. Obesity is now at an all-time high among adults, teens, and children. This disorder leads to many health problems, including high blood pressure, cardiac disease, and type II diabetes. Why are these diseases associated with obesity? Recall that in type II diabetes, an insulin resistance prevents the body from regulating glucose, the form of sugar that fuels the body, which then accumulates in the blood. Type II diabetes was once so rare among young people that it was called "adult-onset diabetes." Now, about half of the new cases of type II diabetes in the United States are young people. Why do you think this is so?

Benefits of Physical Activity to the Body

Many types of physical activity help your body become stronger and more flexible. Certain physical activities also increase your *endurance*, your body's ability to continue exertion.

Benefits of Physical Activity for Your Body
• Can increase your endurance
• Can build muscle mass
• Can strengthen your bones
• Can strengthen your cardiovascular system
• Can increase your body's flexibility

Increases Endurance

Physical activity increases your endurance, or stamina. **Endurance activities** raise your heart rate and breathing and include activities such as walking, running, or swimming. Endurance can be gained through slowly increasing the amount of exercise over time.

Builds Muscles

The more work muscles do, the stronger they become. You can exercise your muscles by pushing against an immovable object, such as a wall, or by repeatedly contracting your muscles, such as by doing push-ups or sit-ups. What are other ways to exercise your muscles?

You need to be physically active.

Why do you think many teens are not physically active?

Strengthens Bones

Just as working your muscles strengthens them, making your bones lift weight strengthens your bones, too. Starting at about age 30, people start to lose bone tissue. In time, their bones can become fragile and break easily. Walking, climbing stairs, playing tennis, and ballet are the kinds of weight-bearing exercises that help to strengthen bones. What other factors increase bone strength? Although swimming, bicycling, and similar activities are great for your muscles and heart, they do not build your bones. Why not?

Strengthens the Cardiovascular System

Your heart is a muscle, so as it works harder during physical activity, it builds muscle fibers. A stronger heart can pump more blood with less effort because it pumps less often, reducing your blood pressure. How will a stronger heart help your body?

Aerobic and Anaerobic Exercise

The cyclists in the photograph are getting a type of physical activity called **aerobic exercise**. This type of activity uses large muscles and is rhythmic and repetitive. Aerobic exercise is especially good for lowering blood pressure. It helps build endurance. To be effective, aerobic exercise should be continued for at least 20 to 30 minutes. **Anaerobic exercise**, another type of physical activity, is made up of brief, strength-related activities, such as sprinting or weightlifing. This type of activity builds muscle and physical strength. An ideal exercise program should include both types of activities.

What are some other examples of aerobic exercise?

Benefits of Physical Activity to Mental and Emotional Health

Have you ever noticed you feel in a better mood after walking or some other physical activity? Physical activity helps you mentally and emotionally in a number of ways. First, vigorous physical activity helps get rid of tension and frustration, so you can relax. Even better, being active stimulates the release of endorphins and serotonin in your brain. These chemicals produce a natural and healthful "high" that can improve your mood and lift feelings of depression. How would physical activity provide an alternate treatment for mild cases of anxiety or depression?

As you begin to exercise regularly, you will feel better, too. Your muscles will be more defined, and you will have more energy to do the things you want—and need—to do. As you use physical activity to help manage stress, you might find it easier to get along with others, including family members. Being physically active can help to provide you with a clear mind and can help you to feel less stress in day-to-day life; this can allow you to communicate more freely with God. How else can engaging in physical activity make you feel better?

Explore-a-Lab

Structured Inquiry

How many steps do you take each day?

1. **Predict** how many steps you think that you walk on an average day. **Record** your predictions.
2. You will be given a pedometer (an instrument that records the number of steps the wearer takes). On a specified weekday, start wearing you your pedometer at 8 A.M., and do not remove it until 8 P.M. Record the number of steps you have taken by noon and the total number of steps at 8 P.M.
3. Repeat step 2 for two other days. Make sure one of the two days is a weekend day.

How did your average number of total steps compare to your prediction? Do some research to see if you can find a recommended number of steps that you and others your age should take to ensure they are getting adequate exercise.

Exercise + Healthful Food = A Healthy You

According to the Harvard School of Public Health 1 out of 3 children (33%) in the U.S. are overweight or obese. In Canada the percentage is lower at 9%. Why are these statistics so important? Overweight children face the same health struggles as overweight adults. What are some diseases associated with obesity? If a person persists in his or her obesity, the attendant health problems can continue for a lifetime. In one study, four out of five children who were overweight at ages 10 to 15 were obese at age 25. As a result, they were more likely to develop health problems that once affected only elderly people.

Check for Understanding

Diet and physical activity go hand in hand with leading a healthful lifestyle. Why is it not enough to either be physically active or have a healthful diet?

A scale cannot tell how much of your body weight is muscle and how much is fat. But that division of body weight is an important indicator of health risks. **Body composition** is the ratio of body fat to lean body tissue (muscles, bone, and connective tissue). A caliper skin test can help determine how much fat is beneath a fold of skin. A qualified technician can also use an electrical device to determine body composition by sending a tiny current through the body. The current travels at a different rate through fat, so its speed indicates the amount of fat in the body.

Concept Check Assess/Reflect

Summary: How do diet and physical activity affect your body? A healthful diet and daily exercise are the means to achieving and maintaining a healthy body. The food you put in your body fuels it. It is important to eat foods with all nutrients present, including carbohydrates, fats, proteins, vitamins, minerals, and water. A balanced diet will include foods from each of the four important food groups: fruits and vegetables, grains, milk products, and protein. Following the seven dietary guidelines is also critical. Physical activity improves your body's endurance and increases bone and muscle strength. Regular physical activity also helps you improve your mental well-being by releasing chemicals in your brain that improve your mood and decrease stress. As obesity becomes an increasing concern in North America, it is important to be aware of how essential physical activity and diet are for avoiding obesity.

1. Write out a day's balanced meals and identify each of the food groups present in the meals.
2. How do vegetarians ensure that they obtain the necessary amount of protein in their diet?
3. What are several physical and mental benefits of being physically active?
4. Suppose that a friend says that he is too tired to exercise. Explain how physical activity can actually give him more energy.

Lesson 4

Essential Question

How Can You Have a Healthy Life?

Being a teenager can be one of the best times of your life, if you choose to make it great. Surrounding yourself with good friends and family helps you to enjoy life. Take advantage of this time in your life. Having a great life involves having fun but still being smart. Think about your actions, and make sure that you are always staying true to yourself and your beliefs. Too often teens worry about what their friends think, and they make poor or unsafe decisions. What are examples of foolish risks you have taken or seen your friends take? The decisions you make can affect not only your life but also your health, with both short-term and long-term consequences. Learning how to make wise decisions is an important life skill.

Objectives

- Identify the major risks (behaviors) that can negatively affect health.
- Explain the roles of heredity and environment in health.
- Describe resources available to provide health-related help.

Vocabulary

anabolic steroid

eating disorder

anorexia nervosa

bulimia

What are some possible outcomes of agreeing to help with a clean-up project such as this one? What are some of the beneficial consequences?

How could sexual abstinence lead to a healthier friendship?

Scripture Spotlight

What risk is described in **1 Chronicles 11:9–19**?

Risks to Life and Health Explain

As a teenager you are confronted with many choices. Some choices are easier to make than others, such as choosing an outfit for school. Others are harder to make, such as deciding where to apply for college. When faced with a decision, think of the consequences of the possible behavior or action. Are the consequences negative or positive? Is the end result worth the risk? Following Jesus makes some tough choices easier; living in this way removes part of the risky elements in life.

One of the greatest flaws and yet a strength of teenagers is their feeling of invincibility. Even the most knowledgeable teens can think that nothing bad will happen to them. Often this feeling leads to teens taking unnecessary and sometimes dangerous risks. A risk is a hazard or dangerous chance. What risks do you face every day? How do you deal with them?

Sexual Relationships

Some major health risks that face teenagers involve poor choices or negative behaviors. Some teens choose not to abstain from sexual relationships and may face consequences for these actions. Possible short-term consequences include losing your parents' trust and feeling used when the relationship ends. One possible long-term consequence includes infection with an STD. Another possible consequence is becoming pregnant or causing someone else to become pregnant, dropping out of school, and becoming a single parent with little or no income and reduced opportunity for going to college. The smart solution is to be sexually abstinent, which means waiting for marriage to become sexually active. Choosing to abstain from sex and living as God wants eliminate these possible consequences in your life. If you or someone you know is infected with an STD or is pregnant, seek medical attention for yourself or that person. If you are uncomfortable talking with a parent about this, talk to a teacher, counselor, or health professional to get the necessary help.

Structured Inquiry

Discover

Record your work for this inquiry. Your teacher may also assign the related Guided Inquiry.

Science Journal

Evaluate and Rank Behaviors

How much risk is involved in life choices?

Procedure

Materials
- large paper
- list of behaviors for teens
- pen or pencil

1. Read the list of common behaviors that you may confront in your ***Science Journal***. On the paper, draw a T-chart. Title the first column *Positives* and the second column *Negatives*.
2. For each behavior, list any positive and negative consequences associated with it.
3. Next, use the chart to evaluate the risk involved in each action and rank it in order from 1 to 10, where 10 indicates the greatest risk. Remember that a risk can be positive or negative.

Analyze Results

Share with a small group the rankings you gave the behaviors. **Compare** your rankings with those of the others in your group. Discuss the items about which there was the greatest difference of opinion. Use your charts to support why you categorized the behaviors as you did.

Create Explanations

1. How much risk is involved in life choices?
2. How can you categorize and evaluate behaviors?
3. How do you think this activity will influence your decision making?
4. Why do you think it is important to evaluate risks?

Using Drugs, Tobacco, and Alcohol

Teens are often confronted with the presence of drugs, tobacco, and alcohol. Peer pressure and the desire to fit in often lead teens to make poor decisions about using these substances. As discussed earlier in this chapter, using any of these substances can have negative health implications and can possibly lead to addiction. Living with temperance and abstaining from using illegal substances or misusing legal drugs keeps your body and mind healthy and sound. What are some of the negative consequences associated with using drugs, tobacco, or alcohol? Some say that it is normal for teens to try these substances. Do you agree or disagree? Why?

Math in Science

Create a bar graph by drug type of the percentage of Colorado teens who abused certain illegal drugs in 2005, 2009, and 2011.

Trends in Illegal Substance Abuse by Teens in Colorado for 2005, 2009, and 2011

	2005 (%)	2009 (%)	2011 (%)
Alcohol	75	72	65
Marijuana	42	43	40
Perscription Drugs	–	–	20
Cigarettes	20	19	16
Ecstasy	6	9	11
Inhalants	10	9	8
Cocaine	6	7	5
Heroin	1	2	3
Methamphetamine	3	1	2

Answer the following questions after evaluating the data.

1. The use of which drug increased the most between 2005 and 2011?
2. Did the use of any drug decrease between 2005 and 2011?
3. The use of which drug decreased the most on a percentage basis?
4. How do you think these values compare with those at your school?

There are numerous negative effects of smoking, many of which are deadly.

Explore-a-Lab

Structured Inquiry

How Do You Rate Consequences?

Find out how others in your class rate the consequences of smoking. Each student will list the six consequences below and will give each one a rating of 1 to 5. (Don't put your name on your ratings. Keep the responses anonymous.) Then compile the ratings in a chart on the board. What, if anything, surprised you about the results of this survey?

In your opinion, how important are these consequences of smoking?

	1 (Least)	2	3	4	5 (Most)
1. Having **bad breath**					
2. **Getting caught** at school/home					
3. **Not fitting in** with friends who smoke					
4. **Coughing**; being short of breath					
5. Getting **cancer**					
6. Developing **heart disease**					

Unsafe Behaviors

Other health risks facing teenagers come from not behaving safely and intelligently. Think before you act. When you are able to start driving put your phone in a glove compartment or someplace where you will not be tempted to use it while driving. Due to lack of experience, teenagers are already higher-risk drivers; adding phone use while driving can make for a potentially lethal combination. Many states have enacted laws regarding cell phone use while driving. What are the laws for cell phones in your state? Do the rules for teenagers differ from those for adults?

Peer Pressure

Choose your friends wisely and remember that a true friend does not need to be impressed or agreed with. Peer pressure is a major issue that often causes teens to make poor or dangerous decisions. While it can be hard, stand up for what you believe is right when confronted with peer pressure. Plato said, "People are like dirt. They can either nourish you and help you grow as a person, or they can stunt your growth and make you wilt and die." How does this quote relate to friendships? How do your friends help you grow? Staying safe and healthy as a teenager can sometimes be as easy as stopping and thinking before acting. Know yourself and God, and trust the path that God wants you to take. This should lead you to a life of happiness and abstinence from dangerous activities.

Heredity or Environment **Explain**

Your body and health are gifts you were given at birth. They are, however, not gifts that are always free of flaws. Some people are born with health conditions that may affect them throughout their entire lives. Some may develop health conditions during their lives. Other people may be extremely healthy and only suffer minor temporary health issues during their lives.

Some health conditions can be the result of heredity. *Heredity* is the transmission of genes from parent to offspring. Often a family resemblance is passed on from parent to child. This may account for hair or eye color, height, or body type. Sometimes health conditions, such as high blood pressure or high cholesterol, can be passed on to a child as well. What are some traits that you have inherited from your parents? How does it help you to know the health conditions of your family that you may be predisposed to?

Other conditions occur due to environmental reasons. Often the habits of parents, such as food choices, are picked up by the children. A person's environment can also lead to health challenges for an individual. The flu and stomach viruses often spread throughout a school or a community because they are contagious. Do you think a person would develop asthma due to heredity or environmental conditions? Explain your reasoning.

As Americans battle with obesity and diabetes wages on, many researchers have discussed whether these characteristics are due to genetics or to environment and lifestyle. Is a child of obese parents born with a gene to become overweight, or does the child become overweight by learning the eating and exercise habits of his or her family and by eating the types of foods made available to the child? Body type is something that seems to be hereditary, but your weight is something that you can manage through diet and exercise, as discussed in the previous lesson.

Trying to Change Who You Are

Some teenagers look in the mirror and are unhappy with what they see. They do not have the body they wish they had, possibly due to heredity, environment, or often a misconceived body image. God made each of us to be different, and God loves us all. Remember this as you look in the mirror each day. Often teenagers cannot embrace themselves and instead work to change their bodies. In this quest for change, sometimes a person's health is ignored. What are some good qualities you have?

Body Image

Some teenagers wish that they were not so skinny and had larger muscles, or that they were in better shape. Occasionally, these teens (both males and females) use **anabolic steroids** to speed up their muscle development. Anabolic steroids are artificially produced hormones that promote the growth of muscle mass. By using these drugs, however, teenagers risk irreversible stunted growth and abnormal sexual development. Here is what steroids offer:

- Bigger muscles right now in exchange for stunted growth for life
- Breast development for men and facial hair for women
- Acne
- Mood swings
- Aggressive behavior
- Hair loss
- High blood pressure
- Heart and liver abnormalities

What are some other side effects of steroids? Considering those consequences, waiting for built muscles through your workouts alone is a whole lot smarter and safer.

Weight loss is a huge industry. It seems that magazine covers are always offering all kinds of fad diets, such as eating nothing but special cookies to lose weight or drinking only liquid protein supplements. Do you think these diets are healthful? Why or why not? Many teens and even adults try these fad diets regardless of the risks associated with them. It is more important for people to feel good about themselves and their weight than it is to attempt to create a body deemed acceptable by society.

Eating Disorders

In an effort to change their bodies, many young people develop eating disorders. An **eating disorder** is any of a group of conditions in which people focus entirely on food and their weight while their health suffers. One eating disorder, called **anorexia nervosa**, involves starving yourself. People suffering from anorexia nervosa develop an irrational fear of gaining weight and a distorted perception of their bodies. What are the negative effects of starvation? **Bulimia**, another eating disorder, involves eating big meals and then vomiting. People suffering from bulimia will eat excessively and then force themselves to throw up the food to avoid gaining weight. This behavior is sometimes referred to as "binging and purging." What are the negative effects of binging and purging? Besides poor nutrition, what other problems might bulimia cause?

Ecclesiastes 4:10 tells us, "If one person falls, the other can reach out and help."

How can you help a friend who is engaging in unsafe behavior?

All eating disorders are harmful and can be life threatening. People who are anorexic will lose fat and muscle tissue so that their bones become visible through the skin. The heart can be severely damaged. Do you think an anorexic person could have strong bones? Does anorexia affect a person's brain? People who are bulimic may actually gain weight but may lose the enamel on their teeth as stomach acid dissolves it. Eating disorders are not a shortcut to healthful weight loss.

If you know anyone who is suffering from anorexia or bulimia, you should tell a trusted adult, such as a parent, teacher, or school counselor. There are many health risks and negative consequences from these disorders, and any person engaging in this behavior needs help. It is important for a person suffering or recovering from an eating disorder to work with a nutritionist and a medical doctor, as well as a counselor. A counselor can help this person to see his or her true self and to banish the extremely negative self-esteem that often accompanies these disorders. There is a common link between eating disorders and depression and possibly suicide. That is why it is of the utmost importance for someone suffering to receive professional help.

The Help of Others Explain

Knowledge is power. The more informed you are as a teenager, the less likely you are to engage in behavior that is detrimental to your health. It is important for you to learn about risky behaviors and consequences, such as sexual activity, use of drugs or alcohol, and eating disorders. Watch your peers carefully for changes in behavior or withdrawal from friends, as these can be signs of depression. If you suspect that a friend is in need of help, let an adult know. If you are in need of help, ask someone for it; there is no reason to go through challenges alone.

Successful people in life have not gotten to that point alone. A successful person knows how to ask for and appreciate the help of others. God has put us on Earth to help one another. When is a time that you have needed the help of others? Many organizations exist for the sole purpose of helping people. What are some such organizations that you know of?

Hospitals and doctors' offices are the most common places for people to go for health care. The Department of Public Health provides health care, training, and disease control for people who may not be able to afford care. It also promotes public awareness about health care. The Centers for Disease Control and Prevention works to prevent disease and protect the public from outbreaks. Many programs promote physical fitness and work to reduce obesity in children. How do you think making the public aware about fitness helps to prevent or reduce obesity?

Concept Check Assess/Reflect

Summary: How can you have a healthy life? As a teenager, you face decisions and choices each day that can end up having a significant effect on your life. It is important to surround yourself with a positive support system and to allow yourself to make good decisions. Peer pressure is a constant in the lives of teenagers. You need to feel confident in who you are and that you can make decisions on your own without worrying about how others will react. Heredity and environment both affect your health. Being informed about family history can be important to prevention and protection against health risks. Having a healthy body image can lead a person to be much happier. Many teens suffer from dangerous eating disorders due to misconceptions about their body image. There are many resources available to provide a person with the help that he or she may need. Government, public, and private agencies are available to provide health care and support for individuals in need. It is important to let a trusted adult know if you or a friend is in need of help or support.

1. List two possible consequences associated with consuming alcohol as a teenager.
2. How can you help a friend who is suffering from an eating disorder?
3. Some states have restrictions for teenage drivers, such as limitations on the time of day that the teenager is able to drive or on the number of passengers allowed to be in the vehicle. Why do you think these restrictions are in place? What are the restrictions trying to prevent?
4. How does living your life by embracing God's words eliminate many of the risky behaviors associated with teens?

Science and Technology

Extend

Stay Smart with Smartphones

What is a smartphone? A smartphone is a combination of mobile phone technology and computing capability. Early smartphones included limited computing capabilities, such as GPS technology, a digital camera, and the ability to play music. In addition to these capabilities, today's smartphones allow consumers to browse the web and use a variety of apps. Many popular smartphones use a multi-touch interface that allows you to interact with the device by using just your finger. This technology has changed the way we communicate, the way we socialize, and the way we do business. Did you know that smartphones can also help you solve your problems?

Being a teen isn't easy, and, although talking to parents, friends, or a trusted teacher or relative about a problem is always the best and first choice, sometimes that is not possible. In those special cases, it is great to know that help is available and as close as your own phone.

If you live in Canada, you've probably heard of Kid's Help Phone (http://org.kidshelpphone.ca/en). It is a phone and online counseling service for kids available 24 hours a day, 365 days a year. Kids all over Canada can call for help with issues such as friends, family, school, and mental and emotional health, as well as physical, verbal, or emotional abuse.

In the United States, several options for help are available. The Boys Town National Hotline (http://www.boystown.org/hotline) provides a service for teens with problems to solve that is available 24 hours a day, 365 days a year. The National Runaway Safeline (http://www.1800runaway.org) has a 24-hour crisis hotline and also offers a live chat option for teens who are thinking about or have run away, or who have a friend considering running away. The hotline can provide counseling and also find a safe place for runaways to stay. The Hopeline (www.hopeline.com) offers crisis hotlines and online crisis chats for teens at risk of suicide.

No matter whether your problems seem small or overwhelming, so many options are available to get help. Whether it's just a sympathetic ear, some useful advice, or a crisis situation, your smartphone can help you find what you need exactly when you need it.

Concept Check

1. What would you do if your friend were considering hurting himself or herself?
2. How do you think smartphone technology has changed society?

Adolescent Psychologist

An adolescent psychologist is a specially trained counselor who is focused on helping teenagers in a variety of ways and in a variety of settings. Typically, a psychologist has completed doctoral-level education and has a Ph.D. An adolescent psychologist is then required to take an additional two years of specialized training in child and adolescent psychology. This training may also require an additional one-year internship. This extra training allows the psychology generalist to focus on the very specific concerns that affect teenagers and how the psychologist may best help teens deal with the issues, conflicts, and concerns that can happen at this time.

Psychologists, unlike psychiatrists, are not medical doctors and therefore cannot prescribe medications. Psychologists may work together with medical doctors to help teens with complex or overlapping concerns, such as addiction or mental health issues.

Adolescent psychologists have a range of options available for how they wish to provide assistance. Some may choose to set up a private practice and offer counseling to individuals in their office. Others may work in hospitals or mental health organizations, rehabilitation centers, or teen hospitals offering specialized assistance to adolescents in crisis situations. Some adolescent psychologists may choose to work in schools as educators, or they may conduct research to find out what techniques and therapies are most effective for teenagers in need.

Concept Check

1. Why is specialized training important for adolescent psychologists?
2. Why might it be important for adolescent psychologists to work with medical doctors?

Chapter 6 Review

Assess/Reflect

Study Guide

Lesson 1

1. Legal drugs can be harmful, such as taking medication not prescribed for you, taking improper doses of a medication, stopping a medication early, and not telling the doctor about allergies and other medications.
2. Stimulants increase metabolism, mental alertness, and energy. Depressants decrease mental alertness, coordination, visual acuity, and memory. Mixed-effects drugs increase metabolism and decrease mental processes and coordination. Hallucinogens cause altered states of perception and hallucinations. Steroids cause hormonal changes and other physiological effects.
3. Some people abuse drugs that make them feel good in some way.
4. In physical dependence, a user has a chemical need for the drug and without it will experience withdrawal. In psychological dependence, a user believes he or she needs a drug to feel good or even just to feel normal.

Lesson 2

1. Alcohol is a depressant. It slows down the body's reaction time and inhibits the function of the central nervous system. Alcohol consumption can cause damage to the liver and other body organs.
2. Food, body size, gender, and the amount and rate of drinking influence the effect of alcohol on the body.
3. Tobacco contains nicotine, a stimulant that causes the central nervous system and internal organs to work faster. Tobacco contains carcinogens, which may cause cancer, emphysema, or other diseases.
4. Smokeless tobacco is just as harmful as other tobacco and can cause mouth cancer and other forms of cancer and disease as well.

Lesson 3

1. The basic nutrients the body needs for health are carbohydrates, fats, proteins, minerals, vitamins, and water. The chart on page 220 provides sources of these nutrients and describes their importance.
2. Important dietary guidelines include: Eat a variety of foods, be at your best weight, the weight at which you look and feel your best, eat fewer fatty foods, eat foods that contain fiber, eat less sugar, eat less salt, do not drink alcohol.
3. Food labels help provide information about the ingredients, calories, fat, sugar, sodium, and other items in a product.
4. Exercise helps you maintain a healthy weight, increase your endurance, improve muscle tone and bone strength, and build a strong heart. It can also help clear the mind and help a person feel less stressed.

Lesson 4

1. Some of the major risky behaviors that teens engage in are sexual activity; use of drugs, tobacco, or alcohol; and unsafe driving.
2. Your health can be affected both by factors you inherit from your parents, such as a high blood pressure or diabetes, and by environmental factors, such as exposure to cigarette smoke or industrial chemicals.
3. There are many public, private, and government agencies available to provide help for mental and physical health issues. If you need help, speak with a trusted adult to guide you.

Vocabulary Check

Fill in the blanks with the correct terms.

1. A disease of the liver called __________ can be caused by excessive alcohol consumption.
2. __________ are the chemicals in food that your body uses for energy, growth, and cell repair.
3. When a person has a chemical need for a drug, this condition is called __________ dependence.
4. A(n) __________ is any of a group of conditions in which people focus entirely on food and their weight while their health suffers.
5. __________ is when a person's mental and physical abilities are impaired because of drug use.
6. Tobacco smoke can cause __________, which is a disease that damages the alveoli in the lungs.
7. When a person feels that he or she needs a drug to feel normal, this condition is called __________ dependence.

Multiple Choice

Choose the best answer.

8. What do withdrawal symptoms indicate?
   A. a psychological dependence on a drug
   B. a physical dependence on a drug
   C. an overdose of a drug
   D. the misuse of a prescription drug

9. Which two drugs slow down your thinking?
   A. steroids and marijuana
   B. cocaine and inhalants
   C. alcohol and heroin
   D. tobacco and alcohol

10. Which is a long-term consequence of using alcohol?
    A. slurred speech
    B. clumsiness
    C. acting foolish
    D. cirrhosis

11. Which of the following is not an effect of using tobacco?
    A. weight gain
    B. emphysema
    C. persistent cough
    D. cancer

Check Point

Answer the following questions.

12. Is taking a drug prescribed by a doctor for someone else safer than taking a drug bought on the street? Explain your answer.
13. What are the differences between anorexia nervosa and bulimia? What are the negative effects of each?
14. **Predict** whether obesity among teens will still be a problem 20 years from now. Explain your prediction.
15. If you wanted to find out how exercising affects a group of teens, how would you **control the variables**?
16. Think about how you could help a friend who you suspect is using anabolic steroids. **Infer** what changes you might see in your friend as steroid usage occurs.
17. What specific benefits do teens get from exercise?

Chapter 7

Making Decisions about Your Body

Scripture Spotlight

As you know God gave us the right to make decisions for ourselves, but he does hope those decisions honor His word. By learning how to become an informed consumer you can honor God. In this chapter you will read the following passages.

Exodus 32 (p. 246)
Matthew 4:1-11 (p. 248)
Proverbs 18:24 (p. 249)

The Big Idea

God gave us free will so we can make our own decisions. To honor God we need to make wise decisions based on accurate, reliable information.

How do you know whether the information on a product label is reliable?

This boy is trying to decide which product to buy. He is carefully reading the labels, but will he choose the best product for his needs?

Inquiry Kick-Off **Engage**

You are bombarded with information every day, including dozens of ads and commercials. Select one health product that is commonly shown during afternoon television. In your ***Science Journal***, you will create a chart to analyze three advertisements for similar products.

Lesson 1

Essential Question

How Can You Make Wise Decisions?

Objectives

- List the steps in effective decision making.
- Describe the effects of peer pressure.
- Explain ways to deal with peer pressure.
- Compare and contrast positive and negative role models.
- Identify situations in which teens need help with decision making.

Vocabulary

refusal skill
support network
role model

Think back to the last time you had a difficult decision to make. What were your choices? What factors were influencing your decision one way or another? Did one of the possible choices involve something that would be considered dangerous, illegal, or unwise in another way? What things did you take into consideration in making the decision? What new information did you need to learn before making the decision? Did you think of the potential consequences to each of your possible choices? How much thoughtful consideration went into your decision? Did you make it quickly or did you take your time? Would you make the same decision the next time you face a similar situation?

Many of the choices this teeenager makes will affect her life as an adult.

What are some decisions you will have to make this year?

The Decision-Making Process **Explain**

Why do smart people make poor decisions? Sometimes people understand the likely consequences of their choices, but still they make the choice that gets them into trouble. They cannot pass up an immediate, short-term reward, even though they know that the choice with the long-term reward would be better for them. What choices will help you excel in school today? How does making good choices now help ensure success in the future?

You may be tempted to take the easy way out, but the easy way can lead to all kinds of trouble. How can you make decisions that you won't regret? Here are some steps for effective decision making.

Effective Decision Making

1. Describe the decision you are about to make so it's clear in your mind.
2. List your choices.
3. Identify the possible consequences of each choice.
4. Consider biblical principles.
5. Pray for wisdom.
6. Consider your personal and family values, such as being honest, doing your best, being responsible, and respecting yourself and others.
7. Select the option that supports your values and has the best possible consequences.
8. Carry out your choice.
9. Evaluate how your decision affected your future and your ability to reach your goals.

These steps make sense and aren't complicated. So, what could stop you from making wise, sensible decisions?

Lesson Activity

Write down three things you could choose to do after school.

Follow the steps of the decision-making process to choose which is the best use of your time. Record your thoughts. Evaluate how well the process worked for you. In what other situations could you use this process? How would the process work if you were making a more difficult decision?

How does the decision-making process work for you?

Peer Pressure

Peer pressure, you will remember, is the influence that people of a similar age have over one another. For example, imagine a friend asks you to go to play ball after school. You say you are busy, but your friend tells you that everyone else in your group of friends is going to be there and you will be the only one missing out on the fun. You really don't want to be left out, but you promised your parents you would do some chores at home. What do you do? Why are some people less affected by peer pressure than others?

Adults also use peer pressure. For example, a mother might pressure a friend to babysit for her "just this once."

Peer pressure is usually thought of as negative and something that should be resisted. When can peer pressure be a positive thing? When have you used peer pressure in this way?

Why do people give in to peer pressure, even when they know they are making a mistake? Saying no to certain people or certain groups is often difficult. Suppose you want to become part of a group, and group members ask you to do something that you're not sure you want to do. Saying no because you fear they won't let you become part of the group can be difficult. Have you experienced a similar situation? What have you done?

The more uncertain you feel, the harder it is to refuse a request. It is best if you make personal decisions about various situations before you encounter them. For example, being trustworthy is an important value. It is likely that it is something your family expects. If two or three of your friends decide to cut class, are they being trustworthy? If you cut class with them, just this once, what else might they pressure you to do to remain in their group? Soon you would lose the trust of school staff and your parents. How long would it take to earn that trust again? Considering your values is a strong tool in resisting negative peer pressure. What other methods can you use to resist peer pressure?

Scripture Spotlight

In **Exodus 32** the people Moses brought out of Egypt gave in to peer pressure. How might their fate have been different if they stood strong against temptation?

Sometimes the best decision is to walk away.

If a group of people all agree to do something, can you be sure that it is the right thing to do? Explain your answer.

Structured Inquiry

Discover

Record your work for this inquiry. Your teacher may also assign the related Guided Inquiry.

Science Journal

Pressure in Advertising

How does peer pressure affect your daily life?

Procedure

Materials

- magazines
- scissors
- 2 sheets of poster board
- glue stick
- markers

1. Working with a partner, cut out 10 advertisements from the magazines your teacher has provided.
2. Discuss each advertisement with your partner to decide the following:
   a. Who is the intended audience is for the product?
   b. What images and words does the advertisement include that would appeal to that audience?
   c. What parts of the advertisement are a form of peer pressure?
   d. What, if anything, would you change about the advertisement to make it more effective?
3. Glue the advertisements to the poster board and write the information listed in Step 2 on the poster board for each advertisement.

Analyze Results

Create a graph to display this information and compare it with your classmates' findings.

Create Explanations

1. How does peer pressure affect your daily life?
2. What seemed to be the most common technique advertisers used to promote their products? How can knowing this help you resist the pressure from the advertisements?
3. Do you think it is ethical for advertisers to target young people with advertisements for certain products to get their business? Explain.

Standing Firm **Explain**

To make wise choices, you may have to use strategies to help you avoid giving in to people who encourage you to do harmful things. **Refusal skills** are strategies that can help you resist negative peer pressure. The following steps will help you develop refusal skills:

1. Say no and sound as though you mean it. You can seem uncertain by making statements such as, "Well, no, I don't think so" or "Maybe later. I don't know." Seeming uncertain encourages the other person to keep pushing. Instead, give a simple, definite reason, such as, "That goes against my values" or "I am not interested."
2. Suggest positive alternatives. For example, if a friend wants to cut class, suggest doing something fun after school instead. If someone offers you a cigarette, suggest getting something to eat. If the other person lights up anyway, leave. If you are there when the smoker gets caught, you both may be in trouble.
3. Stand up for yourself. Repeat your refusal and don't give in. If the other person keeps pushing, leave.

What can refusal skills do for you? How are refusal skills helpful in maintaining your self-respect?

Would a true friend use negative peer pressure to try to influence you? Although friends might suggest questionable activities from time to time, if you refuse to go along, a true friend will respect your decision. How can you avoid the stress of a group that has a negative influence on your life?

You can make strong decisions by identifying a **support network**, which consists of people who share your values and who will support you when you decide not to participate in dangerous or questionable activities. This network might include family members, friends, teachers, pastors, and others. You probably can name people who could be in your support network right now. Are people in your support network true friends? They do not have to know each other, but they should be people you can talk to when you have to make a difficult decision. How can this sort of support be a help to you?

Scripture Spotlight

In **Matthew 4:1–11**, the devil tries to tempt Jesus. How does Jesus resist this temptation? What can you do to be like Jesus?

Your support network can be one or more special friends.

Which values do you think these friends share?

You might also choose a **role model**, a person you consider an example to be imitated. Are role models always positive? How might a role model be good for you? Can you think of someone who would be a good role model for you? You might already have someone in your life who serves as a positive role model. Perhaps your role model is an adult family member, an older sibling, a pastor, a teacher, a character from the Bible, or someone from history, an individual whose decisions and actions seem to be positive, healthful, and admirable. How might you be a role model for someone? How can role models be negative? The connection to a negative role model is just as strong as the connection to a positive role model. How can you identify and avoid negative role models?

A positive role model can also be a mentor or life coach. A mentor or life coach is someone who teaches or gives help and advice to a less experienced and often younger person. What qualities would you want in a mentor? How would you go about finding a mentor?

Scripture Spotlight

Read **Proverbs 18:24**. How can a friend be closer than a brother?

Lesson Activity

With your parent's permission, watch 30 minutes of a popular TV program. During this time, record the number of positive and negative role models in the program. After the program, list the characteristics of each positive and negative role model identified. Write a one-page summary of your findings. Form small groups in class and discuss your findings.

How are role models portrayed on TV?

Help Needed Explain

Think back to when you started this school year. Were you a little nervous as you started taking classes with new people and perhaps a new teacher? Why did you feel this way? Now you are more confident and at ease being in the classroom with your classmates. What happened to make the change? What things did you do that made being in a new classroom better?

It is easy to think that you know all you need to know, but almost every day, year after year, you learn more. For example, many four-year-olds cannot resist giving advice to the adults around them. These youngsters don't realize how much they do not know, but over the years, they will learn more than they could ever have imagined. How has your knowledge of things changed since you were a four-year-old? What have you learned?

How could this father help his son with a tough decision?

Lesson Activity

Brainstorm with a partner for ideas about when to ask for help or when you should help someone else who is struggling to make the right choice. Speak to a school counselor or other trusted adult and look in a local newspaper to find out what reliable resources are available in the community to support young people who need help making important decisions. Record the results.

What resources are available to young persons who need help?

The adults in your life are still learning, too, but they do have years of experience in making decisions and resisting peer pressure. Along the way, they have learned some valuable lessons, sometimes by facing unpleasant consequences. When you are faced with a decision that really troubles you, especially one that directly or indirectly affects your health, ask your parents or a trusted adult for some guidance. Asking for help does not mean you are foolish or childish, but it does show that you recognize the importance of some decisions. Asking for help when you need it is actually a sign of maturity.

As you already know, some decisions are difficult and seem to have no good choices. You can still choose the option that best fits your values and goals. Let the trusted adults in your life help.

Concept Check Assess/Reflect

Summary: How can you make wise decisions? Using the steps of effective decision making can help you make wise decisions. You can use these decision-making skills to decide how to deal with difficult situations before they occur. People use both positive and negative peer pressure to influence others' decisions. Refusal skills, a support network, and a role model can help you stand firm when confronted with negative peer pressure. In addition, having a mentor or life coach can provide a young person the help and advice he or she needs to make difficult life decisions. Asking for help from a parent or trusted adult is a mature strategy in many situations.

1. What are the steps in the decision-making process?
2. Which step in the decision-making process do you think is hardest for teens? Why?
3. Imagine that someone at school offers you a cigarette. Write a paragraph on how you could use refusal skills to successfully make the right decision.
4. Why do people use negative peer pressure on others?

Lesson 2

Essential Question

How Do You Know Information Is Correct?

The U.S. **Food and Drug Administration (FDA)** is a government agency that has the task of ensuring that fruits, vegetables, and shell eggs sold in the United States are safe and healthful. This agency also works to ensure that medicines, medical products, and medical devices are safe and work as they were intended to. Why is this kind of monitoring important? How does one agency monitor all of these products to keep you safe? Drugs must have strong evidence that they are safe and effective for their intended use before they can be sold. How does the FDA ensure that medications are safe?

Like the FDA in the United States, **Health Canada** is a Canadian government agency that uses scientific research to modernize the health-care system, regulate food safety, promote the health of Canadians, and communicate the importance of disease prevention. What process do you think each agency goes through to ensure that food and medicines are safe to consume?

Objectives

- Describe the function of national health agencies.
- Explain how to evaluate health information.
- Describe reliable sources of health-related information.

Vocabulary

Food and Drug Administration (FDA)
Health Canada
dietary supplement

The FDA has warned consumers not to buy or use toothpaste made in China because it may contain a toxic ingredient found in antifreeze. The FDA also prevented more of the toothpaste from entering the United States.

It takes a while for the FDA and Health Canada to identify dangerous products, especially those imported into their respective countries. How can you protect yourself from products such as this toothpaste in the meantime?

The FDA, Health Canada, and Your Health Explain

The FDA and Health Canada are responsible for regulating food and drugs. These agencies cannot check every tube of skin cream and every box of cereal that is produced in or enters the country. Why is this not possible? Instead, each agency has developed laws that limit the claims that can legally be made about products. How do such laws make these products safer?

Under certain conditions, the sellers of a product may claim that the product reduces the risk of a health problem, but they may not claim that the product is meant to treat or cure a disease. Under law, packages must indicate clearly what contents are beneficial to promote good health. For example, a label may claim that over-the-counter calcium tablets reduce the risk of the bone disease osteoporosis, but the label may not claim that the tablets treat or cure this disease. How does this information help you? How can you use this information to make wise decisions?

These tips will help you evaluate the claims made by health-care products and services:

Evaluating Health-Care Product Claims

- Remember that only drugs that are tested and approved by the FDA in the United States or by Health Canada in Canada may support a claim to treat or cure diseases.
- If a product sounds too good to be true, its claims most likely are not true.
- Do not be swayed by apparently true testimonials from delighted customers who have experienced unbelievable results from a product.
- A product that seems to be related to a recent medical discovery may just be taking advantage of that discovery without actually being based on it.
- Any product may claim to have a secret ingredient or be all natural, but such claims do not mean the product works in the way you imagine it should.
- Watch out for limited offers: "Hurry! This price will not last! Place your order today—before supplies run out!" These are simply ways to pressure you into making a purchase.
- Money-back guarantees sound good, but once you make a payment, it is sometimes difficult to get your money back, no matter how poorly the product works.

Structured Inquiry

Discover

Record your work for this inquiry. Your teacher may also assign the related Guided Inquiry.

Science Journal

Reading Medication Labels

How can you use the data on an over-the-counter medication label?

Materials

- 6 different kinds of over-the-counter labels—three brand-name medications and three generic medications

Procedure

1. Use the chart in your ***Science Journal*** to **record** the following information for each brand of medicine your teacher has provided: recommended dose for different age groups, how often to take the medication, active ingredient, amount of active ingredient, price per dose, warnings, and side effects.
2. **Compare** the data for similarities and differences in the active ingredient, dose, and warnings for the paired brand-name and generic medications.
3. Prepare a **graph** that compares each pair of medication on recommended dose, cost per dose, and amount of active ingredient.

Analyze Results

List all the ways you can use the data that you recorded. Include information you would like to know about the products that was missing from the labels.

Create Explanations

1. How can you use the data on an over-the-counter medication label?
2. What were the differences between the brand-name and generic medications?
3. What are the advantages of generic medicine? What are the disadvantages of these medicines?
4. Based on this exercise, what information will you pay closer attention to when you take an over-the-counter medication in the future?

The Regulation of Health-Care Products **Explain**

The regulation of health-care products by the FDA and Health Canada is very limited. One agency cannot test every product in commerce. You must learn how to be a wise consumer. What are some steps you can take to become a wise consumer?

Do you have a big test tomorrow? Just take these I.Q. pills!

Would these pills make you smarter? How do you know?

The FDA and Health Canada ban certain toxic ingredients from products such as skin-care creams. Beyond that, it's up to the manufacturer to make sure the product is safe to use. Instead of a guarantee of safety, many skin-care products contain this note: "Warning—The safety of this product has not been determined." Consumer complaints, however, may lead the health agencies to analyze a skin-care product. What kinds of complaints could be made about a product? If the FDA or Health Canada find harmful ingredients, the agencies may request a recall. Because one agency in each country is responsible for monitoring every drug and many foods in commerce, the agencies typically do not analyze such products unless a consumer complains.

A **dietary supplement** is a product taken orally that contains one or more ingredients that are intended to add to your diet and that are not considered food. Dietary supplements include vitamins, minerals, herbs, enzymes, concentrates, and other products. These products supposedly improve the user's health in some way. They may be sold as tablets, capsules, gels, liquids, powders, or in other forms. What kinds of safety concerns might people have about dietary supplements? What do you think is the major concern about these products?

Unlike prescription or over-the-counter drugs, dietary supplements do not have to be approved by the FDA in the United States or Health Canada in Canada for safety or effectiveness before they are offered for sale. Why do you think the FDA and Health Canada do not have to approve dietary supplements? Is this wise? In fact, manufacturers do not need either health agency's approval to produce or sell these supplements. They do not register their dietary supplements. It is up to the manufacturers to determine whether their products are safe to use and to be

Lesson Activity

As a class, identify one type of over-the-counter medication. Go to a local store and survey at least five products. Write down the warnings on the manufacturer's label. As a class, list the different warnings. Analyze the warnings and the intended audience.

How should you use the warnings provided on the label of over-the-counter medications?

sure that the claims they make about their products are true and supported by evidence. The manufacturers do not have to provide this evidence to the FDA or Health Canada. So, when you buy a dietary supplement, you might be risking your health. Who has the responsibility for ensuring that a dietary supplement is safe for you to use?

Because the FDA and Health Canada are responsible for so many products, they focus their limited resources first on public health emergencies, such as contaminated foods. These agencies also examine products that have been reported to have caused injury or illness. The next priority is products that might be unsafe or violate labeling laws. The agencies use any remaining staff or funds to monitor products that they pull from store shelves or collect during inspections at manufacturing plants. How does each agency make consumers aware of a danger in food or medication? Can you think of a recent problem with a food product? How did you learn about it? What actions did your family take? Because these agencies are unable to check the safety of many health products sold, what should consumers do to avoid ineffective or potentially dangerous health products? What steps can consumers take to ensure that products such as fruits and vegetables are safe and wholesome?

This label is from a product sold in the early 1900s.

What does this product claim to do?

Lesson Activity

Listed below are statements that you might find on labels of different products. With a partner, identify the missing or misleading information in each claim.

1. This product will curb your appetite! You can eat as much as you want and still lose weight!
2. Designed as a complete skin-care system, Wrinkle-B-Gone effectively delivers key anti-aging ingredients deep into the skin. You'll enjoy refined skin texture, increased hydration, and a reduction in the appearance of fine lines and wrinkles around your eyes in just seven days.
3. According to a recent study, novice weight trainers who took MuscleOn for two weeks experienced a 200% increase in testosterone levels and significant lean muscle growth.

Why do some people believe such claims and buy these products?

Finding Reliable Health Information **Explain**

Where do you go to find information about health-related issues? The best source of reliable health information is a medical professional. Many people, however, turn to the Internet to find health information. How would you phrase your questions during an online search? If a statement appears on the Internet, does that mean it is true? Some of what you find on the Internet is reliable and up to date information, but much of what is online is not. Anyone can set up a website where no one checks whether the information it offers is reliable. Some sites, including chat rooms, offer advice from people with strong opinions but no medical training.

Lesson Activity

Form small groups. Each group should develop five questions that can be entered into an Internet search engine to learn more about an illness. Select an example of an illness from your teacher. Refine your questions to match the illness. Conduct an Internet search. List the top five results that come up for each question that you type into the browser. Describe what factors you took into consideration to select one of the top five search items. Open that item and describe the type of information provided.

How reliable was the information that you found online? How can you use these skills to find reliable information in the future?

The tips below will help you find information you can trust on the Internet.

Analyzing Internet Information

- Find out who is sponsoring a website by looking for an "About Us" page. Reliable sites include government agencies, such as the Centers for Disease Control (CDC), the National Institutes of Health (NIH), and Health Canada.
- Look at the end of the URL address. URLs that end with .gov and .edu are generally more reliable than .com sites.
- Be sure the site is not selling anything. For example, if a drug manufacturer sponsors the site, you should suspect that it will favor its own products. If the site has an online shop, get your information from another source. The name of a site might sound official, such as the "American Institute for Healthy Weight Loss," but its real purpose could be to sell you diet pills.

When searching for information online, it is important to understand that there is a vast amount of information available. It is best to start your search with reliable sources. Review several sites and you will begin to see that they provide similar information. When you see the same information on different websites, you can have greater confidence that the information is reliable and true.

Math in Science

The dietary supplement industry is thriving, despite concerns about the worth of these products and possible interactions with medications. Between 1994 and 2000, the sales of these supplements nearly doubled, reaching $17.1 billion. Experts predict that these sales will continue to increase by at least 10% a year. If that estimate is correct, how much will Americans spend this year on dietary supplements?

Concept Check Assess/Reflect

Summary: How do you know information is correct? The FDA and Health Canada try to ensure that food, medicines, and medical products are safe, but one agency in each country cannot evaluate every health-care product and food item in commerce. You play an important role in evaluating the labels on pharmaceutical and over-the-counter medications to determine proper usage and risk. Other products, like dietary supplements and other health-care products, undergo little regulation. It is important to treat the claims of these products with caution. Always try to use reliable sources of health information to make health decisions. Reliable sources include medical professionals and government and academic websites.

1. Why should you read the labels on health-care products carefully?
2. How can you verify that the information you find on the Internet is true and accurate?
3. Why might it be dangerous to take dietary supplements?
4. What are three sources of valid information about your health?

Lesson 3

Essential Question

How Can You Be an Informed Consumer?

Objectives

- Explain the basic FDA and Health Canada drug-approval process.
- Describe common advertising techniques.
- Explain the purpose of disclaimers in advertising.

Vocabulary

media
testimonial
bandwagon
disclaimer

How do you know when to take medicine and what kind of medicine to take? Whom do you trust to give you sound advice about taking medicine? Is a friend a good source of information even if he or she has the same symptoms? Do you think it is better to get help from a pharmacist or to find a natural remedy on the Internet? Medications can improve your health or make you sick if used improperly. How did scientists determine that the medications you can buy at the store are safe to use? What regulations do you think should be in place to keep consumers safe?

Identifying Safe and Effective Drugs Explain

Modern researchers have isolated many plant compounds that help treat or cure illness. Although researchers have been able to create some of these substances in laboratories, they still use plants for medications. Why might we want to create medicinal chemicals and compounds in a laboratory? For example, the Pacific yew (*Taxus brevifolia*) grows naturally in the Pacific Northwest. The indigenous people of that region used the bark, foliage, and fruits to treat wounds, calm an upset stomach, and produce healthful sweating. Today, medical research has extracted from the Pacific yew an anticancer compound, called taxol. Doctors now use this compound to treat ovarian and breast cancers. Why is it important to preserve natural areas?

The Pacific yew grows naturally along the Pacific coast of northern California, Oregon, and British Columbia, as well as in northern Idaho and southeastern British Columbia.

How is this plant an example of the helpful things found on Earth?

Scientists continue to test thousands of plants for medicinal extracts and derivatives. They are seeking more substances that might help treat illnesses. In fact, about one in four prescription medicines contain an active ingredient extracted or derived from plant material. Of course, scientists also are testing thousands of other compounds to see whether they can treat or cure different diseases.

Indigenous healers discovered the plant and animal parts they used to treat their patients by trial and error. What do you think the outcome was from such methods? Would you have been willing to be treated in such a way? Why or why not? Fortunately, today's doctors can be confident that the drugs they prescribe have undergone an extensive testing process. Despite these efforts, medications are still subject to recall if they cause serious health effects. What could happen to cause the FDA or Health Canada to recall a medication?

Check out your ***Science Journal*** for a Structured Inquiry that investigates health risks.
Extend

The FDA and Health Canada require rigorous studies on drugs before they allow the commercial use of those drugs. This process can take several years and cost millions of dollars. If the FDA approves a drug for use in the United States, may Canadians use it? Why or why not?

Drug Approval Process

Process Step	Description
1. Preclinical (animal) testing	This testing shows whether a drug is ineffective or too dangerous to be tested on humans. Many drugs fail this test.
2. Drug company applies to health agency to test drug on humans	The health agency decides whether a drug is safe for human tests and evaluates the tests proposed. It determines whether the possible benefits of the drug appear to outweigh its risks.
3. Phase 1 studies	These tests usually involve 20 to 80 healthy people and determine the drug's side effects.
4. Phase 2 studies	If there are no serious side effects, up to 300 people with the target disease test the drug's effectiveness. Some people take the drug, but others take a placebo, which is something that looks like the drug but has no effect (sometimes called a "sugar pill"). Test results indicate whether the drug is effective.
5. Phase 3 studies	If the drug seems to be effective, it moves to large-scale testing, involving up to 3 000 people.
6. Drug company submits new drug application	After all this testing, the health agency either rejects the drug or assigns a team to review its safety and effectiveness.
7. Health agency reviews label information and inspects manufacturing plant	If everything is in order, the health agency approves the drug for sale.

What do you think about this process? Is it thorough enough, or should it be more stringent? Are there instances when this process is too laborious and when medications should be "fast-tracked" through the approval process? Explain. How does this process help consumers make wise decisions? Should unfavorable research be required to be released?

Structured Inquiry

Discover

Record your work for this inquiry. Your teacher may also assign the related Guided Inquiry.

Science Journal

Pain Relief

How do the tablets react in water and vinegar?

Materials

- safety goggles
- marker
- 9 plastic cups
- 600 mL water
- stopwatch
- regular aspirin tablets
- buffered aspirin tablets
- coated aspirin tablets
- 300 mL vinegar

Procedure

1. Add 100 mL of water to each of three labeled plastic cups.
2. **Record** the time. Simultaneously add a regular aspirin tablet to one cup, a buffered aspirin tablet to the second cup, and a coated aspirin tablet to the third cup.
3. Record any changes in the tablets at 30-s intervals for 10 min or until no further changes are evident.
4. Repeat Steps 1–3 using vinegar in each cup instead of water.

Analyze Results

Create a table. Add a column for each type of tablet and one column for observations. Add a row for each type of solution. How does each treatment affect each type of pill?

Create Explanations

1. How do the tablets react in water and vinegar?
2. Which solution—water or vinegar—best represents the human stomach?
3. Which type of aspirin dissolved the fastest? Why?

Hidden Messages Explain

Over the years, you have seen hundreds of ads selling prescription medicine and over-the-counter health products. These ads are inescapable in all forms of *media*: television, newspapers, magazines, websites, billboards, mailed flyers, and other sources. You might be able to repeat a popular ad word for word, but do you understand its hidden message? What is the ad really telling you? As you consider which products to buy, especially health products, you need to know how the ads can affect your decisions.

Companies use advertising techniques to persuade people to buy their products. If you begged your parents to buy cereal with a free toy in the box, you were responding to an advertising technique. If you bought an energy bar because an athlete said it was good for you, your purchase was influenced by an advertising technique. Should advertising that delivers hidden messages be allowed? Why or why not? Which of the common techniques below sound familiar to you?

This might look like an ad you have seen.

Which technique does this ad use?

Advertising Techniques

Technique	How the Technique Is Implemented	Hidden Message
Free rewards	The product includes a coupon, prize, or gift.	You can't pass up this good deal!
Testimonial	A satisfied customer or a famous person praises the product.	The product works well for this person, so it'll work for you.
Bandwagon	Groups of attractive people are shown using the product.	Everyone is using this product. Don't be left out!
Image	A rich, popular, and/or attractive person is shown using the product.	Buying this product will make you rich, popular, and/or attractive, too.
Good health	An attractive, healthy person stops jogging long enough to discuss the product.	Using this product will make you healthy, too.
Scientific evidence	Surveys and lab tests prove the product works, or a new discovery makes this product the best of its kind.	You can be confident that this product will work for you.
Humor	A funny ad gets people talking.	Buy this product because everyone is talking about it.

The Role of Medical Advertising Explain

How can a carefully tested drug sometimes do more harm than good? If used incorrectly, prescription and over-the-counter drugs can cause problems. For example, a drug that is effective in treating an illness when taken at the recommended dose may be very dangerous if someone takes an overdose. Many medications are also dangerous when combined with other drugs or alcohol. Never mix drugs or take a prescription medication that was not prescribed by your doctor. As you know, a prescribed drug may have many harmful effects, even if prescribed and used as directed by a physician. A drug may be fatal if someone combines it with other medication because drugs may have unpredictable interactions.

While adults may think carefully about the medications they take, many adolescents rarely trouble themselves with these matters. Medicines can have serious, long-term effects on one's body. So why do people sometimes insist that their doctors prescribe a certain medicine? Perhaps it is because they heard about the drug on television, and the ads tend to make the medication sound much more effective than it is. Is this a wise course of action to take? Why?

The average American watches at least 16 hours of drug commercials a year. That is more time than most people spend talking to their doctors. Drug companies spend billions of dollars annually on medication advertisements. Why do you suppose they spend so much on advertising? Many of these advertised drugs are prescription-only medications meant to be taken for chronic conditions such as insomnia, high cholesterol, and weakening bones. Drug companies must summarize the medicines' possible harmful side effects in their ads or refer viewers to a website or a toll-free number for more information.

Lesson Activity

Break up into groups. As a member of your group, monitor the number of hours you watch television each day for one week. During this time, write down every medication (prescription and over-the-counter) advertisement that you see. At the end of the week, get together with your group and analyze the results. If more than one member of your group watched the same program, remove duplicate advertisements.

How many drug commercials did the group view? For what ailments were the drugs intended?

Medications involve risks. For example, a recent magazine ad assured readers that a certain prescription medication would help them with lung disease. The ad showed a man playing with his grandchildren. Part of that page and the entire next page, however, identified important safety information. That information included many situations in which the drug should not be used. The ad also suggested "discussions to have with your health-care provider." It ended with a long list of possible side effects, in tiny print. Why would a person risk taking such medications when there are so many possible side effects?

Sometimes we forget that people on television are almost always paid actors.

Why do drug commercials often use actors who pretend to be ordinary people?

Lesson Activity

Watch several drug commercials on television, or review medical ads in a magazine or newspaper. On a chart, record any information you learned about the medical condition with which the ad is concerned, what benefits the manufacturer claims the drug has, and any warnings about possible side effects.

Do consumers benefit from medical advertisements? Explain your answer.

Drug companies claim their ads help people understand an illness, encourage them to visit their doctors, and give them a more active role in their treatment. Yet doctors say the ads encourage patients to request medication that they should not take. The resulting discussions lengthen office visits. In a survey of physicians, four out of five felt that drug ads encourage patients to request specific medication. What does this say about the effectiveness of these kinds of ads? Have you or a family member been swayed by an advertisement for a prescription medication? What factors caught your attention? How can you use the information in the ads in a positive way to make smart decisions? The doctors considered about 50% of the requests generated by the ads to be inappropriate. Still, they filled nearly 70% of those requests to satisfy their patients. Is this really the best course of action for the patient? What do you think doctors should do instead?

Disclaimers Explain

A **disclaimer** is a legal statement that attempts to absolve a company or other source from responsibility for any consequences resulting from using its product or advice. Some disclaimers say that the advice provided in an article or other source should not take the place of a professional diagnosis or treatment. You should consult a medical professional because the advice is for informational purposes only. Many websites with information about an illness or disease include disclaimers. What kinds of disclaimers have you noticed in ads you have seen?

Check for Understanding

How effective do you think the influence is of users' testimonials or actors in various types of advertising?

Lesson Activity

In small groups, use the Internet to find a commercial for a prescription medication. Identify the type of medication and its purpose. Record the total length of the commercial. Record the length of the part of the commercial when the side effects are listed. Calculate the percentage of time that promotes the product and the percentage of time that lists side effects. Describe the overall mood of the commercial. Write a short presentation detailing your findings. Regroup as a class and take turns sharing the information you learned.

What type of information is being communicated in the ad?

You might also find disclaimers on over-the-counter medication labels. The disclaimer might say, "Ask a doctor before use if you have liver disease." This means that if you have liver disease and you take the medication and suffer harmful side effects, you cannot hold the drug company responsible. Do you think this is an adequate warning? What else could be done?

Commercials and ads for prescription and over-the-counter drugs also often include disclaimers. On TV, though, you must listen carefully to understand them, as the announcer rattles them off at top speed. The drug company spends most of the commercial convincing viewers that the drug will help them. However, it spends as little time as possible warning about harmful side effects and other reasons not to use the drug. What can you do to make sure you are properly informed about such products?

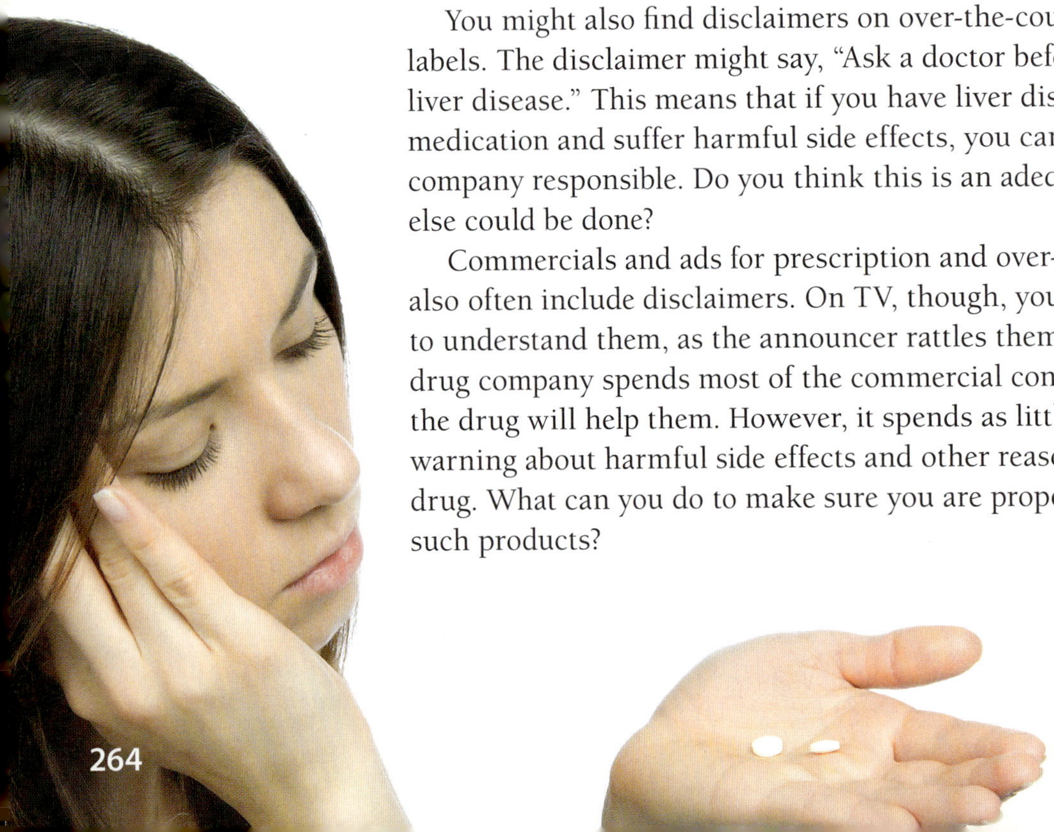

Before taking a medication, be sure you are aware of all of the side effects.

Math in Science

What do you think of the advertising done by these five major industries: soft drinks, fast foods, pharmaceuticals, motor vehicles, and financial services? As a class, find the number of students who believe each type of company to be the most trustworthy. Convert this number into a percentage. Compare your data with similar data from the United States below.

Most Trustworthy

	Total %	Age 18–34 %	35–44 %	45–54 %	55+ %
Soft-drink companies	34	39	40	32	27
Fast food companies	22	17	22	27	24
Pharmaceutical companies	18	20	14	15	19
Auto companies	14	12	13	14	16
Financial service companies	13	13	10	12	14

Note: Percentages may not add up to 100% due to rounding. Source: The Harris Poll Feb. 2010

Concept Check Assess/Reflect

Summary: How can you be an informed consumer? The FDA and Health Canada drug-approval process involves several phases of testing in an attempt to ensure the safety and effectiveness of new drugs. Being aware of common advertising techniques can make you a more informed consumer by alerting you to hidden messages in ads. Medical advertising often includes a considerable amount of information to help consumers become aware of the properties of certain drugs, but advertising also has the aim of influencing consumers to think that a drug will cure their illness. A medical disclaimer attempts to absolve a drug company or other source from responsibility for any consequences resulting from taking their drug.

1. Why do health agencies have complex and expensive drug-approval processes?
2. Do you think the number of medical ads will increase or decrease in the future? Why?
3. What should you remember when viewing or reading ads?
4. Which advertising technique do you think is hardest to resist? Why?

Science and Technology

Extend

Specialized Packaging

If you have purchased any electronics, video games, foods, or medicines recently, there is a good chance your purchase was packaged in specialized packaging. Many manufacturers use smart/intelligent packaging and active packaging to enhance the quality of the packaging and the product. How do you think packaging can make a product better?

Smart packing is a general term that describes several types of specialized packaging that help keep food fresh longer, monitors freshness, provides information on quality, improves safety, prevents damage, and makes using the product easier. There is a difference between smart/intelligent packaging and active packaging.

Smart/intelligent packaging senses and informs the customer about the condition of the product. This type of packaging may include time and temperature history (time, temperature indicators, or TTI), microbial growth indicators, light protection, physical shock, and leakage or spoiling indicators. It has the ability to detect or record changes in the external and internal environment. How do you think tracking these changes benefits the supplier or maker? Many food products and medications use intelligent packaging, including meats, produce, and dairy products.

Active packaging can extend the shelf life of the product, increasing the amount of time it stays a high-quality product. Active packaging can absorb moisture, is antimicrobial, controls oxygen and ethylene, maintains proper temperature, and absorbs and releases odors and flavors. It can reduce the need for additives and preservatives. What types of products do you think would benefit from active packaging?

Specialized packaging is transforming health care with packaging that improves readability, color-changing strips that monitor expiration, and packaging that records when medication is taken. Radio frequency indicators (RFID) used in health-care packaging can even alert the user when it is time to take the medication. These advances reduce the risks of errors and promote healthful habits, especially for senior citizens.

Concept Check

1. In what ways can smart packaging promote healthier lifestyles?
2. Explain the difference between smart/intelligent packaging and active packaging.

Marketing

"Melts in your mouth not in your hands." "Think smart." "Breakfast of champions." "Got milk?" You have probably heard these advertising slogans countless times. Can you think of a few more?

You are exposed to millions of advertisements every day on social media sites, television, blogs, email newsletters, webinars, magazines, and even in person. How many advertisements did you see today? How many of those products do you want to try?

Advertising is one part of the role of a person who works in marketing. Marketing, like science, is really a process. It involves many tasks that start with the initial idea of a product, followed by steps involved in selling and post-sale activities. Selling and post-sale activities can be complex. Companies spend millions of dollars on assessing needs, marketing research, product development, distribution, pricing, selling, advertising, and promotion for just one product. How many people do you think it takes to complete all of these tasks?

Marketing is focused on the customer's needs or creating a need. By identifying a customer need, creating a product to meet the need, and getting the customer's attention, the customer's needs can be satisfied. How do the companies determine what a customer's needs are? Would they go to this trouble for just one person?

The cell phone is a good example of creating products to meet customers' changing needs. The ultimate goal of marketing is matching a company's products and services to the people who need and want them.

Many colleges and universities offer two-year associate's degrees, four-year bachelor's degrees, and master's programs in marketing. Marketing students may take courses in finance, business, communications, multimedia marketing, ethics, public relations, business law, and advertising.

People with marketing degrees often work in advertising as account managers, art directors, product developers, market researchers, and media buyers, as well as in public relations, sales, and promotions. Corporations depend on people with excellent marketing skills to connect their products to the customer.

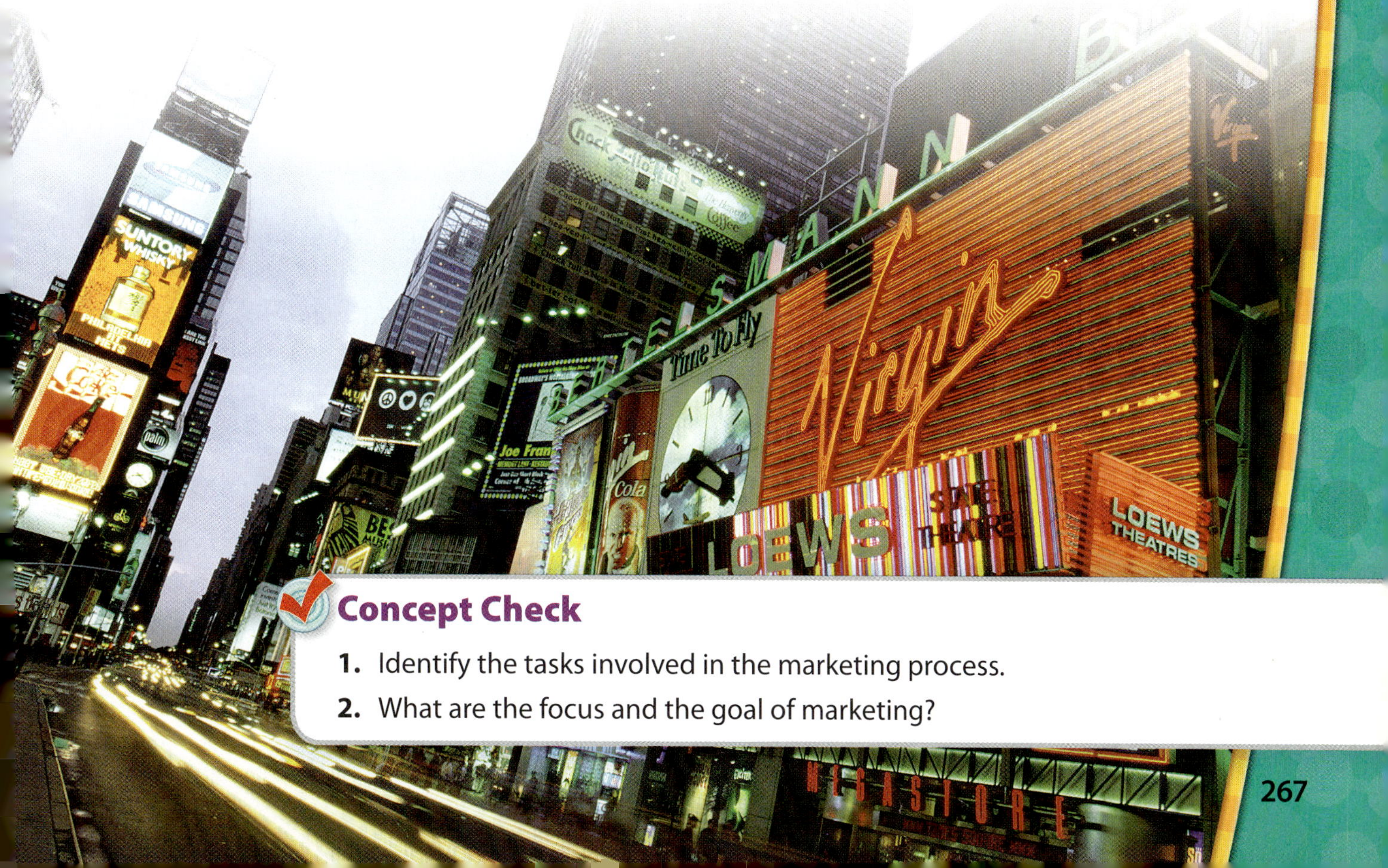

Concept Check

1. Identify the tasks involved in the marketing process.
2. What are the focus and the goal of marketing?

Chapter 7

Review Assess/Reflect

Study Guide

Lesson 1

1. Steps in making wise decisions include: describe the situation, list your choices, identify consequences, consider values, select the option that supports your values, carry out the choice, evaluate your decision.
2. Peer pressure is the influence that people of a similar age have over one another. It can be positive or negative. Negative peer pressure can influence you to do things that you know are a mistake. Positive peer pressure can give you support to make wise choices.
3. Refusal skills, a support network, and a role model can help you stand firm when confronted with peer pressure.
4. A positive role model is a person whose decisions and actions always seem to be positive and healthful. A negative role model is a person who makes poor choices and whose actions can harm health and well-being.
5. Teens may need help with decisions involving skipping class, smoking, or doing activities that are harmful or illegal.

Lesson 2

1. The FDA and Health Canada work to ensure that food and drugs are safe, but these agencies cannot test everything in commerce.
2. The best source of reliable health information is a medical professional. The list on page 252 provides other strategies for evaluation.
3. Good sources of reliable information include doctors or other health professionals. On the Internet, use sources that are associated with government agencies or universities.

Lesson 3

1. New drugs that are proposed for use must go through extensive testing before the FDA and/or Health Canada will approve them. This process is designed to ensure the safety and effectiveness of each drug. This process is summarized on page 259.
2. Advertising techniques, such as testimonials, bandwagon, image, free rewards, good health, scientific evidence, and humor, are ways that companies persuade people to buy their products.
3. A disclaimer is a legal statement that attempts to absolve a company or other source from responsibility for any consequences resulting from taking its product or advice.

Vocabulary Check

Fill in the blanks with the correct terms.

1. The influence that people of a similar age have over one another is ______________.
2. A ______________ is a person looked to by others as an example to be imitated.
3. The ______________ is a government agency in the United States that has the task of ensuring that medicines are safe and work as intended.
4. The government agency that uses scientific research to promote the health of Canadians is ______________.
5. Television, newspapers, and magazines are all sources of ______________.
6. A ______________ is a legal statement that attempts to absolve a company from responsibility for any consequences resulting from taking its product.

Multiple Choice

Choose the best answer.

7. Which action is negative peer pressure?
   A. nagging a friend to go play basketball at a park
   B. nagging your parents to let you go to a concert
   C. nagging a friend to let you borrow his homework
   D. nagging your little sister to feed the dog for you

8. You are making an important decision and have identified your possible choices. What should you do next?
   A. Think about the probable consequences of each choice.
   B. Consider your values.
   C. Select the best choice.
   D. Carry out the choice you like best.

9. If someone offers you a cigarette, which response should not be part of a refusal?
   A. No thanks. I don't smoke.
   B. No, not right now.
   C. No, it's against the coach's rules.
   D. No, I don't want to get suspended.

10. Which website is the most likely source of valid information about heart disease?
    A. www.uptodate.com/information-aspirin-and-cardiovascular-disease
    B. www.smithdrugs.com
    C. www.wikipedia.org/wiki/heart_disease
    D. www.cdc.gov/heartdisease

11. What is a disadvantage of medical advertisements?
    A. The ads encourage people to visit their doctors.
    B. The ads encourage people to ask for medication they might not need.
    C. The ads give people information they otherwise might not have.
    D. The ads encourage people to think about the medicine they take.

12. In a commercial for medication to treat depression, several smiling people report that they were able to resume normal lives after taking this drug. Which technique does this ad use?
    A. testimonials
    B. bandwagon
    C. image
    D. scientific evidence

Unit 3

Earth and Space Science

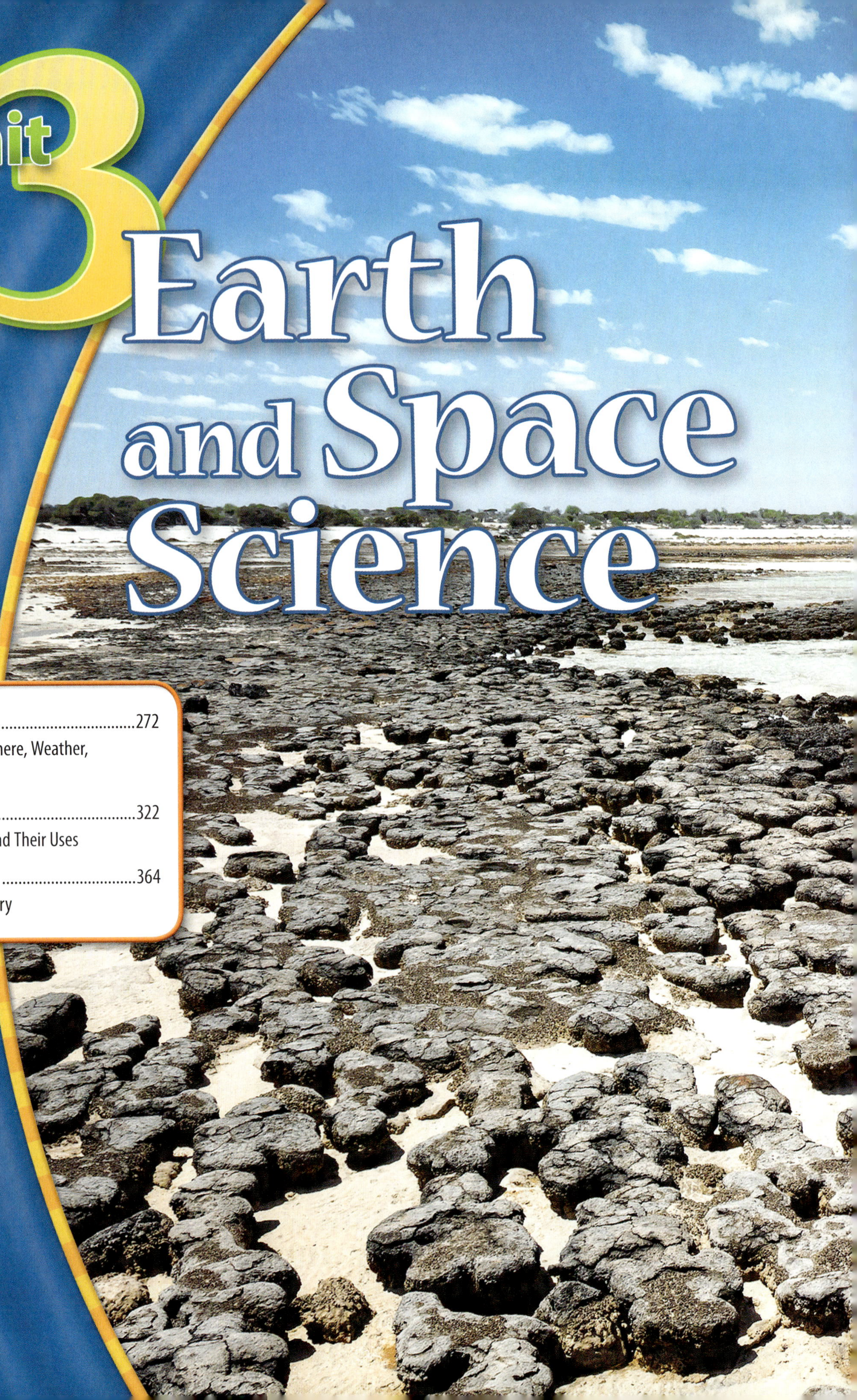

Unit Overview

In the beginning, God gave us air, land, and water with all the beauty and resources they provide. How do you enjoy these resources? How do people affect these resources? In this unit, you will learn about how Earth's air, land, and water act together to create Earth's weather and climate. You will find out about Earth's resources and its history.

Chapter 8 focuses on composition and structure of Earth's atmosphere. You will also find out how clouds form and how air pressure and wind are linked. How does water in the atmosphere cycle to provide all of Earth's living things with the water they need? What factors affect the formation and movement of winds?

In Chapter 9, you will find out how people have affected Earth's resources. How does pollution enter Earth's water and air? What affect does pollution have on living organisms? What have people done to restore and improved the environment?

The focus of Chapter 10 is Earth's history. You will learn about how Earth changes over time and what evidence scientist use as clues to Earth's past. Natural occurrences such as earthquakes and volcanoes have changed the way Earth looks and, in some cases, its climate. What do you think the fossils scientists find tell us about Earth's past? How old do you think Earth is? Why is this such a difficult question for scientists to answer? How do you think Earth as changed since its beginning?

These natural structures, called stromatolites, are formed of layers of blue-green algae that trap sediments in their sticky surface coatings.

Your teacher may assign an Open Inquiry lab and a Lifestyle Challenge activity. Use your *Science Journal* to record your work.

Chapter 8

The Atmosphere, Weather, and Climate

Scripture Spotlight

God created an atmosphere, weather phenomena, and climate conditions that support living things. Many passages in the Bible describe weather and climate. You will study the following passages in this chapter.

Jeremiah 10:12–13 (p. 276)
Ecclesiastes 1:7 (p. 282)
John 3:8 (p. 290)
Mark 4:35–41 (p. 304)
Genesis 41:1–37 (p. 311)

The Big Idea

The uneven heating of the Earth's atmosphere and its surface by the Sun causes air and water currents and the movement of energy across the globe. God designed these processes when He created Earth to make the planet a suitable home for living things.

How do you think interactions between air and water affect weather?

A hurricane is a powerful storm that forms from interactions between air and water.

Inquiry Kick-Off Engage

Winds are caused by uneven heating of the Earth's atmosphere and surface. Why is the heating uneven? Do you think the temperatures of these surfaces contribute to your local winds? In your ***Science Journal***, you will check the temperatures of some of Earth's surfaces.

Science Journal

Lesson 1

Objectives

- Explain how the atmosphere makes life possible on Earth.
- Describe the composition of Earth's atmosphere.
- Explain how air pressure and temperature change with altitude.
- Compare the characteristics of Earth's atmospheric layers.

Vocabulary

cyanobacteria

particulate matter

altitude

Essential Question

What Makes Up Earth's Atmosphere?

Take a deep breath. What did you just do? You just inhaled a tiny part of a very important Earth system—the atmosphere. God designed the atmosphere to keep temperatures on Earth just right for life. How did He design it to protect life? What would Earth be like without the atmosphere? How do you think Earth's atmosphere has changed over time? What processes affect Earth's atmosphere?

Life and Earth's Atmosphere Explain

In **Genesis 1**, the Bible says that God created the atmosphere on the second day of Creation week. You know that the *atmosphere* is a mixture of gases that surrounds Earth. The atmosphere God created allows the right amount of sunlight, heat, and other energy to affect Earth while protecting Earth from harmful energy from the Sun that could injure or damage life. The atmosphere provides the perfect balance of nitrogen, carbon dioxide, and oxygen to support the abundant and diverse life that was created during Creation week. This perfect design of Earth's atmosphere provides strong evidence of God's Design.

Scientists who do not believe in the Creation account in the Bible have a different explanation of how Earth's atmosphere formed. They propose that the Universe, and subsequently Earth, formed as a result of the big bang. According to this theory, early

in Earth's formation, volcanoes covered its surface. The frequent eruptions of these volcanoes emitted gases such as water vapor, carbon dioxide, and ammonia. These scientists theorize that chemical reactions that were powered by energy from the Sun broke down the ammonia into nitrogen gas. According to this theory, oxygen did not become part of the atmosphere until early organisms, called **cyanobacteria**, developed and began to photosynthesize and produce oxygen.

One of the questions asked regarding this theory is whether the right mix of nitrogen, oxygen, carbon dioxide, and trace gases could develop in the proper balance if enough time passed. Scientists who believe in the Bible record do not believe that long periods would result in the development of the complex atmosphere that supports life on planet Earth without a Designer.

Volcanoes have often affected Earth's atmosphere.

The Composition of Earth's Atmosphere Explain

Earth's atmosphere is made mostly of a variety of gases, including water vapor. The amount of water vapor in the air varies with place and time. It can be lower than 1% and as high as 4%. Why doesn't water vapor reach higher than 4% in the atmosphere? In which regions of the world would you expect the highest and lowest percentages of water vapor? Why?

Lesson Activity

Using the Internet or other resources, research what gases are present in the air you breathe. Create a colorful pie chart to display your findings. Were you surprised by the answers you found? Why or why not? How might pollution affect the composition of air?

What gases do you breathe?

Scripture Spotlight

Read **Jeremiah 10:12–13**. How does it relate to Earth's atmosphere?

In addition to gases, the atmosphere contains a small, varying amount of solids, or **particulate matter**, including dust, ash, pollen, and salts. Where do these particulates come from? How long do these particles remain in the atmosphere? Why have scientists found soot from the Industrial Revolution in samples of ice from Antarctica? Why was the soot found in ice samples that were taken from deep beneath the surface?

Atmospheric Pressure and Temperature

The atmosphere extends upward from Earth's surface for hundreds of kilometers. Each molecule of air in the atmosphere has properties of mass and weight, just like all other forms of matter. Recall that *air pressure* is the force exerted by the weight of molecules of air on a particular place or object.

As you sit at your desk, air pressure is exerted upon you equally in all directions—up, down, and sideways. However, air pressure is different at higher and lower altitudes. **Altitude** is the height above sea level. Remember that gravity causes air molecules in the atmosphere to be pulled downward. This causes compression of air molecules near Earth's surface. As a result, air pressure is higher at sea level than it is at higher altitudes. For example, at sea level, the air above pushes down on your skin with a force equal to about 101.3 kilonewtons per square meter (kN/m^2) but with only about 80 kN/m^2 at 1 km above sea level. Why does air pressure decrease with altitude?

Atmospheric pressure is greatest where air molecules are the most compressed.

What would happen to a balloon as it rises through the atmosphere?

You can feel changes in air pressure when your ears "pop" in an elevator due to the decrease in air pressure as the elevator rises. Your ears may also pop in an airplane as it takes off and lands because the cabins are pressurized to account for changes in air pressure with altitude. Recall that *density* is the measure of mass per unit volume. The atmosphere near Earth's surface has more molecules of air at higher altitudes. Therefore, it is denser at sea level than at higher altitudes.

A plane's ability to take off and land is also dependent on the external air pressure and density. Planes need longer runways at higher altitudes than at sea level. Why do you think this is so?

Now imagine driving up into the mountains. In the layer of atmosphere closest to Earth's surface, the temperature generally decreases with altitude. Why does the temperature decrease? However, the opposite can happen in upper layers of the atmosphere. What would explain an increase in temperature with increasing altitude? Why? What protects us from reaching the temperatures experienced in the uppermost layers of our atmosphere?

Explore-a-Lab

Structured Inquiry

What causes a burning candle to stop burning?

Put two candles on a flat surface. Carefully light the candles. Put a glass jar over one candle. Leave the other candle uncovered. Observe the candles for several minutes. What happened to the candles? Why do you think this happens?

Structured Inquiry
Discover

Record your work for this inquiry. Your teacher may also assign the related Guided Inquiry.

Journal

Can Crusher

What happens when air pressure changes?

Procedure

1. Fill the large bowl with cold water and ice. Fill the can with water, and then pour the water into a graduated cylinder to find the volume of the can. **Record** your measurements.
2. Pour 10 mL of water into the can. Use the full volume of the can to find the volume of air remaining in the can.
3. **Measure** and record the temperature of the air inside the can.
4. Read through the rest of the procedure and predict what might happen before you conduct the experiment. Record your prediction in your ***Science Journal***.
5. Place the can on the stove or hot plate while it is off. Then carefully turn on the stove or hot plate.
6. Bring the water to a boil. Use the thermometer to measure the temperature inside the can when it boils and record it.
7. Before the water boils away, use the tongs to get a good grip on the can near its base and so that your palm is facing up. In one swift motion, turn the can upside down as you push the top of the can into the bowl of ice water.
8. Carefully **observe** what happens throughout the experiment. Record your observations in your ***Science Journal***.
9. When the reaction stops, remove the can from the water and quickly measure and record the temperature of the air inside the can and the volume.

Materials
- bowl, large
- water, cold
- clean, empty soda can
- ice
- graduated cylinder (100 mL)
- stove or hot plate
- cooking tongs
- thermometer

Analyze Results

Compare the temperature and volume of the can before and after the experiment. What caused the changes you observed?

Create Explanations

1. What happens when air pressure changes?
2. How did warming the can affect the air pressure in the can?
3. Explain what happened to the soda can when it was plunged into the ice water. Why did this happen?

The Structure of Earth's Atmosphere

Recall that air pressure decreases with altitude. However, temperature can increase or decrease with altitude because it is determined by how well the gases in that layer absorb energy from the Sun and from Earth. Scientists use the boundaries at which the temperature changes to divide Earth's atmosphere into five layers.

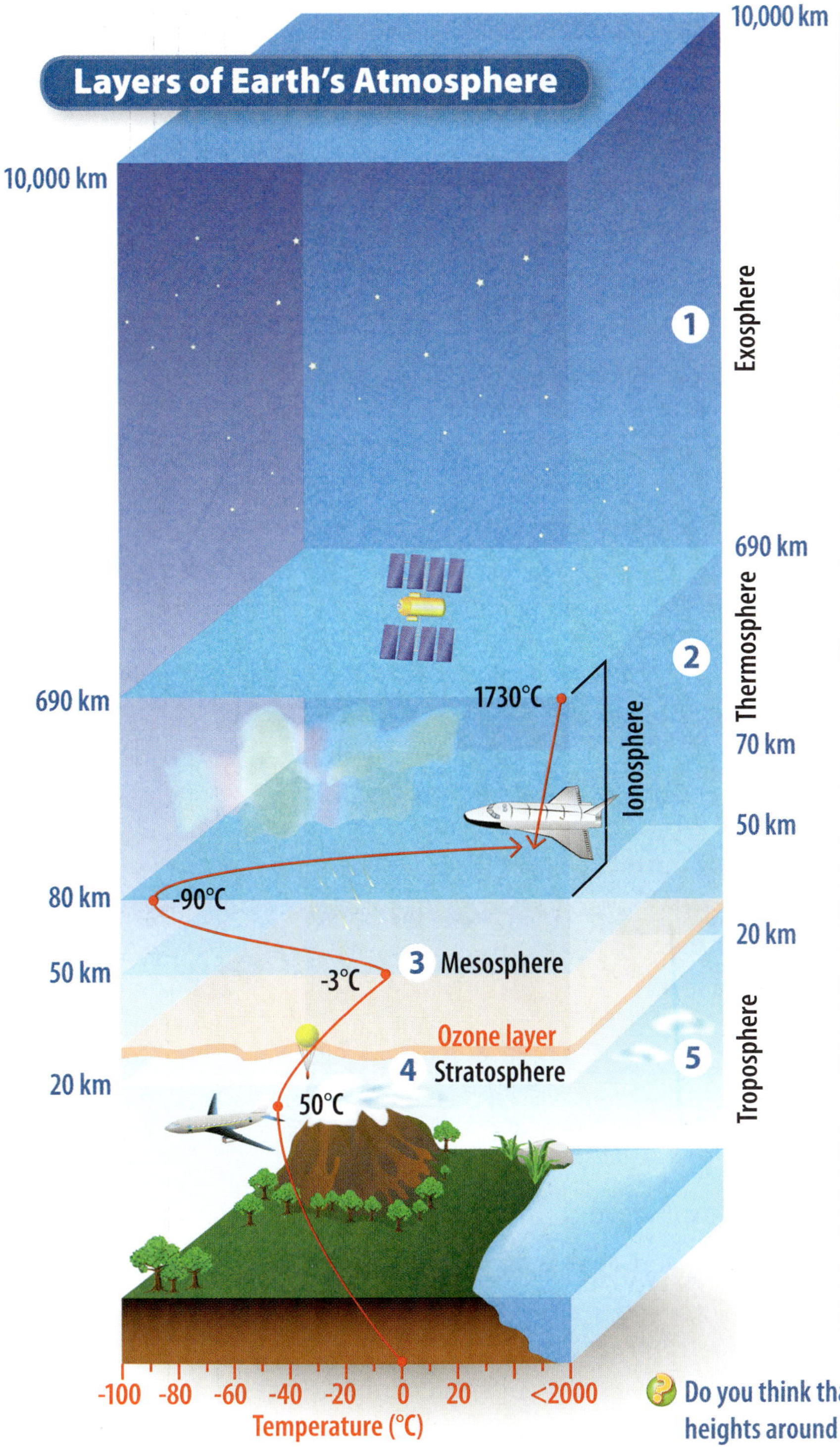

1 Exosphere
- Layer gradually fades into outer space
- Very few gas molecules in this layer can absorb high levels of thermal energy.

Why would you feel cold in this layer?

2 Thermosphere
- Temperature increases with altitude.
- Few air molecules receive intense solar radiation.
- Contains separate layers of different gases with heavier gases at lower levels

What is the relationship between low density and high temperatures in this layer?

3 Mesosphere
- Temperature decreases with altitude.
- Contains little ozone

What might explain the temperature trend in this layer?

4 Stratosphere
- Temperature increases with altitude.
- Contains the *ozone layer*, which absorbs UV rays from the Sun and releases heat

Why can supersonic jets fly faster in this layer than in the troposphere?

5 Troposphere
- Temperature decreases with altitude.
- Contains most of the gases of the atmosphere

What factor of this layer would explain why most of Earth's weather occurs here?

Do you think that the atmospheric layers remain at uniform heights around the world? Explain your reasoning.

Explore-a-Lab

Structured Inquiry

How can you demonstrate changing air pressure?

Blow up two balloons. Attach a piece of sting to each balloon. Hold one string in each hand so the balloons are at nose-level, 15 cm apart. Blow hard between the balloons and watch them move. The air pressure is lowered as you blow between the balloons (think of the air molecules as ping-pong balls, the balls don't have enough time to touch the balloon surface as they zoom by). The air surrounding the balloons that is not really moving is now at a higher pressure, and pushes the balloons together. Draw a labeled diagram of what happens when you blow between the balloons and indicate where the air pressure is lowest and highest.

Next, hold a regular sheet of paper horizontally below your bottom lip (you may have to experiment a bit to find the exact location) and blow hard across the sheet. What happens to the paper? What happened as you blew across the top of the sheet?

Concept Check

Assess/Reflect

Summary: What makes up Earth's atmosphere? Life is made possible by the presence of an atmosphere composed of nitrogen, oxygen, carbon dioxide, argon, and other gases on Earth. The layers of Earth's atmosphere have varying temperatures and air pressures due to the density of the gases in each layer and how well the gases in each layer can absorb energy from the Sun or from Earth. The five layers of Earth's atmosphere are the troposphere, stratosphere, mesosphere, thermosphere, and exosphere.

1. What gases make up Earth's atmosphere?
2. Compare and contrast the troposphere and the stratosphere.
3. Give examples of how the atmosphere affects life with how life affects the atmosphere.
4. The International Space Station orbits Earth at 370 km above the surface. Which of the atmospheric layers is this in? What kinds of space-suit features are necessary to keep astronauts safe when working outside the space station?

Lesson 2

Essential Question

Where Is Earth's Water?

Water is essential to life. You know that most of Earth's surface is covered by water. The vast majority of water on the planet—about 97%—is made up of the saltwater in oceans. Since very little of the water on Earth is fresh, knowing where to find freshwater is crucial for our survival. Water has shaped the development and history of humans just as it has shaped Earth itself. What are some ways in which we are affected by water today? How would our lives be different if we had all the freshwater we needed?

Objectives

- Describe the distribution of freshwater on Earth.
- Explain the processes of the water cycle.
- Classify clouds according to shape, height.
- Compare different types of precipitation.

Vocabulary

condensation nuclei

Freshwater on Earth Explain

As you can see in the diagram, only 3% of the water on Earth is freshwater, but most of this water is not available for human use. Why is most of Earth's freshwater not easily available for use by people? How does your use of water differ from the ways in which other organisms use water?

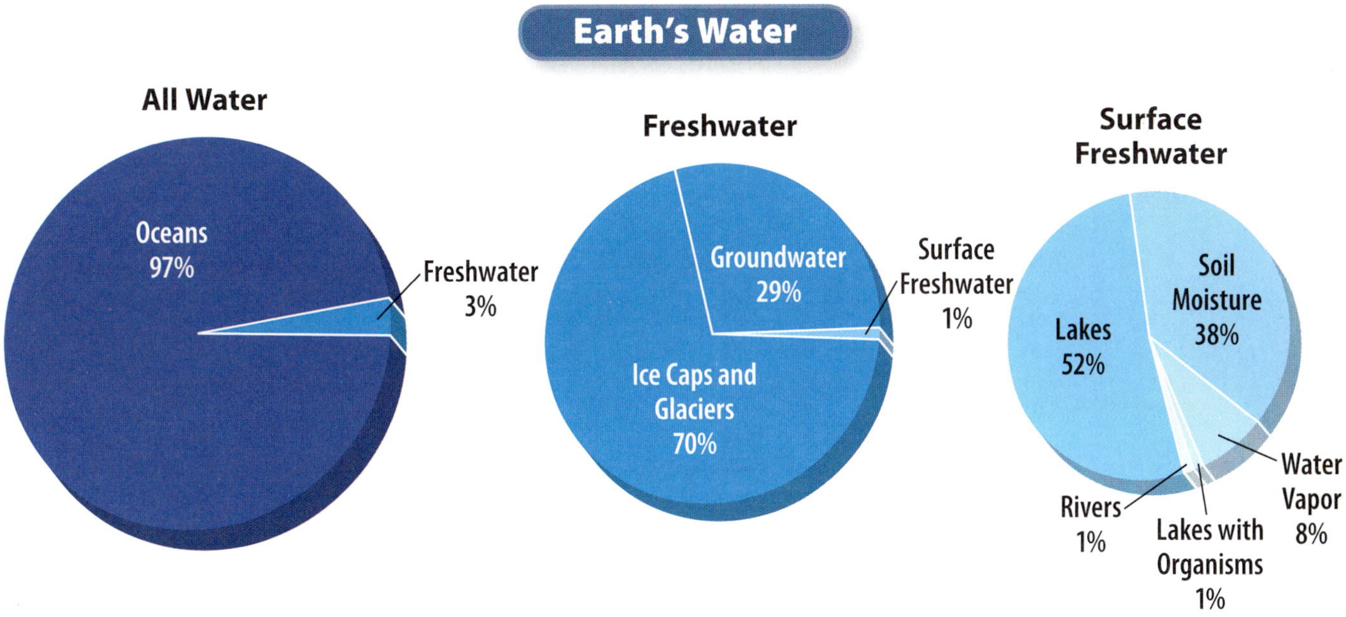

Only a small percentage of Earth's water is liquid freshwater.

Why is most of Earth's water considered hard to use by humans?

The Water Cycle Explain

God knew that plants and animals on land would need freshwater, so He designed the water cycle. The *water cycle*, as you know, is the continuous movement of water from Earth's surface to the atmosphere and back again.

The water cycle includes several processes that work together to recycle water on Earth. Some of these processes involve the transfer of water from one state of matter to another. Such processes also involve the movement of energy. As water changes state, energy is absorbed or released. During what processes of the water cycle might energy be absorbed? When might it be released?

Scripture Spotlight

Read **Ecclesiastes 1:7**. Explain how the "rivers flow into the sea, but the sea is not full."

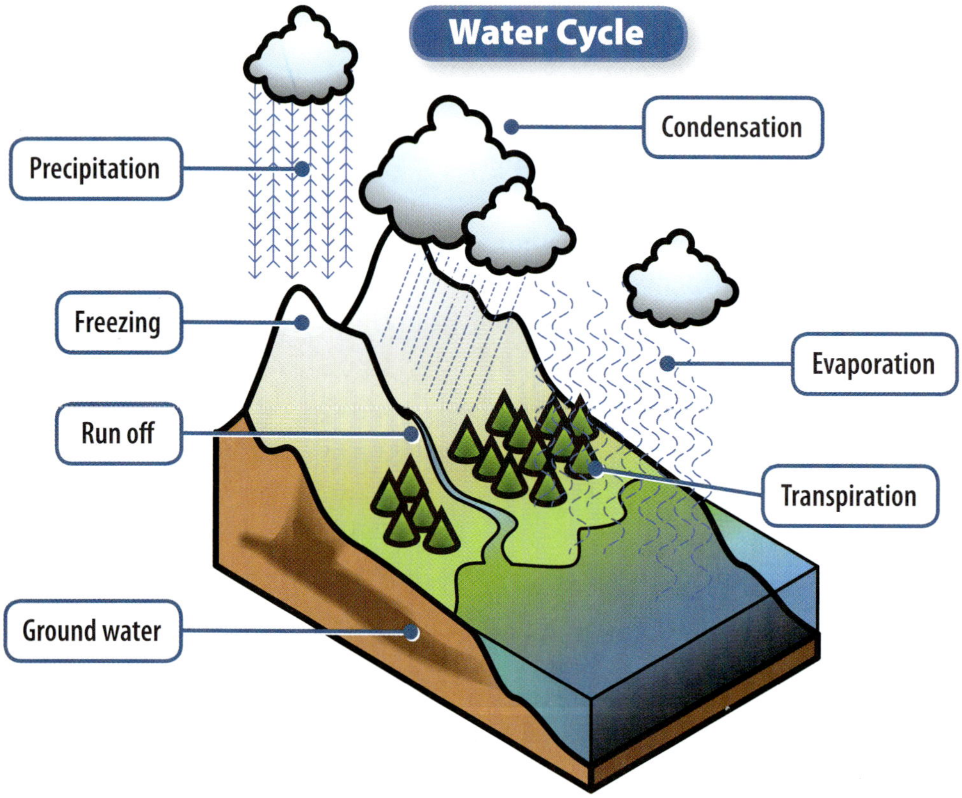

The water cycle shows the movement of water from Earth's surface to the atmosphere and back again.

Which processes directly add molecules of water to the atmosphere?

The water cycle is primarily driven by energy from the Sun. Solar energy causes water from oceans, lakes, rivers, soil, and other surface locations to *evaporate*. During the process of evaporation, the water molecules absorb huge amounts of solar energy. What gas is formed in this process? Water also enters the atmosphere through *transpiration* and sublimation. Recall that *sublimation* is the process in which solid water (ice or snow) evaporates directly into a gas without first becoming a liquid. Under what conditions do you think that can happen? In what other ways does water enter the atmosphere? Which source of atmospheric water do you think provides the most water? Why?

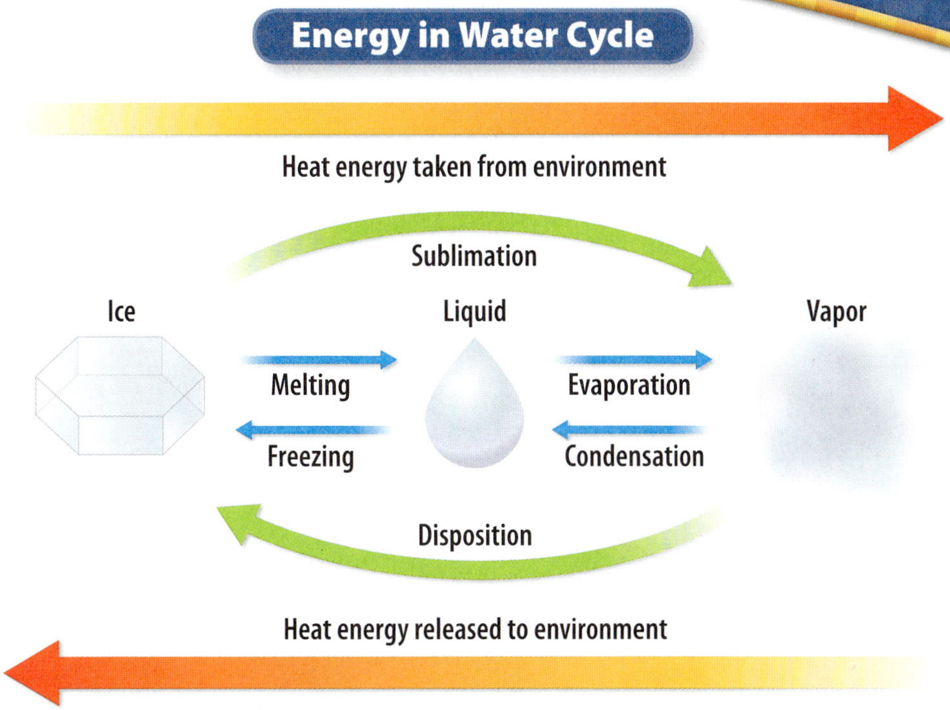

Energy changes take place all around you as water moves through the water cycle.

Cloud Formation

Recall that temperatures generally decrease with altitude in the troposphere. Clouds form when *convection currents* cause warm, moist air to rise. As the air rises, it encounters cooler temperatures. Heat energy moves from the warmer water vapor molecules to the surrounding cooler air. With less energy, the water vapor condenses as liquid droplets around particles of dust, pollen, and other materials. These tiny particles are called **condensation nuclei**. Millions of these droplets join to form a cloud. Air that encounters a mountain can also form a cloud. Why do you think that happens? In what other situations might clouds form?

When condensation happens and heat energy is released into the atmosphere, towering cumulonimbus clouds can result as the energy is released. What type of weather are cumulonimbus clouds associated with? At what altitudes are these clouds found?

Cloud Types

What types of clouds are in the sky today? How do they differ from other clouds you may have seen in the past? Recall that prefixes in the names of clouds indicate the altitude of clouds. Recall also that clouds are classified by their shape, height, and size.

Common Types of Clouds

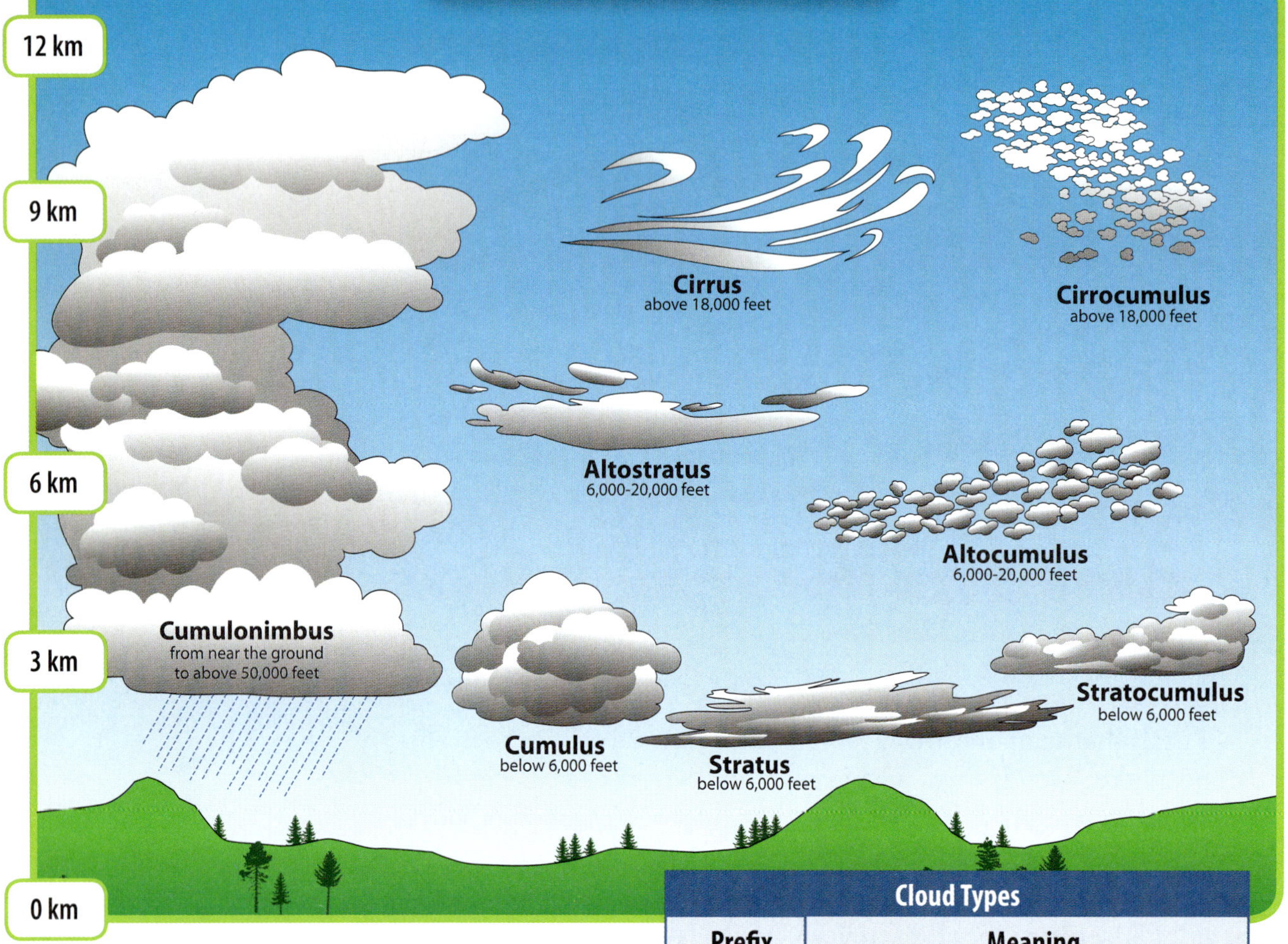

Cloud Types	
Prefix	**Meaning**
cirri-	High cloud; base above 6000 m
alto-	Middle cloud; base between 2000–6000 m
strato-	Low cloud; base below 2000 m
nimbo-	Produces rain (can also be suffix *-nimbus*)

It is possible to predict the type of weather to expect based on the types of clouds that are present.

What type of clouds would you want to see if you were going on a picnic?

Explore-a-Lab

Structured Inquiry

How can you form a cloud?

You can create a cloud in a bottle. Pour just enough warm water into a 1-L bottle to cover the bottom. Carefully light a match and drop it into the bottle. Allow the bottle to fill with smoke and then cap the bottle. Squeeze the bottle hard a couple of times. When you release the bottle, you should see the cloud form. Why did you need to let the bottle fill with smoke? Why did you have to squeeze the bottle to help the cloud form? Did the cloud disappear at any time? If so, can you explain why?

Math in Science

Go outside or look out a large window. If there aren't any clouds, you can use the image of clouds in this lesson for this activity. Estimate the amount of cloud cover in the sky. Write your estimate as a fraction and as a percentage. Identify the different types of clouds. Write each cloud type as a fraction and as a percentage.

Effects of Clouds on Weather

As you know, some clouds bring heavy precipitation, affecting weather in direct ways. Clouds also affect weather in indirect ways. Because winds in the atmosphere can move clouds, the clouds may form in one place and drop precipitation hundreds of kilometers away. In this way, clouds transport matter in the form of water from place to place. How do clouds transport energy from place to place?

Clouds also affect Earth's surface temperatures. Heavy cloud cover can reflect incoming solar radiation back into the atmosphere. As a result, surface temperatures might be cooler on cloudy days. At night, clouds can absorb the heat energy from Earth's surface and reradiate it back to Earth's surface, keeping temperatures warmer. What effect can cloud cover have on the temperature range of a location? How does this help explain the typical temperature ranges of a desert?

Precipitation Explain

As air cools and more condensation occurs, the tiny droplets clump together when they bump into each other and grow in size. When the droplets are large enough, they can fall as *precipitation*. What are some common types of precipitation? What determines the type of precipitation that will fall? Why can more than one type of precipitation fall during the course of a storm?

What can happen to the precipitation that falls from these clouds?

Types of Precipitation

When water droplets in clouds reach approximately 0.2 mm in diameter, they are too heavy to remain suspended in the cloud. They fall as precipitation. The main types of precipitation are rain, freezing rain, snow, sleet, and hail. The kind of precipitation that falls to Earth depends largely on air temperature.

Type of Precipitation	Characteristics
Rain	• Liquid precipitation that falls when air temperature is above freezing • Can fall as a heavy downpour, a steady rainfall, or a light drizzle
Freezing rain	• Caused when rain falls and hits the ground, which is at a temperature that is below freezing • Forms a layer of ice on the ground, trees, and power lines
Snow	• Solid precipitation that falls when air temperature is below freezing • Water vapor in cold air that changes into ice crystals instead of liquid water • The crystals fall as snow.
Sleet	• Solid, small pellets of ice • Forms when drops of rain freeze before hitting the ground
Hail	• Icy precipitation most commonly associated with thunderstorms • Ice crystals that cycle through cumulonimbus clouds, forming layered balls of ice • Falls to Earth's surface when it becomes heavy enough

The hailstone with the largest recorded diameter in the United States was 20 cm across.

Predict what you would see if you took a cross-section of a hailstone.

Structured Inquiry

Discover

Record your work for this inquiry. Your teacher may also assign the related Guided Inquiry.

Science Journal

Moving Water

How does size of the plant affect its transpiration rate?

Materials

- balance
- 3 plant stem cutting (different lengths)
- 3 cardboard (10 cm^2)
- sunlight
- 3 cups, clear plastic
- transparent tape
- graduated cylinder (100-mL)
- water (untreated)
- marker
- scissors
- petroleum jelly

Procedure

1. **Measure** out 3 lengths of plant stem. Make sure each stem has ample leaves. **Record** the length of each stem in your ***Science Journal.***
2. Use the marker to write the length of one of the plant stems on each cup.
3. Fill each cup with 250-mL of water.
4. Determine the mass of each cup of water and record.
5. Use the scissors to make a small hole (just big enough for the plant stem) in the center of the each piece of cardboard.
6. Pull one of the plant stems through the hole and seal around the hole with petroleum jelly.
7. Tape the cardboard to the cup so it does not move.
8. Repeat steps 6–7 with the two remaining plant stem cuttings.
9. Put the cups with stems in a sunny window (or under a lamp).
10. After 24 hours remove the plant stems and cardboard from each cup making sure to let all of the water drip off the stem.
11. Use the balance to determine the mass of each cup of water.
12. Reassemble the cups and stems and leave the cups for 48 hours and repeat Steps 10 and 11.

Analyze Results

Graph your data to show the relationship between stem size and transpiration.

Create Explanations

1. How does the size of the plant affect its transpiration rate?
2. Is the transpiration rate of plants higher in regions of high humidity than it is in regions of low humidity?
3. In this activity what could you have done to decrease the transpiration rates of the plants? What could you have done to increase the transpiration rates?

The water on Earth today was placed here when God created the world. The water cycle is evidence of God's care. He knew the living things He created on land would need freshwater. He designed the water cycle to provide freshwater to the land, filling streams, lakes, and rivers. If the water cycle did not exist, Earth would not receive the life-giving precipitation it needs.

Explore-a-Lab

Structured Inquiry

What relationship can you observe between cloud type and amount of precipitation?

Each day for one week, observe and sketch the clouds in the sky. Identify the cloud types. In addition, place a rain gauge at school. Over the course of the week, record the amount of precipitation that falls daily.

Concept Check

Assess/Reflect

Summary: Where is Earth's water? Only 3% of Earth's water is freshwater, but most of that water is not available for human use. This freshwater moves between living organisms, land, water, and air and changes state as it moves through the water cycle. Energy from the Sun drives the processes of the water cycle.The water cycle plays an important role in weather as it transports matter and energy around the planet. One of the ways it does this is through clouds that can cross large distances. Clouds are formed when water vapor condenses. The shapes of clouds and the altitudes at which they form can be used to predict the weather, including different kinds of precipitation, such as snow and hail.

1. Compare and contrast the shapes of different cloud types.
2. What happens to energy during evaporation and condensation?
3. Describe the formation of hail and what can be learned from studying the cross-section of a hailstone.
4. If freshwater is so limited on Earth's surface, why does it have such a significant effect on weather?

Lesson 3

Essential Question

How Are Air Pressure and Wind Linked?

Like gravity, air is invisible, but you can see and feel its effects. Look outside. What evidence do you see that wind is blowing? You know that the movement of air, called wind, forms because the Sun does not heat all parts of Earth equally. How unequal were the surface temperatures you collected in the Inquiry Kick-Off at the beginning of the chapter?

Wind can be destructive. Strong winds have been known to tear roofs off houses, move large vehicles from one place to another, and force a blade of grass into a wooden telephone pole. How can winds be beneficial? How have you benefited from the wind?

Objectives

- Explain how wind is caused.
- Describe how the Coriolis effect influences wind movement.
- Explain global wind systems.
- Compare and contrast different types of local winds.

Vocabulary

convection cell

jet stream

local winds

valley breeze

mountain breeze

Why Air Moves Explain

The density of warmer air, you will recall, is lower than the density of cooler air. What causes the molecules of warmed air to be less dense? As a result of the lower density, warm air tends to rise. What happens to air molecules when air is cooled? How is the density of the air affected when the air is cooled?

What do you think caused these trees to grow toward one direction?

How do changes in air density relate to winds? Think about a balloon. Is the air pressure in the balloon higher or lower than the air pressure surrounding it? What happens to the air inside a balloon when you puncture the balloon? Likewise, the air in Earth's atmosphere flows from areas of higher pressure to areas of lower pressure—if no other forces act on the air. The greater the differences in the air pressures, the stronger the wind that blows between them.

Consider how this happens. You know that the driving force of winds is energy from the Sun. Recall that the equator receives more intense solar radiation than the poles. Which part of Earth gets the least intense radiation? As a result, the air at the equator is warmed and rises. The warm, rising air creates areas of low pressure. Why does the warm air create less air pressure?

When the warmer air rises and moves away from the area, it will start to cool, become denser, and sink. The downward movement of air creates areas of high pressure. Then the air will move from areas of high pressure to the areas of low pressure. The entire rising and sinking, cyclical movement of air is called a **convection cell**. Today, scientists use a model with three convection cells in each hemisphere, as shown in the diagram. Air rises at the equator and sinks at latitude 30° north and latitude 30° south. This sinking partially drives the adjacent convection cell that rises at latitude 60° north and latitude 60° south. This convection cell partially drives the adjacent cell that ultimately sinks at latitude 90° north and latitude 90° south.

Scripture Spotlight

In what ways is wind like the Spirit of God? See **John 3:8**.

Earth's Convection Cells

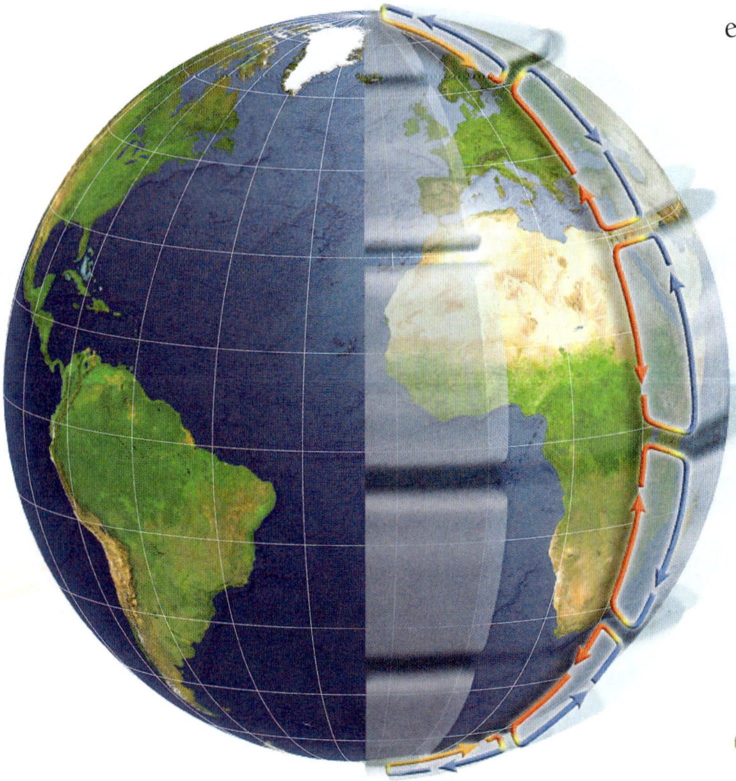

The tropical convection cell shown indicates that air rises at the equator and sinks at latitude 30° north and latitude 30° south.

Is rain more common in areas of low pressure or high pressure? Why?

The Coriolis Effect **Explain**

The diagram of convection cells is a simplified version of how wind flows in Earth's atmosphere. You know that Earth rotates on is axis. This rotation affects winds so that they do not flow directly north or south. Instead, the paths of the winds curve. Recall that Earth's rotation causes winds in the northern hemisphere to turn to the right, and winds in the southern hemisphere turn to the left. The tendency is called the *Coriolis effect*.

Explore-a-Lab

Structured Inquiry

How can you model the Coriolis effect?

Working in groups, gather a sheet of poster board and two different colored markers. Find an open space on a wall or whiteboard. One student should hold the poster board on the wall using a finger or the nonwriting end of the marker in the center of the paper. The paper needs to stay in place but be loose enough for another student to spin it. Another student should practice spinning the paper at a steady rate. A third student should practice drawing a straight line from the top to the bottom, using the nonwriting end of the marker. When all group members feel they have synchronized their movements, draw a line with the marker. After completing the line, mark the beginning and end points and indicate the direction the marker moved. Now turn the paper over and repeat the lab, but turn the paper in the opposite direction. Based on your observations, why doesn't wind blow in straight lines? What did turning the paper in the opposite direction model? What effect does this have on air travel?

The Coriolis effect is responsible for the spiral rotation of storm clouds on Earth.

Global Winds **Explain**

Convection cells, winds, and the Coriolis effect combine to produce air circulation patterns called global winds. Recall that *global winds* are large wind patterns that are present in particular latitudes in each hemisphere. Global winds distribute heat around Earth's surfaces. How do these winds influence ocean currents? How do they influence weather? The chart and diagram below show the major global wind patterns.

Global Wind Patterns	Characteristics
Doldrums	• A zone of little to no wind near the equator • Caused by the upward movement of air as it is heated at the equator
Trade winds	• Produced by the pressure difference between the low-pressure system at the equator and the high-pressure system at 30° latitude • Blow toward the equator from east to west
Prevailing westerlies	• Produced by the pressure difference between the high-pressure system at 30° latitude and the low-pressure system at 60° latitude • Blow toward 60° latitude and move from west to east • Responsible for moving weather systems across much of the United States and Canada
Polar easterlies	• Produced by the pressure difference between the low-pressure system at 60° latitude and the high-pressure system over the poles • Begin in the east and blow toward the west • Often not as strong as the prevailing westerlies

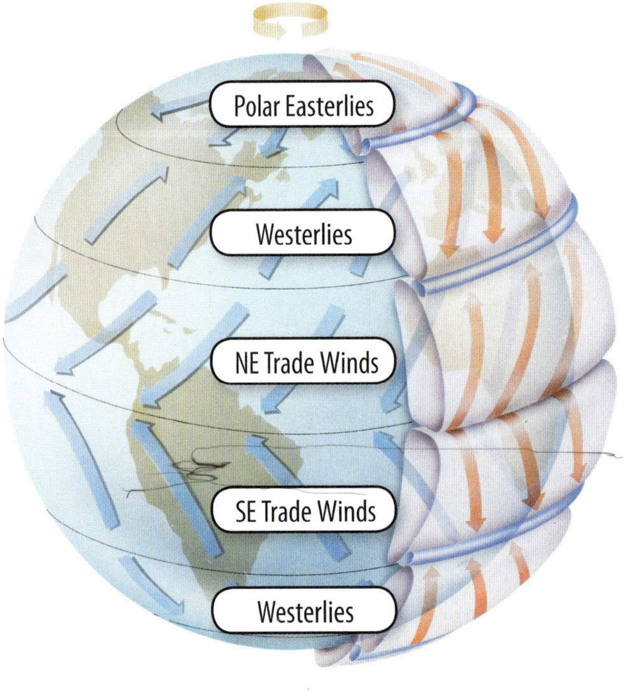

Earth has three distinct, global wind systems.

Which wind system has the greatest effect on the weather in your area? Why do hurricanes form near the equator, and what variables can affect their paths in the northern hemisphere?

Jet Streams

The global wind systems are found near Earth's surface. Other global winds flow high in the troposphere. A **jet stream** is a fast-flowing wind between 10 km and 15 km above Earth's surface. Like all global winds, jet streams move weather systems from place to place. They flow in an easterly direction. Why do eastbound airplanes try to fly within the jet stream? What advantage would there be from doing this? What about westbound airplanes? Do they fly in the jet stream also?

The factors that influence the flow of the jet stream are landmasses and the Coriolis effect. Landmasses interrupt the flow of the jet stream because of friction and temperature differences, while Earth's rotation accentuates these changes. So the jet stream meanders across Earth like a river meanders before it reaches the sea. The meandering sections of the jet stream continue to change as they interact with landmasses once again, creating ever-changing temperature differences.

In winter, the temperature of the stratosphere can also have an effect on the strength and position of the jet stream. The cooler the polar stratosphere, the stronger the polar/tropical differential becomes, encouraging the jet stream to gain in strength. The warmth of the landmasses and oceans (such as the El Niño Southern Oscillation) can also have a bearing on the strength and amplitude of the jet stream.

Jet Streams

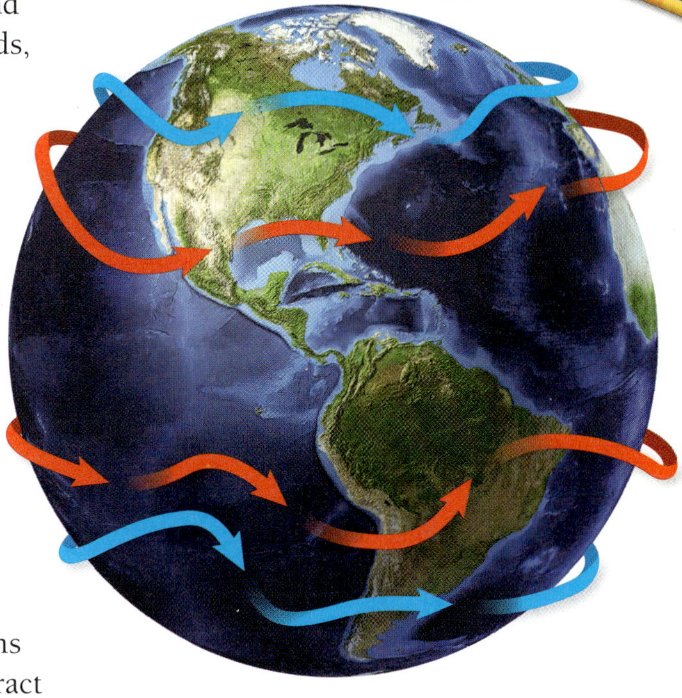

The polar jet stream is located between cold, polar air and warmer, midlatitude air. The subtropical jet stream is located farther south. The location of each jet stream can vary by season.

What other factors might affect whether a jet stream moves northward or southward from its usual location?

Local Winds

Recall that global winds are caused by pressure differences that result from the unequal heating of Earth's surface. Most **local winds** also result from differences in pressure caused by the uneven heating of Earth's surface. However, these pressure differences result from a different process from those that produce global winds. Local winds result from pressure differences that are caused by the differences in how quickly matter heats up. For example, rocks heat up more quickly than water. Areas of low pressure form over matter that heats up quickly. Local winds can blow from any direction, and they generally move short distances.

Sea Breezes and Land Breezes

You already know that one kind of local wind that forms near coastal areas is a *sea breeze*. Sea breezes blow from the sea over land. At what time of day do sea breezes form? How do they form? What generates the energy that drives this process?

You also know that a *land breeze* is a local wind that blows from land toward the sea. What process forms these breezes?

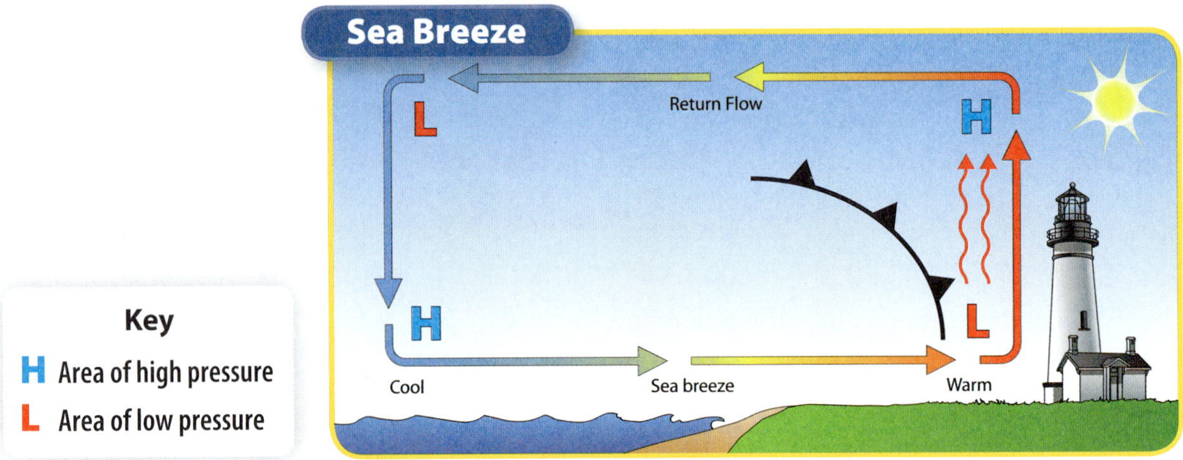

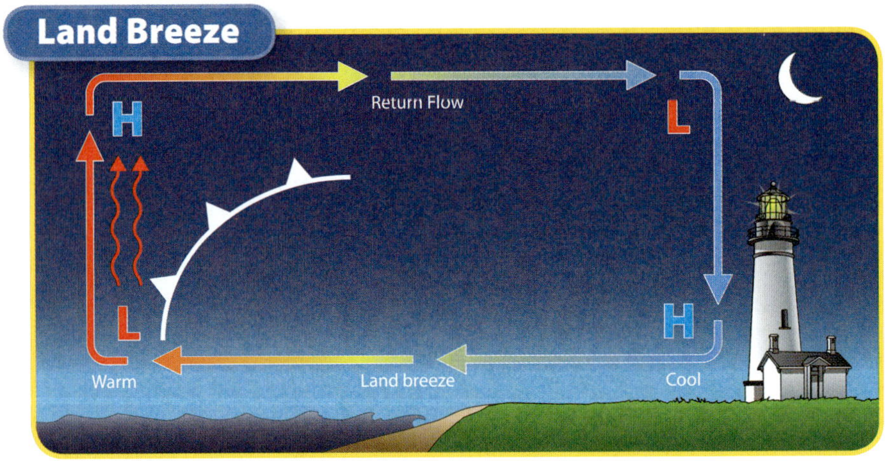

Explore-a-Lab

Structured Inquiry

How do the heating and cooling rates of land and water compare?

Measure and record the temperatures of a cup of water and a cup of soil. Place the cups under a lamp. Measure and record the temperatures of the materials every 20 min for 1 hr. Then turn off the lamp and measure and record the temperatures of the materials every 20 min for 1 hr. Graph your data and analyze the results. Which material heated up more quickly? Which material cooled down more quickly? How would wind form between the two cups during the day (when the lamp is on) and at night (when the lamp is off)?

Structured Inquiry

Discover

Record your work for this inquiry. Your teacher may also assign the related Guided Inquiry.

Science Journal

Wind and Evaporation

How does wind affect the rate of evaporation?

Procedure

1. Fill *all* bowls with the same amount of room-temperature water. **Record** the amount of water in each bowl and the starting temperature of the water in each bowl.
2. Label one of these bowls *Control*. Place it where the fan will not affect it but where it otherwise experiences the same conditions as the bowls that you will use with the fan.
3. Place a fan on a flat surface. Place one bowl 30 cm from the fan and label it *A*. Place the other bowl 1 m from the fan and label it *B*.
4. Turn on the fan so that it blows on the water in bowls *A* and *B* for 30 minutes.
5. Record the temperature of the water in each bowl every two minutes.
6. **Measure** and record the water left in each bowl after 30 minutes.

Materials

- 3 equally sized bowls
- 3 Celsius thermometers
- water
- fan
- graphing paper
- colored pencils

Analyze Results

Draw a line **graph** showing how the temperature of the water changed in each bowl. Use a different colored pencil for each bowl's data.

Create Explanations

1. How does wind affect the rate of evaporation?
2. Compare your results with those of other groups. What factors could account for any differences? What was the importance of the control?
3. Did the thermal energy of the water in each bowl change during this investigation? Explain.
4. What part of the water cycle is modeled in this investigation? What role does the wind play? Explain.

Valley Breezes and Mountain Breezes

Local breezes can also form in mountainous regions. A **valley breeze** forms during the day, when the Sun heats the valley floor. The warm air in the valley becomes less dense and gently rises up the mountainside. The warm, rising air cools and sinks, and the cycle repeats.

At night, a **mountain breeze** forms when the air on the mountaintop cools down more quickly than the air on the valley floor. The cooler, denser air moves down the mountainside into the valley. The air warms and rises, then cools and sinks again. Why might this be important to search-and-rescue helicopter pilots? What other activities would be affected by these kinds of breezes?

Check for Understanding

How does the uneven heating of the atmosphere by the Sun cause air movement, or wind?

Concept Check Assess/Reflect

Summary: How are air pressure and wind linked? Wind is the result of uneven heating and cooling of Earth's surface and its atmosphere. The rising of warm air and the sinking of cool air form three global convection cells in both the northern and the southern hemispheres. The convection cells are also affected by the turning motion of Earth. These factors affect the strength and the direction of flow of global wind systems on Earth. The effects of unevenly heating Earth's surface and the atmosphere above it can also create differences in local air pressures. Examples include the daily sea and land breezes as well as the valley and mountain breezes.

1. Describe the movement of air in a convection cell.
2. Compare a land breeze with a sea breeze.
3. Suppose a storm forms over Kansas. In which direction would the storm most likely move? Explain.
4. Imagine you are flying from the equator to the North Pole. How would you plan your route to account for the Coriolis effect? Draw your expected flight path.

Lesson 4

Essential Question

How Are Weather Patterns and Severe Storms Related?

If you hear the sound of an approaching train but don't live near train tracks, you should take cover in the lowest room of your house that doesn't have any windows. This sound often precedes a tornado. Meteorologists do their best to warn people about weather dangers, but they can't always predict when tornadoes might develop. Why do you think it is so hard to accurately predict severe weather? How do you use the information in weather forecasts? What professions rely on accurate weather forecasts? You want to plan a picnic for the weekend and need to check on the weather. Where could you find weather information that is the most accurate?

Objectives

- Identify factors that affect weather.
- Describe the characteristics of major air masses.
- Compare weather conditions associated with pressure systems and fronts.
- Explain the formation of different types of severe weather.

Vocabulary

storm surge

blizzard

visibility

Instruments used to describe the weather can be high tech or relatively simple.

Wind vanes indicate the wind direction. In which direction would a wind vane measuring the prevailing westerlies point?

Weather Factors Explain

Weather is the result of many different factors that interact and produce a variety of results. Most of these factors are measured using different instruments. Recall that weather stations include many of these instruments to provide the data used to describe and forecast the weather. Some of these are shown in the table below.

Weather Factor	Instrument
Air pressure	Barometer (N/m^2)
Wind speed	Anemometer (km/hr)
Wind direction	Wind vane (degT)
Air temperature	Thermometer (°C)
Humidity	Hygrometer (%)
Precipitation	Rain gauge (mm/hr)

Science Journal

In your ***Science Journal***, find out how human hair can help you measure humidity.
Extend

Humidity

You know that *humidity* describes the amount of water vapor in air at a given place and time. Why do people care about the humidity of the air? How can you tell whether the humidity is high or low?

Hygrometers are used to measure the humidity. The first hygrometer used human hair to measure the humidity. When the humidity was high, the hair shortened. When the humidity was low, the hair stretched out. *Humidity* is a general term. To be more specific when describing humidity, scientists often describe the relative humidity and the dew point.

Lesson Activity

Before people had modern technology to predict the weather, they relied on observations to spot weather patterns. They incorporated some of these patterns into sayings, such as "Red sky in morning, sailor take warning. Red sky at night, sailor's delight." Choose an old-time weather saying. Design an investigation to see whether the saying is accurate.

Are weather sayings as accurate as weather forecasts? Defend your answer.

Relative Humidity

A given volume of air can contain only so much water. At a certain point, it becomes saturated. Recall that *relative humidity* is a ratio of the amount of water vapor in air to the amount of water vapor needed to saturate the air at a given temperature. As the air temperature increases, more water vapor is needed to saturate the air. How might this affect athletes who train in areas that are usually very humid? Under what conditions might athletes want to be in a high-humidity environment for short periods? Relative humidity is written as a percentage. People are normally comfortable when the relative humidity is at 45%. What percentage indicates that the air is saturated?

Math in Science

To calculate relative humidity, you need the actual vapor density and the saturation vapor density. Vapor densities are measured in g/m^3. Use the equation shown to fill in the missing information in the table below.

$$\text{Relative humidity} = \frac{\text{Actual vapor density (g/m}^3\text{)}}{\text{Saturation vapor density (g/m}^3\text{)}} \times 100\%$$

A Mass of Air		
Relative Humidity (%)	**Actual Vapor Density (%)**	**Saturation Vapor Density (%)**
	8	20
56		25
72	24	

Dew Point

You know that *dew point* is the temperature at which a mass of air becomes saturated and condensation occurs. When air reaches its dew point, clouds may form. Dewdrops on grass are also a result of air reaching its dew point temperature. During what part of the day does dew often form on grass? Why?

As the chart shows, relative humidity decreases as temperature increases. Why does the relative humidity change if no water is being added or removed from the air? Air with a temperature of 37°C and a dew point of 24°C would feel hot. Yet the relative humidity of the air would be only 48%. At what temperature would condensation most likely occur?

Dew Point and Relative Humidity for a Mass of Air		
Air Temperature (°C)	**Dew Point (°C)**	**Relative Humidity (%)**
24	24	100
32	24	63
37	24	48

Directed Inquiry

Discover

Record your work for this inquiry. Your teacher may also assign the related Guided Inquiry.

Science Journal

How Humid Is It?

How do you use a psychrometer to measure the moisture in the air?

Materials

- 2 Celsius thermometers (with increments of at least 1°)
- clamp stand
- gauze
- small fan
- wood or plastic foam, solid piece, for mounting
- stopwatch
- tape
- water

Procedure

1. To make the psychrometer, use the tape to firmly secure the thermometers to the same side of the wood or plastic foam. The thermometer bulbs should protrude slightly off the end of the wood or plastic foam, and the scale on the thermometers should be easy to read.
2. Wrap the gauze around one of the bulbs. This will be your "wet bulb."
3. Securely mount the psychrometer in the clamp stand.
4. **Record** the room temperature at the start of the experiment.
5. Now wet the gauze and turn on the fan so that it is blowing directly on the thermometers from about 15 cm away.
6. Over a period of 10 min, use the stopwatch to measure 1-min intervals and check the temperature from each thermometer. Then record the data. By this time, the temperature on the wet bulb should be steady. If not, continue to take readings until the temperature reading on the wet bulb becomes steady.
7. To figure out the relative humidity, subtract the wet-bulb reading from the dry-bulb reading. Then use the table in your ***Science Journal*** to obtain the relative humidity in your room. Record the relative humidity.

Create Explanations

1. How do you use a psychrometer to measure the moisture in the air?
2. How would you expect the relative humidity to change if a storm were approaching?
3. Would you expect all of the rooms in a building to be at the same relative humidity? Explain your answer.
4. Under what conditions might the relative humidity vary in a natural environment?

Air Masses **Explain**

Scientists use temperature and moisture content to describe air masses. Recall that an *air mass* is a body of air with nearly the same temperature and moisture characteristics throughout its reach. Air masses get their properties from the source regions over which they form. What characteristics will an air mass have if it lies over warm, tropical water for long enough? If an air mass lies over cold, dry land long enough? What kind of an air mass is over your area right now? Did it form there, or did it move in from elsewhere?

The map below shows the main air masses that affect weather in North America. How would winds form between continental tropic and maritime polar air masses? Explain.

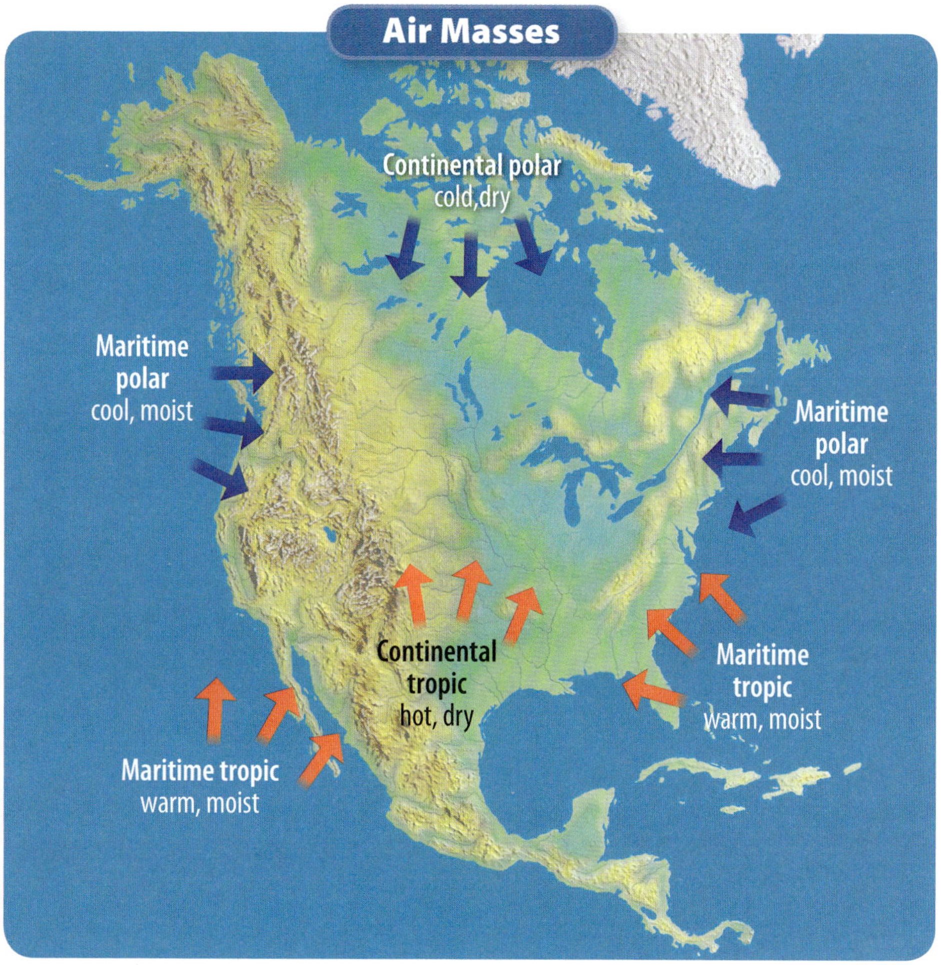

Air masses form over land and water. Their characteristics range from cold and dry to warm and wet.

How do the seasons affect the formation of these air masses?

Fronts Explain

Much of the weather you experience happens when winds move air masses, causing air masses of different characteristics to clash. A front is the boundary between two air masses with different characteristics. What happens when fronts collide? Some fronts are associated with violent weather. Others bring gentle rains.

Types of Fronts

Cold front	• Forms when a cold, dry air mass overtakes a warm, moist air mass • Clouds form. • Heavy precipitation and thunderstorms are common.	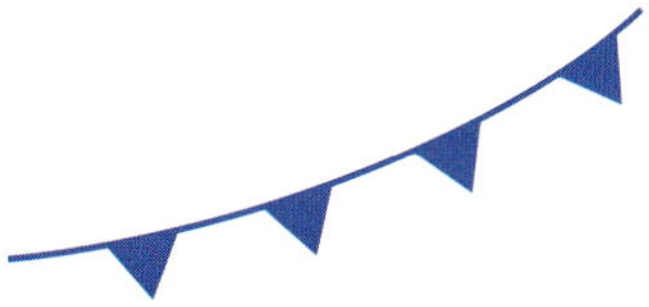
Warm front	• Forms when a warm, moist air mass overtakes a cold, dry air mass • Light, steady precipitation is common.	
Occluded front	• Forms when a fast-moving cold front overtakes a warm front • Can bring strong winds and heavy precipitation	
Stationary front	• Forms when two air masses meet and neither moves • Usually brings light precipitation	

Pressure Systems

Steady and changing air pressure can affect the weather in an area. High pressure is usually associated with fair weather, and low pressure is usually associated with stormy weather. Based on the weather, what type of air pressure is over your school? How would you know whether it was changing?

High-Pressure System

When air is dense, it sinks and creates an area of high pressure. The Coriolis effect causes air to move clockwise around a high-pressure center in the northern hemisphere. In the southern hemisphere, the opposite occurs—air moves counterclockwise around a high-pressure center. Sinking air is cooler and, usually, relatively dry. So high-pressure systems often bring fair weather to an area. They are represented by the capital letter *H* on weather maps. What factors do you think affect the speed at which these systems move? How do you think the speed at which pressure systems move can affect the weather?

Low-Pressure System

As you have learned, less dense air rises and creates an area of low pressure. In the northern hemisphere, air moves counterclockwise around a low-pressure center. Again, the opposite occurs in the southern hemisphere. As the air rises, the water vapor in the air condenses. This can lead to cloud formation and precipitation. For this reason, low-pressure systems often bring stormy weather. They are represented by the capital letter *L* on weather maps. Under what conditions would you expect a low-pressure storm to spin clockwise in the northern hemisphere?

High-Pressure System

Low-Pressure System

Air sinks and spirals outward in a high-pressure system. The movement is in a clockwise direction in the northern hemisphere. In low-pressure systems, air rises and spirals inward in a counterclockwise direction.

What makes air rotate in pressure systems?

Severe Weather Explain

All the factors you have read about so far—air masses, fronts, and pressure systems—combine to create the weather you experience each day. Sometimes, conditions are just right for a warm, sunny day. Other times, severe weather strikes.

Thunderstorms

A thunderstorm forms when cold air lies atop a layer of warm, moist air near the ground. Once the warm, moist air reaches its dew point, a cumulus cloud forms and begins to rise vertically through the cold overlying air. This motion creates updrafts. If strong enough, the updrafts can cause the cloud to continue to grow in height and become a cumulonimbus cloud.

Heavy precipitation falls from the tall, cool tops of the cumulonimbus cloud. As the precipitation falls, it cools the air, making it denser. The denser air sinks, causing downdrafts to occur next to the updrafts. Sometimes, the downdrafts can be so strong that they develop into a *downburst*. This is when rain-cooled air falls toward the ground and then spreads out in all directions as straight-line winds. These winds can reach more than 240 kph and cover more than 300 km in extreme cases. What pattern would you expect to see if a downburst occurred over a field of crops? As precipitation continues to fall, lightning and strong winds also occur at this stage of the storm. Hail may also fall. Why doesn't hail form with all thunderstorms? What other dangers might develop because of severe thunderstorms?

Scripture Spotlight

Read about Jesus's experience in severe weather in **Mark 4:35–41**.

Check for Understanding

What can you infer from the movement of air masses?

Thunderstorm

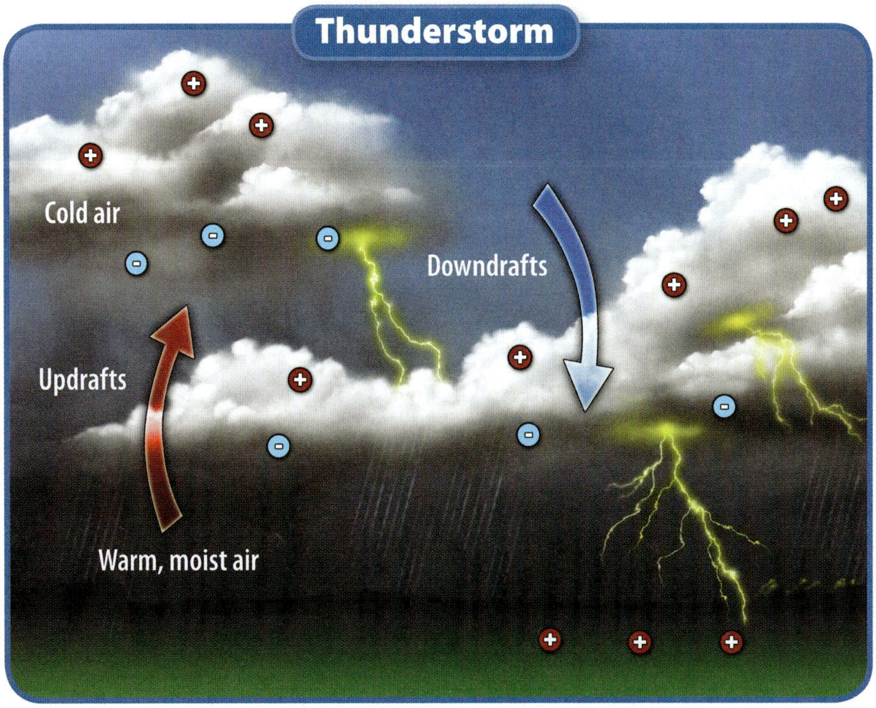

Thunderstorms usually go through three stages.

What determines how much rain falls during a thunderstorm?

Eventually, the updrafts stop because little warm, moist air is left. At this point, the storm begins to break up. Light rain may continue for an hour or more. Which time(s) of the year would you expect to have the most thunderstorms? Why?

Tornadoes

Do you have tornadoes where you live? A *tornado* is a powerful, rotating column of air that touches the ground. Often, it begins as a strong thunderstorm.

1. The air from the cold front changes speed and direction, moving more or less horizontally across an area.
2. The air nearest the surface begins to roll along the ground.
3. It encounters the upward-rising warm air that has formed the cumulonimbus clouds.
4. This causes the horizontally blowing air to tilt vertically while still rolling.
5. This turns into the rotating column of air that becomes the tornado.

Most tornadoes last only a few minutes and affect a relatively narrow area. However, some tornadoes can affect much wider areas and last much longer. On May 31, 2013, a tornado with a width of 4.2 km and a damage path of 26 km ravaged El Reno, Oklahoma. At the time, it was the widest tornado on record, with winds that reached 474.75 km/hr, easily making this tornado one of the most powerful known. Eight people were killed, but this storm could have been far more tragic. What saved so many people's lives that day? Since tornadoes are so dangerous, how do we study them? What kinds of instruments are used to measure conditions within the tornado itself?

Sometimes the rotating column of air produces a funnel-shaped cloud that just hangs below the clouds. It is not called a tornado until the funnel cloud touches the ground.

Hurricanes

A *hurricane* is a low-pressure storm that forms over tropical waters and has wind speeds of at least 118 km/hr. What symbols would be used to indicate a hurricane on a weather map? Hurricanes are large storms—they can be more than 500 km wide! A hurricane gets its energy from the warm ocean water. It begins when warm, moist air over the ocean begins to rise and create a rotating low-pressure system. If conditions are right, the size and the wind speed of the storm increase. When wind speeds reach at least 118 km/hr, the storm is classified as a hurricane. What will happen to a hurricane that moves over land? How might the size of the hurricane and the direction in which it moves change as it moves over land?

Storm surges can flood coastal habitats.

After the water from a surge recedes, what sorts of damage must people deal with?

Hurricanes can cause tremendous damage when they reach land. For example, a **storm surge** is primarily caused by the winds of a hurricane pushing a huge wall of ocean water onshore. The wall of water may be more than 6 m above sea level! This wall of water floods coastal areas, destroys habitats, and can wash away trees, buildings, and even entire sandy islands. How might such a wave affect agricultural land? What do you think would happen as the storm surge reached an estuary?

Hurricanes are the most powerful of the severe storms. A hurricane generally weakens after it reaches land because it loses its source of energy—the warm ocean water. Before it dies out, however, it can spawn tornadoes, thunderstorms, and flooding rains that can reach far inland and cause a lot of damage. For example, in 2005 Hurricane Katrina killed more than 1800 people and caused more than $80 billion in damage.

Lesson Activity

Use the Internet to research a historic hurricane. Write up several bullet points that detail the strength of the storm, the path of the storm, the area affected, the damage it caused, and the effectiveness of emergency responses. You can take turns presenting your information to the class. After the presentations, compare and contrast the storms as a class.

How have hurricane emergency responses changed over time?

Blizzards

A **blizzard** is a winter storm that is characterized by low temperatures, winds that gust or blow at 56 km/hr or more, and low visibility because of blowing snow that lasts for at least three hours. **Visibility** is the greatest distance at which objects can be seen. For example, a visibility of 100 m means that objects beyond 100 m cannot be seen. In blizzards, visibility is often zero or close to it, which makes any sort of travel difficult. What other hazards are associated with these storms? How are the jet streams related to the development of a blizzard?

Called to Serve

The Seventh-day Adventist church does relief work in many areas of the world where natural disasters occur. See what your class can do to help.

Explore-a-Lab

Guided Inquiry

How does wind speed affect storm surge?

SAFETY: *Caution: Be careful that the hair dryer does not fall into the water.*

In groups of four, use playdough to form land on one side of a rectangular, plastic container so that it grades gently from the bottom of the container to about 3 cm below the edge. Then pour in water until it covers half the "land." Now place sugar cubes on the land to simulate houses. Put a couple of drops of blue food coloring into the water. Then turn the hair dryer on low to blow the water from the ocean toward the land. What happens? Turn the hair dryer to high. What do you observe? How does this simulate a storm surge? What materials would help simulate the effects of a storm surge more accurately?

Concept Check

Assess/Reflect

Summary: How are weather patterns and severe storms related? Factors including humidity and the formation and movement of air masses have a strong effect on weather. Air masses obtain their temperature and moisture from their source regions. These characteristics mean that air masses also have different air pressures. The Coriolis effect determines the direction of air movement in air pressure systems. When air masses with different characteristics such as high-pressure, fast-moving, and cold, dry air come into contact with a low-pressure, warm, humid air mass, severe weather can arise. Thunderstorms, tornadoes, hurricanes, and blizzards are examples of the kinds of damaging severe weather that can result at the boundaries of different air masses.

1. Describe the characteristics of a continental polar air mass.
2. How does a storm surge form?
3. Compare the weather conditions associated with occluded fronts and stationary fronts.
4. Air is sinking and moving outward in a counterclockwise direction around a pressure center. Identify the pressure system and the hemisphere in which it is found.
5. As global warming continues, how could it affect the severity and frequency of weather events?

Lesson 5

Essential Question

What Is Climate?

Imagine waiting for 30 years before finding the answer to a question. That's what you would have to do if you wanted to accurately describe the climate for a specific area. This also means that once you have more than 30 years of data, you can track whether the climate in that area is changing. All organisms, including humans, are adapted to the climates in which they live. How would you adapt to a climate that seems to be turning colder than what you are used to? What kinds of changes might you expect to see in local plant and animal communities?

Objectives

- Describe how climate and weather are related.
- Identify the factors that affect climate.
- Compare climate zones.
- Explain how biomes are dependent on the local climate.

Vocabulary

topography

polar days

Climate Versus Weather Explain

Climate, you will remember, is the long-term weather pattern for an area. To determine an area's climate, scientists study weather factors, such as precipitation, temperature, wind, and humidity. They look at data covering 30 years or more. What is the climate like where you live? What factors determine the climate where you live? Do you think the climate in your area is the same as it was 100 years ago? Why or why not?

This tree changes with the seasons. Its leaves begin to grow in spring, continue to grow during summer, and change color and fall off in autumn. During winter, the tree's branches are bare.

What do you think triggers the leaves to change color and fall off the tree? What triggers the leaves to begin growing in the spring?

Structured Inquiry

Discover

Record your work for this inquiry. Your teacher may also assign the related Guided Inquiry.

Science Journal

Latitude and Temperature

How does latitude affect a location's average temperature?

Procedure

1. Locate the following cities on the world map.
- Escuela, Antarctica 63°S 57°W
- Quito, Ecuador 0°N 78°W
- Kyoto, Japan 36°N 135°E
- Sandy Bay, Australia 43°S 147°E
- Chalatenango, El Salvador 14°N 89°W
- Detroit Lakes, MN 47°N 96°W
- Carltonville, S. Africa 26°S 27°E
- Guangzhou, China 23°N 113°E
- Kodiak, AK 58°N 152°W

Materials
- pencils, colored
- graph paper
- ruler, metric
- world map

2. Research the average high and low temperatures each month for one year for each city. **Record** the data.

3. Make a line graph that displays the monthly temperature data for all nine cities. Use a different color for each city. Create a key to show which lines correspond to each city.

Analyze Results

Are there any patterns in your graph? What are the similarities and differences between higher latitudes and lower latitudes?

Create Explanations

1. How does latitude affect a location's average temperature?

2. Which city had the greatest difference between its lowest and highest temperatures? Why?

3. Which city had the smallest difference between its highest and lowest temperatures? Why?

4. A few of the cities at the same latitude and elevation most likely have very different climates from what you would expect. Can you explain this difference?

God's Perfect Climate **Explain**

Earth's climate has not always been the same as it is today. Unlike the climate and weather extremes we experience today, after Creation the newly created Earth had a mild, temperate climate. A mist watered Earth daily. Rain, snow, hail, and deadly storms were unknown. Freezing winter nights and scorching summer days were not part of God's perfect plan. With the entrance of sin, this mild, uniform climate began to change. Dramatic changes in climate occurred after the Flood, which Creation scientists believe brought on the ice ages.

Scripture Spotlight

Read **Genesis 41:1–37**. Why do you think the climate of Egypt would be susceptible to famine?

Explore-a-Lab

Structured Inquiry

How do the sun's rays warm Earth?

On a globe, tape one thermometer (P) so that its bulb is near the North Pole and tape another thermometer (E) so that its bulb is near the equator. Record the temperatures. Next, place a lamp about 30 cm from the globe and so that the lamp is pointing directly at the globe's equator. Take the temperature every 2 min for 20 min. What can you conclude from your results? What do you think are the similarities between your set-up and how Earth is heated by the Sun? What other factors affect how the Sun warms different parts of Earth?

Factors That Affect Climate

Why is Hawaii warm and northern Canada cold for much of the year? What other areas have climates that are similar to the climates of Hawaii and northern Canada? What do those areas have in common? Different factors affect an area's climate. These factors include latitude, elevation, bodies of water, ocean currents, and topography.

Northern Winter / Southern Summer

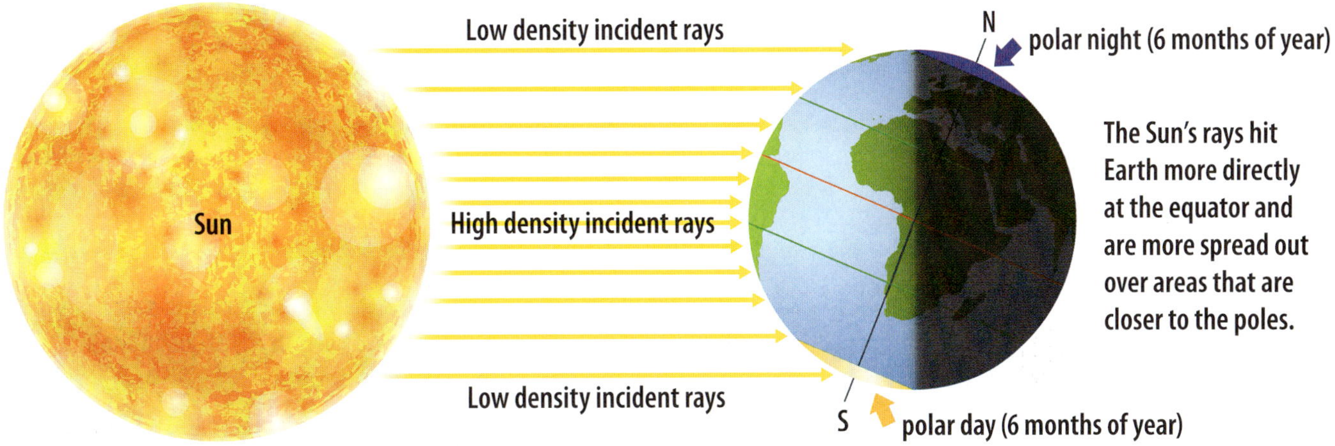

Factors Affecting Climate		
Factor	**Effect**	**Think About It**
Latitude	Areas near the North and South Poles have cooler temperatures than areas near the equator.	Some places have climates similar to that at the North and South Poles. What do you think living in such a place would be like?
Elevation	Higher elevations have cooler climates than do areas of lower elevation.	What is the elevation where you live? What locations might have the same climate as where you live?
Bodies of water	Coastal areas have cooler summers and milder winters than do inland areas.	What bodies of water might affect the climate of your area?
Ocean currents	Warm ocean currents can carry warm waters from the equator toward the poles, warming climates. Some ocean currents carry cooler waters from high latitudes toward the equator, cooling the climate.	In the northern hemisphere, currents from the equator flow up the western boundary of the ocean and back to the equator along the eastern boundary in a circular motion known as a gyre. How could we harness the energy in these currents?
Topography	**Topography** describes the features of an area in terms of altitude and slope. The leeward side of a mountain often has a drier, warmer climate.	What deserts in North America are affected by this *rain-shadow* effect?

Rain-Shadow Effect

A rain-shadow effect on the leeward side of a mountain occurs when the air has already lost its moisture on the windward side of the mountain.

How will this topographic feature affect climate?

Climate Zones Explain

Scientists use data about temperature and precipitation to classify regions of Earth into different climate zones. A **climate zone** is a region that shares similar climate characteristics. The broadest classification schemes divide Earth into three main climate zones. Each of these zones can be further divided into smaller, more specific climate zones.

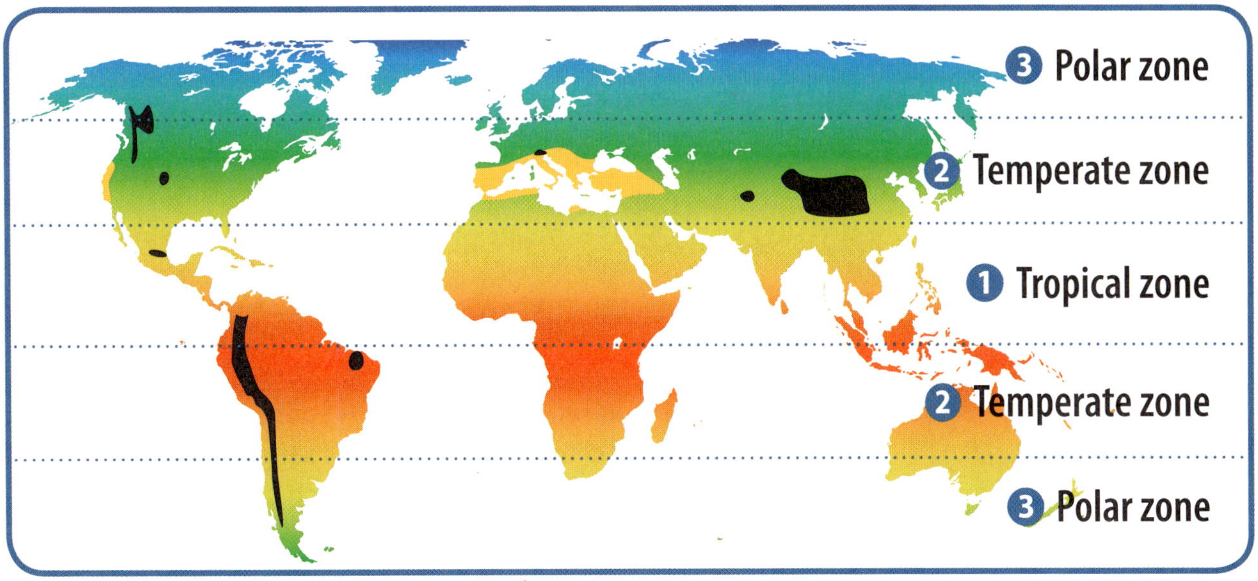

Earth can be divided into broad climate zones.

In which climate zone do you live?

1 Tropical Zones

- Temperatures high year round; high levels of precipitation
- Tropical wet: hot and very rainy all year
- Tropical dry: hot all year with wet and dry seasons

2 Temperate Zones

- Temperatures and precipitation vary throughout year
- Temperate: moderate and moist
- Desert: very dry
- Mountainous: varying precipitation depending on elevation
- Mediterranean: hot, dry summers; mild rain winters

3 Polar Zones

- Always cold; low rates of precipitation
- Tundra: always cold with short, warmer summers
- Ice cap: freezing temperatures year-round

Climate and Biomes Explain

Check for Understanding

How do you think we can use the distribution of plants and animals to assess how the climate zones are changing?

As you have learned, different factors determine the climate of an area. Climate, in turn, is one factor that helps determine a biome. Recall that a *biome* is a large ecosystem made up of distinct plant and animal communities and distinct environmental conditions in a large geographic region. The climate determines which plants and animals make up a biome. For example, a desert is a biome with a dry climate. It receives very little precipitation. God gave the plants and animals that live in the desert the ability to adapt to these dry conditions. For example, the waxy stems of cacti help conserve water, and kangaroo rats have the ability to thrive without drinking water.

Lesson Activity

Use the Internet to research times in Earth's history when the global climate was cooler and warmer. Then choose a cooling-Earth or a warming-Earth scenario and use a map to illustrate how the climate zones would change. Predict what would happen to the organisms in each of these climate zones. How do you think modern society would be affected? Present your findings to the class and discuss the differences and similarities.

How does a changing climate affect life on Earth?

The arctic fox has a white coat in the winter and a dark coat in the summer to blend into its surroundings.

What other animals might benefit from the ability to change colors with the seasons?

Like the plants and animals of the desert, the Creator gave plants and animals of the tundra the ability to adapt. Many tundra plants grow in clumps to be protected from the extreme cold and wind. Animals of the tundra often have thick fur to keep them warm. The arctic fox has the added adaptation of fur that changes color so it can better hunt and hide during different seasons.

Concept Check Assess/Reflect

Summary: What is climate? Climate is the weather patterns of an area over a long period of time, while weather is the different atmospheric conditions happening every day. Climate is affected by factors such as latitude, altitude, air masses, global winds, bodies of water, and topography. In general, Earth is broadly divided into three main climate zones—polar, temperate, and tropical. Climate also determines the kinds of biomes that occur in an area.

1. Similar climate changes occur as altitude increases and latitude decreases. Assess whether the preceding statement is true or false and provide your reasoning.
2. Use animal and/or plant adaptations to illustrate the conditions in each of the three main climate zones.
3. Which factor is most likely to affect the climate of an area in central Canada? Explain.
4. Shrubs grow in tundra biomes. What might happen to shrubs if the climate of the tundra became cooler?

People in Science

Extend

Get to Know
Dr. Bob Sheets

Dr. Bob Sheets has been the calm center of frenzied activity during three of the worst storms to hit the United States and the Caribbean. Born in 1937, Sheets began his weather career when the U.S. Air Force offered him a chance to go to meteorology school, where he studied under a Japanese expert on tropical typhoons. In 1965, he was hired as a meteorologist for the National Hurricane Center in Miami, eventually becoming its director. He is well remembered for numerous interviews given from the Center during Hurricane Andrew.

In 1992, when Dr. Sheets was then the director of the National Hurricane Center, one of the strongest hurricanes ever to hit the United States forever changed the face of South Florida. Hurricane Andrew was a small and fast-moving hurricane, tiny compared to slow giants like the more recent storms Katrina and Rita. Most residents had less than a day to prepare their homes and to evacuate coastal areas. The storm's wind speeds were later estimated at more than 160 miles per hour; the wind even broke the gauges used to collect such data! Perhaps a hundred thousand people were left with no power, no water, no food, and no security. Hurricane coverage was a combination of reporters in the field, viewers' questions, and conversations with Dr. Sheets.

When giving updates on Hurricane Andrew, Dr. Sheets had far fewer tools than exist today. Forecasters relied on observations from a single satellite and wind speed measurements from aircraft flying through the storm at 10,000 feet. Since Hurricane Andrew, the government has learned a lot about how to prepare for and respond to hurricanes. It has also improved its ability to predict the storms' paths. Today, there are many more satellites and additional devices that are dropped from airplanes to take readings from within a hurricane's core. They provide much more data for the computer models that forecast hurricane tracks.

Since retiring in 1995, Dr. Sheets has continued his relationship with the media, becoming a special-situation hurricane analyst with ABC network affiliates in Florida. Nontelevision work includes lectures and workshops on hurricane threats and preparedness and serving on advisory boards at Florida International University and the University of Oklahoma.

Concept Check

1. What made forecasting storm situations more difficult during Hurricane Andrew than today?
2. Since Dr. Sheets retired, what has he been doing to continue educating the public about storm forecasting and preparedness?

Science and Technology

Extend

SMART-R

SMART-R, or *Shared Mobile Atmospheric Research and Teaching Radar*, originated from a group of weather institutions' efforts to improve weather research and education. This unique partnership has created two C-band mobile Doppler weather radars for storm-scale research that is able to accurately measure both clear-air circulations and damaging winds in heavy rain. The units are used to study atmospheric processes to help improve forecasts of significant weather events such as flash floods, hurricanes, and tornadoes. Weather radars have improved quantitative rainfall estimates and help predict hail, strong winds, and heavy rains. SMART-R is also used to provide education in radar meteorology.

Doppler radar is a tool used to forecast atmospheric conditions and processes. The Doppler effect was named after Christian Doppler, who first discovered it in 1842. The Doppler effect is an increase or decrease in the frequency of sound, light, or other waves as the source and participant move toward or away from each other. A stationary transmitter is located in a weather station and the moving object being studied is a storm system. Radio waves are emitted from a weather station at a specific frequency. The waves are large enough to interact with clouds and other atmospheric objects. The waves strike objects and bounce back toward the station. If the clouds or precipitation are moving away from the station, the frequency of the waves reflected back decreases. If the clouds or precipitation are moving toward the station, the frequency of the waves reflected back increases. Computers electronically convert Doppler data into pictures showing precipitation, wind speeds, and wind direction.

SMART-R is a mobile observation system—in the past meteorologists have relied on stationary and airborne systems. A stationary system can operate continuously, providing rapid sampling of storm systems in lower altitudes. Airborne systems require lead time for launch, cannot operate continuously, and have longer times between samples of the same part of the storm (due to time to turn the aircraft around), and sampling in the lower altitudes is difficult. Mobile radar systems, such as SMART-R, combine many of the strengths of these two systems. They are flexible in deployment and can operate continuously with rapid sampling of the lower altitudes and, with new technology, can extend the range of sampling up to tens to hundreds of kilometers into large-size areas of multiple storm systems.

Concept Check

1. In addition to forecasting atmospheric conditions and research, what else is SMART-R designed to do?
2. What makes the SMART-R a better radar than other traditional radars?

Chapter 8 Review Assess/Reflect

Study Guide

Lesson 1

1. Earth's atmosphere makes life possible on Earth by providing the perfect balance of nitrogen, carbon dioxide, and oxygen.
2. Earth's atmosphere is made mostly of gases, including water vapor, which varies with place and time. Varying small amounts of particulates are also present in the atmosphere. The atmosphere is categorized into layers—troposphere, stratosphere, mesosphere, thermosphere, and exosphere—based on composition and temperature trends.
3. In the atmospheric level closest to Earth, temperature generally increases with altitude. The opposite can happen in the upper layers of the atmosphere. Air pressure is higher at sea level than at higher altitudes.
4. The boundaries between these layers are the tropopause, stratopause, mesopause, and thermopause, all of which occur at the extreme of the temperature trend in that atmospheric layer. The chart on page 279 summarizes the characteristics of each layer.

Lesson 2

1. Only 3% of the water on Earth is freshwater, and of that only a small percent is available for human use.
2. Freshwater on Earth is recycled through the processes of the water cycle. The water cycle is primarily driven by energy from the Sun. Several processes move water through the cycle, including evaporation, transpiration, sublimation, condensation, and precipitation.
3. Clouds are classified according to shape, altitude, and the precipitation they produce. See the illustration and table on page 284 for a summary of the characteristics of cloud types.
4. The type of precipitation that falls depends on air temperature, relative humidity, and air movement. See the chart on page 286 for a summary of the characteristics of different types of precipitation.

Lesson 3

1. Wind is the movement of air between high- and low-pressure systems that form from the uneven heating of Earth's surface by the Sun.
2. The Coriolis effect influences the direction in which global winds flow. Winds deflect to the right in the Northern Hemisphere and to the left in the Southern Hemisphere.
3. Global wind systems near Earth's surface include the trade winds, prevailing westerlies, and polar easterlies. See the chart on page 292 for a summary of the characteristics for each type of wind. High in the troposphere, global winds known as jet streams also occur. These winds can also have a strong effect on the weather.
4. Sea breezes blow from the sea over land. Land breezes blow from land toward the sea. A valley breeze forms during the day when the Sun heats the valley floor. The warm air rises, cools, and sinks, and the cycle repeats. A mountain breeze forms when air on a mountain cools more quickly than the air on the valley floor. The warmer air rises, cools, and sinks, and the cycle repeats.

Lesson 4

1. Air pressure, wind speed and direction, temperature, humidity, precipitation, and other factors affect local weather.
2. When an air mass lies over a region long enough, its temperature and humidity become characteristic of that source region. These characteristics also contribute to the air pressure of that air mass and the formation of winds. See the map on page 301 for a map of the major air masses and their characteristics.
3. See the chart on page 302 for a summary of the types of fronts and the weather conditions associated with each. High-pressure systems often bring fair weather to an area. Low-pressure systems often bring stormy weather.
4. Thunderstorms form when cold air lies atop a layer of warm, moist air near the ground. Tornadoes often begin as strong thunderstorms when cold air changes speed and direction and moves horizontally across an area. Air nearest the ground begins to roll until it meets the upward-rising warm air. this causes the horizontally blowing air to tilt vertically, forming a tornado. A hurricane is a low-pressure storm that forms over tropical waters and has wind speeds of at least 118 km/hr. A blizzard is a winter storm that forms winds that gust at 56 km/hr or more.

Lesson 5

1. Climate is long-term weather in a region.
2. Climate is affected by latitude and altitude. Generally, increasing latitude and altitude results in cooler, drier climates. Bodies of water and topography can also affect the climate of a region. Oceans and lakes help maintain a more moderate climate. Mountain ranges can cause rain-shadows behind them.
3. The plant community in a large geographic region is determined by the local climate and supports a unique animal community.

Show What You Know

Visualize It Complete the map of the water cycle to identify either the process or its definition.

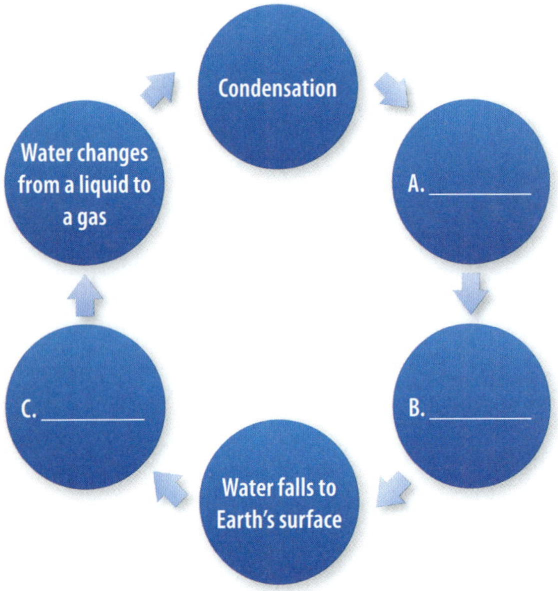

Vocabulary Check

Explain how the following pairs of terms are related.

1. condensation nuclei—cloud
2. convection cell—high pressure
3. storm surge—hurricane
4. local wind—sea breeze
5. topography—rain shadow
6. latitude—altitude

Multiple Choice

Choose the best answer.

7. Which of the following is **not** a gas in the Earth's atmosphere?
   A. oxygen
   B. carbon dioxide
   C. sulfur
   D. nitrogen

8. The __________ of air molecules is greater at the bottom of a mountain than at the top.
   A. mass
   B. volume
   C. density
   D. altitude

9. During which process of the water cycle do water molecules absorb a great amount of solar energy?
   A. evaporation
   B. condensation
   C. precipitation
   D. disposition

10. Which wind system is present at the highest altitudes?
    A. prevailing westerlies
    B. jet streams
    C. trade winds
    D. sea breezes

11. Which correctly sequences the atmospheric layers, beginning with the layer closest to Earth's surface?
    A. exosphere, mesosphere, troposphere
    B. troposphere, stratosphere, thermosphere
    C. stratosphere, exosphere, mesosphere
    D. thermosphere, exosphere, stratosphere

12. What can happen when air reaches its dew point?
  A. The air becomes warmer.
  B. Water vapor evaporates from it.
  C. Water vapor condenses from it.
  D. The air becomes more saturated.

13. Which of these explains the formation of large hailstones?
  A. There were strong updrafts in the cumulonimbus cloud.
  B. There were strong, straight-line winds in the nimbostratus cloud.
  C. The air had low relative humidity.
  D. A strong downburst was beginning to build up.

14. The weather forecast is calling for a day of gentle, steady rain. What type of front is approaching?
  A. warm front
  B. cold front
  C. occluded front
  D. stationary front

15. What determines which plants and animals make up a biome?
  A. precipitation
  B. air pressure
  C. atmosphere
  D. climate

Check Point

Answer the following questions.

16. Which type of cloud is shown below? At what altitude are these clouds usually found?

17. **Compare** two cities at the same latitude. Under what conditions might they have different climates?

18. What determines the boundaries of each of Earth's atmospheric layers? How is it possible that in some layers of the atmosphere the temperature increase with an increase in altitude?

19. Patagonia is a region of South America that is known for being dry. What role do the Andes, which lie to the west of this region, play in Patagonia's climate?

Chapter 9

Resources and Their Uses

Scripture Spotlight

In Genesis, God gave man the Garden of Eden to cultivate and to keep. As Christians, we should be excellent stewards of the Earth. You can strengthen your faith using science by studying about resources. You will read the following passages in this chapter.

Exodus 23:10–11 (p. 344)
Genesis 26:19–22 (p. 348)
Psalm 42:1 (p. 347)
Genesis 2:15 (p. 355)

The Big Idea

Interactions among living things and nonliving things affect Earth's climate and its land, water, and air resources.

Which of your daily actions might affect Earth's water resources?

Alternate energy sources, such as wind and solar energy, can be used in place of traditional energy sources that are more harmful to the environment.

Inquiry Kick-Off Engage

How do you use natural resources every day? Are the objects you use made from one resource or multiple resources? Are these resources renewable? In your *Science Journal*, you will explore what natural resources were used to make the items in your classroom.

Lesson 1

Essential Question

How Can We Use Resources Wisely?

Objectives

- Describe how people use natural resources.
- Distinguish between material resources and energy resources.
- Compare and contrast renewable and nonrenewable resources.
- Describe ways to conserve natural resources.

Vocabulary

material resource

energy resource

In ancient times, the only natural resources available to people were those that were close at hand, such as trees, blocks of ice, sod to build houses, animal skins for clothes and shoes, and wild game for food. How much do you think people in the past processed these natural resources before using them? In today's world, natural resources are used to produce everything we need and use in our lives. What led to this increased use of natural resources? Do you think this increased use of natural resources is a good thing? Make a list of pros and cons about natural resource use to discuss with your class.

How People Use Natural Resources Explain

Make a list of products made with natural resources. Compare your list with the lists your classmates. Did you include petroleum products? Petroleum is used for thousands of products, including gasoline, perfumes, house paint, DVDs, plastic food wrappers and bottles, and most sunglasses. How much would your life change if petroleum products were no longer available? What would your life be like without any manufactured products at all? Was there ever a time when there weren't any manufactured products?

All life on Earth depends on water to grow and thrive.

What other natural resources do you see in this photo?

Natural resources also include the air you breathe, the ground you stand on, and the water you drink. In addition, the food you eat—vegetables, wheat, fish, chickens, and cattle—are also natural resources. God gave us these gifts to show He cares for us. It is our job to care for those resources by using them wisely and keeping the air, land, and water on Earth clean.

How Natural Resources Are Classified

Natural resources are classified by how they are used. A **material resource** is one that is used to make other products. The desks, books, windows, and all the other items in your classroom are made of material resources. Material resources include rocks, minerals, sand, metals, trees, and food crops. Plants such as tomatoes and animals such as cattle and chickens are also considered material resources because they are used to make other products we use. What products in your home are made from these material resources?

Common Material Resources and Their Uses

Product	Application	Product	Application
Sand and gravel	Roads, driveways, foundations, concrete	Salt	Food seasoning, roadway deicing, an electrolyte in our blood
Bauxite ore (aluminum)	Cars, utensils, aluminum foil, beverage cans	Iron ore	Steel, building materials, knives, cars
Clay	Pottery, floor tiles, bricks, porcelain	Plants	Food, clothes, furniture, salad oil
Copper	Electrical wire and cables, pennies, cookware	Animals	Food, leather, fertilizer, pets
Diamond	Drilling bits, jewelry	Silver	Jewelry, coins, metal products, wire
Titanium	Paint, metal parts of spacecraft, medical uses	Petroleum	Fuel, plastics, cosmetics, packaging

Another classification of resources is energy resources. An **energy resource** is a source of energy that is used to produce other types of energy. Energy resources such as gasoline and heating oil are used to power vehicles, heat buildings, and generate electricity. Energy resources you may not have thought of include the Sun, wind, and water. What energy resources does your family use every day?

Lesson Activity

Rank the material resources above in order of importance to your daily life. Compare your list with the lists of two other students and take turns defending your choice for the most important resource.

Which of the resources in the table do you think you could do without for long periods?

Renewable Resources

Resources can also be classified according to their availability. You know that a *renewable resource* can be replaced by natural processes as fast as, or faster than, it is used. Air, water, animals, and plants are renewable resources if conserved and used wisely. How are renewable resources, such as land, water, and air, threatened?

Solar Power

The Sun, for example, is an inexhaustible energy resource. Solar panels convert the energy from the Sun to generate electrical energy. Many businesses and homes have solar panels installed. Where else do you see solar panels being used? Why might solar energy be less costly than the energy produced by fossil fuels?

Wind and Tidal Power

Wind energy and tidal power are also sources of renewable energy. Energy generated by wind turbines now supply states such as Texas, Iowa, and California with an important part of their energy needs. Tidal power has the potential to provide energy as well. What states and provinces do you think would have the best location for this kind of renewable energy?

Forests

When new trees are planted to replace those used for paper or other products, the new trees may require 10–20 years to mature. Properly managed forests that are carefully replanted can be considered a renewable resource. Is this kind of resource management feasible? Explain.

The Tehachapi Wind Farm is a collection of wind generators in California, comprising approximately 5000 wind turbines.

Which type of alternative energy resource is best suited to your area? Explain.

Hydroelectric Power

As you know, *hydroelectric energy* is electricity produced by harnessing the energy of moving water. Niagara Falls is one of the best-known hydroelectricity locations. Where else would you expect hydroelectric power to be a significant part of the energy supply? Is hydroelectric power available everywhere? Explain.

Geothermal Energy

Recall that *geothermal energy* is energy that comes from Earth's core. At certain places on Earth, this thermal energy can be tapped and used to provide hot water for heating or steam to turn turbines to produce electricity. What geologic features would be characteristic of regions that utilize this kind of energy? What states and provinces do you think would have geothermal plants?

Biomass Energy

Biomass energy comes from using plant life to generate electricity. Biomass, such as wood or waste, can be burned directly for energy. Other forms of biomass, such as corn and willow, can be used in the process of fermentation. The fermentation reaction produces carbon dioxide and the fuel ethanol, a substance that is added to gasoline to power trucks and automobiles.

This facility produces fuel from corn.

What might be some negative effects of using farmland to produce fuel?

Explore-a-Lab

Guided Inquiry

How can you use renewable energy?

If you did not have access to a stove, microwave, or fire, how could you heat 100 mL of room-temperature water (20°C) to 70°C? Using a pizza box, aluminum foil, plastic wrap, and any other materials your teacher supplies or you ask for, use the Sun to heat your water. Record the starting temperature of your water and how long it takes to change the water temperature. Compare your results to those of your classmates. Which solar oven design worked best? How would you change your design?

Structured Inquiry

Discover

Record your work for this inquiry. Your teacher may also assign the related Guided Inquiry.

Science Journal

Evaluating Biofuels

Which biomass provides the best fuel?

Procedure

1. Label the water bottles *Fuel Stock 1, Fuel Stock 2*, and *Fuel Stock 3*.
2. Add 250 mL of warm water and one package of dry active yeast to each of the plastic foam cups. Use the plastic spoon to mix the yeast with water.
3. Add 30 mL of molasses to the bottle labeled *Fuel Stock 1*, add 30 mL of fruit juice to the bottle labeled *Fuel Stock 2*, and add 30 mL of finely chopped or blended lettuce or spinach to the bottle labeled *Fuel Stock 3*.
4. Add 1 cup of the warm water and yeast to each of the bottles containing fuel stock.
5. Stretch a balloon over the neck of each bottle, and use a rubber band to secure each balloon in place.
6. In your ***Science Journal***, **record** the time that each bottle was covered with the balloon.
7. After 10 minutes **observe** each bottle and balloon. Record your observations.
8. Let the bottles sit for at least two hours.
9. Determine the volume of carbon dioxide produced by each fuel stock, following the steps below:
   - Determine the diameter of the balloon at its widest point.
   - **Calculate** the radius of the balloon by dividing the diameter in half.
   - Use the radius and the formula below to calculate the volume of carbon dioxide produced by each fuel stock. Record this measurement.

$$V = \frac{4}{3} \pi r^3$$

Materials

- 3 yeast packets
- plastic foam cups
- 3 1-L plastic water bottles
- balloons (large balloons to fit over the top of bottles)
- rubber bands
- funnels
- graduated cylinder (100 mL)
- plastic spoon
- stopwatch
- molasses
- fruit juice (non-acidic)
- lettuce or spinach

Analyze Results

Construct a graph of your data. Which fuel stock produced the most CO_2? Which produced the least?

Create Explanations

1. Which biomass provides the best fuel?
2. What was the function of the yeast in this activity?
3. Why are some fuel stocks better than others?

Nonrenewable Resources

Nonrenewable resources, you might remember, are resources that are being used faster than nature can replenish them. What are these resources? Why does it take so long to replenish them? Are there some that will never be replaced?

Fossil Fuels

Most of the fuels we use in our daily lives—fossil fuels—are nonrenewable resources. What are three types of fossil fuels? How do they form? Why do we continue to rely on these fuels if they are nonrenewable? What do you think is the best thing to do regarding the use of these fuels?

Besides fossil fuels, what are other nonrenewable resources do you depend on in your daily life? Are there nonrenewable resources that you can see around you right now? What are they?

This natural gas power plant produces electricity.

What type of fuel resources does your community use to provide electricity?

Nuclear energy

Nuclear energy is generated by splitting atoms through the process of fission in nuclear reactors. They are powered by a rare and nonrenewable resource—uranium. Like coal and oil, uranium is mined from the ground. What are the advantages of nuclear energy over fossil fuels? What are the disadvantages?

Minerals

Most minerals are considered nonrenewable. What are examples of nonrenewable mineral resources? What might be some renewable mineral resources? Look around; what mineral resources do you see in use?

Energy Use in North America

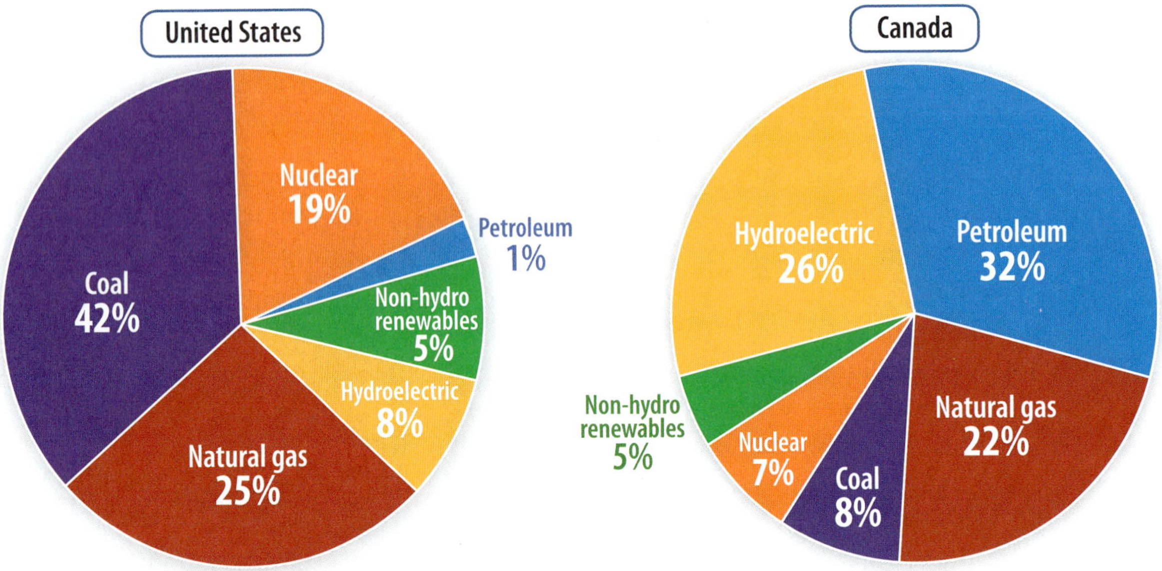

These statistics from a recent study show the breakdown of energy use by type.

What are the major differences in energy consumption between the United States and Canada?

The human population has risen dramatically in the past few hundred years. As the human population grows, the demand for resources grows as well. Some nonrenewable resources may be depleted in the not-too-distant future. In addition, some renewable resources can become nonrenewable if they are not used wisely. For example, many people get water from wells drilled into reservoirs of groundwater. What factors might contribute to reduced water supplies for these people? What steps can individuals and communities take to help reduce the risk of this problem?

How Can We Conserve Natural Resources? Explain

God made us stewards of His Creation. That means it is our responsibility to care for the resources God lovingly provided us. Every one of us can help protect important natural resources. What can you do to be an agent of resource conservation? Learn the 3Rs: reduce, reuse, recyle. Learn how reducing, reusing, and recycling can help you, your community, and the environment by saving money, energy, and natural resources. Recycling programs are managed at the state and local level. Find information on recycling in your community and put this information into action at home and at school. Complete the Lesson Activity on the next page to get started.

Faith Connection

How could a worldwide flood contribute to the huge coal and oil deposits found in the world today?

Lesson Activity

Make a list of things you can do to protect or conserve the natural resources you rely on every day. Use the Internet and other sources to add more items to your list. Compare your list with those of classmates and add any of their ideas to your list. Your list should contain at least 10 items. Choose at least three of those items that you can commit to doing to care for God's Creation. Explain why you are making these changes and ask your family to join your commitment as well.

What can you do to protect Earth's natural resources?

Replacing incandescent light bulbs with fluorescent light bulbs can help reduce energy use.

How do these two types of lighting differ? Why is fluorescent lighting more energy efficient?

Concept Check Assess/Reflect

Summary: How can we use resources wisely? A natural resource is something used to meet the needs of humans and other living things. A material resource is used to make products. An energy resource is energy or is used to produce energy, such as electricity. A renewable resource, such as energy from the Sun, can be replaced. A nonrenewable resource, such as fossil fuels, is used faster than it can be replaced. Conservation is the wise use of resources. Everyone is responsible for the proper care of the resources that God gave us.

1. How did you use natural resources today?
2. Compare material resources with energy resources.
3. It can take a forest ecosystem thousands of years to develop. Is a forest renewable or nonrenewable? Defend your answer.

Lesson 2

Essential Question

How Have People Affected the Land?

Objectives

- Describe the effects of urbanization on the environment.
- Describe how industry and agriculture affect land resources.
- Identify ways to protect land resources.

Vocabulary

urbanization
urban sprawl
crop rotation
organic farm
contour plowing
strip farming
no-till farming
mine reclamation
sustainable logging
stewardship

How much do you think the world's population has grown since your grandparents were born? How much has it grown since you were born? How much bigger is the world's population today than it was yesterday? From the beginning of recorded history to the start of the Industrial Revolution, the human population on Earth grew at a very slow rate. It took thousands of years for the population to grow to 1 billion people in 1804. Why do you think it grew so slowly?

Over the last 200 years, though, the population has grown much faster. The human population reached 2 billion in 1927, a period of only 123 years. Better sanitation, improved medical science, and lower infant mortality rates led to increased population growth. What might be another reason for the rapid population growth? A little more than 30 years later, the world's population reached 3 billion. The Earth's population now exceeds 7 billion. How does increasing world population affect you and your family? Is it possible for the population on Earth to get too big? How many people do you think Earth can sustain?

World Population Growth to 2000

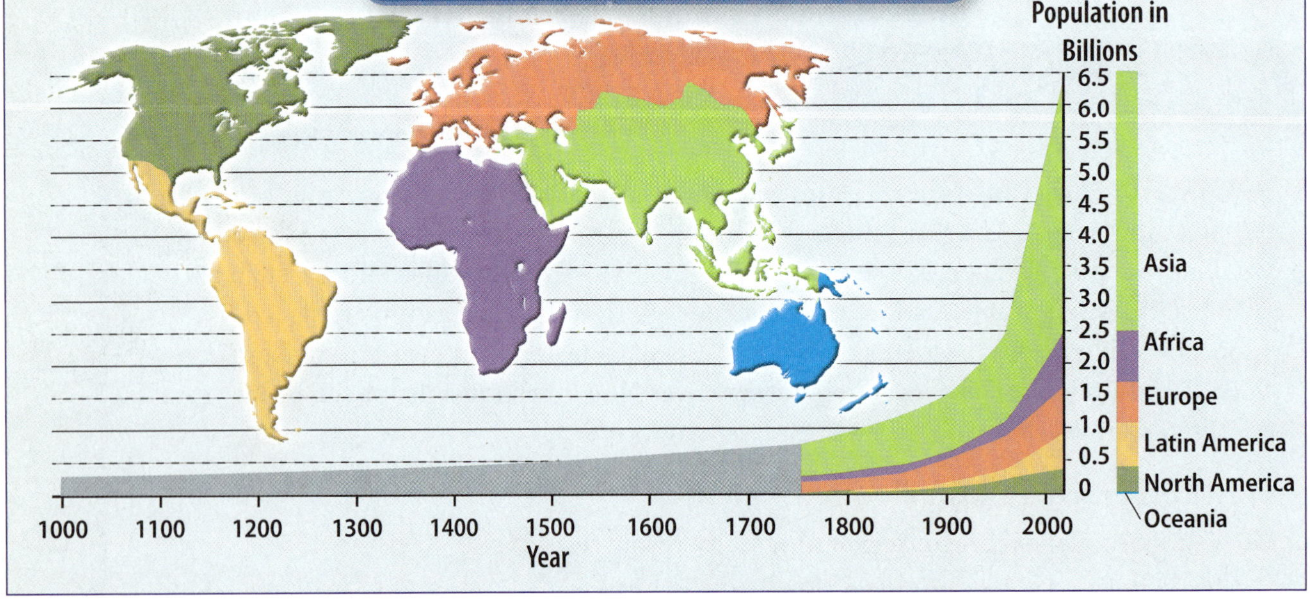

Which continent's population has grown the most over the last 200 years? The least?

Urbanization Explain

One of the most dramatic effects of human development on the global ecosystem is urbanization. **Urbanization** is the process by which towns and cities form and become larger as more people begin working and living in central areas. Where are some areas that you have seen urbanization taking place? In 1790, only 5% of Americans lived in cities. In 1890, 35% of U.S. citizens lived in cities, largely due to the large number of factories created during the Industrial Revolution. Today in the United States and Canada, about 80% of the population lives in urban areas. On a global basis, in 1900 only 15% of the world's population lived in cities, but today the figure has grown to more than 50% and is still increasing.

As a result of these changes, more and more land has been used for cities, roads, industrial areas, and housing. Urbanization results in higher population density, the number of people living in an area. How does increased population affect the natural resources of the area? What natural resources do you think are affected the most? Why?

Look at the graph below that shows the past and future predictions for the migration of people from rural to urban areas. What explanation can you give for the trends shown in the graph?

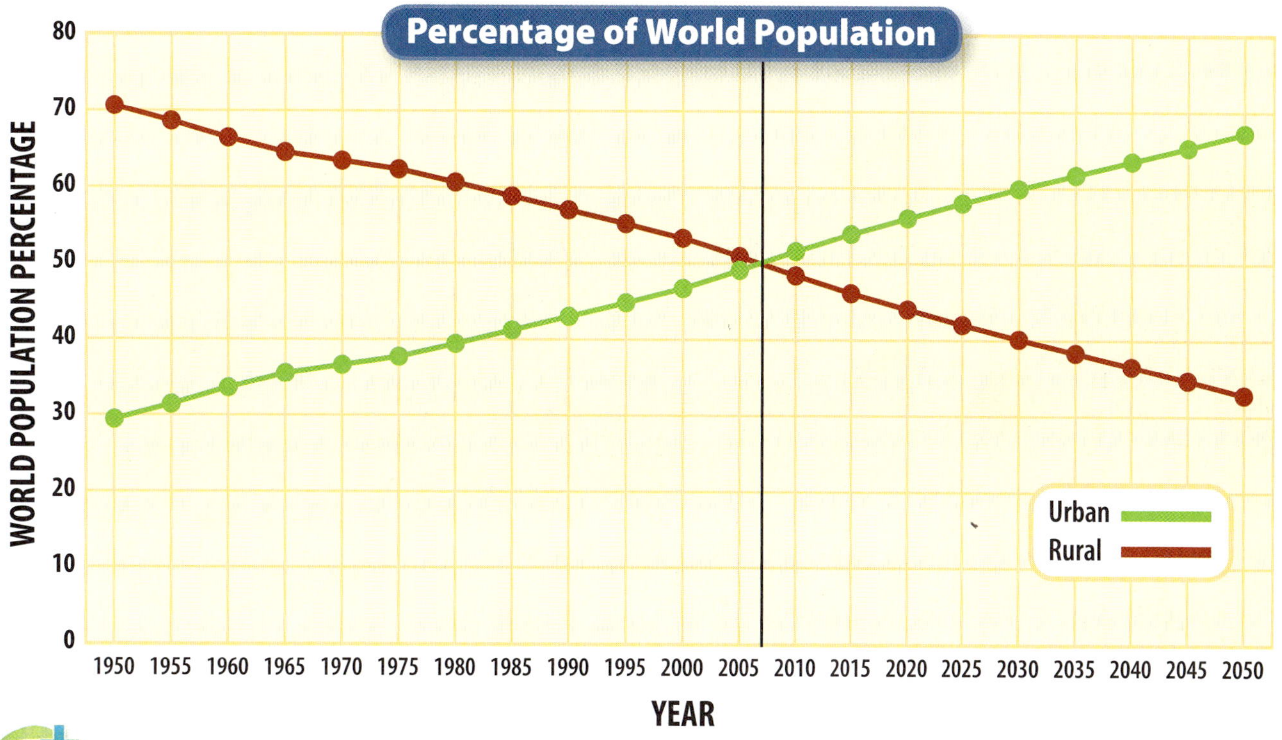

Math in Science

What is the projected rate of growth (% per year) in urban population that is predicted for the world population between 2010 and 2050?

Effects of Urbanization

This uncontrolled spread of urban development into surrounding countryside (and agricultural land) is called **urban sprawl**. What can result from this kind of urbanization? Water that falls on paved surfaces does not seep into the ground, depleting the groundwater levels. This puts a strain on sanitation systems and increases flash flooding. Paved surfaces absorb more heat than grasslands, forests, and other unpaved surfaces. So, cities experience warmer temperatures than nearby rural areas. Urbanization also results in increased use of vehicles, which causes additional pollution.

Several decades ago, this land was used for farming. Today, it is used for suburban homes. The people who live here commute to work in the cities, sometimes more than an hour each way.

What impact does commuting have on natural resources? On the environment?

Explore-a-Lab

Structured Inquiry

How do different types of surfaces affect heat absorption rates?

On a sunny day, place one thermometer on a paved area, such as asphalt. Place a second thermometer on a nearby grassy area. Make sure both thermometers are in direct sunlight. Record the initial temperature of the thermometers. Wait 20 min. Record the temperatures again. If time permits, record the temperatures a third time, 20 min later. What was the difference in the temperatures of the two areas? Based on your results, what recommendations would you make to a group planning development of a new section in a town? How could you use this information to improve conditions in an existing large city?

What impact does this slum have on its environment?

Landfills

Large cities must dispose of the trash generated by millions of people. Solid waste may be burned, buried in landfills, or even dumped at sea. How much garbage does your family produce in one day? One week? Landfills take up enormous amounts of area and must be carefully lined to avoid waste seeping into groundwater. Once landfills are filled to capacity, another landfill, farther away, must be created. This uses up more land, and the additional travel means more pollution as garbage trucks travel back and forth to the dump each day. How far does your garbage have to travel to reach the landfill for your community? Some trash is illegally discarded by people, dumped into fields, forests, or by the side of the road. In addition to being unsightly, how can trash have negative impacts on plants and wildlife?

Lesson Activity

The United States uses 3 million kg of paper per year. Of that, about 66% is recycled. If your school doesn't have a paper-recycling center, start a program by placing recycling bins in each classroom. Collect the paper for one week and weigh it. Place a poster in each classoom suggesting what students and teachers can do to use less paper. Wait a few days and then collect paper for another week and weigh it. Did your conservation effort succeed?

How can you use less paper?

The Effects of Industry and Agriculture **Explain**

While our urban centers are growing, 40% of land in the United States has been developed for agricultural purposes, including growing crops and grazing animals. This crucial land use provides about 30% of the world's corn, 10% of its wheat, and 20% of various meat products. How has the formation of large and small farms affected the environment?

Science Journal

Check out your *Science Journal* for an activity to see who can create the best fertilizer. **Extend**

Agricultural Land Use

The amount of land set aside for agriculture has actually decreased over the years. At the same time, changes in farming techniques have increased crop yields. However, many of these changes can cause environmental problems. Chemical fertilizers and pesticides used to increase crop yield can seep into the ground and pollute soil and groundwater. Chemicals can also be washed into nearby streams and lakes when heavy rain occurs. How could these chemicals affect the microorganisms in the soil? What other effects could this runoff have in the environment?

Some agricultural land is used to raise cattle. Overgrazing can damage plant roots and leave the soil bare. Bare soil is more easily eroded than soil that is covered with vegetation. Eroded topsoil can wash into waterways, causing changes that can have negative effects on animals that depend on the ecosystem. But managed grazing techniques can be used to prevent overgrazing. Moving animals so they can graze in different areas means they'll have access to different pastures while pastures they have been feeding on grow back.

Today's agriculture uses chemicals to control insect pests and diseases on food crops.

How does spraying chemicals on land affect our water and air resources?

What are practices that a good farmer can use to protect the environment? Farmers who practice *sustainable agriculture* use nonrenewable resources effectively while growing nutritious foods rather than focusing only on economic gain. Food produced on sustainable farms is more healthful for consumers, although the cost at the supermarket is somewhat higher. Some sustainable-agricultural methods include rotating crops, choosing insect-resistant crops, attracting the natural predators of insects such as birds and bats, and removing weeds by hand instead of using herbicides.

What are some minerals that plants take in from the soil to support their life processes? What happens to the soil in a farm field over time when the same kind of crop is planted there every year for many years? **Crop rotation** involves planting different crops on a field to improve the quality of the soil. For example, farmers might alternate crops that add important nutrients to the soil with those that need the same nutrients to grow healthy. Crops such as beans, lentils, peas, and peanuts add nitrogen to the soil. Farmers might also plant deep-rooted crops one season and shallow-rooted crops the following season to improve soil structure. How would alternating deep-rooted and shallow-rooted plants improve soil quality? What would farmers who do not use these practices have to do to keep their soil healthy enough for crop growth?

Huge commercial farms grow most of the fruits and vegetables you eat. To increase yield, these farms often use chemical fertilizers that contain nitrates and phosphates. But small, local farms often engage in the practice of organic farming. **Organic farms** use only naturally occurring substances to fertilize their crops and protect them from weeds and insect invaders by using natural substances instead of toxic pesticides. What natural substances do you think organic farmers might use to control insect pests? How might they control weeds?

Other agricultural practices include contour plowing and strip farming. **Contour plowing** is the practice of plowing along the natural contours of the land. As a result, the land is not overly disturbed, and soil erosion is reduced. **Strip farming** involves growing alternating rows of a cultivated plant, such as corn, and a sod-forming plant, such as alfalfa, to reduce the rates of soil erosion. How do sod-forming plants help prevent erosion?

Farmers can also engage in a process called no tilling. **No-till farming** is a way of growing crops from year to year without disturbing the soil through plowing and tilling. Practicing this technique increases the amount of water and organic matter in the soil and decreases erosion.

Structured Inquiry

Discover

Science Journal

Record your work for this inquiry. Your teacher may also assign the related Guided Inquiry.

Preventing Hillside Erosion

What effect can planting a hillside garden have on soil erosion?

Procedure

Materials

- 3–9 × 13 aluminum baking pans
- alfalfa seeds (one teaspoon)
- two small ground cover plant starts
- spray bottle
- a variety of small rocks, gravel, or pebbles - 2 cups
- 5 cm × 10 cm × 15 cm wood block

1. Using only the materials listed (include both kinds of plants and pebbles), **develop a plan** for your garden that will best keep soil in place when your garden is on a slope. Draw a diagram of your garden plan in your ***Science Journal***.
2. Fill each pan with potting soil. Use your hand to gently pack the soil. Add more soil as necessary to bring the level of the soil to the top of the pan.
3. In Pan 1 plant a garden following the garden plan you developed in Step 1. Add pebbles and rocks as desired. When you have finished planting your garden, make any necessary modifications to the drawing of your garden you made.
4. Water the garden in Pan 1 as needed. Record the amount of water used at each watering. Water Pan 2 (soil only) with the same amount of water each time you water Pan 1.
5. Once the garden has grown to the point where the roots have been established, two weeks should be plenty, test your garden for erosion.
6. Place your garden pan on a block so that one end of the tray is elevated about 10 cm.
7. Place the opposite end of the garden pan in the third aluminum pan in order to catch the dirt and water runoff.
8. Sprinkle the garden with 200 mL of water and collect the runoff in the catch pan. If there is no runoff continue sprinkling the garden with 100 mL of water at a time until you get enough runoff to test. Be sure to keep track of the total amount of water used in this procedure.
9. Now **measure** the amount of soil and water runoff and **record** this amount in your ***Science Journal***.
10. Repeat Steps 5–8 with the second pan (soil only).

Industrial Land Use

What manufactured products do you use routinely? How do these products influence your daily life? Could you live without these products? What would you be willing to give up to keep these products on the market?

Manufactured products have invaded the lives of everyone in North America. Many of the products make our lives easier and safer. However, many industrial processes generate hazardous wastes. As you know, *hazardous wastes* are toxic by-products of industrial processes. A nuclear power plant, for example, generates radioactive wastes that must be safely stored for thousands of years. Oil, household cleaners, paint, and other chemicals are also hazardous wastes. What hazardous wastes do you have at your home? How should you get rid of these hazardous wastes? What safe products could you use in place of these hazardous materials?

Industries that manufacture items, such as vehicles, furniture, computers, and plastic containers, and factories that process food products all require large pieces of land. Air and sound pollution sometimes result in the manufacturing process. Industrial farming, the raising of thousands of chickens or other livestock, also has a negative effect on the environment. What industry practices cause these negative effects? What could industries do to prevent the negative effects?

By-products of mining processes can spread and pollute land and water that are away from the mine.

What steps could be taken to keep these pollutants out of the water?

Grassy Narrows: A Case Study

In Canada, many people are familiar with the case of Grassy Narrows, the home of several First Nation communities about 80 km north of Kenora, Ontario. The traditional lands of these people have been located for a long time on the Wabigoon River, downstream from a local paper mill that dumped toxic chemicals, including mercury, into the river.

In the 1970s, extensive mercury contamination was discovered in the river. The toxic levels of mercury had poisoned the ecosystem's organisms, including the river's fish and the people who lived there. As a result, commercial fisheries were closed, and the government of Ontario ordered that no more mercury could be dumped into the river system. However, the emission of airborne mercury continued until 1975.

Although the mills now have been shut down for more than 50 years, the residents of the area are still dealing with the long-term impacts of mercury poisoning. What are the symptoms and effects of mercury poisoning on health? What social and economic impacts might the mercury poisoning have caused 50 years ago and today?

What must individuals, communities, and governments do to prevent impacts such as what occurred at Grassy Narrows?

The people of Canada have learned from the Grassy Narrows disaster. Today the environment is an important concern of Canadians. Waste materials are recycled. Jobs are created by cleaning up and protecting the environment. Companies increasingly recognize that switching to clean and green technology leads to greater profits.

In the United States, the Clean Air Act and other regulations have had a role in reducing the pollution emitted by industrial processes. In 20 years, the levels of six common pollutants decreased 51%. Estimates are that the reduction of ozone and fine particles in the air will prevent more than 160,000 deaths.

To protect the environment and the health of people, all hazardous wastes must be disposed of properly. Pouring chemicals down a drain or dumping them on the ground pollutes the land, and the chemicals are likely to end up in our drinking-water supply. What household substances are safe to pour down the drain? Which are not? How are the unsafe substances properly disposed of?

Mining

How many products can you think of that are made from iron, aluminum, copper, gold, sand and gravel, granite, and marble? What characteristics make these materials so popular? Where are these materials found?

Some of these materials are removed from the ground in a process called strip mining, or surface mining. *Strip mining* disturbs large portions of land by removing the top layers of soil and rock. You can see an example of an area of land that has been strip mined in the picture below. How has the mining process changed the land? What effect do you think this process has had on the organisms that lived on the land before the mining process began? What effect might the process have on other ecosystems that are not directly mined? How can people use the land as it appears in the picture?

Mined land can become usable again, as shown in this picture.

When is the best time to plan land reclamation?

The strip-mined area may appear to have limited use. However, **mine reclamation** can restore land that has been mined to a natural or usable state. In the United States and Canada, mine reclamation is a regular part of the mining process. Plans for converting the mined land to a usable form begin before the mining process begins. After the mining is completed, the land is restored to a condition that was approved in the premining planning. The restored land can take many forms, often depending on community needs. The end result could be a forested area, a golf course, or a park. What roles do you think the local community and the mining company should have in the land reclamation process? What factors other than the removing of soil and rock do you think mining companies should consider before they begin the mining process?

In recent years, new mining technologies and stricter regulations have significantly improved mining efficiency and reduced the negative impacts of mining on the environment. Overall, mining techniques have become much more environmentally sensitive. A few of the new strategies being employed are shutting down illegal and unregulated mines, choosing environmentally friendly mining processes, implementing recently discovered green mining technologies, and cleaning up the sites of shut-down mines. These strategies have actually improved the efficiency of mining companies since fewer waste products are being produced.

Mine runoff pollutes rivers.

What other problems might an open mine like this cause?

Logging

Another large-scale use of forested land is logging. The United States and Canada are two of the world's largest timber producers. What are some uses of forest trees? Which of these products do you consider essential? What materials could be used instead of trees to produce these products? Are those alternate materials more "Earth friendly"?

Although tree planting can restore forests, restoring a complex forest ecosystem can take years. For this reason, a good conservation plan often involves cutting down a few selected trees throughout the forest rather than entire tracts of forests. Today many North American forests are carefully managed in multi-year cycles to ensure a continuous supply of timber. Trees are logged in a rotating grid pattern or are selectively cut within a large forest.

In addition, in 1992 the United Nations adopted the Forest Principles, which lay out strategies for sustainable forestry. **Sustainable logging** is a practice in which loggers balance the use of the forests for wood production with protecting the ecosystem and the environment. The timber industries take only one third of the trees eligible to harvest in each area while preserving habitat for fish and wildlife and sites of cultural and historical importance. How do you think these principles affect the availability and cost of lumber? How do they affect the economies of regions that rely on logging?

A history of logging in four regions of the United States.

Why are the logging history curves for the South and North so different from those of the Pacific Coast and Rocky Mountain regions?

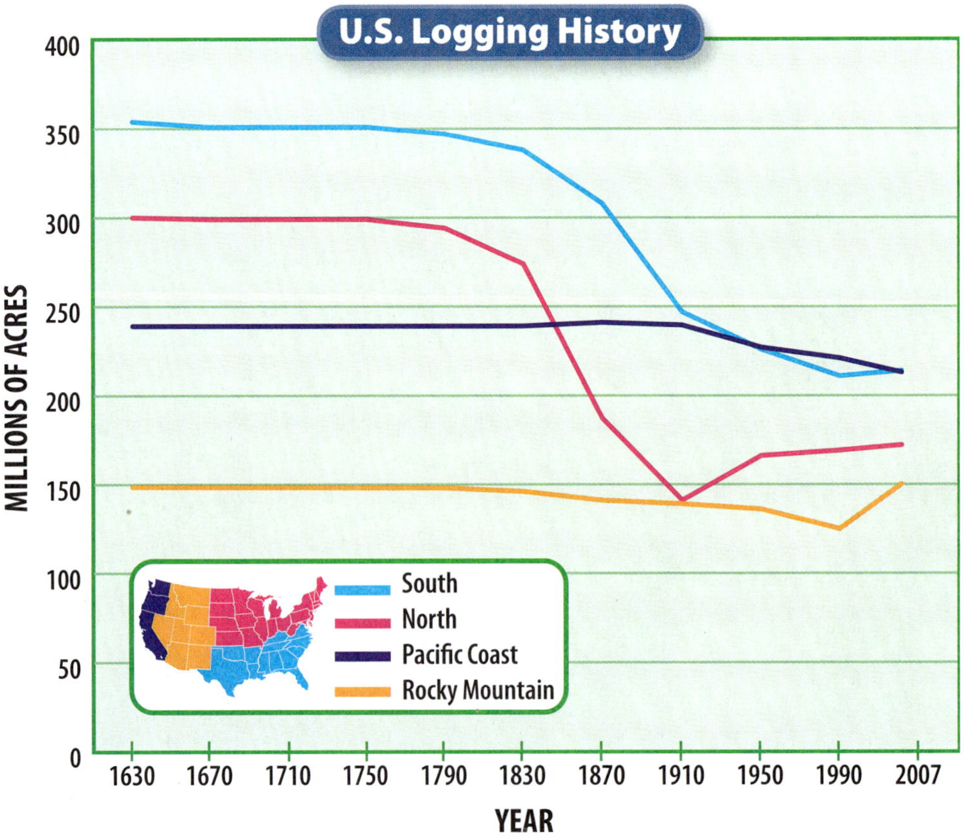

Created in 1853, Central Park in New York City was the United States' first landscaped public park. It includes about 843 acres.

Why is parkland important to residents in large cities such as New York City?

Environmental Stewardship Explain

In the mid-nineteenth century, virtually no thought was given to preserving natural beauty and ecosystems. Although throughout time some people, especially the native peoples, have looked to nature and wildlife with awe and reverence as God's creations and sought to leave sites of natural beauty untouched, most people were only concerned with their jobs and personal lives.

A few champions of conservation in the 1800s sought to preserve large tracts of land for the enjoyment of the people and to protect one of our most important natural resources—the land. Today an important part of being a good citizen is stewardship. **Stewardship** is taking action as an individual to improve the quality of the environment. Anyone can be an environmental steward. All it takes is being aware and knowledgeable about the world around you and doing as much as you can to avoid having a negative impact on the world. What can you do at school to practice environmental stewardship? What are the benefits of being a good steward of the land?

To help reduce problems associated with urbanization, governments can set aside land as national parks and preserves. These areas are largely protected from development. In addition to providing recreational activities for people, they are important habitats for wildlife.

In 1872, Congress established Yellowstone National Park "as a public park or pleasuring-ground for the benefit and enjoyment of the people." Since then, the federal government has created more than 1200 parks or preserves. Each park is accessible to citizens, but each is also carefully controlled to preserve the natural beauty of the land and the animals that live there. The first of Canada's 44 national parks, located in Banff, was established in 1885. What national park is closest to where you live? What are the features that attract people to it?

Scripture Spotlight

God gave the Israelites specific farming instructions to help preserve the land. See **Exodus 23:10–11**.

Each state and province in the United States and Canada preserves locations because of their natural beauty, historic interest, or recreational potential. Niagara Falls (1885) and Mackinac Island (1895) were the first state parks in the United States. Now more than 6800 state parks are spread over the 50 states. Thousands of provincial parks have been established in Canada for its citizens to enjoy nature at sites where natural beauty is virtually untouched.

Cities can expand green spaces, such as parks, within their boundaries. Grass-covered parks help reduce rates of heat absorption. Trees filter some pollutants from the air, help make the air cleaner to breathe, and provide important habitats for birds and other animals. Planting trees along city streets and encouraging gardens on city rooftops are other examples of good stewardship. How many of these measures are used in your community?

Densely populated European cities have found a solution for adding green space. Roof gardens are increasingly popular in Germany, Spain, and the United Kingdom. In fact, nearly 50% of German cities have building incentives for roof gardens.

Lesson Activity

Use the Internet or other resources to locate the state or provincial park nearest to your school. What is its size? When was it created? What recreational facilities are available? Does the area preserve important natural resources or protect endangered species?

How does the park help the environment and the people living in the area?

What effect will this rooftop garden have on the environment?

Lesson Activity

Make a plan to become a steward of the land and follow it. Report back once a month on how your plan is working.

What can you do personally in your home and neighborhood to protect Earth's environment?

Concept Check Assess/Reflect

Summary: How have people affected the landscape? Human population growth is increasing exponentially and is becoming more urbanized. The Industrial Revolution began the migration of people from rural to urban areas, which continues today. Much of the original landscape in North America has been changed for other uses. These include urbanization, farming, industries, grazing, and logging. Agricultural land is shrinking, leading to increased use of chemical fertilizers and pesticides. Overgrazing and deforestation lead to soil erosion. Industrial land use and mining generate toxic waste that pollutes the surrounding countryside. Mining disturbs large areas of land. Grazing and logging also result in the clearing of large tracts of forest. To preserve land, national and state park systems have been developed in the United States and Canada. To help protect the environment, each one of us can be a steward and do our part in protecting our valuable natural resources and the beauty of nature that surrounds us. Reclamation efforts by mining companies and logging companies have helped restore damaged areas back to their natural states. Techniques such as crop rotation and no-till farming can reduce soil erosion and increase the amount of water and organic matter in the soil.

1. How can urbanization affect the environment?
2. Compare the effects of agriculture and industry on land resources.
3. What is organic farming? How does it differ from commercial farming?
4. Give three examples of how you can help protect the environment in your home or school.
5. Describe two ways that land can be preserved for people to enjoy and appreciate.
6. What are some techniques that loggers can use to promote sustainability in forestry?
7. How can farmers increase yields while at the same time helping to protect the environment?

Lesson 3

Objectives

- Describe how people use water.
- Compare and contrast point-source and nonpoint-source pollution.
- Demonstrate how water resources can be conserved.

Vocabulary

point-source pollution

nonpoint-source pollution

Essential Question

How Have People Affected Water Resources?

In most parts of North America, clean water is a given. If you're thirsty, you turn on the tap and clean water comes right out. All kinds of drinks are at our fingertips. But nearly 800 million people on Earth do not have access to clean, safe drinking water. Dirty water kills. In developing countries, waterborne illnesses are responsible for 80% of all illnesses. Why do you think it is so difficult for some people to find clean water? Whose responsibility do you think it is to get clean water to these people? How would these groups go about getting clean water to these areas? What can you do to help?

These Ethiopian women gather water from a rainwater pool. Purification tablets make the water safe for use.

How People Use Water **Explain**

Your body needs water to survive. The most important way you use water is for drinking. In what other ways do you use water? How much water do you use in a day? How much does your family use?

Residential water use makes up a small percentage of the overall water use in North America. Another small percentage is used by public facilities, factories, mines, and ranches. Most water in the United States and Canada, however, is used for agriculture and industry.

You know that only a small fraction of freshwater is available for us to use in our homes and businesses. After the water is used, it must often be filtered and cleaned to remove waste products and other chemicals before it is returned to the environment. In cities and towns, water is cleaned in water treatment plants. In rural areas, many homes use septic tanks to dispose of wastes and depend on well water for their needs.

Scripture Spotlight

Psalm 42:1 compares the need of a deer for water with our need or longing for God. How is your need for water like your need for God?

Looking at the categories of water use, what do you think would be the best way to conserve water on a daily basis?

In addition, people use water in ways that do not involve its removal from the environment. For example, oceans, rivers, and lakes are used for transportation. Dams are built across rivers to provide hydroelectric power. Bodies of water behind these dams are also used for recreational purposes, including swimming, boating, fishing, and other uses.

Explore-a-Lab

Structured Inquiry

How much water can you save by taking a shorter shower?

How many gallons of water could you conserve by cutting the time it takes to shower by 50%? Turn the shower on and hold a bucket under the head. Have a friend (or parent) time the flow for 30 sec. Calculate how many gallons of water you would use for a 10-min shower. How many gallons of water would you save by taking a 5-min shower?

Water Pollution Explain

How do your activities affect the quality of water resources? Water resources can be impacted by both point-source pollution and nonpoint-source pollution.

Scripture Spotlight

Water was a precious resource in Bible times. Read **Genesis 26:19–22**.

Point-Source Pollution

Point-source pollution is water pollution that comes from a single point. This type of pollution can come from a sewage-outflow pipe, a storm drain, or factory discharge. Other examples include any ditch, well, channel, container, or vessel that releases pollutants. What are some examples of point-source pollution in your community?

Pouring motor oil, paint, solvents, herbicides, and pesticides down drains causes point-source pollution as well. These chemicals contaminate water resources and must be disposed of properly. Where should these materials be disposed of? Another example of point-source pollution is oil spills. These environmental disasters kill tens of thousands of marine animals and contaminate beaches and groundwater over an astonishingly large area. What are some examples of massive oil spills over the past 30 years?

Explore-a-Lab

Structured Inquiry

Which cleaning agent does the best job of removing the oil residue from the feather?

Fill an aluminum tray with water. Pour a thin layer of motor oil over the top of the water. The oil will float on the water. Your teacher will give you a feather. Drop it into the tray and submerge it. You will be given four different cleaning agents. Which did the least damage to the feather?

Pollution and Water Quality

Point-source contamination can be traced to specific points of discharge from wastewater treatment plants and factories or from combined sewers.

Air pollution spreads across the landscape and is often overlooked as a major nonpoint source of pollution. Airborne nutrients and pesticides can be transported far from their area of origin.

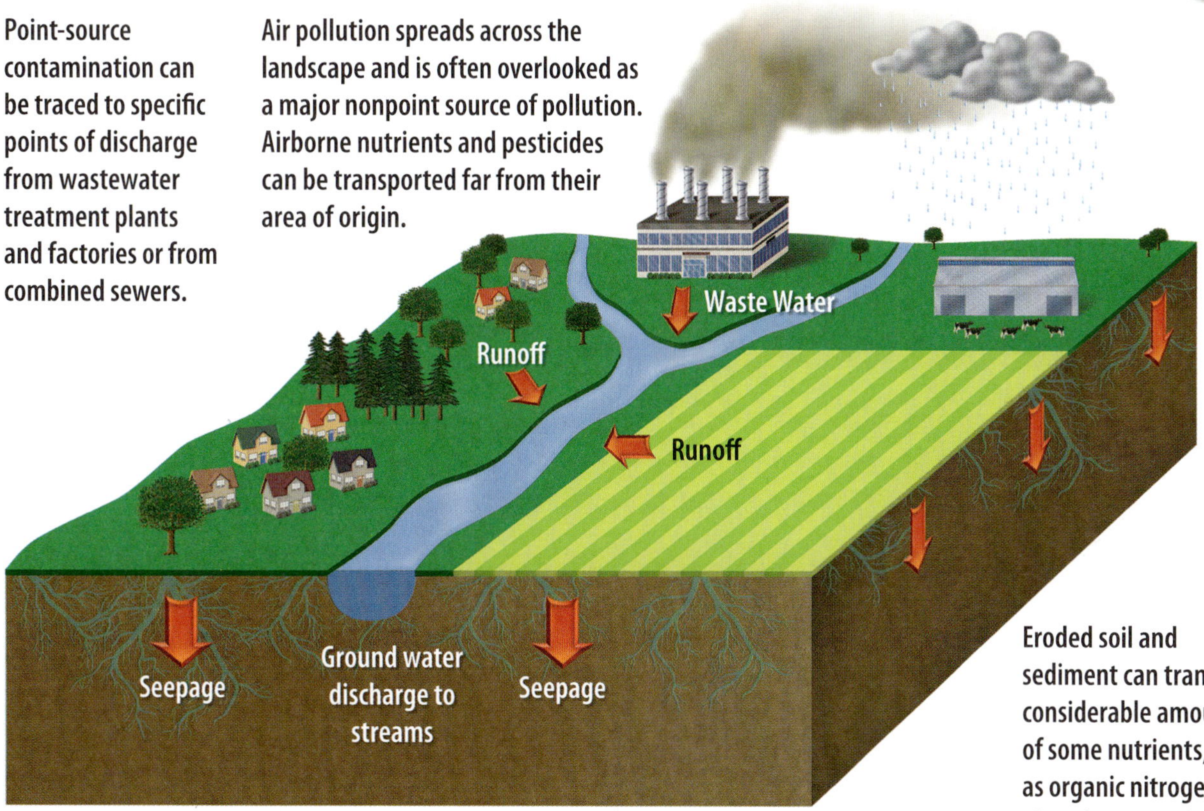

Eroded soil and sediment can transport considerable amounts of some nutrients, such as organic nitrogen and phosphorus, and some pesticides, such as DDT, to rivers and streams.

Both point-source pollution and nonpoint-source pollution affect water quality.

Identify sources of point-source pollution in the diagram.

Nonpoint-Source Pollution

Nonpoint-source pollution is pollution that is difficult to trace to a single source. It may come from several sources. Examples of nonpoint-source pollution include chemicals sprayed on farms, lawns, and mines. Because water is a good solvent, many substances can dissolve in it. Water can carry these dissolved substances into the ground or into streams and rivers. This pollutes groundwater supplies and disrupts aquatic ecosystems. What are some examples of nonpoint-source pollution in your community?

An astonishing example of nonpoint-source pollution can be seen in the Pacific Ocean. Ocean currents in the North Pacific form a large, rotating area of water. Litter washed from the land, as well as garbage dumped in the ocean, ends up traveling in a large, rotating surface current called the North Pacific Gyre. This area of wastes is estimated to be thousands of square kilometers in size.

The North Pacific Gyre is one of five main oceanic gyres. Exceptionally high amounts of debris that have been trapped by the gyre have created the Great pacific garbage patch.

What do you think might be the source of the debris?

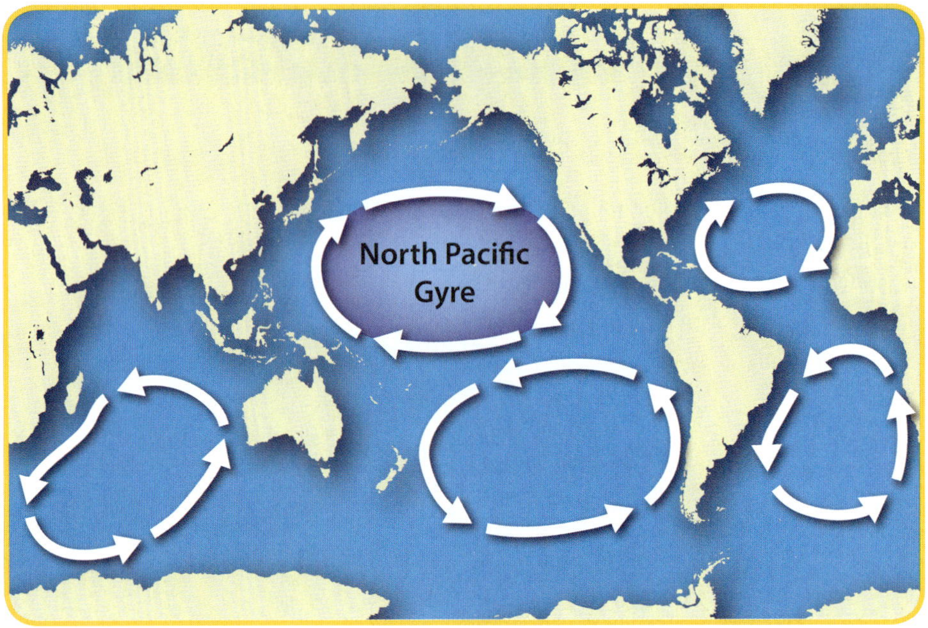

Another example of nonpoint-source pollution is acid rain. Recall that *acid rain* is precipitation that is highly acidic due to airborne pollutants that dissolve in water droplets in the sky. These chemicals, including sulfur dioxides and nitrogen oxides, are released by burning fossil fuels. The acidic precipitation can fall far from the factories, power plants, and vehicles that release the chemicals into the atmosphere. In water, they form sulfuric acid and nitric acid.

Power Plant Scrubbers

Power plants can use scrubbers to reduce the amount of pollutants released into the atmosphere.

1. Dirt and smoke particles move through a smoke stack.
2. The particles pass through a negatively charged metal plate and receive a negative charge.
3. The charged particles are attracted to positively charged collecting plates.
4. Collecting plates are knocked to remove the dirt and smoke particles.
5. Clean waste gas continues through the smoke stack.

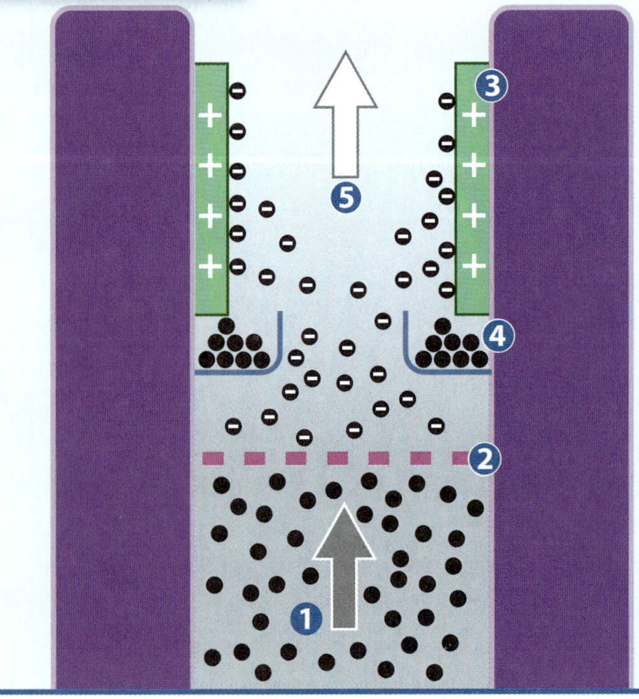

Structured Inquiry

Discover

Record your work for this inquiry. Your teacher may also assign the related Guided Inquiry.

Science Journal

It's Raining

What is the effect of acid rain on plant growth?

Procedure

1. Pour 5 mL of vinegar into 100 mL of distilled water, stir well, and check the pH. **Record** the pH in your ***Science Journal***. **Measure** and record the pH of the spring water.
2. Put one of the following labels on each cup:
   - Water philodendron
   - Acid philodendron
   - Water coleus
   - Acid coleus
3. Pour about 230 mL of spring water into the *Water-philodendron* and *Water-coleus* cups.
4. Pour about 230 mL of the vinegar/water mixture into the *Acid-philodendron* and *Acid-coleus* cups.
5. Measure and record the mass of each cutting.
6. Put one philodendron cutting into each philodendron labeled cup, covering the stem and part of the leaf with the liquid. Repeat with the coleus.
7. Add root hormone to each cup.
8. Set the cups where they are not likely to be spilled and where they will receive some daylight.
9. About every two days, check to be sure that the plant cuttings are still in the liquid. You may need to add more liquid if the cups become dry.
10. After one week, **compare** the new root growth of each plant in spring water with the new root growth of its corresponding plant in acid water. Measure the mass of each cutting. Record the results.
11. After two weeks, again **observe** the plant cuttings for new root growth, measure their mass, and record the results.

Materials
- 4 cups
- spring water
- distilled water
- white vinegar
- 100-mL graduated cylinder
- plastic spoon
- 2 cuttings of a philodendron plant (1 leaf and small amount of stem)
- 2 cuttings of a coleus plant (1 leaf and small amount of stem)
- pH paper
- safety goggles
- balance
- root hormone

Analyze Results

Place your measurements of root growth and mass of cuttings into a table for comparison. **Create a graph** of your data.

Create Explanations

1. What is the effect of acid rain on plant growth?
2. Which plant cuttings had the fastest root growth?
3. How might farmers reduce the damage to their crops resulting from acid rain?

Explore-a-Lab

Structured Inquiry

How can you model the effect of runoff on groundwater?

You will need two glass jars, landscape gravel, aquarium gravel, sand, potting soil, red food coloring, a spray bottle, and water. Fill two clean glass jars about one-third full of landscape gravel. Pour a cup of water into each jar. Add a layer of aquarium gravel. Add a layer of sand and then soil over the gravel. Label one of the jars *No Chemicals*. Label the other jar *Treated with Chemicals*. Place 5 drops of red food coloring on the soil in the chemical jar. Fill a spray bottle with water. Slowly spray an equal amount of water on the soil in the two glass jars. Stop when the water begins to seep into the pebbles and rocks. Observe and compare the groundwater that has accumulated in the bottom of each jar. Record your observations. Why should we be concerned about chemicals being released in the environment? How can we prevent this?

Water Conservation Explain

In general, it is easier to prevent pollution than to remove pollution from the environment. That's partly why governments pass laws to reduce pollution. For example, the Clean Water Act limits the amount of pollutants that can be released into the environment. The Safe Drinking Water Act in Canada helps protect supplies of drinking water by providing standards for water treatment plants. Why might they want these laws?

Farms and power plants in the United States use the most water. What steps can they take to conserve water? Farmers can use new drip irrigation technology that reduces the amount of water used to grow crops. They can also use drought-tolerant crops and seeds.

What can you do on a personal level? How can you help conserve water by properly disposing of household substances, such as paints and motor oil? Be sure to contact your local waste-disposal agency for guidance.

Other ways to conserve water include turning off the faucet when you brush your teeth or do dishes. Take shorter showers. Use nontoxic cleaning items and lawn products to help reduce harmful runoff. Buy energy-efficient appliances, such as dishwashers and washing machines. Share these tips with your friends and family to help educate others about water conservation. What other things can you do to help conserve water use?

Explore-a-Lab

Structured Inquiry

How much water is lost per day by a leaking faucet?

Gather a 100-mL graduated cylinder, large nail, small nail, plastic pan, two plastic water bottles, and water. Use the small nail to punch a hole in the side of a water bottle about 1 cm above the bottom. Leave the nail in the hole for now. Fill the bottle with water. Position the bottle so that as the water drips out of the hole, you can collect it in the plastic pan. Remove the nail from the hole, and allow the water to drip out of the bottle into the plastic pan for 5 minutes. Use the graduated cylinder to measure the amount of water that was collected. Record the data. Repeat the procedure but add a second hole that is the same diameter as the first to the water bottle. Repeat the procedure one more time with a new water bottle, but use a large nail to punch the hole in the bottle. If this were a leaky faucet, how much would your family have to pay for the water lost if water cost as much as gasoline? Why do you think many people are so reluctant to fix a leaky faucet?

Concept Check Assess/Reflect

Summary: How have people affected water sources? Only about 0.3% of Earth's freshwater is easily accessible surface water. Water is used residentially, industrially, agriculturally, and in generating electricity. Point-source pollution is pollution that is easily traced to a single source, such as a factory pipe that discharges pollutants into a stream. Nonpoint-source pollution comes from a variety of hard-to-trace sources, such as runoff and acid rain. New irrigation systems, scrubbers to remove pollutants, and conserving and protecting water resources in our daily lives help keep our water safe.

1. Identify four ways that you use water in a typical day and at least two ways to conserve this resource.
2. How does acid rain form?
3. Almost three-fourths of Earth is covered with water. Why is it necessary to conserve water resources?
4. A company is releasing smoke into the atmosphere. Is the smoke a form of point-source pollution or nonpoint-source pollution? Explain your reasoning.

Lesson 4

Essential Question

How Have People Affected Earth's Air Resources?

Objectives

- Describe how human activities affect the atmosphere.
- Describe ways to reduce air pollution.

Vocabulary

smog

chlorofluorocarbon (CFC)

The average person takes 28,800 breaths a day. Our lungs filter the air we breathe, removing dust and other particulates that hitchhike in with air ... to a point. Coal miners in the nineteenth and twentieth centuries worked underground, breathing in large amounts of coal dust every day. As a result, many died from black lung disease. What other health problems do you know about that are a result of air pollution? How do you think scientists measure levels of pollutants in the air? What pollutants do they look for?

Have you ever heard a weather report that included information about air quality? Often such information is given in a chart called an Air Quality Index (AQI). If air quality in a particular city is poor, the AQI might recommend that people limit their outdoor activities. Can you recall a day when air quality was poor? If so, how did that make you feel?

U.S. Air Quality Index

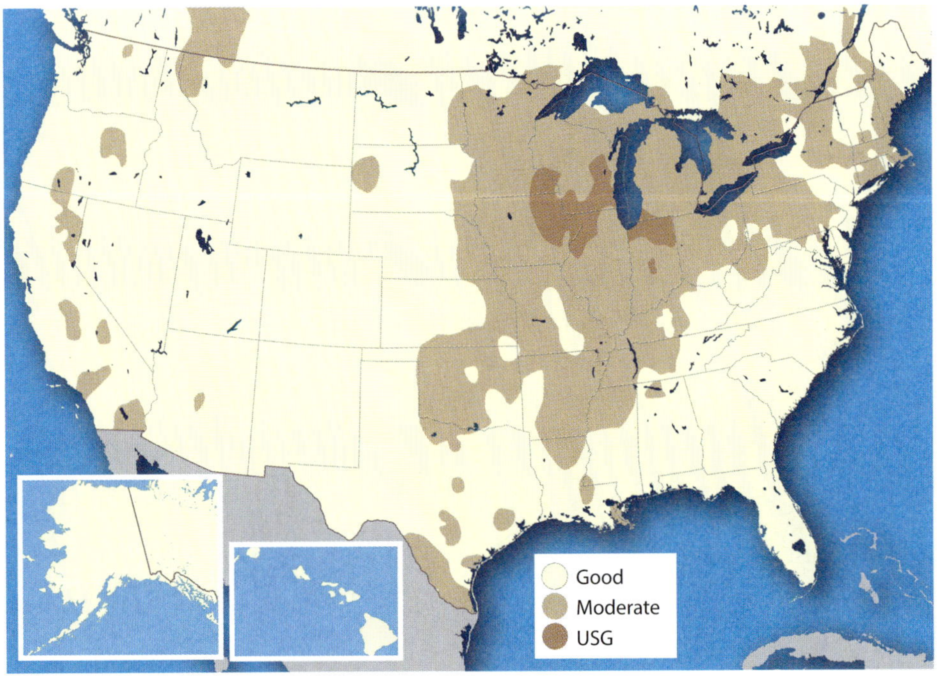

An AQI gives information about an area's air quality.

What is the AQI today in your community?

Air Pollution and Its Effects **Explain**

As you know, air pollution is any particle or gas in the air that affects the health or well-being of living organisms. During the past century, human-caused air pollution has increased and has become a growing threat to our health. What types of particles or gases make up air pollution? Which are human made? Which have natural origins? Is human-made air pollution more harmful than natural air pollution? Why or why not?

Smog

The picture shows smog hanging over a major city in the United States. **Smog** is a yellow-brown haze formed when sunlight reacts with pollutants that are released from the burning of fossil fuels into the air. What do you think caused the smog in this city? Where do you think you would be most likely to find smog? How would you feel if you had to breathe this air?

Ozone Depletion

One of the primary components of smog is ozone. You will recall that *ozone* is a molecule made up of three oxygen atoms. It forms when the ultraviolet (UV) rays of sunlight split a regular oxygen atom. While the ozone layer in our atmosphere protects Earth from harmful UV radiation, ozone on the ground is hazardous to human health. Ground-level ozone is formed when vehicle exhaust and organic compounds interact in the presence of sunlight. Why would some of the sunniest cities be some of the most polluted?

Chlorofluorocarbons, also known as **CFCs**, are chemical ingredients used in the manufacture of aerosol sprays, packing materials, and refrigerants. CFCs are major contributors to the destruction of the ozone layer in Earth's atmosphere. Without the ozone layer, Earth's average temperature would drop from the present-day 16°C to a very chilly –18°C. How do you think your daily life would change if you lived where the average temperature was much colder?

Scripture Spotlight

Read **Genesis 2:15**. How does this passage relate to air-quality issues today?

Atmospheric conditions can trap pollution near the ground, causing thick smog such as this.

Global Warming

Using tree ring and ice core data, scientists have documented that Earth's climate has changed dramatically many times in the past. Periods of very warm temperatures have been followed by periods of bitter cold. These cold periods are known as ice ages. These climate changes have occurred over significant amounts of time, over many thousands of years in some cases. What kinds of events do you think caused these changes?

Since the beginning of the Industrial Revolution, the warming of the planet has accelerated. Scientists estimate that the average temperature of Earth may increase by as much as 3°C over the next 100 years. This may not seem like a lot, but even a small rise in temperature can have widespread effects. Glaciers and ice caps are already melting at rates never before recorded, causing a rise in sea level and habitat loss in arctic ecosystems. In addition, the change in global temperature is warming the surface of the ocean, causing water to expand and increasing evaporation. How might this increase of water vapor and energy in the atmosphere affect the frequency of severe storms? How might an increase in global temperatures affect species on Earth as a whole and in specific regions of Earth?

Many reasons have been suggested as the cause of this warming. Some scientists think that this has happened because of increased release of *greenhouse gases* due to increased use of fossil fuels. What are some greenhouse gases? How do greenhouse gases raise temperatures in our atmosphere?

Explore-a-Lab

Structured Inquiry

What does air pollution look like?

Your teacher will assign your group one item that emits energy. It might be a long fireplace match, paraffin wax candle, soy-based candle, alcohol or kerosene lamp, or Bunsen burner. You will also need a flat glass plate or test tube. Light your energy source and hold the glass plate or test tube above the flame for exactly one minute. You may want to hold the glass plate or test tube with tongs or an oven mitt to avoid burns. Extinguish your flame and place your plate or tube next to your energy sources. Record how much soot you collected and compare it with the other energy sources tested by classmates. Which source gave off the most pollution? Why do you think this source had the most pollution? To examine your pollution more closely, scrape some soot onto a microscope slide and view the particles under low power. Draw what you see. Why should pollution be a concern for all of us?

Structured Inquiry

Discover

Record your work for this inquiry. Your teacher may also assign the related Guided Inquiry.

Science Journal

It Feels Like a Sauna!

How does the greenhouse effect change temperature?

Procedure

1. Tape the thermometers in the bottom of the shoeboxes so they can be easily read without picking them up.
2. **Record** the beginning temperature reading of each thermometer in your ***Science Journal***.
3. Cover one shoebox with the acrylic sheet and place it in direct sunlight.
4. Place another uncovered box next to the box with the acrylic sheet. Be sure that it also is in full sunlight.
5. Place the third box (uncovered) in a shady area.
6. Record the temperature of each shoebox every 2 min for 30 min.
7. **Construct a graph** that shows your data.

Materials
- acrylic sheet (20 cm × 40 cm)
- 3 plastic shoe boxes
- tape
- 3 thermometers

Analyze Results

According to the data, how does the temperature change of each box compare?

Questions

1. How does the greenhouse effect change temperature?
2. Which shoebox had the highest temperature?
3. What human activities contribute to an increase in greenhouse gases?
4. How does the acrylic sheet covering the shoebox act like Earth's atmosphere?

Explore-a-Lab

Structured Inquiry

What will be the increase in global temperature between 2010 and 2050?

The link between increasing amounts of carbon dioxide (CO_2) in the atmosphere and increasing temperatures on Earth is shown on the graph. What was the increase in Earth's temperature between 1880 and 2010? Sketch the graph and extend it to 2050.

Global Temperature and Carbon Dioxide

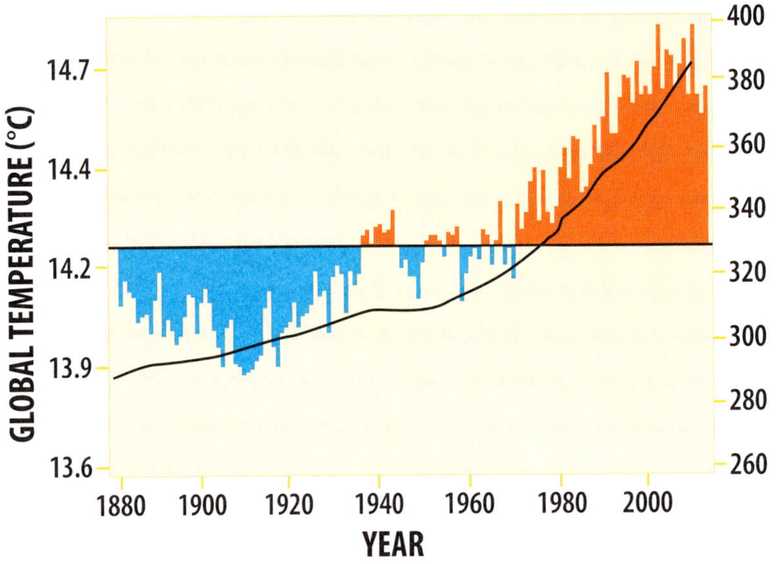

Reducing Air Pollution Explain

A concerted effort by the U.S. federal government to regulate the automobile industry and to demand the production of more energy-efficient cars is helping to reduce air pollution problems in some cities of North America. In California's Los Angeles Basin, levels of some vehicle-related air pollutants have decreased by about 98% since the 1960s, even as area residents now burn three times as much gasoline and diesel fuel. Between 2002 and 2010 alone, the concentration of air pollutants emitted by cars dropped by half.

Two other government regulations have contributed to cleaning up the atmosphere. In the United States, the Clean Air Act, passed in 1990, works to protect and improve air quality. In 2006, the Canadian Clear Air Act became law. Levels of many air pollutants in North America have been reduced since these acts were implemented.

International agreements have been implemented to clean up Earth's air worldwide. In 1987, nations signed the Montreal Protocol to Reduce Substances that Deplete the Ozone Layer, a global environmental treaty that had a provision to reduce the 1986 levels of CFCs by 50% before the year 2000. An amendment approved in 1990 called for elimination of the production of CFCs by 2000. The production of this chemical ended for the most part in 1996. The chemistry that was the basis for the Montreal Protocol was developed by several scientists, who received the 1995 Nobel Prize in Chemistry for recognition of their work. As a result of these activities, the ozone layer has begun to recover. What other international efforts have countries worked on to protect the planet?

All UN recognized nations have ratified the Montreal Protocol, which prohibits the use of CFCs in products such as aerosol sprays.

Find out about recent developments under the Protocol that have been proposed by the United States, Canada, and Mexico.

Concept Check Assess/Reflect

Summary: How have people affected Earth's air resources? The quality of air is affected by human activities, such as burning fossil fuels. Smog is a brown haze formed when sunlight reacts with air pollutants that are released when fossil fuels burn. CFCs are chemicals that contribute to the destruction of the ozone layer. CFCs are found in aerosol sprays and other products. Both domestic and international governments create and agree to regulations to reduce pollution and protect the environment.

1. What human activities affect the atmosphere?
2. Identify three ways to reduce air pollution.
3. A friend says that there is no global warming. How would you respond? What evidence would you present?

People in Science

Extend

Get to Know John Muir

John Muir was a Scottish-born American naturalist, author, and advocate for the preservation of wilderness in the United States. Today he is referred to as the "Father of the National Parks." Born in 1838, he moved from Scotland with his family to a Wisconsin farm. He went on to study geology and botany at the University of Wisconsin. In 1867, after recovering from a factory accident that left him temporarily blinded for months, he left a promising career in industry to walk from Indiana to Florida, creating botanical sketches on his way. From there he sailed to California and then walked from San Francisco to the Sierra Nevada—the place that would change his life—and surrendered to nature.

Muir's letters, essays, and books about his adventures in nature, especially in the Sierra Nevada, continue to be read today. Muir became particularly concerned about natural landscape preservation. He focused his attention on the destruction of natural resources in areas surrounding the state-administered Yosemite Land Grant. After inviting a magazine editor onto the land, Muir convinced him that the area could only be saved if it were incorporated into a national park. The magazine published an article that initiated a bill in the U.S. Congress, making the Yosemite National Park a reality in 1890. Muir's activism helped to preserve Sequoia National Park and other wilderness areas. The spiritual quality and enthusiasm toward nature expressed in his writings inspired readers, including presidents and congressmen, to take action to help preserve large nature areas. He taught people the importance of experiencing and protecting our natural heritage. In addition to Yosemite and Sequoia National Parks, his activism contributed greatly to the creation of Mount Rainier, the Petrified Forest, and Grand Canyon National Parks. Dozens of places are named after Muir, including the Muir Woods National Monument, the John Muir Trail, and Muir College.

Called to Serve

Muir felt a spiritual connection to nature; he believed that humankind is just one part of an interconnected natural world, not its master, and that God is revealed through nature.

In 1892, John Muir and a number of his supporters founded the Sierra Club. Muir served as the club's president until his death in 1914. Today, the Sierra Club has more than 1.3 million members and supporters and is the oldest, largest, and most influential grassroots environmental organization in the United States.

Concept Check

1. John Muir's activism has had a major impact on wilderness preservation today. What was one of Muir's greatest concerns?
2. How did John Muir influence presidents and congressmen to preserve large nature areas?

Land Planning Coordinator

Careers in Science

Extend

If you are dedicated to the intelligent use and protection of our natural resources, you might want to investigate a career in land planning coordination. You will consult with specialists in wildlife, archaeology, and recreation, as well as citizens and community leaders, engineers, and architects as you prepare plans for land use. Your plans will establish a community's vision for the future and outline a roadmap for achieving that future by guiding land use and development.

The early stages of a plan involve thinking about natural areas and wildlife. By considering natural resources during the initial planning stages, planners have the opportunity to create ways to conserve vulnerable habitats while planning for future development.

Some land-use plans do not address natural areas and wildlife, making it difficult for communities to know what actions to take to protect these areas. It is important to have good natural area and wildlife information, with a clear outline of how the community can conserve those resources. A good plan will reflect local conditions, concerns, and resources and can be easily understood by all community members, including citizens, developers, builders, and public officials.

To accurately focus on natural resources in an area, planners should consider some basic principles:

- Natural features are part of the community's character, and they provide ecosystem services, including water purification, that contirbute to the area's quality of life.
- Planning ahead protects natural resources better than a reactive response to environmental concerns.
- Planning helps reduce conflict by raising awareness of environmental concerns before the project begins.
- Natural resources do not always follow community boundaries.

Land planning coordinators have to relate natural areas and wildlife to other important issues in the community. Showing that these relationships have been considered will help decision makers. For example, actions that protect natural resources can be considered together with goals to enhance housing availability. Highly populated areas can be directed away from sensitive habitats. Transportation corridors can have a tremendous effect on natural areas and wildlife. Placement of new roads can be coordinated with resource-protection goals.

Including natural resource protection in a community's vision is the first step in creating protective programs, policies, and legislation. A plan is the basis for constructing ordinances and laws as it guides leaders in making good decisions to follow the community's vision for development.

Education requirements for land planning coordinators vary greatly. Some coordinators learn their skills on the job as assistants. Others obtain two-year and four-year degrees.

Concept Check

1. Why is it so important to consider natural areas and wildlife at the start of land planning?
2. Describe one of the basic principles that land planners use to focus on the area's natural resources.

Chapter 9 Review Assess/Reflect

Study Guide

Lesson 1

1. People use natural resources to make products such as house paint, DVDs, and sunglasses. We also use air to breathe, land to stand on, and water to drink.
2. A material resource is one that is used to make other products. An energy resource is a source of energy that is used to produce other types of energy.
3. A renewable resource can be replaced by natural processes as fast as, or faster than, it is used, such as sunlight. A nonrenewable resource cannot be replaced by natural processes in a reasonable amount of time, such as fossil fuels.
4. Overuse of resources can adversely affect the environment. The wise use of resources can help reduce some of these effects. A government, for example, can pass laws to reduce the amount of pollutants released by vehicles. A business can recycle used paper. A person can turn off lights when he or she leaves a room. These measures help conserve resources.

Lesson 2

1. Urbanization can result in urban sprawl, in which urban areas expand into surrounding countryside. Urbanization can increase heat absorption rates in cities and increase rates of energy use.
2. Agricultural practices can result in pollution from fertilizers and pesticides, bare soil from overgrazing, and nutrient depletion in the soil. Industrial wastes from manufacturing processes can pollute land. Mining processes can disturb large portions of land by removing the top layers of soil.
3. Good farming practices, such as crop rotation, contour plowing, strip farming, and no-till farming, can protect the environment. Mine reclamation can restore land to useful purposes. New techniques and stricter regulation have significantly improved the efficiency of and reduced the negative impact of industry and agriculture.

Lesson 3

1. Most water in North America is used for farming and generating electricity.
2. Human activities cause point-source pollution (such as wastewater treatment plants and factories) and nonpoint-source pollution (such as chemicals sprayed on crops) of water resources.
3. Turning off the faucet when you are brushing your teeth or doing dishes and taking shorter showers are ways to conserve water resources.

Lesson 4

1. Air pollution can result in smog, ozone depletion, and global warming. Ozone depletion reduces the ability of the atmosphere to filter out the harmful UV rays of the Sun. Global warming can cause melting ice caps and rising sea levels.
2. Government regulations are one way to reduce environmental pollution.

Vocabulary Check

Fill in the blank with the correct vocabulary word.

1. Lumber is used to make other products so it is a(n) __________ resource.
2. The uncontrolled spread of urban development into neighboring areas is __________.
3. Acid rain is a type of __________ that cannot be easily traced.
4. The restoration of land that has been mined to a natural or usable state is __________.
5. A practice in which loggers balance the use of the forests for wood production with protecting the ecosystem and the environment is __________.
6. __________ is a yellow-brown haze formed when sunlight reacts with pollutants that are released from the burning of fossil fuels into the air.
7. A discharge pipe from a factory that pours polluted water into a stream is an example of __________.
8. __________ are chemical ingredients used in aerosol sprays that are major contributors to the destruction of the ozone layer in Earth's atmosphere.

Multiple Choice

Choose the best answer.

9. Which is a nonrenewable energy resource?
   A. geothermal energy
   B. nuclear energy
   C. solar energy
   D. wind energy

10. Which is an example of stewardship?
    A. A man drives to work every day.
    B. A girl tosses her soda can out the window of her car.
    C. A farmer takes action to improve the quality of her soil.
    D. A boy keeps his television on during the night to help him sleep.

11. The farming practice that involves growing alternating rows of cultivated plants and sod-forming plants to reduce the rates of soil erosion is __________.
    A. crop rotation
    B. contour plowing
    C. no-till farming
    D. strip farming

Check Point

Answer the following questions.

12. **Communicate** a way to use water energy to do work.

13. Your task is to reduce pollution in a lake. Which type of pollution will most likely be harder to control: point-source pollution or nonpoint-source pollution? Explain your answer.

14. Choose a renewable resource that could replace fossil fuels as an energy source for heating homes. Tell why the resource you chose would be a more environmentally friendly choice.

Chapter 10

Earth's History

Scripture Spotlight

God created a dynamic Earth, so it has undergone many changes since Creation. By studying these changes, you can better understand God's plan. You will read the following passages in this chapter.

Proverbs 8:27–29 (p. 367)
Genesis 7–11 (p. 370)
Psalm 29:10 (p. 376)
Genesis 7:21–22 (p. 391)
Genesis 1 (p. 393)

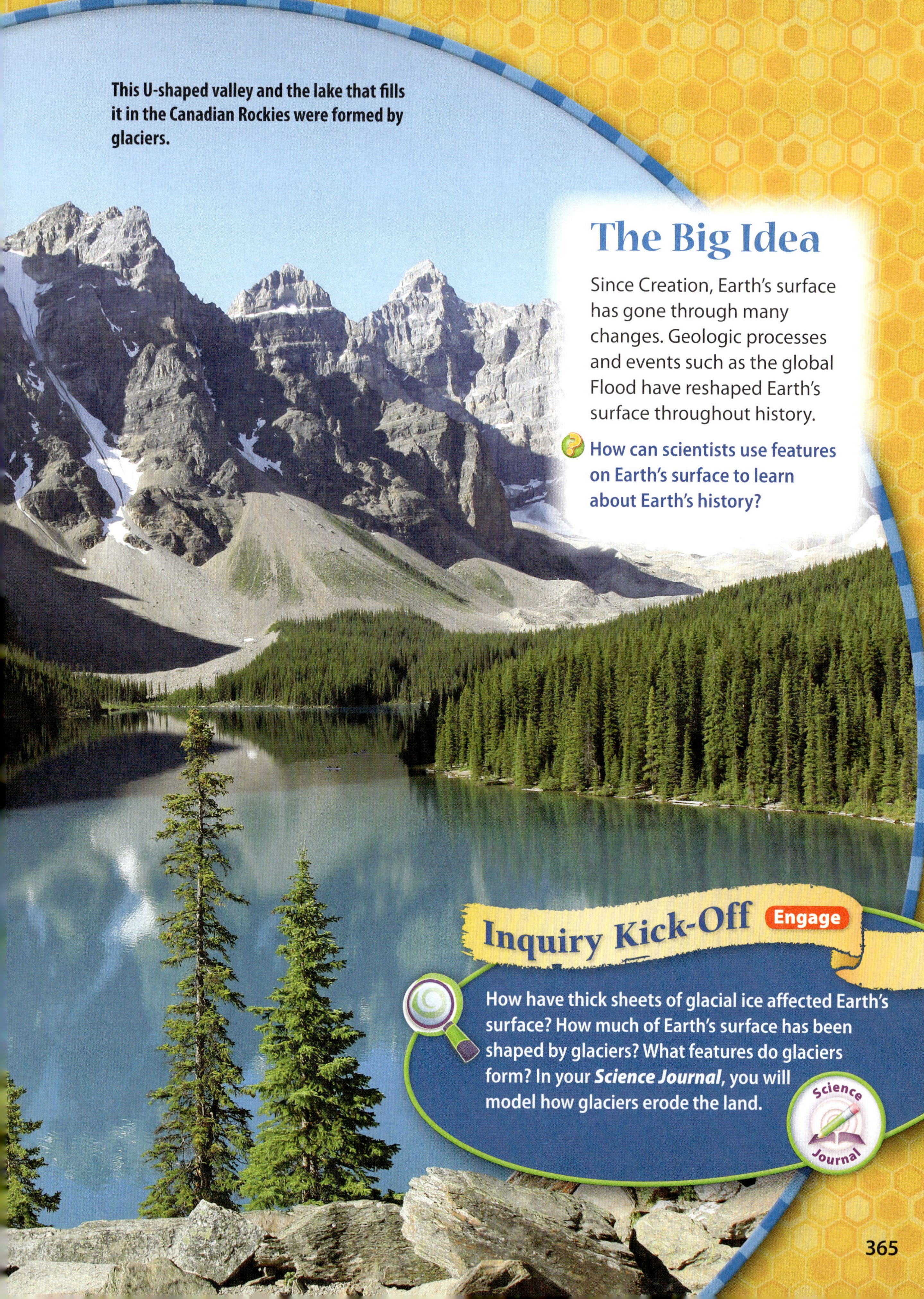

This U-shaped valley and the lake that fills it in the Canadian Rockies were formed by glaciers.

The Big Idea

Since Creation, Earth's surface has gone through many changes. Geologic processes and events such as the global Flood have reshaped Earth's surface throughout history.

How can scientists use features on Earth's surface to learn about Earth's history?

Inquiry Kick-Off Engage

How have thick sheets of glacial ice affected Earth's surface? How much of Earth's surface has been shaped by glaciers? What features do glaciers form? In your ***Science Journal***, you will model how glaciers erode the land.

Science Journal

Lesson 1

Essential Question

How Does Earth Change?

Objectives

- Describe the theory of plate tectonics and how plates move relative to each other.
- Describe the geologic changes that can occur as a result of plate movements.
- Explain how weathering, erosion, and deposition change Earth's surface.
- Explain how the evidence at Mount St. Helens supports rapid geologic processes.
- Summarize the evidence that supports the worldwide Flood described in **Genesis**.

Vocabulary

tectonic plate
mass movement
deflation
abrasion
intrusion
unconformity

Do you think Earth looks the same today as it did when God created it? In what ways do you think our planet has changed throughout history? What, if any, features of Earth's structure have remained the same?

Changes in Earth's Lithosphere Explain

Recall what you have learned about Earth's layers. The outermost layer is called the crust and consists of rocky materials that form Earth's surface. Below the crust is the mantle, which consists of an upper part that is rocky like the crust. Together, the crust and the upper part of the mantle form the lithosphere. The lower part of the mantle, called the asthenosphere, is not as solid or dense as the upper part. The material in the asthenosphere is molten because of the intense heat and pressure. Beneath the mantle is the core. The outer part of the core consists of molten metals, such as iron and nickel. The inner part of the core consists of a dense ball of solid iron and nickel.

Summary of Earth's Layers

Crust	Thin, rocky outermost layer of Earth that includes: • Continental crust (less dense) • Oceanic crust (more dense)	Lithosphere (crust + upper mantle)
Mantle	Upper: Made mostly of solid rock	
	Lower: Made of partly molten rock	Asthenosphere
Core	Outer: Made of liquid iron and nickel	
	Inner: Made of solid iron and nickel	

Plate Tectonics

As you know, Earth's lithosphere is broken into huge slabs called **tectonic plates**, which rest on top of the plastic-like and easily molded asthenosphere. Plate tectonics refers to the theory that these plates move across Earth's surface. Plates move at different rates and in different directions. For example, the Pacific Plate is moving to the northwest at a speed of between 7 cm and 11 cm per year. In contrast, the North American Plate is moving toward the southwest at a rate of 2 cm per year. Many scientists believe that this plate movement is responsible for major geologic change during Earth's history. Which plate would result in greater geologic change—the Pacific Plate or the North American Plate? Explain the reason for your choice.

Scripture Spotlight

Which of God's activities described in **Proverbs 8:27–29** might relate to this section?

Earth's Tectonic Plates

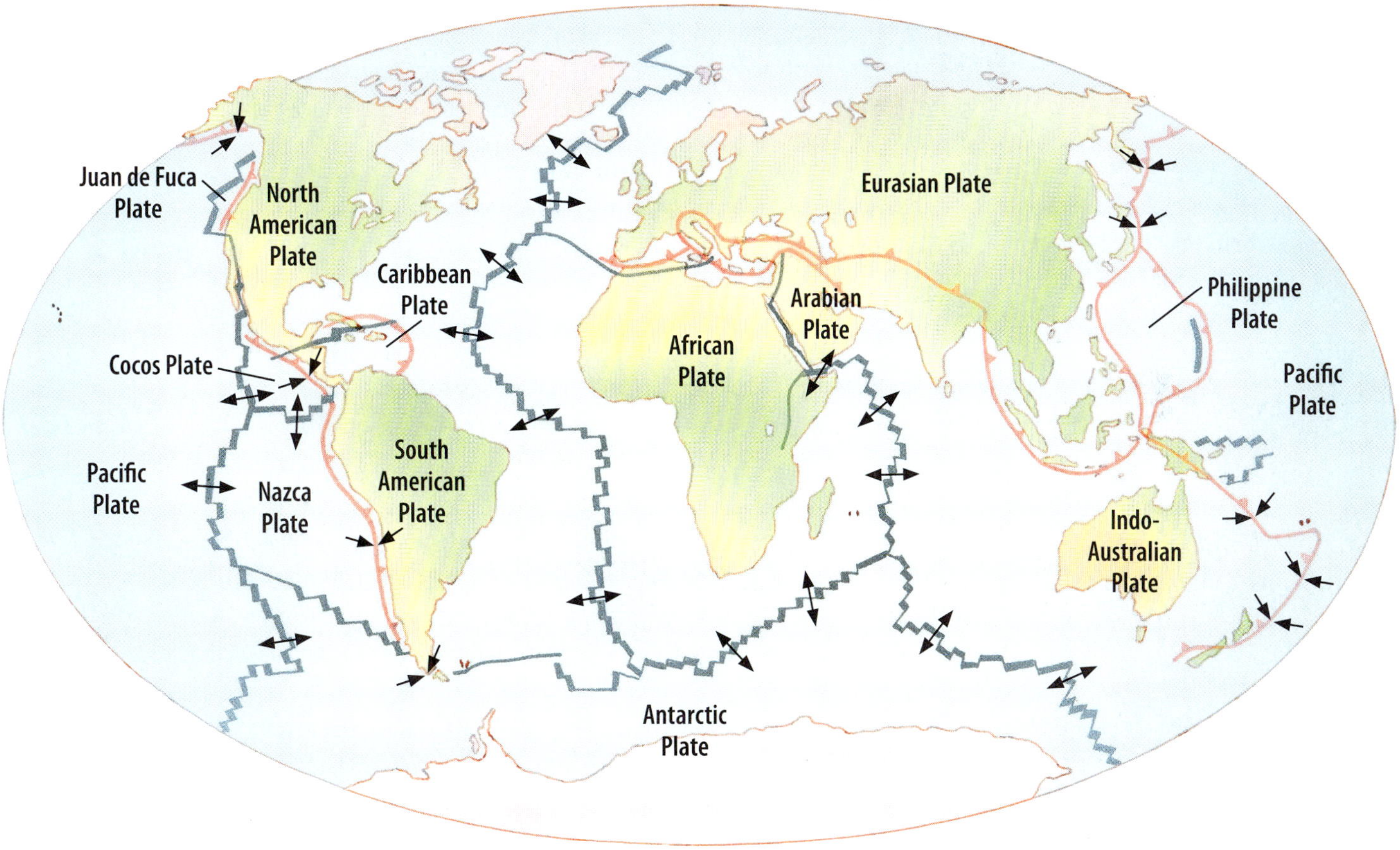

The lines in the diagram show the edges of tectonic plates. The arrows show the directions in which they are moving.

If a plate moves an average of 4 cm/yr, how long will it take to move 1 m? What might speed this process up?

Plate Boundaries

Tectonic plates move in three different ways. They can slide past one another, diverge (pull apart from one another), or converge (come together or collide).

Types of Boundaries

Type	Description
Transform 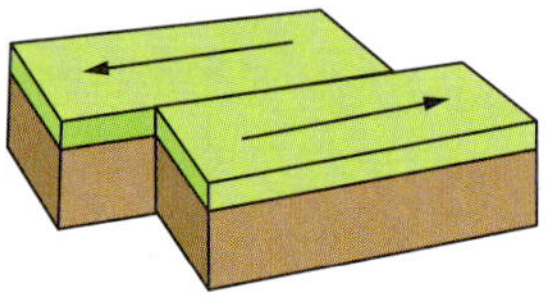	Earthquakes are common when tectonic plates slide and grind past one another. Example: The San Andreas fault in California is a 1300-km fault zone where the Pacific and the North American Plates slide past each other, causing fairly frequent earthquakes.
Divergent 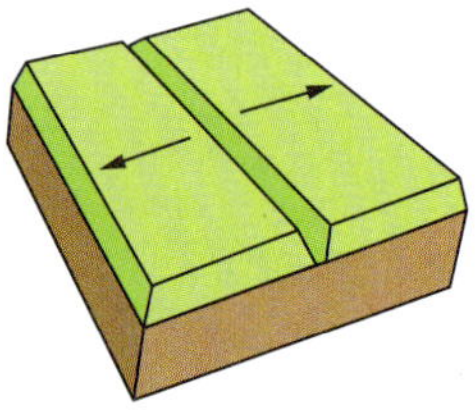	Mid-ocean ridges can form when two oceanic plates pull apart, allowing magma to push through and create new oceanic crust. Example: The Mid-Atlantic Ridge is part of the longest mountain range in the world, separating the North and South American Plates on the west from the Eurasian and African Plates on the east. The current average spreading rate is 2.5 cm per year.
Convergent (oceanic crust and continental crust) 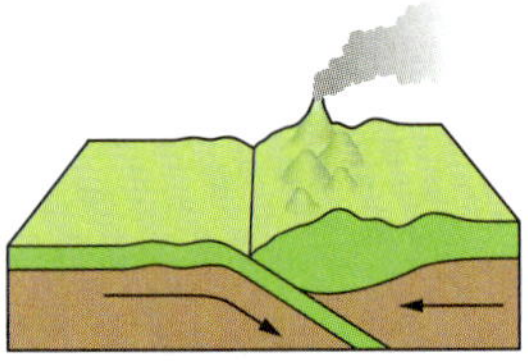	When an oceanic and a continental plate collide, the oceanic plate, because of its greater density, bends down and sinks under the continental plate and eventually becomes incorporated into the mantle. This process is called *subduction*. Earthquakes and volcanoes are common in subduction zones, along the boundary between oceanic and continental crust. Mountain building can sometimes occur as well. Example: Along the Peru-Chile trench, the Nazca Plate is being subducted under the South American Plate, which responds by crumpling to form the Andes.
Convergent (two oceanic plates) 	When two oceanic plates collide, the denser of the two is subducted. A curved volcanic mountain chain forms above the subducting plate and rises out of the ocean as a chain of islands known as an *island arc*. An example of an island arc is the Philippines, which consist of a chain of volcanic islands. Earthquakes and tsunamis can originate in these collision areas. Example: The subduction of the Pacific Plate under an oceanic portion of the North American Plate caused the formation of the Aleutian Peninsula of Alaska.
Convergent (two continental plates) 	When two continental plates collide, mountains can form due to folding and buckling of the crust. Example: The formation of the Himalayas, located along the border between India and Tibet, can be explained as the result of the collision of the Eurasian and Indian Plates.

Structured Inquiry

Discover

Record your work for this inquiry. Your teacher may also assign the related Guided Inquiry.

Science Journal

A Faulty Candy Bar

How do Earth's plates fit together?

SAFETY: If you have a food allergy such as peanuts, use precautions when working with the candy bar.

Materials
- candy bar
- colored pencils

Procedure

1. Look at the diagram in your ***Science Journal*** and think about how the structure of a candy bar can be used as a **model** for the structure of the Earth. On each cross-section, label the layers that represent the crust, the mantle, and the core.
2. Use your fingernail to make a few breaks across the "crust" or top of the candy bar.
3. Illustrate tension forces by gently pulling on the ends of the candy bar. Describe what you see and draw a picture of what the candy bar looks like when you pull the ends. **Record** your **observations** and make a drawing in your ***Science Journal***. Push the candy bar back together.
4. Illustrate shear forces by sliding one half of the candy bar forward (away from you) and the other half backward (toward you). Observe what the candy bar looks like when one side is pushed forward and one half is pushed backward. Record your observations and make a drawing in your ***Science Journal***. Move the candy bar back together.
5. Illustrate compression forces by pushing both ends of the candy bar together. Describe what you see and draw a picture of what the candy bar looks like when the two ends are pushed together. Record your observations and make a drawing in your ***Science Journal***.

Create Explanations

1. How do Earth's plates fit together?
2. What do the "cracks" you made in the surface of the candy bar represent in Earth's crust?
3. Tension is a force that pulls on the plates of Earth's crust, causing them to move apart. What are two things you observed when you created tension forces on the candy bar?
4. What are two things you observed when you created shearing forces on the candy bar?
5. What are two things you observed when you created compression forces on the candy bar?

Scripture Spotlight

What is described in **Genesis 7–11**? How might this be related to what you have been reading?

Supercontinents

Related to the theory of plate tectonics is the idea that supercontinents have formed and broken apart throughout Earth's geologic past. Some scientists think that a supercontinent named Rodinia formed about 1.2 billion years ago and broke up about 750 million years ago. They think a second supercontinent called Pangaea formed about 250 million years ago and broke up over many millions of years, forming the continents that exist today. Do you think these time estimates are based on catastrophism or uniformitarianism? How do you think these time estimates were calculated? Can you tell anything about the worldviews that may have influenced these conclusions? What other possibilities could explain the formation of supercontinents?

Changes in Earth's Landscape Explain

In addition to the changes caused by Earth's moving plates, Earth's landscape is also changed by weathering, erosion, and deposition. Weathering is the process by which rocks and other Earth materials are broken into smaller pieces by physical processes or chemical changes.

Erosion and deposition are related processes that move sediments to a new location. Once a rock has been physically or chemically broken down by weathering, it can be eroded from its original location by gravity, glaciers, wind, or water. What can cause erosion to happen faster? Often these agents transport the eroded materials for very long distances. Deposition happens as these agents lose energy to carry the sediments farther and drop them in a new location. Why are some sediments transported for only a short distance?

Explore-a-Lab

Structured Inquiry

How long do sediments take to settle out of water once they stop moving?

You will need aquarium gravel, coarse sand, fine sand (washed), unwashed sand, an empty plastic 0.5-L bottle, water, and a stopwatch. Add about 2–4 cm of each solid material to the empty bottle. Use a funnel if needed. Then add water to fill the bottle completely. Put the top on securely. Shake the bottle and contents vigorously for 30 sec. Use the stopwatch to time how long each sedimentary material takes to settle out of the water. Record the times. Graph your data. In what order did the sediments settle? Did all the sediment settle out? Why or why not? How could you decrease the time for the turbidity to clear? How long do you think the water will take to become as clear as it was at the beginning of the activity? Where would you expect to find the finest sediments on the ocean floor: near the edges of the continents or out away from the edges of the continents? Explain.

The processes of weathering, erosion, and deposition create amazing structures both above and below the ground.

How do you think water might play a role in the formation of these caves?

Forces That Change Earth's Landscape

Erosion Moving weathered materials	**Deposition** Dropping eroded sediments
Gravity	
Gravity can cause rocks and soil to slide downhill. Landslides, mudslides, and rock falls—collectively called **mass movements**—can be triggered by earthquakes and often slide downhill more quickly following heavy rains.	When mass movements can go no further, sediments are deposited.
Glaciers	
Glaciers, large masses of slowly moving ice, can act like giant bulldozers by picking up and carrying dirt, rocks, and boulders. As they move, they flatten hills and carve out valleys.	When a glacier begins to melt, the materials that were picked up by the ice are deposited, forming *eskers* and *moraines*.
Wind	
Wind can carry bits of sediment as it blows across land. This is called **deflation**. **Abrasion** occurs when these bits of sediment strike other rocks, slowly weathering and reshaping them. Is abrasion an example of mechanical or chemical weathering?	Like water, wind drops its sediment load when it loses its energy. These deposited sediments can sometimes form features on Earth's surface. Dunes in deserts, for example, are built up as sand is deposited by wind.
Water	
Flowing rivers erode Earth's surface. Rapidly moving water has enough energy to move large rocks and even boulders. Heavy rains can wash sediment downhill or into rivers. Ocean waves cause erosion along coastal areas. Crashing waves can carry broken bits of sediments to new places.	As water slows down, it has less energy and deposits larger rocks. As it slows further, smaller rocks, sand, and silt are deposited. When a river enters a lake or an ocean, the water slows and the sediments are deposited in a triangular-shaped *delta*. When rivers flood, they cover surrounding land with sediment, forming a *floodplain*.

Clues about Past Changes **Explain**

The geologic record is filled with fascinating and complex rock formations. If you know what to look for, you can figure out a lot about their history by examining the clues they contain. These clues offer information about the order of events that created the formation and the rate at which these geologic processes occurred. Certain principles are valuable in unraveling the mysteries of rock formations.

Geologic events can cause layers of rock to become tilted.

What geologic processes might explain why these layers are no longer horizontal?

Principle of Original Horizontality

The principle of original horizontality states that layers of sediment are deposited in horizontal layers. If you see rock layers that are no longer horizontal, that is generally a clue that something dramatic has happened to them in the past. For example, the bedrock on one side could have been pushed upward, or uplifted. What processes during or after the Flood might have caused certain rock layers to be uplifted?

Principle of Lateral Continuity

The principle of lateral continuity says that as sediment is deposited, it is deposited over a large area and spreads out to the sides, or laterally. Some sediments are spread out over tens of meters, while others cover hundreds of thousands of square kilometers and multiple states and provinces. Could these geographically widespread layers be formed by the kinds of geologic processes we see at work in the world today?

Principle of Superposition

The principle of superposition explains that if the layers of rock are not disturbed by forces beneath Earth's surface, the oldest layer will be on the bottom and the youngest layer will be on top. Recall that geologists use this principle to determine the relative ages of the rock layers. The word *relative* means "when compared with something else." So relative age describes the age of a rock layer when compared with other rock layers. Geologists might describe the relative age of a particular rock layer as older or younger than another. Relative age does not tell the exact number of years since the layer was formed or tell how many years passed between the deposition of the various layers.

Most sediments are deposited in horizontal layers.

How does the photograph show an example of the principle of original horizontality?

Principle of Cross-Cutting Relationships

The principle of cross-cutting relationships explains how a sequence of rock layers can be disturbed by something that cuts through the layers. For example, an earthquake might create a fault through layers of rock, or magma may push up through or between the layers. The rock layers being cut must be older than the feature that cuts through them. Notice the order of events represented by this diagram. Four successive layers were deposited prior to the magma pushing up through them. As the magma approaches the surface, it cools and hardens into a rock. This rock is called an **intrusion**. Is an intrusion always younger or older than the rock layers around and beneath it?

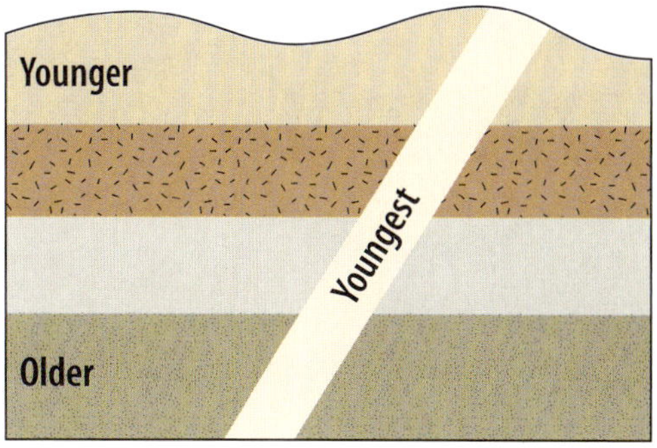

The magma intrusion cuts through the existing layers of rock.

What do you think is represented by the top line of the diagram being wavy instead of flat?

Sometimes the process of deposition in an area stops for a period of time. The layers of rock that had been formed are eroded before deposition starts again. This event, known as an **unconformity**, leaves a gap in the geologic record. In one type of unconformity, an entire layer of rock is missing. In another, the layers of rock are tilted during the pause in deposition. When deposition starts again, new horizontal layers are formed on top of the tilted layers. In a third type of unconformity, erosion exposes deep layers of metamorphic or igneous rock. When deposition starts again, layers of sediment are formed on those rocks. What type of events in Earth's history might lead to large amounts of rock or sediment being eroded?

Explore-a-Lab

Structured Inquiry

What's the Oldest?

Redraw the figure below at a larger scale on a sheet of paper. Number each layer to show its relative age. The granite layer is oldest, so mark it with a *1*. Mark the next oldest feature with a *2*, and so on. Were any layers of rock deposited after the igneous dike formed? What type of unconformity is shown? Which layers show significant erosion? Which layers show no apparent erosion? Is it possible that originally there were more layers of rock than shown in the diagram? Explain.

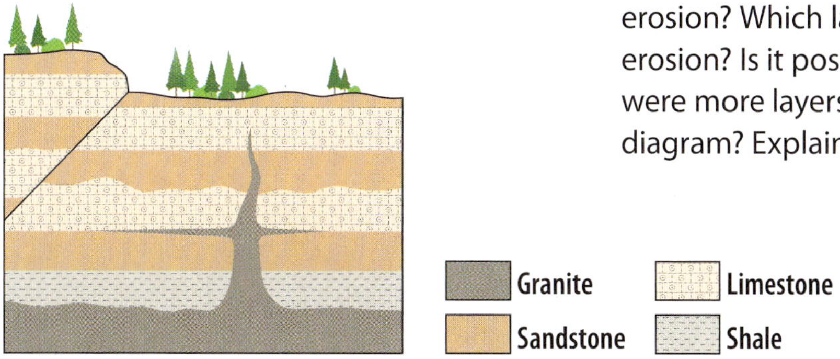

Mount St. Helens is a volcanic mountain located in Washington State.

How is a volcanic mountain such as this formed?

Rate of Geologic Processes **Explain**

Geologic processes can occur very slowly or very rapidly. Those processes that occur very slowly are usually not observable because of the very long time that they take to occur. How can scientists conclude that these processes occurred if they cannot be observed? Those geologic processes that occur very rapidly may be observed and their impact noted. For example, a major geologic event happened very rapidly and in plain sight on May 18, 1980. This was the eruption of Mount St. Helens in the state of Washington.

Mount St. Helens

For weeks before May of 1980, scientists had been watching a bulge growing on the side of Mount St. Helens. What might cause a growing bulge on the side of a volcanic mountain? On May 18, the Pacific, North American, and Juan de Fuca tectonic plates moved, causing a moderate earthquake below Mount St. Helens. The earthquake triggered the largest recorded landslide in history along one side of Mount St. Helens. The sound was heard all the way into Canada.

Some of the debris from the landslide flowed more than 20 km. The rest of the avalanche debris flowed into Spirit Lake, which is just north of the mountain. As the debris fell into the lake, it displaced water from the lake, creating a giant wave of water that washed up its northern shore. This huge mass of water washed hundreds of thousands of trees into Spirit Lake.

With part of the side of the mountain gone, the magma contained within the volcano erupted. Snow and ice melted by the hot magma formed a giant mudflow that washed away dozens of bridges, logging camps, and homes.

Rapid Strata Formation

Mount St. Helens provides scientists with direct evidence of geologic processes. One such process is the rapid formation of rock layers, or strata. Near Mount St. Helens is a sequence of strata 183 m tall. Because of eyewitnesses and photographic evidence, scientists know precisely when these layers were deposited. The bottom section formed from ash and pumice deposited during the eruption on May 18, 1980. Above that are layers of volcanic ash 8 meters thick known to have formed in only three hours about a month later on June 12. Above that is a deposit from a mudflow on March 19, 1982. Why is this important? These deposits demonstrate that the formation of strata does not require millions, thousands, or even hundreds of years. They can form very rapidly.

These layers also show that instead of mixing up all the sediments as you might expect, a catastrophe can actually separate the coarse from the fine particles and lay them down in very thin layers. Some successive layers in the geologic record are thought to have required a long time to be deposited. For example, the sediments that make up the Grand Canyon are thought by many scientists to have been deposited over a time span of 250,000,000 years. The observations made at Mount St. Helens could help provide alternative models for a much more rapid deposition of sediments.

Math in Science

The average depth of the Grand Canyon is 1.6 km. Suppose that the sediments of the Grand Canyon were laid down at the same rate as the deposition recorded at Mount St. Helens (8 m in 3 hrs). How much time would it have taken for the deposits of the Grand Canyon to form?

Mount St. Helens after the eruption

What does Mount St. Helens reveal about the rate of some geologic events?

In a single day, mudslides carved out a system of canyons near Mount St. Helens.

How does the formation of the Little Grand Canyon provide an alternate explanation for the formation of other canyons?

Rapid Erosion

Another geologic process exhibited by the Mount St. Helens eruption is rapid erosion. Recall the initial avalanche that deposited 62 km/sq^2 of debris to an average depth of 46 m in the valley. Today that valley is filled with evidence of major erosion that scientists know has happened since the eruption in 1980. One canyon has been carved out to a depth of 30 m. Another is more than 180 m deep. In the system of canyons visible there today, there is a 43-m deep canyon—known as the Little Grand Canyon—with a stream in the bottom. It might appear that the river eroded the canyon over thousands of years, but we know it happened on March 19, 1982, when a major mudflow breached the giant debris dam deposited by the eruption of Mount St. Helens in 1980 and carved out the canyons.

These erosional features demonstrate that millions of years are not needed to erode large canyons. If the Little Grand Canyon was carved out in a day, how quickly might the real Grand Canyon have been carved out?

Worldwide Flood

The eruption of Mount St. Helens in 1980, which caused the catastrophic devastation you read about, was a local event described by volcanologists as a small- to average-sized eruption. Imagine for a moment what it might be like if multiple earthquakes and volcanic eruptions happened in a short period of time all around the world. The Bible describes a worldwide Flood that began when the "fountains of the great deep" were broken up. This was followed by 40 days and 40 nights of rain. The Bible describes the fountains of the deep and the windows of heaven being stopped and the rain from heaven being restrained after five months. During the months that followed, the waters decreased. If we know a canyon can be carved in a day, what kind of erosion might have been caused by months of receding floodwaters? Evidence for catastrophic geologic events is found around the world that is consistent with the story of the worldwide Flood.

Scripture Spotlight

What does **Psalm 29:10** say about the Flood?

Worldwide Flood Evidence	
Fossils	Millions of fossils—individual creatures as well as larger populations, some buried rapidly, others buried slowly—provide evidence of destruction on a worldwide scale.
Widespread sedimentary deposits	Many examples of widespread sedimentary deposits provide evidence that geologic processes on a much larger scale than we see today happened sometime in the past.
Flat gaps	The geologic column contains examples of places where, in spite of missing layers, the rocks above and below the absent layers are in flat contact with each other, showing little or no evidence of erosion. This is consistent with rapid deposition rather than deposition over millions of years.
Turbidites	Turbidites are formed from massive amounts of underwater sediments that slide downhill like an avalanche and spread out in deep ocean waters. Thousands of turbidites, which form rapidly and only underwater, are found on the continents.
Ocean sediments on the continents	Marine fossils on the continents, some of which appear to have been deposited as water flowed across the continents, are plentiful.
Continental-scale currents	Evidence indicates that many sedimentary layers were deposited by water that was flowing in the same direction all across the continent. This would require geologic processes on a much larger scale than we see today.

Concept Check Assess/Reflect

Summary: How does Earth change? The theory of plate tectonics states that pieces of Earth's rigid layers in the crust and mantle form plates that are in constant, slow motion. This plate motion is the result of a molten layer found deeper in the mantle. Tectonic plates can move in different ways. Two plates can slide past each other. Two plates can also converge, or collide. Two plates can even diverge, or move apart from each other. As a result of tectonic plate movement, geologic events, such as earthquakes, mountain formation, and subduction, can occur. Weathering, erosion, and deposition change Earth's surface. Rocks weather or erode as a result of exposure to the air and water. The sediments formed by weathering can be transported by wind and water and deposited in locations far from their original site. As a result, weathering, erosion, and deposition cause changes in Earth's features. Such geologic changes can occur rapidly or very slowly.

1. Explain three ways that tectonic plates move and describe the results.
2. Identify four agents of weathering and erosion. Choose one and describe how it changes Earth's surface.
3. Explain the four principles related to deposition.
4. Explain two lessons we can learn from Mount St. Helens.

Lesson 2

Essential Question

How Old Are Earth and Its Layers?

Objectives

- Distinguish between relative dating and radiometric dating.
- Describe how a geologic time scale is created.
- Identify concerns related to the reliability of radiometric dating.
- Describe evidence that is inconsistent with long geologic ages.

Vocabulary

absolute age
isotope
radioactive decay
half-life
geologic time scale
bedding
bioturbation
lamination

How old do you think Earth is? Many scientists currently estimate that Earth is about 4.6 billion years old. What kind of evidence do you think is used to reach that conclusion? Compare that estimate with other ideas about the origin of Earth. Is the origin of Earth part of empirical science or historical science? How might a scientist's worldview affect his or her interpretation of scientific data related to the history of Earth? This lesson will examine data and interpretations related to time.

Radiometric Dating Explain

In the last lesson, you learned how geologists use certain principles about deposition to determine the relative ages of rock layers. Recall that the principle of superposition states that, in undisturbed rock layers, the oldest rock layer is located at the bottom, while the youngest rock layer is closest to Earth's surface. Consequently, a fossil found embedded in the bottom layer is older than any fossil found in the top layer. This process, known as relative dating, can tell you only which fossil is older. However, relative dating cannot tell you how old a fossil actually is. To do this, scientists use a process called radiometric dating. As you know, radiometric dating is a process used to date the age of an object, such as a rock or fossil. This process reveals the **absolute age** of an object, such as a fossil. The absolute age of an object gives the number of years that have passed since the object was formed.

Radiometric dating uses **isotopes**, atoms of the same element that have different numbers of neutrons. For example, some isotopes of carbon have six neutrons, while other carbon isotopes have seven neutrons. What do all isotopes of an element have in common?

Scientists have long tried to determine Earth's exact age.

Why are there different estimates of Earth's age among scientists?

Structured Inquiry

Discover

Record your work for this inquiry. Your teacher may also assign the related Guided Inquiry.

Science Journal

Radiometric Decay

How can you model the decay of an unstable isotope?

Materials
- 100 pennies
- shoebox with lid

Procedure

1. Put 100 pennies heads up in the box and shake.
2. Remove all the pennies that are now tails.
3. Count the remaining pennies and **record** this data in your ***Science Journal***.
4. Repeat until you have no pennies left.
5. **Graph** your data by placing the number of pennies on the *y*-axis and the number of shakes on the *x*-axis. You should start by placing 100 pennies at 0 shakes.

Analyze Results

Compare your graph with those made by your classmates. Describe similarities and differences between the graphs.

Create Explanations

1. How can you model the decay of an unstable isotope?
2. What does each shake of the shoebox represent?
3. Explain why all the groups did not take the same number of shakes to get rid of all the pennies.
4. Is it possible that all the pennies would be gone after just one shake? Explain.
5. How would your graph look if you had started with 200 pennies?

Some isotopes are stable, which means that they do not break down. Others are unstable, which means that they break down into other isotopes during a process called **radioactive decay**. Unstable isotopes, or parent isotopes, decay at known, steady rates and form stable daughter isotopes. The isotope's **half-life** is the time it takes for one half of the amount of an unstable isotope to decay into daughter isotopes. Scientists determine the ratio of parent isotope to daughter isotope and, by using the half-life value for that radioactive isotope, calculate an age for the material being tested. For dating purposes, the half-life must be short enough to have produced a measurable amount of daughter isotope. It must be long enough that a measurable amount of parent isotope is still present.

Several different elements are commonly used for radiometric dating. The table lists the half-lives of different unstable isotopes.

Half-Life of Isotopes

Parent Isotope	Daughter Isotope	Half-Life
Uranium-238	Lead-206	4.5 billion years
Uranium-235	Lead-207	704 million years
Thorium-232	Lead-208	14.0 billion years
Rubidium-87	Strontium-87	48.8 billion years
Potassium-40	Argon-40	1.25 billion years
Samarium-147	Neodymium-143	106 billion years

When scientists have used this process to date Earth's rocks, the results suggest that Earth is over 4 billion years old. Because of the activities associated with plate tectonics and the rock cycle, however, scientists realize that Earth's rocks have been altered repeatedly by geologic processes and so would not provide an accurate estimate of Earth's age. Assuming that Earth and the other solid bodies in the Solar System would have formed at the same time, scientists have used radiometric dating on meteorites that have been found on Earth's surface. Because radiometric dating suggests that the meteorites are about 4.6 billion years old, scientists conclude that Earth must be about the same age. How does this compare with other estimates for the age of Earth that you may have heard?

Radiometric dating methods are also used to date specific layers of rock on Earth. They are used mostly to date igneous and metamorphic rocks rather than sedimentary rocks such as sandstone, shale, and limestone. The dates associated with the geologic column—referred to as the **geologic time scale**—were calibrated based on radiometric dating. A geologic time scale relates life forms to geologic events in Earth's history. For example, the diagram shows that a gigantic extinction of life forms occurred between the Permian and Triassic periods. According to the geologic time scale, how many years ago do scientists think that extinction happened?

Accuracy

How accurate is radiometric dating? For this method to be accurate, careful procedures must be followed when samples are collected and when data is measured and analyzed in the laboratory. In addition, four assumptions must be true:

1. The decay rate (half-life) has always been constant.
2. The rock has experienced no chemical exchange with the rocks around it. (Scientists describe this as a closed system.)
3. The initial concentration of daughter products at the time of formation of the rock can be determined.
4. The parent-to-daughter ratios were caused *only* by radioactive decay.

In spite of some questions about accuracy that exist and the fact that some individual dates have been shown to be wrong, the overall pattern of dates suggested by radiometric dating methods is generally

The geologic time scale separates Earth's history into parts, depending on how long ago they occurred.

How can the diagram be interpreted in terms of relative age?

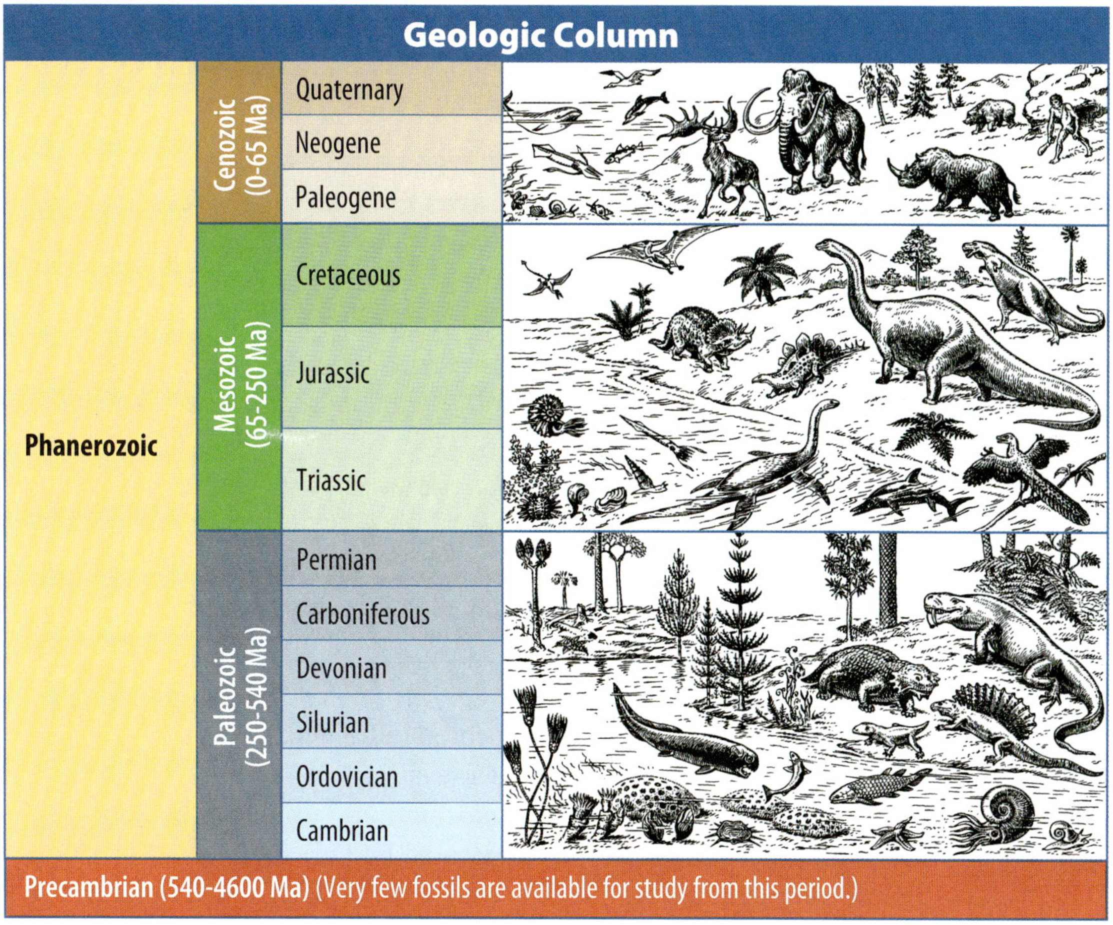

"Ma" is an abbreviation for mega-annum, or one million years.

quite consistent. It does seem to be a reliable method at least for relative dating. Even if the ages determined are not entirely accurate, they can be used to compare rock layers because they are consistent. For example, if radiometric dating assigns an age of 50 million years to certain rocks in North America and to others in Africa, the rocks in both places are probably about the same age, and both are probably younger than rocks from India that are said to be 60 million years old.

Rock layers in different locations can be compared according to their ages found by radiometric dating.

How is radiometric dating useful for comparing rock layers in different locations?

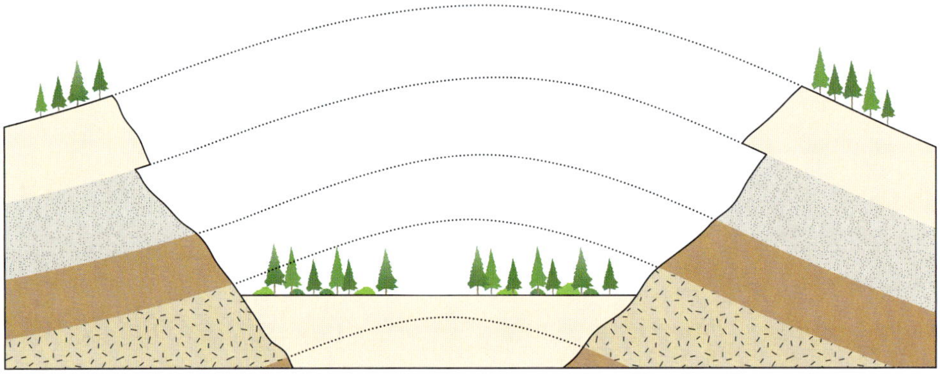

Some scientists, however, have raised questions about the absolute ages suggested by radiometric dating. Look again at the assumptions of radiometric dating. Then read about the concerns that some scientists have about some of these assumptions.

Concerns about Radiometric Dating

Assumption	Comments/Concerns
The decay rate has always been constant.	There seems to be little doubt at the present time about the theory of the decay process.
The rock has experienced no chemical exchange with the rocks around it. (Scientists describe this as a closed system.)	If the minerals to be dated were cracked or partially heated, some of the isotopes could have escaped the system. In this case, the ratios would not accurately reflect the age of the rock.
The initial concentration of daughter products at the time of formation of the rock can be determined.	If scientists did not know some old daughter product was left over when a new rock formed, the rock would date as much older than it really is.
The parent-to-daughter ratios were caused *only* by radioactive decay.	Some scientists wonder whether any other factor could explain this ratio.

While many scientists have accepted radiometric dating as a given truth, others still consider it a hypothesis to be tested. One reason for this is that another source of information—the Word of God—tells a different story than that suggested by radiometric dating. The Bible is very clear about the Creation of life on Earth in six literal days with a seventh added for worship and rest. The genealogies in

Genesis 5 and **11** seem to indicate that this happened several thousand—rather than million—years ago. The Bible is less clear about whether the entire Universe was created during that same six-day period or whether it, and perhaps the raw materials for Earth, was created sometime before that. Even if Earth's raw materials are older than several thousand years, it doesn't really solve the problem of the geologic time scale.

Although the Creation story in Genesis is not something that science can prove or disprove, some scientists believe it to be a reliable history of Earth's origins and have chosen to accept it as accurate. These scientists look for alternate explanations of radiometric dates that would be consistent with the biblical record. More study is needed, and scientists who believe in the biblical Creation story predict that there are discoveries yet to be made about the process of radioactive decay that will shed light on this issue.

Inconsistent Evidence Explain

It is important to realize that there are questions on both sides of the discussion that are difficult to answer. How are unanswered questions related to the scientific process? What are some different reactions people might have to unanswered questions? You have read how radiometric dating is hard to explain in light of the timing suggested in the Bible. Next we will note several pieces of geologic evidence that are difficult to explain in ways that are consistent with millions of years of time.

Science Journal

Check out your ***Science Journal*** for a Guided Inquiry to investigate whether extinct species could be brought back to life.

Extend

Rates of Erosion and Sedimentation

Rates of erosion of the continents and sedimentation in the ocean are inconsistent with the geologic time scale of millions of years. As rivers flow, they erode parts of continents and carry the sediments to the oceans, where they are deposited. Based on current rates of erosion, scientists have estimated that it would take 10 million years for the continents to be eroded to sea level. In the billions of years claimed by the geologic time scale, this has not happened. Assuming that in the past, rivers have always carried approximately the same amount of sediment they currently transport to the oceans, it has also been estimated that in over 100 million years the ocean basins should be completely filled with sediment. What observation can you make about this prediction? Radiometric dates suggest that certain mountain ranges, such as the Appalachians in North America, have been here for about 300 million years. In such a long time, they should have been eliminated completely by erosion, but they are still here.

Bedding

Many of the sedimentary rock layers found within the geologic column have very distinct boundaries that do not blend gradually one into another. This distinct layering is called **bedding**. If the layers are a record of millions of years of gradually changing climates and environments, we should expect to see a much more gradual shift from one kind of rock to another. For example, a slow change in sea level might produce a gradual shift from clay to sandstone but would not be expected to produce very distinct layers of clay and sandstone.

Another reason why the boundaries between layers should be blurred if they were formed over long periods is the normal burrowing activity of animals. Creatures that live within sediment continually burrow, and over time this activity, which is called **bioturbation**, would blur or destroy the original layering. So the bedding in the geologic column is difficult to explain if the layers were deposited slowly over millions of years.

Explore-a-Lab

Guided Inquiry

How can you model bioturbation?

Bioturbation can also occur in the seas. For example, sea cucumbers are responsible for bioturbation in coral reefs. Coral reefs are extremely diverse ecosystems that support enormous biodiversity. The coral in a reef consists mostly of calcium carbonate. Use eggshells that contain calcium carbonate to model a coral reef. Use vinegar that dissolves calcium carbonate to model the action of sea cucumbers. Design and carry out an experiment that illustrates the impact of bioturbation by sea cucumbers on a coral reef. What factors can slow down this process?

Much of the geologic column exhibits bedding.

How do the obvious layers contradict the geologic time scale?

Flat Gaps

If the various rock layers of the geologic column represent many millions of years, there should be abundant evidence of erosion of valleys and even canyons. In many places, we do see this kind of evidence. In many other places, however, we see flat gaps. Flat gaps are consecutive sedimentary layers that, in spite of missing layers in between, are in flat contact with each other, showing little or no evidence of erosion. If the missing layers represent millions of years, it is difficult to explain the absence of erosion before the next layer was deposited.

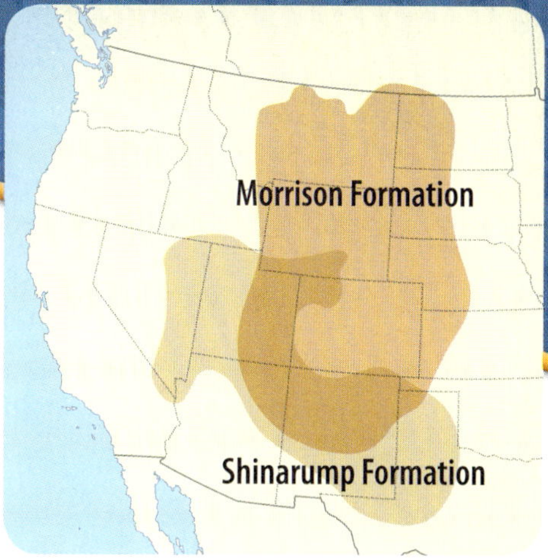

The Morrison and Shinarump Formations are quite extensive and cover large areas.

How do the widespread sedimentary deposits contradict the geologic time scale?

Widespread Sedimentary Deposits

Recall how an erosional agent such as water carries sediment along, dropping the largest pieces first and then dropping the smaller ones as the water slows down. Recall also the principles of original horizontality and lateral continuity. An interesting feature of the geologic column is the presence of many individual rock layers that spread across multiple states. The Shinarump Conglomerate covers more than 250,000 km^2 in multiple states in the southwestern United States. The Morrison Formation is even larger, extending from Texas to Canada and including parts of over a dozen states. Other unique deposits, including chalk and coal, cover extensive geographic areas around the world. How such widespread deposits could have been formed over many millions of years is difficult to explain.

Rapid Geologic Processes

Based primarily on radiometric dating, the Pisco Formation in Peru is thought to have formed over 10–12 million years, but certain evidence suggests that it was formed much more quickly. By looking at fossils, scientists can often get a pretty good idea about whether the animal was buried rapidly or not. If an animal was buried quickly, its fossilized bones are found in a much more orderly arrangement than if scavengers had been able to eat parts of the body or if it had been moved around by water or wind. The well-articulated and well-preserved vertebrate fossils indicate rapid burial. In addition, evidence in the sediments indicates that they were formed by storms and other fairly rapid processes.

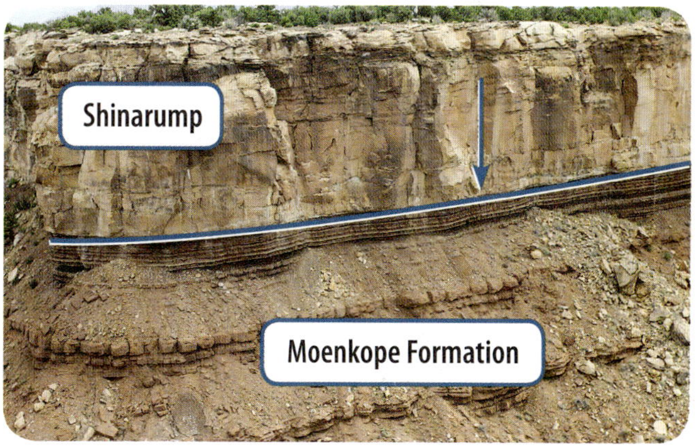

The Moenkope Formation and the Shinarump exhibit a flat gap.

How many layers are missing? How many years are represented?

A similar situation exists in the Green River Formation in southwestern Wyoming. Thousands of thin layers that are full of exquisitely preserved fossils suggest that the formation was deposited in much less than the several million years described by the geologic time scale. These thin layers are known as **laminations**. Lamination is a sedimentary rock structure in which thin layers, or laminae, are recognizable. A lamina is generally less than a centimeter in thickness. In contrast, bedding layers are usually greater than 1 cm. Laminations are more visible in sedimentary rocks that have undergone weathering.

Imagine trying to preserve a 3-m-long crocodile with 0.1 mm of sediment per year. How likely is it that the animal would be preserved at all, let alone be preserved exquisitely like the animals found in this formation? How these formations could have been laid down over millions of years is not an easy question to answer.

These examples do not prove that the geologic time scale is wrong, but they raise questions that are difficult to answer if the geologic time scale is accepted. There are plenty of questions left on both sides of the discussion, and science is all about seeking answers to difficult questions. Maybe someone in your class will become a scientist and discover the answer to some of these questions. Maybe it will be you!

Concept Check Assess/Reflect

Summary: How old are Earth and its layers? In undisturbed rock layers, relative dating can be used to determine which layer is oldest. Radiometric dating is a process that compares the ratio of parent isotopes with daughter isotopes to determine the absolute ages of rock, fossils, or other objects. Radiometric dating involves the use of unstable isotopes that decay at a constant rate. Radiometric dating has been used to date the absolute ages of rocks to create a geologic time scale that is a record of the geologic events and life forms in Earth's history. Concerns have been raised about the reliability of radiometric dating. For example, some of the daughter isotopes may have escaped from the sample before it was tested. Evidence collected from rates of erosion, beddings, and flat gaps is inconsistent with long geologic ages.

1. Compare relative dating with absolute dating.
2. List two assumptions that must be true for radiometric dating to be accurate.
3. Compare two ideas about the age of Earth with the scientific evidence consistent with each one.
4. Review the pieces of evidence that supports and detracts from the theory of the geologic time scale. Decide which ones you think offer the most compelling evidence. Arrange them in order, beginning with the evidence you think is best. Explain why you chose that order.

Essential Question

What Does the Fossil Record Show?

What is the basic claim of evolution? What things have you learned that would be necessary for that claim to be true? How are microevolution and macroevolution different? Evolutionary scientists claim that the fossil record is the record of the evolution of living organisms. What kind of evidence in the fossil record would be consistent with this idea? Scientists who have a biblical worldview claim that the fossil record is a record of the catastrophic destruction of created organisms. What kind of evidence would be consistent with this idea? In this lesson, we will look at the observable data found in the fossil record and ask ourselves which interpretation is best supported by the evidence.

Objectives

- Describe how fossils provide information about Earth's history.
- Describe how an index fossil can be used to correlate rock layers.
- Explain why carbon-14 is limited as to what it can provide absolute ages for.
- Explain how the fossil record supports the **Genesis 1** account of Creation.

Vocabulary

carbon dating
phyletic gradualism
Cambrian explosion
morphology
morphological intermediate
hominid
stasis

Order in the Fossil Record Explain

Fossils are found in the same basic order all around the world. Using this predictable sequence, scientists have identified certain fossils—called *index fossils*—that are helpful in determining the relative age of rock layers found in different places. You will recall that an *index fossil* is a fossil of an organism that is found within a limited portion of the geologic column but is relatively widespread and abundant in that limited portion. Index fossils are used to identify the layers of the geologic column.

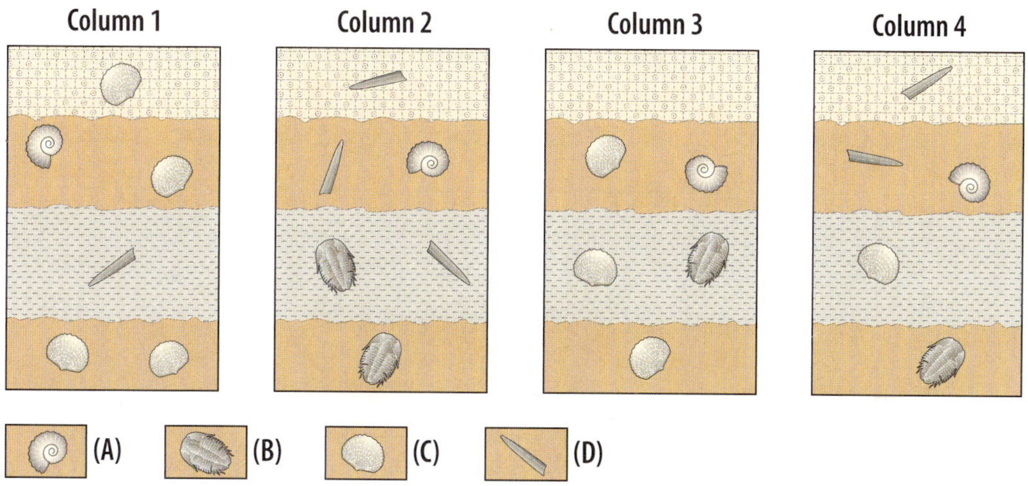

Fossil A could serve as an index fossil.

What characteristics make it a useful index fossil?

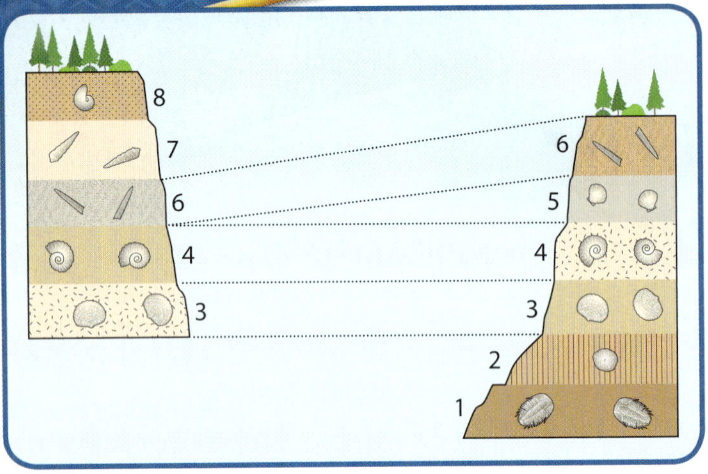

Layers of rock in different locations can be matched according to their fossils.

Why might the corresponding layers be at different depths in each location?

To understand how index fossils can be used to correlate layers in different locations, look at the diagram. It shows two rock outcrops separated by several miles. Each rock outcrop contains index fossils and distinct rock layers. By comparing the index fossils in the two different locations, scientists can infer which layers from one location correspond with layers in the other location. Notice that this kind of correlation is done using the fossil content of the layers, not the kind of rocks that make up the layers.

Presence of Carbon-14 Explain

In the same way that scientists use radiometric dating methods to try to determine the absolute age of a rock, they can also use radiometric dating to try to date organic materials such as wood. Living organisms continuously take in carbon. Most of that carbon is stable carbon-12, but some is the unstable isotope carbon-14. Carbon-14 is continually decaying into nitrogen-14. While an organism is alive, the ratio of carbon-12 to carbon-14 is the same as it is in the atmosphere and in other living organisms. Once the organism stops living, it no longer takes in new carbon. As carbon-14 decays, the ratio of the two carbon isotopes changes. In a process called **carbon dating** or radiocarbon dating, scientists compare the ratio of carbon-14 with carbon-12 to estimate the age of a fossil. The half-life of carbon-14 is 5730 years. Thus, after 10 half-lives, so little carbon-14 remains that it cannot be detected. Therefore, carbon-14 dating can be used only to date fossils that are a maximum of 50,000–60,000 years old. Because of this, scientists use it to date fossils only in the very top layers of the geologic column. Also, scientists cannot use carbon-14 to date a fossil

The graph shows the percentage of carbon-14 remaining in a sample over time.

How much of a 100-g sample remains after three half-lives?

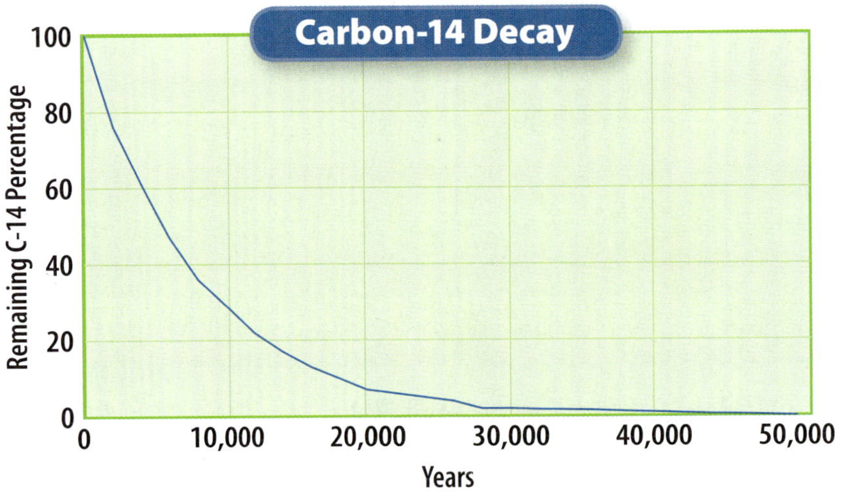

Lesson Activity

Select another two unstable isotopes other than carbon-14. Using the decay graph of carbon-14 as a model, draw a decay graph for each of these two isotopes. How do these two graphs compare with the decay graph of carbon-14? Be sure to describe both similarities and differences. Prepare bar graphs that show the decay of the unstable isotopes you selected.

What are the advantages of showing the decay with the use of a bar graph?

in which minerals have replaced the organic matter. Why must a fossil contain organic matter for it to be dated using carbon-14? Interestingly, though, small amounts of carbon-14 are found in fossils throughout the geologic column. Some scientists claim that this is because of contamination. What other possibilities might there be to explain the presence of carbon-14 in fossils that are supposed to be millions of years old?

Cambrian Explosion Explain

Remember that animals are classified into domains, kingdoms, phyla, classes, orders, families, genera, and species. There are over 30 different phyla, which are groupings of animals based on their body plans. How many different body plans can you describe?

Macroevolution refers to the idea that all living organisms evolved from a single-celled, common ancestor over billions of years. If this is so, macroevolution would predict that the fossil record should show organisms with new body plans appearing gradually in the fossil record. This process is called **phyletic gradualism**.

Notice on the geologic column diagram that there is a name for all the layers above the Precambrian. It is called the Phanerozoic, which means "visible life." It was given this name because of the abundance of fossils found in these layers in comparison with the Precambrian layers below. In the lowest section of the Paleozoic layers, called the Cambrian, we find the abrupt appearance of most known phyla from the animal kingdom. Instead of phyletic gradualism, paleontologists describe the same body plans visible in nature today, as well as some that have gone extinct—all with no apparent ancestors.

Geologic Column

Eon	Era	Period
Phanerozoic	Cenozoic	Quaternary
		Neogene
		Paleogene
	Mesozoic	Cretaceous
		Jurassic
		Triassic
	Paleozoic	Permian
		Carboniferous
		Devonian
		Silurian
		Ordovician
		Cambrian
Proterozoic (Precambrian)		
Archean (Precambrian)		
Hadean (Precambrian)		

In the lowest section of the Phanerozoic—the Cambrian layers—we find the abrupt appearance of most known phyla from the animal kingdom with no apparent ancestors.

Why is it significant that most known body plans, as well as some that have gone extinct, are found so early in the geologic column?

To understand what this means, imagine the hundreds of millions of years evolutionists suggest for the history of life compressed into one 24-hr day. According to the fossil record, nothing but single cells would have existed for the first 18 hr. At about the twenty-first hour, in about 2 min, most major animal phyla would have appeared in their current forms. The abrupt appearance of most animal phyla in the Cambrian layers is called the **Cambrian explosion**. Why do you think the word *explosion* was included in the name?

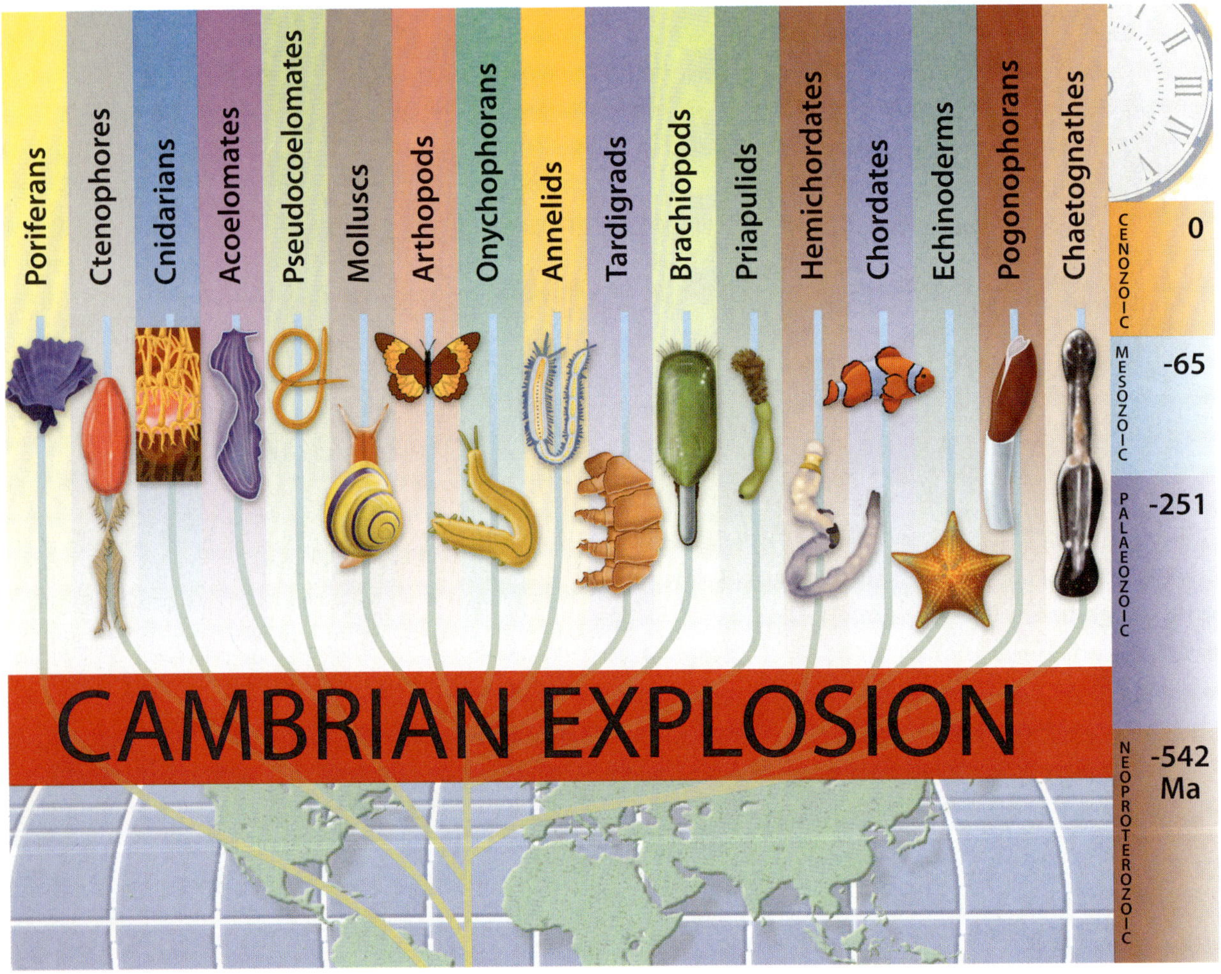

According to the fossil record, most major phyla appeared rapidly during the Cambrian explosion.

How does the Cambrian xxplosion contradict the theory of phyletic gradualism?

Charles Darwin was aware of the Cambrian explosion and the challenge it presented to his theory of evolution. He predicted that as more fossils were discovered, evidence supporting phyletic gradualism would be discovered. Today the fossil record is much more complete, but the evidence that Darwin anticipated has not been found.

What do you think? Is macroevolution's prediction of phyletic gradualism verified by the fossil record? Is the evidence found in the Cambrian explosion consistent with the theory of evolution? Is it consistent with the biblical stories of Creation and the Flood?

The Cambrian explosion is more consistent with the biblical account of specific creatures being created by God and later buried in a worldwide Flood than it is with the claims of macroevolution.

Complexity throughout the Fossil Record Explain

Another prediction of macroevolution would be that the fossil record should show a sequence of organisms from simple to complex. For example, we might expect to find a layer of simple creatures, such as simple sponges, near the bottom. A little farther up, we might expect to find a layer with simple sponges, more complex sponges, and corals. Later we might find a layer with simple and more complex sponges, simple and more complex corals, plus brachiopods, which are extinct organisms that lived in a bivalve shell but were not clams. New creatures, as well as more complex creatures, continuing to appear in a simple-to-complex sequence like this would be consistent with evolution, but that is not what we find in the fossil record.

In addition to finding most body plans in the Cambrian explosion, paleontologists have found fossils of highly complex creatures in the lower Phanerozoic layers. Consider the trilobite, for example, which is found all throughout the Paleozoic layers. As complex as a modern crab, it had a brain, gut, heart, and compound eyes. Each organ would have been made from specific types of cells with specified proteins and genes. It appears to have had body processes as complex as those of modern arthropods.

Scripture Spotlight

How might **Genesis 7:21–22** be related to the fossil record?

Macroevolution would predict that simple organisms, such as these, would gradually give way to more complex organisms. This prediction is not supported by the fossil record.

What does the fossil record actually show regarding simple and complex organisms?

Trilobites were hard-shelled organisms with multiple body segments. They are considered arthropods.

Why would macroevolutionists not expect these types of organisms before the Cambrian explosion?

Furthermore, paleontologists have discovered fully integrated, complex communities in succession instead of finding increasing complexity when moving up the geologic column. For example, the fossil marine communities of the Paleozoic (including the Cambrian) are just as complex as the marine communities that we find in the modern world.

So, although the fossil record does show a definite sequence—invertebrates, fish, amphibians, reptiles, mammals—it does not show the expected sequence of simple creatures evolving into complex creatures or simple communities evolving into complex communities. Complex creatures and complex communities at the beginning of the sequence are more consistent with the biblical story of Creation and the Flood than with evolution.

Morphological Intermediates Explain

Have you ever heard of "missing links"? According to the theory of evolution, a series of intermediate forms between major groups of organisms existed in the past. In other words, if fish evolved into amphibians, we should be able to find in the fossil record an animal that was somewhere between a fish and an amphibian. If dinosaurs evolved into birds, we should be able to find fossil evidence of the transition. Scientists have suggested that *Archaeopteryx* is an example of this transition. This suggestion is based on examining an organism's structure or morphology. **Morphology** refers to what an animal looks like—its structure or shape. Sometimes the creatures that share characteristics with different animal groups are called **morphological intermediates**. An example of a morphological intermediate would be a creature with some characteristics like a fish and others like an amphibian. Macroevolution predicts that morphological intermediates should be abundant in the fossil record.

Paleontologists acknowledge that the fossil record is no longer as incomplete as Darwin believed it was in his day. In spite of this, they have found only a relatively small number of fossils that they consider to be transitions, or morphological intermediates. None exist between kingdoms. None exist between phyla. At the class level, scientists believe they have found evidence of intermediates between reptiles and birds, but there are none between the jawless fish and jawed fish and none between fish with skeletons made of cartilage (sharks and rays) and those with bony skeletons. At the order level, scientists suggest that the four-legged creatures such as *Ambulocetus* may be morphologically intermediate between artiodactyls (hoofed animals such as pigs and camels) and whales, but no intermediates have been found between other orders, such as insectivores and bats. While most mammalian families lack morphological intermediates, scientists suggest that there is a sequence of horse transitions and hominid transitions. **Hominids** are ape-like animals that walked on two legs.

Instead of abundant transitional fossils that actually demonstrate one creature evolving into another, paleontologists usually find the abrupt appearance of new creatures in the fossil record, followed by **stasis**, which means "no change for long periods," followed by the replacement of one species with another.

Creationism would predict that no evolving intermediates should be found in the fossil record. Since distinct kinds of organisms were created by God, we would expect to find evidence of only separate creatures and not intermediate forms in the fossil record. How should creationists respond to the claims of scientists regarding these morphological intermediates? When thinking about this question, it is important to remember some things you have already learned:

- First, for macroevolution to occur, there must be a way for living organisms to acquire the new genetic material necessary to create new body parts and processes. This new genetic material comes from mutations, or mistakes, that happen as DNA is duplicated. Most mutations are harmful.
- Second, remember how the process of natural selection weeds out organisms that are less fit. Organisms with harmful mutations would be less likely to survive.
- Finally, remember what you learned about irreducible complexity. Living organisms contain many systems made up of multiple parts that all must be present and in working order for the system to work at all.

How likely would it be for an entire system to evolve all at once from mutations? If it evolved only part of the way, how likely would it be to survive while the rest of the system evolved? It might be easy to look at a series of living organisms and imagine that they evolved in that order. But remember that lining things up in a sequence doesn't demonstrate that they possessed the biological resources for such dramatic change.

Scientists have created this image of what *Archaeopteryx* may have looked like based on fossil evidence.

How is this creature an example of a morphological intermediate?

Scripture Spotlight

Review the Creation account in **Genesis 1**. How many times does the phrase "after its kind" or "after their kind" appear?

If these animals are not evolving intermediates, what other explanations might there be for these creatures that share physical characteristics with more than one grouping of animals? Do you think God might have created some animals with shared characteristics that have since gone extinct? How might a scientist's worldview affect his or her explanations of morphological intermediate animals?

In spite of a relatively small number of fossils considered by evolutionary scientists to be intermediates, the majority of the fossil record does not show evidence of macroevolution. Instead of phyletic gradualism, we see the Cambrian explosion. Instead of a simple-to-complex sequence, we see highly complex creatures and communities at the bottom of the fossil-bearing layers that continue to be present throughout the fossil record. Instead of abundant transitions, we usually see the abrupt appearance of new creatures in the fossil record followed by stasis and replacement.

Science Journal

Check out your ***Science Journal*** to learn more about fossils to determine their relative ages.

Extend

Notice on the image on the next page that there is a detailed order of first and last appearances in the fossil record that can be correlated around the globe. Fossils appear abruptly, followed by stasis (no significant change), and ending with abrupt disappearance. Some fossils such as the tyrannosaurs and trilobites exist unchanged through several layers of the geologic column, other fossils such as the horseshoe crab exist from early in the geologic column to the present with little change, still other fossils such as mastodons came and went in a relatively brief time of geologic history with no change. In the diagram the green line indicates when the first example of the fossil is found, the red line indicates where last fossil is found (it is assumed that the fossil went extinct at this time), the vertical purple line in between the two horizontal lines indicates the period of time the animal lived on Earth with little change in its appearance—stasis.

Is this data more consistent with macroevolution or Creation and the Flood? Recall what you have learned about the limitations of historical science. Can either macroevolution or Creation be proven scientifically? What role does faith play in accepting either idea?

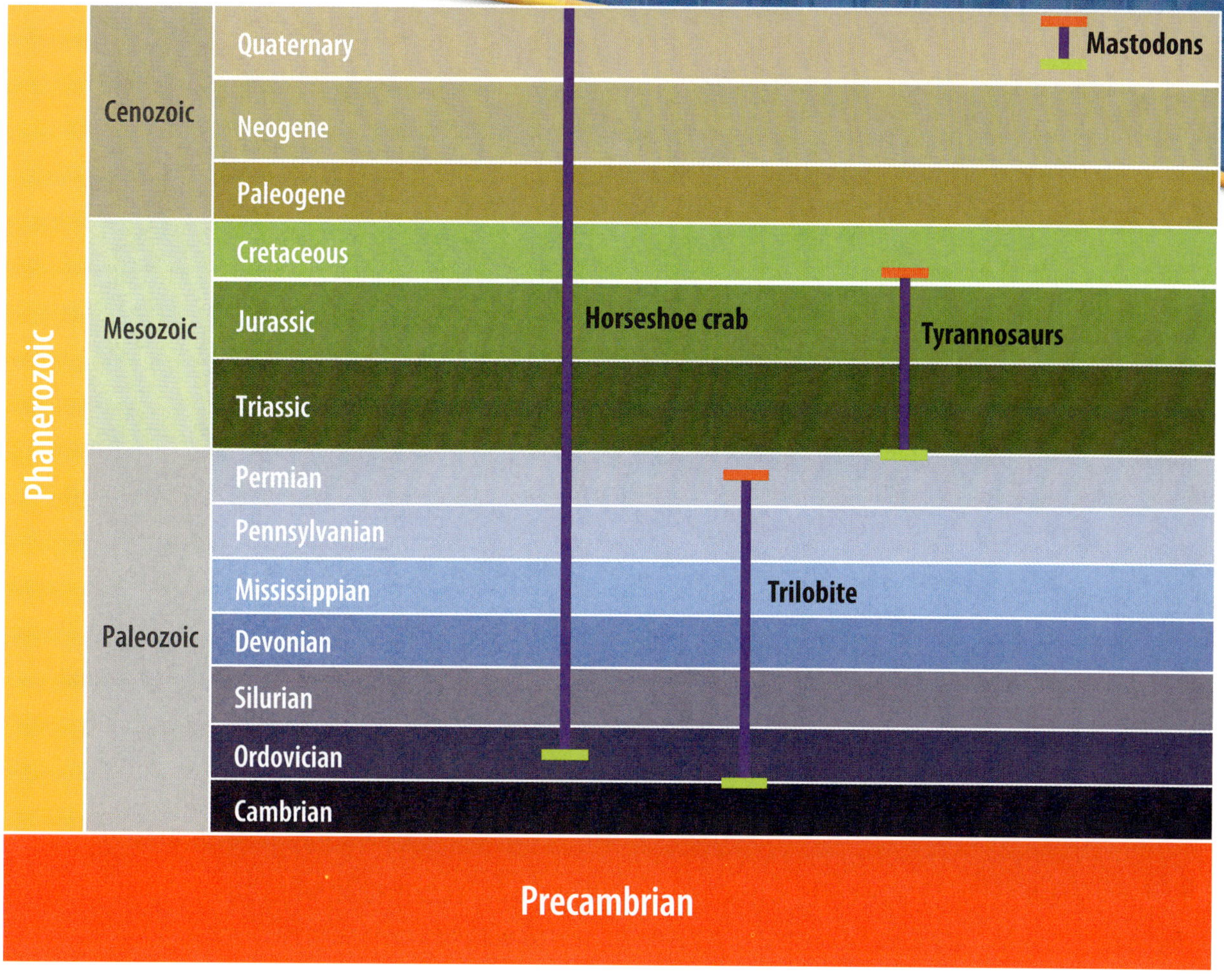

Concept Check Assess/Reflect

Summary: What does the fossil record show? Fossils are found in a definite order all around the world. The fossil record can be compared with the expectations of Creation and evolution. The presence of small amounts of carbon-14 in fossils throughout the geologic column, the Cambrian explosion, and the complexity of both individual creatures and communities throughout the fossil record are more consistent with Creation than evolution. Morphological intermediates would be more consistent with evolution, but there are far fewer than evolution would predict, and it has not been demonstrated that such large changes could happen through evolutionary processes.

1. Can carbon dating be used to date igneous rocks? Why or why not?
2. Contrast phyletic gradualism with the Cambrian explosion.
3. Describe the sequence of fossils that can be found in the fossil record. Is this a simple-to-complex sequence?
4. Give an example of a morphological intermediate. Explain how evolutionists and creationists might interpret such a fossil.

People in Science

Extend

Get to Know Dr. John Baumgardner

Dr. John Baumgardner is a geophysicist and a young Earth scientist. For the past 25 years, his main focus of research has been to explain how the global Flood occurred. Dr. Baumgardner uses parts of the theory of plate tectonics to explain this catastrophic event.

The Bible describes the time when God destroyed the surface of Earth and almost all living things. Dr. Baumgardner believes that the action of plate tectonics was the vehicle through which God accomplished this destruction. He uses the presence of rock formations all over Earth as evidence. The fossils contained in layers of rock are the remains of the animals and plants that died when the Flood occurred.

The theory of plate tectonics is accepted by most scientists. However, the way plate tectonics works has been the topic of study for many years. Dr. Baumgardner has designed a computer program that models how plate movement occurs. He uses it to demonstrate how this movement could have caused the destruction of Earth's crust. His model shows how the cold, heavy, ocean crust plates could have sunk below Earth's surface just before the Flood occurred. The force of gravity pulled the plates downward. They moved as if they were on a conveyor belt—one after another sinking below the surface. With no ocean basin left for the ocean water to fill, the water then flooded the land.

Called to Serve

Dr. Baumgardner sees that part of his job as a geophysicist is to explain processes on Earth as part of God's work.

Concept Check

1. According to Dr. Baumgardner, how is plate tectonic activity related to the global Flood?
2. What evidence on Earth supports Dr. Baumgardner's ideas?

Finding Dinosaurs Using Satellite Technology

With the technology at our fingertips that maps all parts of Earth and even offers images of all of Earth's landscape, it can sometimes be hard to remember that much of Earth's surface is still incredibly remote and unmapped or that the maps of those areas are highly inaccurate.

For paleontologists, hunting for fossils often means weeks or months of trekking through some of the world's least hospitable terrain, searching for hidden treasures. When a fossil bed is found, researchers often return to the site year after year, hoping that erosion will uncover more of the precious artifacts.

In 2000, scientists started using satellite technology to help in their search. Landsat 5 images are true-color photos of Earth's surface, showing rock formations and some vegetation. Landsat 7 images show shortwave infrared images in red, green, and blue. By combining images from these two satellites, scientists can clearly determine areas with sedimentary rock and little vegetation that are promising sites for fossils.

Using this technology, paleontologists have concentrated their searches in areas that already have shown considerable promise, such as Mongolia's Gobi Desert. Searching by vehicle or on foot in this area, with its 38°C temperatures and frequent sandstorms, makes fossil hunts expensive and unpleasant—and sometimes dangerous. With satellites, researchers can pinpoint likely sites in safety.

Today researchers in other areas of the world are using satellite technology to aid in their studies. Australian paleontologists have used this new approach coupled with helicopters to find additional fossil sites in northwestern Queensland, helping them to locate fossils very different from the ones they have previously seen.

Concept Check

1. How can using new technology help scientists find ancient artifacts?
2. Why is it important for scientists to continue to innovate in their fields of study?

Chapter 10 Review Assess/Reflect

Study Guide

Lesson 1

1. The theory of plate tectonics explains the formation and movement of Earth's plates in the solid lithosphere that float on top of the molten asthenosphere and that move due to convection currents.
2. The movement of tectonic plates can result in earthquakes, mountain formations, and subduction.
3. Weathering can be caused by both physical factors such as wind and chemical factors such as acid precipitation. Weathering results in erosion of rocks to form sediments that can then be transported and deposited elsewhere. Weathering, erosion, and deposition change the landscape by rearranging Earth's surface features.
4. The eruption of Mount St. Helens and the worldwide Flood are examples of geologic processes that occurred very rapidly. Geologic processes can also occur very slowly.
5. See the chart on page 377 for evidence to support the worldwide Flood described in **Genesis**.

Lesson 2

1. Relative dating can be done on rocks in an undisturbed formation to determine which layer is the oldest. In contrast, radiometric dating can be used to determine the absolute age of an object, such as a rock or fossil.
2. The geologic time scale associates each layer in a geologic column with a specific date. According to the scale, Earth is estimated to be over 4 billion years old.
3. The accuracy of radiometric dating has been brought into question because it is based on several assumptions that may not always be true. These assumptions include that the decay rate (half-life) has always been constant, that the decay takes place in a closed system, that the initial concentration of the daughter product can be determined, and that the ratios of parent-to-daughter material were caused only by radioactive decay.
4. Several pieces of evidence are inconsistent with the ages determined through radiometric dating: rates of erosion, bedding, flat gaps, widespread sedimentary deposits, and rapid geologic processes at work (Mt. St. Helens). All suggest more rapid processes than would be associated with longer ages.

Lesson 3

1. The fossil record provides information about Earth's history, including what environments may have been like and how organisms have changed over time.
2. An index fossil is one that is widely distributed but found only in a limited portion of a geologic column. Index fossils are useful because they help us to correlate layers of the geologic column.
3. After 10 half-lives, so little carbon-14 remains that it cannot be detected.
4. Macroevolution predicts a gradual increase in the complexity of organisms found in the fossil record and morphological intermediates that demonstrate phyletic gradualism, but instead the fossil record shows the rapid appearance of most major phyla, a high level of complexity throughout the fossil record, and very few morphological intermediates. These findings are consistent with the Creation account in **Genesis 1**.

Show What You Know

Vocabulary Check

Explain how each pair of terms is related.

1. deflation—abrasion
2. radioactive decay—half-life
3. bedding—bioturbation
4. relative age—absolute age
5. radiometric dating—carbon dating
6. phyletic gradualism—Cambrian explosion
7. tectonic plates—supercontinent

Multiple Choice

Choose the best answer.

8. Which statement might a scientist make when describing the relative age of a rock layer?
   A. This layer is 12,000 years old.
   B. The top layer will erode away in the future.
   C. The bottom layer is older than the top layer.
   D. This layer formed 40 million years after that layer.

9. According to the geologic time scale, about how old do scientists estimate Earth to be?
   A. 46,000 years
   B. 4.6 million years
   C. 4.6 billion years
   D. 46 trillion years

10. Why do flat gaps suggest that Earth is younger than scientists estimate it to be?
    A. They show little evidence of erosion.
    B. They show that mountains wear down quickly.
    C. They show that sediments are deposited unevenly.
    D. They show that canyons can be carved out rapidly.

11. Which of these happened during the Cambrian explosion?
    A. Rock layers were destroyed from the geologic record.
    B. Canyons were formed rapidly.
    C. Many species went extinct at the same time.
    D. Many phyla appeared abruptly.

Check Point

Answer the following questions.

12. In 1910, a German scientist named Alfred Wegener proposed that Pangaea slowly broke apart to form the present-day continents. However, he could not explain how this happened. What evidence did Wegener lack that would have supported his proposal?

13. Some fossils are the imprints left in sediments that have hardened over time. Can carbon-14 be used to determine the absolute age of a fossil imprint? Explain the reason for your answer.

14. Explain why horseshoe crabs that are found early in the fossil record and that have existed on Earth with little change would not be useful as an index fossil.

15. Explain why the lack of transitional forms in the fossil record is used as an argument against macroevolution.

Unit 4 Physical Science

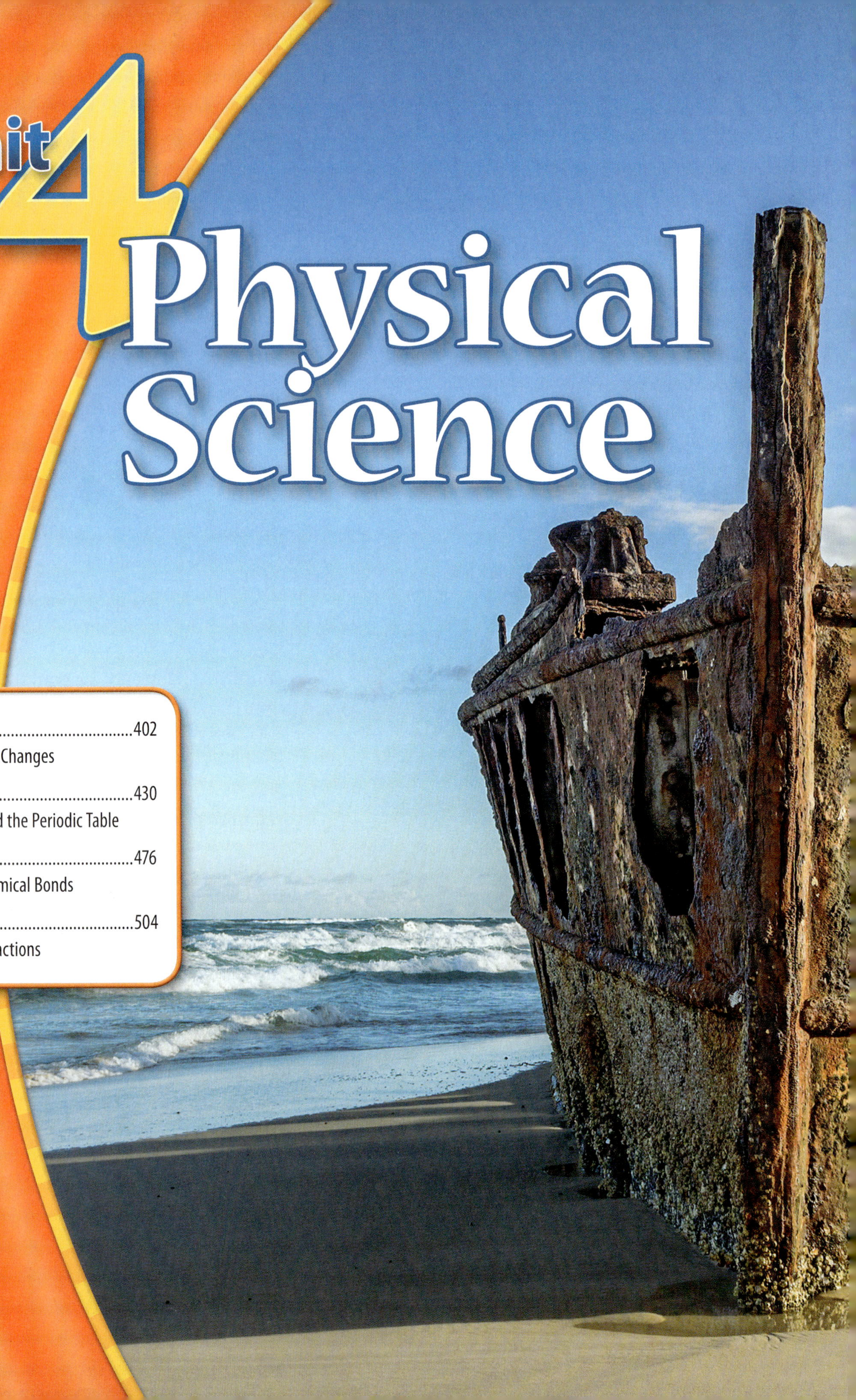

Unit Overview

Most of the objects around you are made from matter that has changed in some way. In this unit, you will explore how matter changes and how the particles that make up matter are involved in these changes. What happens when iron rusts? How does matter with new properties form? How can you describe the processes that produce this new matter?

You will also examine the periodic table of the elements and learn about the characteristics that certain groups of elements share. What properties do gold and silver share? How is oxygen different from helium? What are some other ways that elements differ?

The rust on this boat resulted when the iron on the boat chemically reacted with oxygen in the air. What other examples of chemical reactions are around you?

ADVENTISTS InStep for Life

Your teacher may assign an Open Inquiry lab and a Lifestyle Challenge activity. Use your ***Science Journal*** to record your work.

Chapter 11

Ways Matter Changes

Scripture Spotlight

God created matter with a wide variety of characteristics that we can use to make our lives on Earth easier. By studying these characteristics, we can see how God cares for us. In this chapter, you will read the following passages.

Matthew 25:31–46 (p. 414)
1 Corinthians 1:2 (p. 416)
2 Corinthians 5:17 (p. 416)
Matthew 5:13 (p. 417)
John 17:11, 14–16 (p. 417)
John 13:35 (p. 421)
John 14:15 (p. 421)

This foundry takes solid iron ore and heats it until the iron melts and becomes liquid. The foundry then further processes the liquid and changes it changes it back into a solid. This chapter explores the states of matter and how they change.

The Big Idea

Physical properties can be used to classify matter and explain how it changes. These changes make it possible for us to use matter in many different ways.

What is happening to the atoms in the melted iron? Is the liquid iron the same substance as solid iron? What is different?

Inquiry Kick-Off

Engage

Many scientists classify matter based on its properties. Does all matter fit neatly into standard categories? What do you do if matter cannot be classified in a standard group that already exists? In your *Science Journal*, you will examine two types of matter that might be hard to classify.

Lesson 1

Objectives

- Distinguish between physical and chemical properties of matter.
- Distinguish between physical and chemical changes.
- Identify the four main states of matter.
- Describe the properties of each state of matter.
- Explain the changes in energy that occur during changes in state.

Vocabulary

extensive property
intensive property
crystalline solid
amorphous solid
viscosity
surface tension
compression
endothermic
exothermic
sublimation

Essential Question

What Are Properties of Matter?

Think about eating a juice pop on a hot day. At first, the pop is very cold and solid. But soon, the pop starts to melt and soften. Liquid juice drips from the juice pop. You quickly lick the dripping juice, but some still falls to the ground. What causes these changes in the juice pop? What will happen to the liquid on the ground over time? In your everyday life, what materials are able to change from one state to another? How are these changes useful to you?

Properties of Matter Explain

How would you describe your science book? Which properties of the book could you use to describe it? Which properties could be used to distinguish your book from those of your classmates? Properties of your book include its mass and volume. Would these properties be the same for every book, or would they be different? Why? Find out. Use a balance to determine the book's mass and a metric ruler to determine its volume. How do the science books in your class compare?

Physical Properties

Properties of matter can be classified as either physical properties or chemical properties. You have learned that *physical properties* of matter are characteristics that can be observed using our senses. These properties include color, shape, smell, mass, density, and others. Physical properties can be further categorized as intensive or extensive properties. **Extensive properties** are those that depend on the amount of matter present, such as mass and volume. **Intensive properties** are properties that do not depend on the amount of matter present. Color, density, hardness, and temperature are examples.

Would "tasty" be a physical property of this juice pop?

What are some extensive and intensive properties of a golf ball? What about the pen you use to write with? How many of these properties can you list?

Gold is a soft, yellow metal that can be hammered easily into thin sheets. It also has a high density and keeps its yellow color. These descriptions are the physical properties of gold. These properties of the gold stay the same even if you cut the gold into two pieces. On the other hand, what physical properties of the gold nugget change when it is cut in half?

Chemical Properties

Chemical properties, you will remember, are those properties that describe how different kinds of matter combine and react with each other to form new substances. What examples of chemical properties can you remember? What would be a chemical property of a sheet of paper? What are some chemical properties of steel wool? Look at the chart below to review chemical and physical properties.

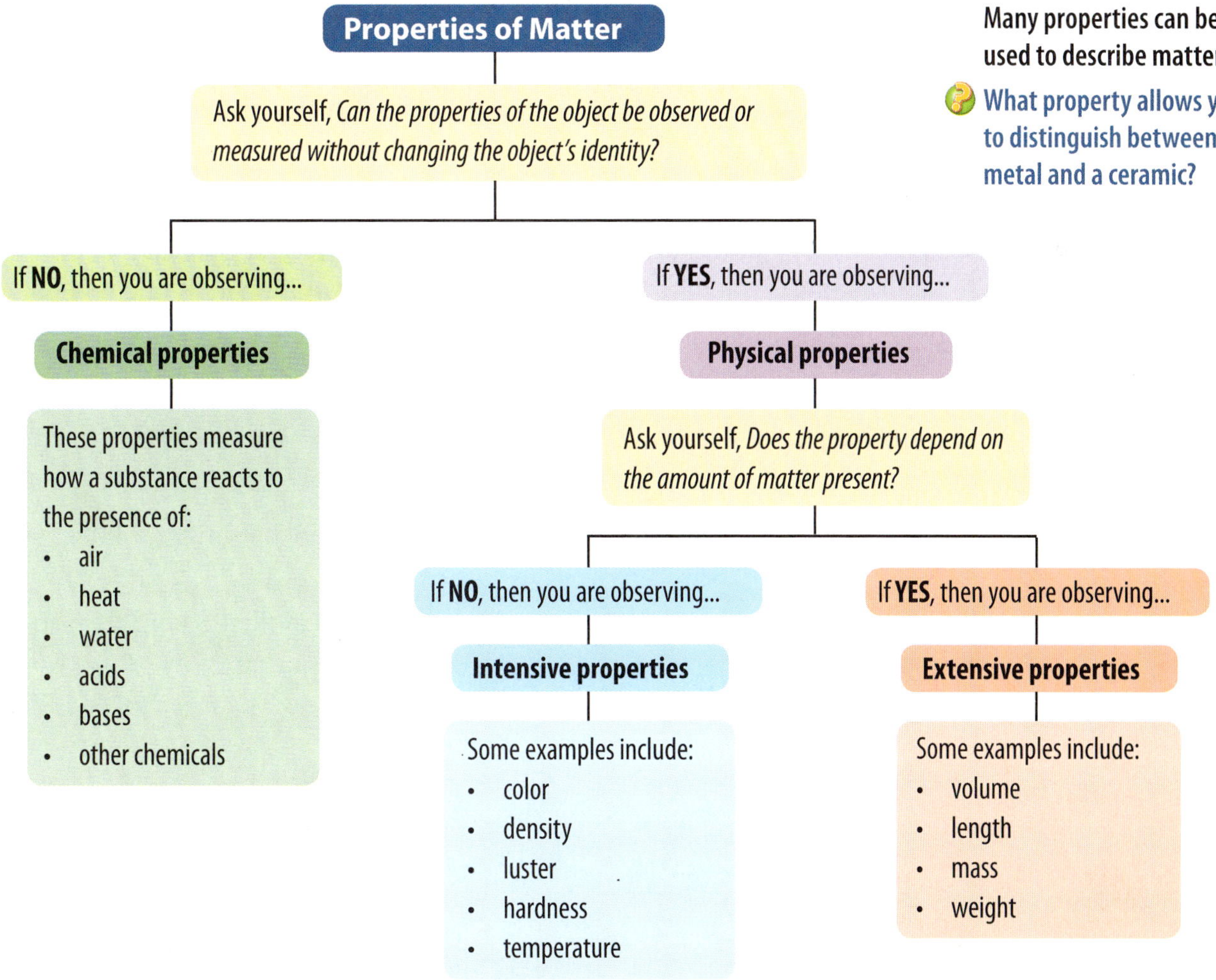

Many properties can be used to describe matter.

What property allows you to distinguish between a metal and a ceramic?

Structured Inquiry

Discover

Record your work for this inquiry. Your teacher may also assign the related Guided Inquiry.

Science Journal

How Hard Is This?

How can the hardness of a mineral be determined?

Materials

- mineral assortment
- penny
- putty knife
- glass plate

Procedure

1. Look at each of the minerals in the assortment you have. Predict the hardness of each sample by ranking them in order from softest (#1) to hardest (#10). **Record** your prediction in the data section.
2. Now begin to test the hardness of each sample.
3. Test Sample A to see whether it can be scratched by your fingernail. Record the results in your ***Science Journal***.
4. Test Sample A to see whether it scratches the penny. Record the results.
5. Test Sample A to see whether it scratches the putty knife. Record the results.
6. Finally, test Sample A to see whether it scratches the glass plate. Record the results.
7. Repeat Steps 3–6 with the remaining Samples (B–J).

Analyze Results

Place the samples in order of hardness from softest to hardest. If two samples appear to be the same, scratch one with the other to determine which one is harder.

Create Explanations

1. How can the hardness of a mineral be determined?
2. What kind of physical property is hardness?
3. Is the hardness of a mineral sample enough to determine what mineral it is? Why or why not?
4. What other physical and chemical properties are helpful in identifying a mineral sample?

Physical and Chemical Changes Explain

You learned earlier that a *physical change* is a change that affects one or more physical properties of matter, but the identity of the matter stays the same. Cutting a long piece of wire into shorter lengths, breaking a rock, melting an ice cube, or blowing up a balloon are examples of physical changes. What are some physical changes that have taken place in your classroom today? What physical changes have you caused? What physical properties of a lump of clay can you change? Which can you not change? Try it.

Recall that a *chemical change* is a change in which one type of matter is changed into another type of matter that has different properties. Many chemical changes are difficult to observe because they occur at the molecular level. Other chemical changes are easy to observe. For example, when a match burns, it is easy to observe that the tip, head, and stick are all changed into different substances. What physical changes are happening to the match?

A seltzer tablet is fizzing in water, showing both a chemical and physical change.

What part of this image suggests that a physical change is happening? What about a chemical change?

Lesson Activity

Pick a type of weather event (snow, rain, heat wave, tornado) and make a list of as many physical and chemical changes that you can think of that could be caused by the event. Think about what the weather might do to your house, your yard, or to the plants or animals outside.

Can you think of a weather event that does not cause physical and chemical changes?

States of Matter Explain

Imagine that the weather is cold and that there is snow on the street. As the Sun warms up the street, the snow begins to melt. Most of the water runs off the street, but puddles remain. As the street gets even warmer, the puddles get smaller and smaller. Where does the water go when the puddles disappear? Has the water really disappeared? Explain.

Snow and ice are water in its solid state. The water in the puddles is water in its liquid state. The water that evaporates from the puddles is water in its gaseous state. Solids, liquids, gases, and a fourth form of matter—plasma—represent the four common states of matter. Describe all the ways that water changes state in your home. How are these changes useful, and how can they be problems?

Lesson Activity

Recall what you previously learned about the states of matter and create an illustration for each state. You may either draw or create a physical model. How many of these states of matter can you see in the room you are in?

What do the states of matter look like?

Solids

You already know that a solid has a definite shape and volume. The definite shape and volume of a solid result because the solid's particles have little kinetic energy. Recall that *kinetic energy* is the energy of motion. The particles in a solid do not move from place to place, but they do vibrate while remaining in place.

There are two kinds of solids—crystalline and amorphous. When the atoms in a solid are arranged in a very regular, three-dimensional pattern, the material is known as a **crystalline solid**. Most metals are crystalline solids. So are many rocks, gemstones, and ice. The atoms in a crystalline solid line up into regular rows and flat surfaces.

Amorphous solids, such as plastic and rubber, do not have this orderly atomic pattern. In an **amorphous solid**, each atom has a specific position where it remains, but the atoms do not have the ordered arrangement of a crystalline material. Describe what you would do to make models that show the difference between these two kinds of solids.

How would you describe the viscosity of this liquid?

Liquids

Recall that a *liquid* has a definite volume but no definite shape. In a liquid state, atoms have more kinetic energy than a solid. They move faster and can bounce farther off one another. This movement causes the atoms in liquids to be farther apart. How does this change in spacing affect the atoms? What is the evidence of this change?

Liquids have a property known as **viscosity**, which is a liquid's resistance to flow. The higher the viscosity of a liquid is, the more slowly it will flow. Why does heating a liquid lower its viscosity?

Another special property of liquids is surface tension. **Surface tension** is the attractive force that exists between the particles on the surface of a liquid. Liquids that have a high surface tension, such as water, form spherical drops or beads. In contrast, liquids that have a low surface tension, such as gasoline, form flat droplets. Why does the difference in surface tension cause these two liquids to form drops that have different shapes? How can you model this with your classmates?

Explore-a-Lab

Structured Inquiry

How many drops of water can dance on Lincoln's head?

Use the medicine dropper to put as many drops of water as possible onto the surface of a penny without the water flowing off. Draw a side-view diagram of the penny and the water, showing the penny on the bottom and the water sitting on top. Record the number of drops before the water begins to flow over the penny's edge. Repeat this process two more times. Construct a graph that displays your data. Why do you think there is a difference in the number of drops that fit on the penny? Do you think this is a chemical property or a physical property of water? Think of some other variables that you could test in doing this investigation.

Gases

A gas has neither a definite shape nor a definite volume. The atoms in a gas have the highest kinetic energy of the three common states of matter, and they move the fastest. The atoms in a gas can spread out to fill the container in which it is kept. Gas particles create a pressure on the walls of the container as they bounce against it. The space between particles in a gas is relatively large compared with the space between particles in solids and liquids. Therefore, the particles of a gas can be squeezed together or compressed. **Compression** is a property in which a substance can be forced to occupy a smaller volume. Is compression a physical property or a chemical property? Explain why. Which state of matter is less likely to be compressed? Explain why.

Explore-a-Lab

Structured Inquiry

How does temperature affect a helium-filled balloon?

Fill a balloon with helium gas. Predict what will happen if you place the balloon in a freezer. Then test your prediction by placing the balloon in a freezer for 1–2 hr. What happened to the balloon? Now submerge the balloon in hot water. What happens? Explain your observations.

Plasma

Plasma is a state of matter where many of the atoms have lost their electrons. A plasma is a gas, so it can change its shape or volume. The gas inside of a fluorescent light bulb is a plasma. Most of the gas inside the Sun is also a plasma. What do you think causes the gas inside the Sun to be a plasma?

Other, even rarer, states of matter exist. Scientists think that there are probably states of matter that we haven't discovered yet.

What physical properties do the plasmas in the light bulb and in the Sun share?

Changing States Explain

Suppose you put a paper towel on your desk and then place an ice cube on the paper towel. What will happen? How long do you think it will take for the solid water to change completely to liquid water? What will eventually happen to the water on the paper towel? What is needed to bring about these changes in state? How could you make these changes happen faster?

Energy Changes during Melting and Freezing

You learned in your earlier study of science that any change in state is accompanied by a change in energy. Energy as heat is usually either added or removed during a change in state. This change in energy can be measured by measuring the temperature of a substance as it changes state. If a pot full of very cold ice is placed onto the stove and heated, energy will flow from the stove to the pot and into the ice in the form of heat. This energy will cause the atoms in the ice to gain kinetic energy and move faster, and the temperature will rise. As the temperature reaches 0°C, the ice will start to melt. At first, the pot will be filled with mostly ice with a little bit of melted water, but as the pot keeps heating, the amount of liquid water will increase and the amount of solid ice will decrease. Eventually, there will be mostly liquid water and a tiny amount of ice left, which will finally melt. Objects that absorb energy while they change their state are said to be experiencing an **endothermic** change.

A similar effect occurs while a substance is freezing. From the point where the smallest bit of a substance starts to freeze to the point where the last amount of liquid changes into a solid, the temperature of the whole mixture will stay at the freezing temperature. While a liquid is freezing, the substance is losing energy, which is flowing into the surrounding environment. Objects that give off energy while they change their state are said to be experiencing an **exothermic** change.

Energy Changes during Boiling, Condensation, and Sublimation

The temperature in the pot with ice will only start to rise after all of the water is melted. Once all the water is a liquid, the energy going into the pot will then increase the movement of each atom of the water. This increased movement causes the temperature of the liquid water to rise, until the temperature of the liquid water reaches 100°C. While the water is heating, a small amount of it will evaporate, but most will stay as a liquid in the pot.

When the temperature reaches 100°C, the water will start to boil and turn into a gas. The steam that comes out of the pot also has a temperature of 100°C. What happens to the temperature of the water while it is boiling? Energy keeps flowing into the pot, because it is still being heated. Where does that energy go? Since boiling water absorbs energy from its surrounding environment, it is an endothermic process.

Explore-a-Lab

Structured Inquiry

How quickly does the temperature of ice water change as the ice melts?

Fill a 400-mL beaker with ice. Measure and record the temperature of the ice. Measure 100 mL of room-temperature water. Measure and record the temperature of the water. Pour the water into the beaker of ice. Predict what you think the temperature of the ice and water mixture will be. Record your prediction. Now measure and record the actual temperature of the ice and water mixture. Is the temperature of the mixture what you expected it would be? Predict what the temperature of the ice water mixture will be in 10 min. Record your prediction. Record the temperature of the ice water every 3 min until all the ice has melted. Design a data table and record your data in the table. Construct a graph to display your data. How does the initial temperature of the ice water compare with the temperature just before the last bit of ice melted? Where did the heat that was in the room-temperature water go? What could you have done to make the temperature of the ice water colder? Explain.

If the gas condenses back to liquid water, the process will release energy from its surrounding environment. *Condensation* is an exothermic process. Under the right conditions, some solids can also change their state directly into a gas without melting. This process is known as **sublimation**. How can the properties of giving off heat be used to our advantage? Why might a farmer spray water on his strawberry plants when they are in danger of being frozen by a late spring freeze?

Explore-a-Lab

Structured Inquiry

What happens to the temperature of water as it melts and boils?

Fill the beaker with ice, and place the thermometer in the ice. Heat the ice with the hot plate, and record the time, temperature, and state of the water in the beaker (ice, ice + liquid, liquid, boiling liquid). Turn off the hot plate after the last water has boiled away. Measure the temperature at 15 or 20 different times while the ice is warming and melting and the water is heating up. Use the data to make a plot of temperature versus time. Label the portions of the curve with "ice," "ice + liquid," "liquid," and "boiling liquid" as appropriate. What happens to the energy that is flowing into the water as the state of the water changes?

Concept Check

Assess/Reflect

Summary: What are properties of matter? Each type of matter can be described by its physical and chemical properties. Matter can undergo physical changes, which can change the way the matter looks but does not change the matter itself. Chemical changes change the type of matter. The same matter can exist in different states, such as solid, liquid, gas, or plasma. Matter can change from one state to another by processes such as evaporation, boiling, condensation, sublimation, melting, or freezing. State changes are accompanied by changes in energy. Melting, boiling, and sublimation are endothermic state changes. Freezing and condensation are exothermic state changes.

1. Trees continually draw water from the ground through their roots. Where does this water go?
2. Why does the mirror in the bathroom fog up when you take a really hot shower?
3. Do you think sublimation is an exothermic process or an endothermic process?

Lesson 2

Essential Question

What Are Mixtures and Solutions?

Objectives

- Describe the properties of a mixture.
- Distinguish between homogeneous and heterogeneous mixtures.
- Analyze a solution in terms of solute and solvent.
- Distinguish between a solution, colloid, and suspension.

Vocabulary

heterogeneous mixture
homogeneous mixture
dilute solution
concentrated solution
saturated solution
supersaturated solution
suspension
colloid

When you mix flour, sugar, milk, eggs, baking powder, and vanilla together, what are you making? Cake, of course! But how do these ingredients combine when you put them together? Are some of them getting dissolved? Are some of them just getting physically mixed together? Recall that matter can be a pure substance or a mixture of pure substances. Pure substances include elements and compounds in which every particle is alike. What happens to those pure substances when you combine them in a mixture? After all those cake ingredients are mixed together, could you pull out a single particle of flour? What about the sugar?

Heterogeneous Mixtures Explain

A bag of jelly beans is a mixture that you can separate. If you don't like the lemon ones, you can always pick them out and give them to a friend who likes them. But what if you have a bag of popcorn that is too salty? Is it easy to separate the popcorn from the salt?

The bag of jelly beans is a mixture of beans of different flavors and colors. A mixture is formed when any two elements or materials are blended together. These blended materials touch each other but are not chemically combined. This means that the mixture can be physically separated. The lemon jelly beans can be easily removed because they can be identified by their color. Mixtures that have components that are not uniform or have regions with different properties are called **heterogeneous mixtures**.

Suppose you add a spoonful of salt and a spoonful of pepper together on a paper plate. What would be the best way to separate the salt and pepper from the mixture? If you add a spoonful of salt and a spoonful of pepper to a cup of water, could you use the same technique to separate the mixture? Why or why not? If you were given a sample that was made up of iron and sulfur powders blended together, how would you separate this mixture? What properties of the materials help you to separate them when they are combined? The materials that go into these mixtures have their own distinct identities and properties.

Some mixtures are not easily separated. For example, granite is a rock that is a mixture of several different types of rock. However, the granite formed originally as a liquid mixture and then solidified, so the particles of different materials are "glued" together. This makes granite very difficult to separate, but it is still a mixture of several unique substances rather than a single material. Most steel and cement materials are also examples of mixtures that are difficult to separate.

Scripture Spotlight

What method for separating a mixture is described in **Matthew 25:31–46**? What "distinguishing property" determines which group each mixture will be separated into?

It is easy to see the different kinds of matter in these mixtures. Separating the materials can be easy or very challenging, depending on the mixture.

What kind of a candy is a mixture that would be difficult to separate?

Math in Science

Soil is a very important mixture. Knowing the contents of a soil sample can help scientists understand an ecosystem or determine what kinds of crops can be grown in an area. A sample of soil is analyzed and is found to contain 15 g of clay; 65 g of sand; 35 g of silt; 3 g of roots, leaves, and other plant material; and 1 g of insects and other insect matter. If a farmer needs this soil to have 4% plant and insect matter, does the current soil meet his needs? Explain.

Homogeneous Mixtures Explain

Many mixtures are so well mixed together that they don't look like a mixture at all. Fruit punch is a mixture of water, sugar, and ingredients that add flavor and color. Shampoo is a mixture of water, soap, and other chemicals. Mixtures such as these, where the components are uniformly distributed throughout the mixture, are called **homogeneous mixtures**. It can be difficult to tell whether they are mixtures or pure materials, and they can be very difficult to separate. Can you think of a food in your kitchen cabinet that has many ingredients but looks like a pure material?

Structured Inquiry

Discover

Record your work for this inquiry. Your teacher may also assign the related Guided Inquiry.

Science Journal

Sands of Time

What particles make up a sample of sand?

Procedure

1. **Measure** about 1 teaspoon of sand. Spread out your sample of sand so that a single layer is formed.
2. Use the hand lens and the toothpick to count out 100 grains of sand.
3. **Sort** the 100 grains according to color and features. Use the Particle Key to sort the grains.
4. Count the number of grains of each type. Record.
5. **Calculate** the percentage of composition each mineral type of sand represents in the total sample.

Materials
- measuring spoons
- sand (coarse)
- hand lens
- toothpicks
- Particle Key
- tweezers
- magnet

Analyze Results

Construct a bar graph that shows a comparison between the percentage of each particle you found in your sample with the class average of each percentage. Is there much difference between your sample and the class average?

Create Explanations

1. What particles make up a sample of sand?
2. How did your mixture compare with your classmates'?
3. Which mineral was most common?
4. What type of mixture is sand—homogeneous or heterogeneous?
5. What technique did you find most effective in separating the mixture?

Particle Key
Calcite—pure white
Feldspar—opaque, white to gray, sometimes red
Muscovite mica—white to gray, transparent, paper-like pieces
Quartz—colorless, semitransparent, glassy appearance
Pyrite—light brassy yellow, cubic structure
Biotite mica—brown to black, paper-like pieces
Iron— black, dark brown, magnetic
Hornblende—green to black
Organic material—bits of wood, leaf, etc.

Solutions Explain

When you add a teaspoon of sugar to a hot cup of water, how do you know that the sugar is still in the cup even though you cannot see it? You already know that the sugar and water form a liquid mixture known as a *solution*. Since you cannot tell which part of the solution is the water and which is the sugar, the solution is homogeneous. Solutions have two parts, a solvent and a solute. Recall that a *solvent* is the part that dissolves another material. The *solute* is the material that is dissolved. For sugar water, what is the solvent? What is the solute? How would you separate the components of this solution?

Scripture Spotlight

Think about the definitions of *solute* and *solvent* as you compare **1 Cor. 1:2** and **2 Cor. 5:17**. What spiritual application can you think of?

Explore-a-Lab — Guided Inquiry

How can you remove the carbon dioxide from a carbonated drink?

Develop a plan to remove the carbon dioxide from a carbonated drink. You may want to try several methods with a few cans.

While many solutions use water as the solvent, many other kinds of solutions do not use water as a solvent. Iodine is sometimes used as an antiseptic to help prevent infections. Iodine does not dissolve in water, but it does dissolve in alcohol. When a substance is dissolved in alcohol, the resulting solution is called a tincture. Is it always easy to figure out what is a solvent and what is a solute? How about a solution of equal amounts of water and alcohol?

Gases and liquids can be solvents and solutes. The air we breathe is a solution of nitrogen and oxygen with other gases dissolved in them. Alcohol and water form a solution when alcohol dissolves in water. In this case both the solvent and solute are liquids. When both solvent and solute are gases or liquids, the material that makes up the majority of the solution is usually called the solvent. Why do most fish tanks have a bubbler that moves a constant stream of air bubbles through the water?

Solids can also be solutions. Metal alloys are solutions made by completely mixing together two melted metals, and then cooling the solution so that it forms a solid. Metal alloys are called solid solutions. Brass is a solid solution of copper, nickel, and other atoms.

Concentration of Solutions

If you dissolve more and more sugar into a cup of hot water, what will the solution taste like? What will eventually happen if you keep adding sugar to the water? When a small amount of solute is dissolved in a solvent, the solution is called a **dilute solution**. If a large amount of solute is dissolved in the solvent, the solution is said to be a **concentrated solution**. If you continue to add sugar to the solution, the concentration will continue to increase until a point is reached when no more sugar can be dissolved in the water. A solution that has reached this point is called a **saturated solution**.

If you want to add more sugar to the saturated sugar solution, you have to heat the solution. Why would heating the solution allow more sugar to be dissolved in it? What happens to the atoms in the solution as heat is added to the system? A solution that has more solute dissolved in it than could be dissolved by the solvent under normal conditions is called a **supersaturated solution**. What would you expect to happen if a supersaturated solution cools? Why?

Scripture Spotlight

Compare **Matthew 5:13** with **John 17:11, 14–16** and think of a spiritual application that has something to do with substances mixing together.

Explore-a-Lab

Structured Inquiry

Can you use your senses to determine concentration?

Your teacher will provide you with five sugar solutions, each with a different concentration of sugar. Taste each of the five sugar solutions, and rank the solutions in terms of sweetness from 1 to 5, with 1 being sweetest. As a class, record the rankings and make a plot. What explanation can you give for the differences in ranking?

Suspensions and Colloids Explain

Gold flakes are often stored in water. If the container is shaken, they will float around for a while, but eventually they will settle to the bottom of the container. It is easy to distinguish between the gold and the water. This type of mixture is called a suspension. A **suspension** is a mixture in which the particles in the solvent are so large that they will settle out unless the mixture is constantly stirred. What force causes the particles to settle out of the solution? What are some examples of suspensions?

The particles in a suspension are larger than the particles in a solution—usually larger than 1 micrometer (1 μm). You can see the individual particles in a suspension under a microscope, and often you can see them with the naked eye. Sometimes you can separate the parts of a suspension by filtering. How else might you separate the components of a suspension?

How could you cause the particles in the peanut butter to fall out of the solution? What two very familiar colloids can you see in the picture?

Sometimes particles are larger than those in a solution but smaller than those in a suspension. These mixtures are called **colloids**. The particles in a colloid can be solids, liquid, or bubbles of gas. They will float in the solution for a very long time. In this case, the atoms in the solvent have enough kinetic energy to push the particles around and keep them from settling out. Since the particles in a colloid are often large enough to absorb or scatter light, most colloids are not transparent. Milk is a colloid, and so are paints and inks.

Types of Colloids

Type	Phases	Examples
Gel	Gel	Gelatin
Emulsion (liquid)	Emulsion (liquid)	Milk; mayonnaise
Emulsion (solid)	Emulsion (solid)	Cheese, butter
Foam	Foam	Shaving cream; whipped cream
Aerosol (solid)	Aerosol (solid)	Smoke; auto exhaust
Aerosol (liquid)	Aerosol (liquid)	Clouds; mist; aerosol spray

Concept Check Assess/Reflect

Summary: What are mixtures and solutions? A mixture consists of two or more different substances. It is easy to distinguish the different materials in a heterogeneous mixture. Homogeneous mixtures are very well mixed and appear like a single material, even if they are not. Solutions are homogeneous mixtures. The amount of solute that can be dissolved into a solvent depends on temperature. Solutions can be solids, liquids, or gases. Suspensions and colloids consist of mixtures of particles in a liquid or gas. In a suspension, the particles are large enough that they will not stay mixed. In a colloid, the particles are small enough that they will stay mixed for a very long time.

1. What physical properties of a mixture could help you decide whether it is heterogeneous or homogeneous?
2. What physical force causes suspensions to settle over time?
3. A mixture of raisins and peanuts is a heterogeneous mixture. How could you make it homogeneous?
4. What solutes would you expect to find in a bottle of soda?
5. What is an example of a suspension that does not settle out?

Lesson 3

Essential Question

What Are Acids and Bases?

Why does a lemon taste sour? Why does vinegar add zest to a salad? Why does soap feel slippery? Why does toothpaste help control bad breath? The answers to all these questions depend on certain types of chemicals that are found in each of these substances. These chemicals are known as acids and bases, and they have certain properties that make them useful in a wide variety of applications.

Objectives

- Compare the properties of acids and bases
- Explain how an indicator can be used to identify an acid or base.
- Explain the pH scale.

Vocabulary

acid

base

acid-base indicator

pH scale

Acids, Bases, and Ions Explain

Generally, a substance that is an acid or a base will act as an acid or a base only when it is dissolved in water. These substances produce ions when dissolved in water. Recall that an *ion* is a charged particle.

In a solution, most **acids** produce a hydrogen ion (H^+). A hydrogen ion is made up of a proton and has a positive charge. The diagram shows what happens when hydrogen chloride (HCl) is dissolved in water (H_2O). What two ions are produced when HCl dissolves in water?

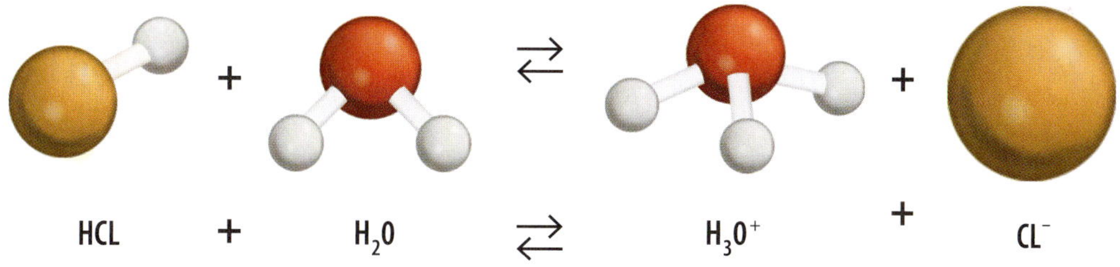

Each HCl molecule can form one hydrogen ion.

According to the diagram, what happens to the hydrogen ion?

When most **bases** dissolve in water, they release a hydroxide ion (OH^-). For example, when potassium hydroxide (KOH) is dissolved in water, it releases potassium ions (K^+) and hydroxide ions (OH^-).

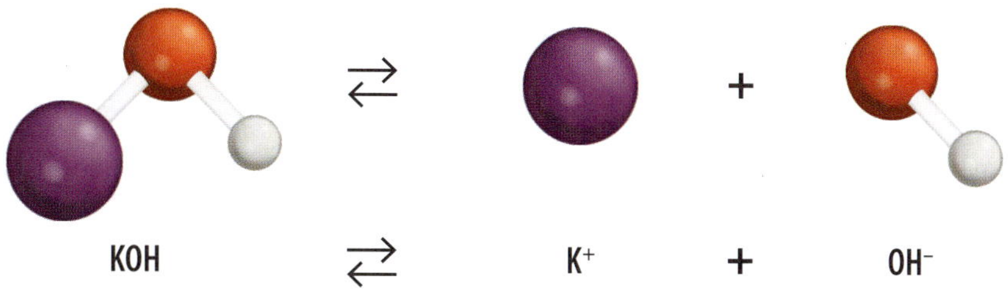

Acids Explain

As you just learned, an acid is a substance that increases the number of hydrogen ions (H_3O^+) when dissolved in water. Lemons and vinegar are acids. Some acids are used to clean surfaces, etch glass, and create laboratory reactions. Acids are also used in the manufacturing of paints, plastics, and some types of fertilizer. One type of acid is used in car batteries. Another acid helps to digest food in your stomach.

Physical Properties of Acids

All acids share certain physical properties. For example, even though it is very dangerous to taste most acids, acids have a sour flavor. In fact, the word *acid* comes from Latin and means "sour." Citrus fruits get their name from the citric acid that they contain. The concentration of the citric acid determines the sourness of the fruit. Which citrus fruits have the highest concentration of citric acid? Vinegar and yogurt are also sour because of the acids that they contain. Acids also exhibit high electrical conductivity.

Even though there are many acids in foods, an unknown chemical should never be tested by tasting it. Some acids are poisonous, and they can be dangerous when concentrated. The table below lists some common acids and their uses.

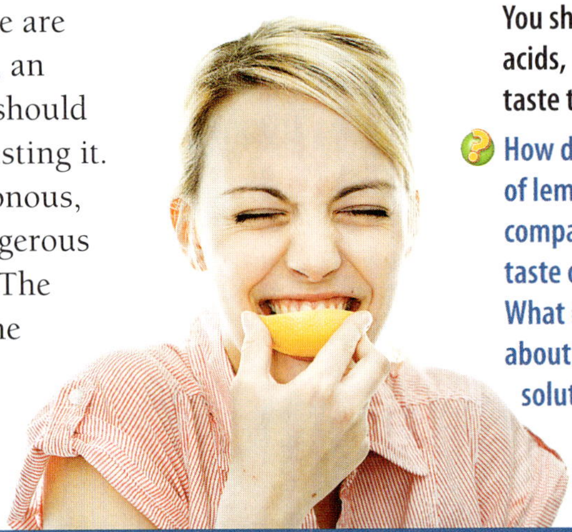

You shouldn't taste acids, but you can taste this one!

How does the taste of lemon juice compare with the taste of lemonade? What is different about the two solutions?

Acid Name	Acid Formula	Use
Acetic acid, or vinegar	CH_3COOH	Making pickles and salad dressing, in some cleansers
Acetylsalicylic acid, or aspirin	$HOOC\text{-}C_6H_4\text{-}OOCCH_3$	Fever reducer and pain reliever
Ascorbic acid, or vitamin C	$H_2C_6O_6$	Vitamin
Carbonic acid, or seltzer water	H_2CO_3	Adds carbonation to sodas
Hydrochloric acid, or muriatic acid	HCl	Cleaning pools and brick walls, stomach acid
Nitric acid	HNO_3	Fertilizer production, industrial chemical
Phosphoric acid	H_3PO_4	Ingredient in detergents and some cola drinks
Sulfuric acid	H_2SO_4	Car batteries, industrial chemical
Citric acid	$C_6H_8O_7$	Adds tart taste in jellies and jams

You see some of these acids every day.

What is similar about most of these acid formulas?

Chemical Properties of Acids

You just learned that in water an acid releases a hydrogen ion. Another chemical property of most acids is their ability to react with metals. The picture shows the bubbles that form when a metal such as zinc is placed in an acid—in this case, hydrochloric acid. These bubbles are hydrogen gas. However, some metals, such as gold and silver, will not react with hydrochloric acid in this way.

You can easily test whether a substance is an acid by using another characteristic. Acids can react chemically with substances known as acid-base indicators to cause the indicators to change color. An **acid-base indicator** will change to a specific color when exposed to an acid. One common type of acid-base indicator is litmus paper. Blue litmus paper turns red when acid is placed on it.

Many metals react strongly and visibly with acids.

What will happen to the solid metal as this reaction occurs?

Bases Explain

Recall that a base is a substance that increases the number of hydroxide ions (OH^-) when dissolved in water. Bases are used to make soap, detergent, and drain cleaner. Calcium hydroxide is a base that is used to make cement and concrete. Some bases are used in antacids. A list of common bases and their uses is shown below.

Base Name	Base Formula	Use
Aluminum hydroxide	$Al(OH)^3$	Antacids and water purification
Calcium hydroxide, or caustic lime	$Ca(OH)_2$	Ingredient in mortar and plaster, soil additive
Magnesium hydroxide, or milk of magnesia	$Mg(OH)_2$	Antacid and laxative
Sodium hydroxide, or lye or caustic soda	NaOH	Ingredient in soap, oven cleaner, and drain cleaner; used in papermaking
Ammonia	NH_3	Ingredient in fertilizers, industrial chemical

You see some of these bases every day.

What is similar about most base chemical formulas?

Physical Properties of Bases

Bases share some common properties. Bases taste bitter, and they feel slippery. But tasting and touching is not a good way to test a substance to see whether it is a base. In fact, both strong acids and strong bases can cause severe burns to the skin. Bases also conduct electricity.

Scripture Spotlight

What litmus tests are described in **John 13:35** and **John 14:15**?

Most soaps are a combination of oil or fat and a base. What physical characteristics of bases are displayed by soaps?

Chemical Properties of Bases

Like acids, bases react with indicators. When red litmus paper is placed into a base, the litmus turns blue. Phenolphthalein is also an indicator and is normally a colorless liquid. When phenolphthalein is mixed with an acid, the solution remains colorless, but the solution will turn pink or fuschia when exposed to a base.

Strength of Acids and Bases Explain

Acids and bases can be strong, weak, or somewhere in between. You have probably already experienced one big difference between a weak and a strong acid. If you have ever had the unfortunate experience of vomiting, then you know how bad that strong stomach acid can taste. How does this compare with the taste of vinegar in a salad dressing? The strength of an acid or a base also has a big effect on how hazardous it is. Strong acids and bases usually require far more safety precautions than weak acids or bases, even if they have the same chemical formula. You probably know how painful it is to get soap in your eyes, and soap is a weak base. What safety precautions should you take if you need to handle a strong base?

The yellow color of this indicator solution indicates that the solution is slightly acidic.

Structured Inquiry

Discover

Record your work for this inquiry. Your teacher may also assign the related Guided Inquiry.

Science Journal

Cabbage Chemistry

How do chemical indicators help to identify whether substances are acids or bases?

Materials

- cabbage solution
- graduated cylinder (10 mL)
- test tube
- 2 medicine droppers
- vinegar
- household ammonia

Procedure

1. Pour 5 mL of cabbage solution into the test tube. Note the starting color. As you work through this activity, the solution will change color. The following legend can be used when interpreting color changes:
   - Red indicates an acidic solution.
   - Blue indicates a neutral solution.
   - Green indicates a basic solution.
2. Use the medicine dropper to add vinegar to the test tube of cabbage solution. Add one drop at a time until the color changes. **Record** how many drops it takes for the color change to appear.
3. Add ammonia with the second medicine dropper, one drop at a time, to the test tube until the color changes back to the original blue. Record how many drops it takes for the color to change.
4. Continue adding ammonia until the color changes again. Record how many additional drops of ammonia it takes to change the color again.

Analyze Results

Identify whether vinegar and ammonia are acids or bases.

Create Explanations

1. How do chemical indicators help to identify whether substances are acids or bases?
2. Is vinegar an acid or a base?
3. Is ammonia an acid or a base?
4. If the soltuion is green, what would you do to turn it red?
5. What do you think would happen if you added a seltzer tablet to the solution at the end of Step 2?

The strength of an acid is determined by the concentration of hydronium ions that it produces in solution. The more hydronium ions that are produced in solution, the stronger the acid. Similarly, the strength of a base is determined by the concentration of hydroxide ions it produces in solution. The more hydroxide ions that are produced in solution, the stronger the base. The relative strength of acids and bases can be measured with the use of the pH scale.

Science Journal

Check out your *Science Journal* for a Structured Inquiry that explores the pH of common household substances.

Extend

The pH Scale

The **pH scale** measures the concentration of hydronium ions in a solution. Values for pH range from 0 to 14. A low pH value means that the substance is acidic, and the concentration of hydronium ions is high. On the other side of the scale, a high pH value means that the substance is basic, and the concentration of hydronium ions is low. The pH of a solution can be measured using indicator paper. When a typical indicator is exposed to an unknown solution, the paper will turn a color that indicates the pH of the solution.

Strong acidic and basic substances are at the ends of the scale, and mild substances are close to the middle, around a pH of 7. Pure water has a pH of 7 and is considered neutral, meaning that it is neither an acid nor a base. Why does tap water not have a pH of 7? What would the pH of milk be?

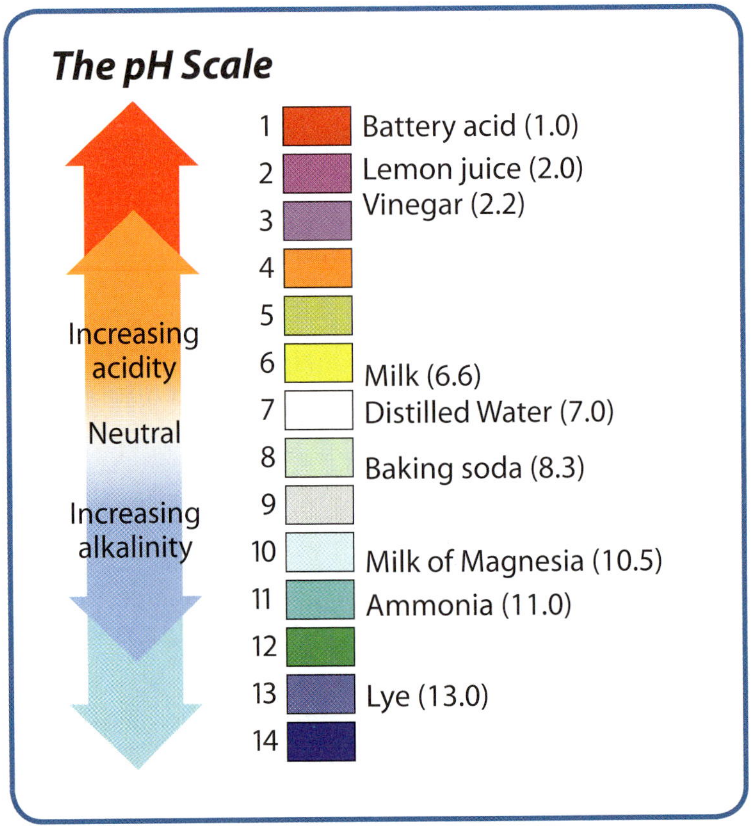

The indicator paper color shows the pH of the solution. The pH measures hydrogen ion concentration and can identify the strength of an acid or a base.

What is the pH range of liquids that are commonly considered "safe"?

Explore-a-Lab

Structured Inquiry

How can you find the concentration of an acid?

Make 5 different solutions of distilled vinegar and water with target pH values of 2.5, 4, 5, 6, and 7. You will need to mix the solutions together in differing amounts to achieve these pH values. Record your initial guess as to the water-to-vinegar ratio. Then use trial and error to achieve the target pH. Be sure to monitor the volumes of water and vinegar necessary to achieve the pH targets. How can you use the data to calculate the actual concentration of vinegar?

Concept Check

Assess/Reflect

Summary: What are acids and bases? Acids and bases are solutions with unique chemical properties. Acids have a sour taste, react with metals to form hydrogen, and react with indicators. They are formed when a substance dissolves in water and forms hydronium ions. Bases have a bitter taste and a slippery feel. They are formed when a substance dissolves in water and forms hydroxide ions. The strength of acids and bases is measured using the pH scale. A pH that is below 7 means that a solution is an acid. A pH of 7 means that a solution is neither an acid or a base. A pH above 7 means that a solution is a base.

1. Think of three ways that you could safely distinguish between a strong acid and a strong base.
2. What similarities are there in the chemical formulas of acids and bases?
3. What will happen to the pH of an acid solution if some of the water evaporates?
4. Is a solution with a pH of 13 a weak base or a strong base?

People in Science

Extend

Get to Know

William Thomson, Lord Kelvin

William Thomson was born in 1824 in Belfast, Ireland. Soon afterward, he moved with his family to Glasgow, Scotland. He was considered a child prodigy and attended university from age 10. He later studied at Cambridge and at Paris universities and became a professor of natural philosophy in 1846 at the age of 22. He held this position for more than 50 years.

He was focused on practical applications of science and established the first physics laboratory in Britain. Along with Michael Faraday, he introduced the concept of electromagnetic fields. He also developed marine tools, such as submarine telegraphy, a mariner's compass, a tide machine, and depth measuring devices. He was the first scientist to develop the concept of energy, and he formulated the first and second laws of thermodynamics. He is best known for his development of a thermodynamic scale of temperature beginning at absolute zero. The scale based on this is named after him, and we measure these temperatures in degrees Kelvin.

In 1892, William Thomson was elevated to the barony and given the title Baron Kelvin of Largs, named after the Kelvin River near Glasgow University, and is therefore known as Lord Kelvin.

Lord Kelvin died in 1907 and is buried at Westminster Abbey in London, England.

Called to Serve

Although a tremendous force for scientific understanding in his time, Lord Kelvin remained a strong and devoted Christian and opposed Darwin's concept of evolution. Using his concept of thermodynamics, he calculated the age of Earth as a maximum of 100 million years, too short a time for evolution.

Concept Check

1. Why do you think Lord Kelvin focused on practical scientific applications? Do you agree that science should be practical? Explain.
2. How did Lord Kelvin express his belief in God?

Titanium in Biomedical Technology

Science and Technology

Extend

Titanium has a long list of benefits for different industries. However, due to its biocompatibility, it has become a staple in the biomedical technology field. It resists corrosion, is long lasting, and has a natural ability to join with human bone. From surgical titanium instruments to orthopedic titanium rods, pins, and plates, medical and dental titanium has become an essential material used in biomedical technology today.

Titanium is most often alloyed with molybdenum, manganese, iron, and aluminum. By weight, it's one of the strongest readily available metals. It is lighter than steel yet has comparable strength and is twice as strong as aluminum, while being only 60% heavier. The element itself is located in the minerals found in beach sand. Titanium is the ninth most abundant element in Earth's crust.

Two of the best properties of titanium are its high strength-to-weight ratio and its corrosion resistance. Couple this with its nontoxic nature and its ability to resist all corrosion from bodily fluids, and it's no wonder titanium has become the metal of choice within biomedical technology. Titanium is also incredibly durable and long lasting.

Osseointegration is a unique process where your body's natural bone and tissue actually bond to the artificial implant. This firmly anchors the titanium dental or medical implant into place. Titanium is one of the only metals that allows for this bonding of bone to implant. When titanium cages, rods, plates, and pins are inserted into the human body, they can last for more than 20 years. Dental titanium, such as titanium posts and implants, can last even longer.

Titanium alloys are used to manufacture the following biomedical devices:

- Pins and pegs
- Bone plates
- Valves
- Expandable rib cages
- Wires and screws
- Parts to repair injuries of the face and jaw
- Finger and toe replacements
- Rods
- Spinal fusion cages

Some of the most common medical uses for titanium are in hip and knee replacement surgeries. It is also used to replace shoulder and elbow joints and to protect the vertebrae following complicated and invasive back surgery. Titanium pegs are used to attach false eyes and ears, and titanium heart valves are even competing with regular tissue valves. The uses for titanium continue to evolve as researchers recognize the unique qualities this metal alloy has to offer. What other metal alloys have properties similar to titanium that researchers may want to consider for medical use?

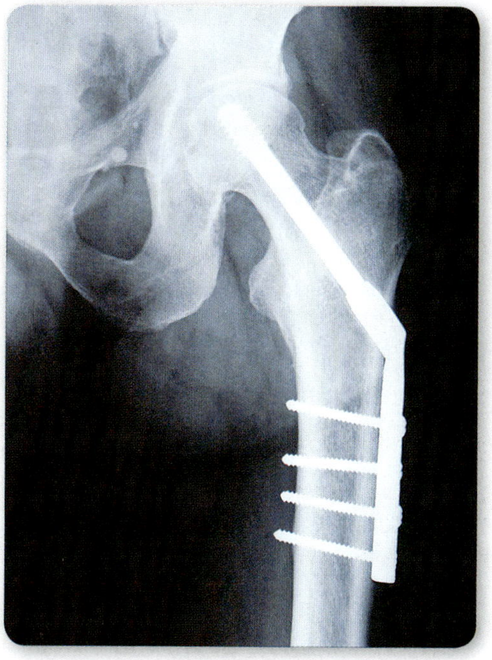

Concept Check

1. Why is titanium such an essential material for use in biomedical technology?
2. What are two of titanium's biggest benefits to biomedical technology?

Chapter 11 Review Assess/Reflect

Study Guide

Lesson 1

1. Physical properties describe substances and can be observed by the senses, while chemical properties describe how a substance reacts with other substances.
2. A physical change does not change the identity of the matter, while a chemical change converts one kind of matter into a different kind of matter.
3. The three main states of matter are solid, liquid, and gas.
4. Solids have fixed volume and shape, liquids have fixed volume and can change their shape, and gases have no fixed volume or shape.
5. While a substance is melting or boiling, it will absorb energy, and when it is freezing or condensing, it will release energy.

Lesson 2

1. A mixture consists of more than one kind of matter physically mixed together but not chemically combined. Many mixtures can be easily separated.
2. Physical forces, such as a pull from a magnet, can be used to separate some mixtures. Sometimes one component of a mixture can be dissolved, leaving the other behind.
3. In a heterogeneous mixture, the different components can be easily identified. In a homogeneous mixture, the components are so well mixed that they cannot be easily distinguished.
4. When a solid is mixed with a liquid, the solid is the solute and the liquid is the solvent. When both components are in the same state of matter (solid, liquid, gas), the substance that makes up the majority of the mixture is the solvent.
5. A solution is a mixture where the different substances are spread evenly throughout each other. Suspensions and colloids are liquid solutions that contain solute particles. In a suspension, the particles are large enough to fall out of the solution. In a colloid, the particles are so small that they stay in solution for a very long time.

Lesson 3

1. Acids have a sour taste, turn litmus paper red, release hydrogen gas when they react with metals, and conduct electricity, while bases have a bitter taste, turn litmus paper blue, and have a slippery feel.
2. The pH scale measures the strength of an acid or a base and ranges from values of 0 for a strong acid to 14 for a strong base. Pure water has a pH of 7.
3. An indicator is a chemical that changes color when exposed to an acid or base solution. The resulting color can be used to determine the pH of the solution.

Vocabulary Check

Explain how each pair of terms is related:

1. Acid—hydronium ion
2. Heterogeneous—homogeneous
3. Colloid—suspension
4. Extensive—intensive

Fill in the blank

Complete the sentence with the correct term.

5. When a metal is made from two different kinds of atoms mixed together, it is called a(n) __________.
6. In a solution of salt and water, the salt is called the __________.
7. If a solution chemically reacts with a metal and forms hydrogen gas, it is probably a(n) __________.
8. If a solution has a pH value of 12, it is called a(n) __________.

Multiple Choice

Choose the best answer.

9. Which solution would have the most amount of solute per solvent?
   A. a supersaturated solution of sugar and water
   B. a concentrated solution of sugar and water
   C. a saturated solution of sugar and water
   D. a dilute solution of sugar and water
10. A bowl of chicken soup would be called a
    A. heterogeneous mixture.
    B. homogeneous mixture.
    C. heterogeneous alloy.
    D. homogeneous alloy.
11. What characteristic do all solutions share?
    A. They are heterogeneous mixtures of different substances.
    B. They consist of only one type of substance.
    C. The solvent is always water.
    D. They are homogeneous mixtures in which the particles of matter cannot be distinguished from each other.
12. What characteristic do milk of magnesia, lye, and ammonia all share?
    A. They are all acids.
    B. They are all heterogeneous solutions.
    C. They are all bases.
    D. They are all solids.

Check Point

Answer the following questions.

13. Pick your favorite food and make a list of all the physical and chemical properties of that food. Think about all of your senses while making this list.
14. Granite is a mixture of different minerals that is very difficult to separate. Describe how you might try to separate the minerals from each other.
15. A dilute solution of sugar is heated on a stove. Describe the changes to the solution as the water is boiled away, using vocabulary from this chapter.
16. When a steel tool rusts, the rust changes both the chemical and physical properties of the tool. Explain.
17. Why are the intensive properties of 1 g of silver the same as the intensive properties of 10 g of silver?
18. Why does rain form drops when it lands on the window of a car?

Chapter 12

Elements and the Periodic Table

Scripture Spotlight

We know that God created matter in an orderly fashion. Scientists have found patterns in His Creation that make the world easier to study and bring us closer to God. In this chapter, you will read the following passages.

Ephesians 4:3 (p. 437)
Colossians 3:14 (p. 456)

The trees, the soil, the lake water, and the air surrounding them are all made up of tiny particles of just a few basic types of matter.

The Big Idea

Everything you can see, touch, smell, and taste is made of atoms. Atoms can be organized by their properties and behaviors. Their interactions are predictable because of the laws that govern their structure.

The clouds and the lake are made of the same atoms, so why are their properties so different?

Inquiry Kick-Off Engage

How can you determine what is inside an "atom" without taking the atom apart to look inside? In your *Science Journal*, you will make a model of an atom to answer the question.

Science Journal

Lesson 1

Essential Question

What Are Atoms?

Objectives

- Identify the contributions of scientists who have aided our understanding of the atom.
- Relate movement of electrons to energy absorption or emission.
- Distinguish between the atomic number, mass number, and average atomic mass of an element.
- Explain what an isotope is.

Vocabulary

energy level
electron cloud
orbital
mass number
average atomic mass
quark

The matter that surrounds us exists on many different scales. Earth itself is a large ball of matter of different forms—some solid, some liquid, some gas. Earth can be broken into smaller parts, such as its land, oceans, and atmosphere. Even smaller are stones, sand, dust, mists of water, and the particles that make up the air you breathe. How can these smaller pieces be broken into smaller and still smaller pieces? What do you think are the smallest pieces of matter that surround us? Can you see those pieces without magnification? Why do you think such small particles are necessary? What would life be like if those particles were larger and all visible?

The Atomic Theory Explain

Use a pair of scissors to cut a strip of aluminum foil in half. Can you detect any difference other that the dimensions of the foil? Cut one of the pieces in half again and again until you can no longer divide the foil. How do the properties of the smallest piece compare with the original? Is there a smallest possible piece that cannot be divided? This very question puzzled the ancient Greeks over 2000 years ago. At that time, a Greek philosopher named Democritus thought that you could keep cutting a piece of matter until eventually you would wind up with something that you could no longer cut into anything smaller. Democritus called this smallest piece of matter *atomos*, a Greek word that means "not able to be divided." Why would Democritus have not been able to investigate this concept experimentally?

These finely ground powders are used to make dyes and paints.

Is there a limit beyond which powdered materials cannot be ground?

Democritus's idea eventually developed into the atomic theory, which states that all matter is made of tiny particles called atoms. How is an aluminum atom different from a copper atom? Are all atoms the same size?

You will remember that atoms that have the same properties make up an *element*. Each chemical element has its own set of properties and cannot be divided into something else. The atom is the smallest particle into which an element can be divided and still retain all of its properties. What elements can you see around you?

Developing a Model of the Atom

Democritus took the first step in developing the atomic theory more than 2000 years ago by predicting that atoms existed. The specifics of the atomic theory still remained a puzzle. In fact, Democritus's contribution remained the only piece of the puzzle for a very long time. Why did the concept of atoms remain unchanged for so long?

John Dalton

By the early 1800s, scientists had broken many materials down into their components. They discovered dozens of elements. John Dalton studied several materials that were made of more than one element. He found that the atoms of elements that combine with one another always did so in specific ratios.

Based on this observation, Dalton proposed the first true atomic theory. There are several points that make up Dalton's theory:

- All matter is made of atoms. Atoms are indivisible and indestructible.
- All atoms of a given element are identical in properties.
- Compounds are formed by a combination of two or more different kinds of atoms. (You will learn about compounds in the next chapter.)
- A *chemical reaction* is a rearrangement of atoms.

The atomic theory as we use it today is more complicated, but Dalton's basic concept of atoms remains valid. What could you use to provide a concrete model of Dalton's atom? Why do you think the atomic theory today is different from the original theory that Dalton proposed?

Dalton's model of the atom was a solid, unbreakable sphere, much like a marble.

Lesson Activity

Using 2 quarters and 2 nickels, make as many combinations of 2 or more coins as you can. Write down the value of each combination. How are all of the values related? Are there amounts of money that you cannot make using quarters and nickels?

If you were given a value, how could you determine the number of quarters and nickels that you would need to make up that value?

Thomson's Model of the Atom

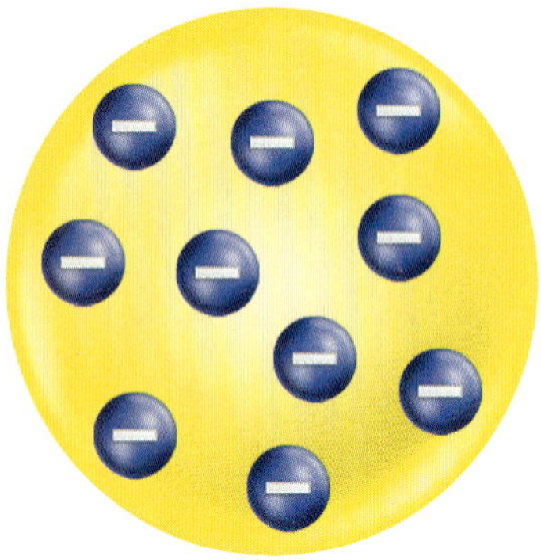

Thomson's observations provided more details about how atoms are constructed.

Do you think we will ever know everything about the structure of an atom? Why or why not?

J. J. Thomson

The discovery of electricity led to observations that an electric current could cause atoms to emit charged particles. In the late nineteenth century, J. J. Thomson reasoned that these very small charged particles (much smaller than an atom) were negatively charged. This was the first evidence that Dalton's atomic model needed to be modified.

Atoms could be divided into smaller, subatomic particles. Based on his observations, Thomson proposed a revised atomic model. In the Thomson model, atoms consist of a positively charged mass embedded with negatively charged electrons. Why would Thomson propose positive charges in the sphere if he only observed negative charges in his experiment? What object can you think of that would make an accurate model of Thomson's atom?

Ernest Rutherford

In the early 1900s, Ernest Rutherford designed an experiment to test the Thomson model. Why would Rutherford have wanted to test Thomson's model? To test the model, Rutherford aimed a beam of positively charged particles at a thin sheet of gold foil. Rutherford placed a detector around the gold foil so that he could tell where a particle went after it passed through the gold foil.

Rutherford reasoned that if the Thomson model accurately described atoms, then all of the particles should pass straight through the foil.

Rutherford's Experiment

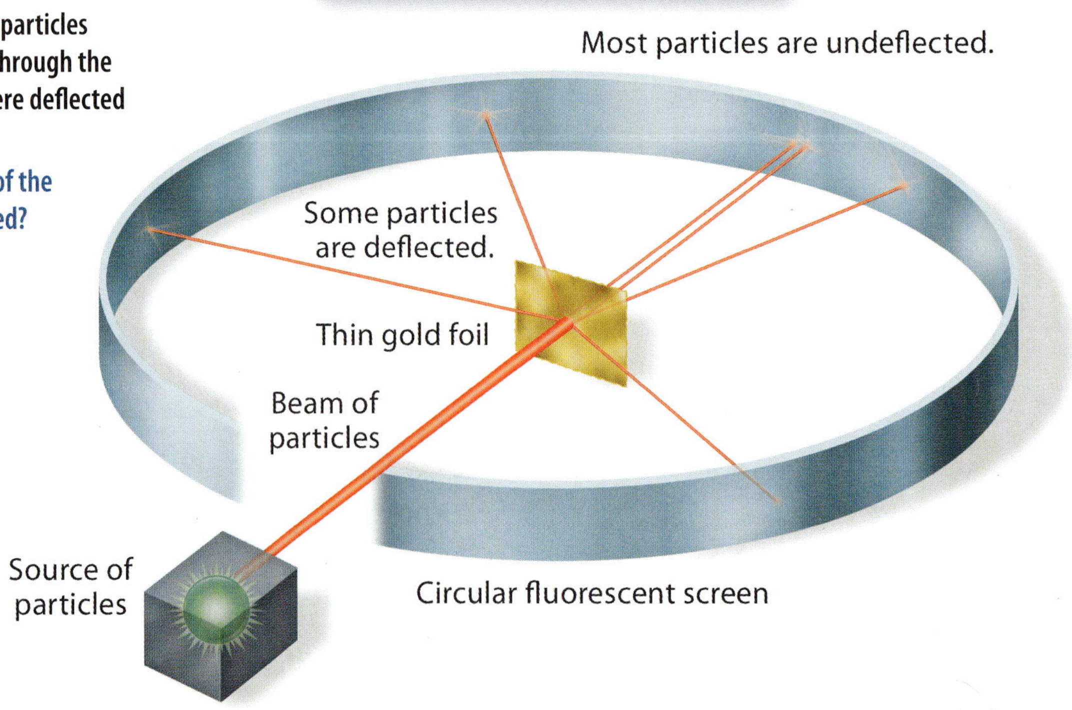

Almost all of the particles passed straight through the foil, but some were deflected sharply.

Why weren't all of the particles deflected?

What reasoning might Rutherford have used to make this assumption? However, Rutherford observed something unexpected. Most of the particles did pass through the foil, just as the model predicted. But Rutherford was very surprised to observe that many particles were deflected, some of them even bouncing straight back at the source. What might cause a positively charged particle to bounce or head off in a new direction?

Rutherford reasoned that these particles struck compact particles in the gold foil that also had a positive charge. Therefore, Rutherford concluded that not only was most of the mass of an atom found in its nucleus, but also that the nucleus contained positively charged particles. Rutherford called these positive charges *protons*, after a Greek word meaning "first." As you know, a *proton* is a positively charged particle that is found in the nucleus of an atom. If an atom is neutral, how do the numbers of electrons and protons compare?

Rutherford's observations suggested that a new atomic model was needed. In Rutherford's model, almost all of the mass of the atom is contained in a tiny, positively charged region in the center of the atom, called the *nucleus*. According to Rutherford, the electrons move in the vast space that makes up most of an atom's size. In fact, the nucleus is so small that if an atom were the size of an athletic stadium, then its nucleus would be only the size of a pea.

Rutherford's Model of the Atom

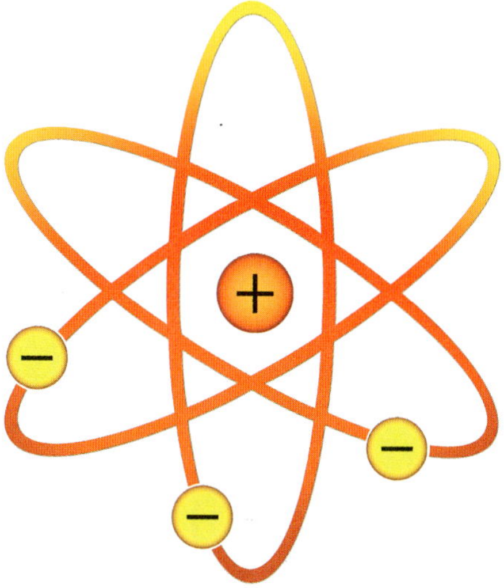

The nucleus of the atom is very small, but it has almost all of the mass of the atom.

How is it possible to know about a nucleus inside an atom that is too small to see with a microscope?

Explore-a-Lab

Structured Inquiry

How did Rutherford's experiment show that the nucleus was small?

Make a circle about 0.5 meters across on the floor or sidewalk using tape or chalk. Scatter a handful of marbles in the circle. Shoot marbles into the circle without aiming at a marble inside the circle. Record the path of each marble. If a marble stops before leaving the circle, predict where it would have exited if it had continued moving. Make a table to show how many marbles followed a straight path and how many marbles were deflected. How do you think your results would change if you replaced the shooting marbles with very light, hollow, plastic balls?

James Chadwick

A third subatomic particle was discovered a few years later by James Chadwick. Chadwick's experiments demonstrated that the nucleus had more mass than the number of protons suggested that there should be. Chadwick made another observation—while the mass was greater than expected, the charge of the atom did not change. This suggested to Chadwick that there was another subatomic particle in the nucleus. Chadwick described this particle as having about the same mass as a proton but having no electrical charge. He called the particle a *neutron*.

Because each proton or neutron has a mass that is about 2000 times as large as the mass of an electron, almost all of the mass of the atom is located in the nucleus. Although many other subatomic particles have been discovered, electrons, protons, and neutrons are the basic structures that make up the modern model of the atom. How do you think the structure of the atom provides evidence of the Creator?

Niels Bohr

Rutherford's model placed the electrons in the area around the nucleus but did not describe any specific locations. As scientists made more observations about atoms and elements, they were able to add more information to the model.

One important observation was that atoms absorb and release energy. Niels Bohr used this observation to develop the revised atomic model. According to the Bohr atomic model, also known as the *planetary model*, electrons move around the nucleus in specific orbits just as the planets move around the Sun in specific orbits. These orbits are called **energy levels**. Electrons can move from one energy level to another but only if energy is involved. An electron moving from one orbit to another absorbs or releases energy.

Bohr's Model of the Atom

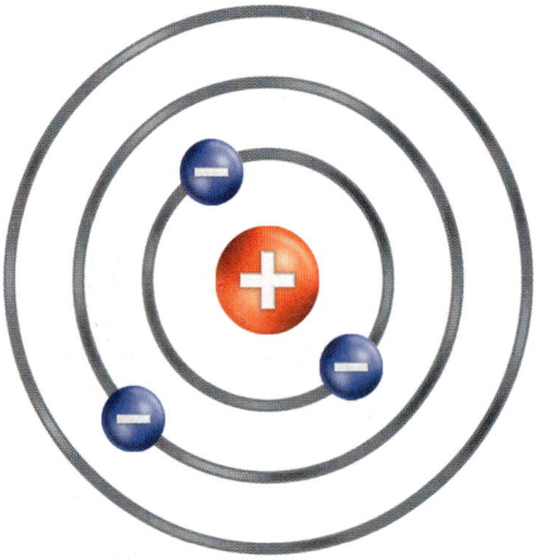

Electron Cloud Model of the Atom

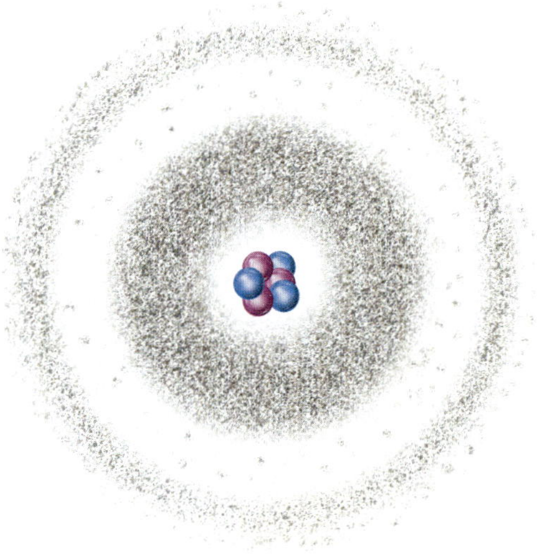

Each new set of observations and measurements provided more detail about the structure of the atom.

Do you think we will learn more about atomic structure in the future? Why or why not?

Current Atomic Model

The current atomic model is similar to the Bohr model, but it describes the positions of the electrons in a way that is more consistent with recent observations about electron energy. Scientists now know that electrons do not follow a fixed orbit around the nucleus. Instead, the current model, the electron cloud model, defines the probability of an electron being in a particular location around the nucleus. These areas of probability are defined as **electron clouds**. Unlike the Bohr model, these electron clouds are not fixed paths around the nucleus. They are three-dimensional regions. In the illustration on the previous page, think of the electron clouds as extending into and out of the page, in the shape of a sphere surrounding the nucleus. In a drawing representing the electron cloud, what would an area with dots close together represent? Similar to the Bohr model, each electron cloud corresponds to a specific level of electron energy. These energy levels are known as **orbitals**.

Scripture Spotlight

Imagine fellow believers as atoms being joined together in a chemical bond. What bonding agent is responsible for the spiritual bond described in **Ephesians 4:3**?

Orbitals and Energy Absorption and Emission

Moving an electron from one orbital to another orbital farther from the nucleus requires the absorption of energy. When an electron moves to a lower energy orbital, energy is released. Scientists use the differences in energy levels as one way to find out whether a particular element is in a material. When a sample containing sodium is burned in the flame from a gas burner, the flame becomes bright yellow. However, if the sample contains copper, the flame is green. The heat of the flame adds energy that causes an electron to move to a higher energy level. When the electron returns to its normal level, it emits light. Would this test be useful for analyzing a sample that contained more than one element?

Fireworks designers use different elements in their rockets to create a colorful display.

Where does the energy to move electrons in the atoms come from?

Looking Deeper into Atoms Explain

You know that each element is defined by the number of protons in its nucleus. Every hydrogen atoms has one proton, every helium atom has two protons, and so on. Recall that the number of protons in the nucleus of the atoms in an element is the *atomic number* of that element. The atomic number of hydrogen is 1. The atomic number of uranium is 92. What does this number tell you about a uranium atom?

Mass Number

The **mass number** of an atom is the sum of the number of protons and the number of neutrons in its nucleus. Although the atoms of an element always have the same number of electrons and protons, the atoms can have a different number of neutrons. Recall that atoms with the same number of electrons and protons but different numbers of neutrons are known as *isotopes*. Most elements have more than one isotope. For example, consider two isotopes of helium. Most helium atoms have two protons and two neutrons, giving them a mass number of 4. However, a tiny fraction of helium atoms has only one neutron. The *atomic mass* of an atom is the total mass of its protons, electrons, and neutrons. What is the atomic mass of this helium isotope?

Atomic mass is used to distinguish different isotopes of an element. For example, a helium isotope, called helium-4, has a mass number of 4. Another isotope is called helium-3. Consider carbon-12. What is its mass number? The atomic number of carbon is 6. How many protons, electrons, and neutrons does an atom of carbon-12 have?

Average Atomic Mass

Elements with large atomic numbers can have 20 or more different isotopes. Many of the isotopes are unstable and last for a very short period of time. The **average atomic mass** of an element is the weighted average of the atomic masses of its isotopes. Often the average atomic mass of an element is different from the atomic masses of any of its individual atoms. For example, bromine has two isotopes with atomic masses of 79 and 81. These isotopes are found in almost equal amounts. The average atomic mass of bromine is about 80 even though bromine atoms with an atomic mass of 80 are extremely rare.

Even Smaller Particles

Electrons, protons, and neutrons don't tell the whole story of the modern model of the atom. There are many other particles that play a role in holding atoms together and that contribute to their properties. Scientists have also discovered that protons and neutrons are themselves made up of even smaller particles, called **quarks**.

Structured Inquiry

Discover

Record your work for this inquiry. Your teacher may also assign the related Guided Inquiry.

Science Journal

Flame Tests for Metal Ions

How can you use changes in energy levels to identify elements?

SAFETY: Wear approved safety goggles and chemical-resistant gloves and apron. After you finish each test and extinguish the flame, place the used wooden splint in the disposal container.

Materials
- 250-mL beaker
- laboratory burner
- colored drawing pencils
- goggles, aprons, and gloves
- wooden splints, soaking separately in each of the following solutions: NaCl, KCl, LiCl, $CaCl_2$, $SrCl_2$, $CuCl_2$, and unknown.

Procedure

1. Put on gloves, an apron, and safety goggles. Fill a 250-mL beaker with tap water and place it at your workstation. This will be used to extinguish wooden splints.
2. Obtain a wooden splint from one of the metal chloride solutions.
3. Carefully light the laboratory burner and slowly pass the wooden splint back and forth through the flame. **Observe** and **record** the color of the flame in the data table in your ***Science Journal***. Immerse the wooden splint in water, make sure it is completely extinguished, and then place it in the disposal container.
4. Repeat Steps 2–4 with another splint and another salt until all ions have been tested.
5. Repeat Steps 2–4 with the unknown salt solution. Determine which ion is present in this solution.

Analyze Results

Discuss some of the limitations of using flame tests to identify ions present in solution.

Create Explanations

1. How can you use changes in energy levels to identify elements?
2. Why do different elements produce different flame colors?
3. How did you determine the identity of the unknown sample?
4. **Propose** at least one possible method for improving the accuracy of results in this laboratory experiment.

Math in Science

The average atomic mass of an element is calculated by multiplying the atomic mass of each isotope by the percentage of that isotope present. For example, chlorine is made up of 75.8% chlorine-35 and 24.2% chlorine-37. The average atomic mass of chlorine is:

$$(0.758)(35) + (0.242)(37) = 26.5 + 9.0 = 35.5$$

1. Rubidium has two isotopes, 72.2% rubidium-85 and 27.8% rubidium-87. Calculate the average atomic mass of rubidium.
2. Bromine also has two isotopes: 50.7% is bromine-79 and 49.3% is bromine-81. Can you estimate the average atomic mass of bromine without using a calculator?
3. What is the average atomic mass of fluorine, whose only naturally occurring isotope has an atomic mass of 19?

Concept Check Assess/Reflect

Summary: What are atoms? All matter is made up of about 100 elements that have different properties. The atomic theory states that all matter is made up of basic particles, known as atoms, and that the particles of different elements are different. The atomic theory is based on the ancient Greek concept of the atom, but the theory developed as a result of many observations about matter. The current model describes the atom as having a tiny, dense nucleus made of protons and neutrons that is surrounded by clouds of electrons. The factor that defines an element is its atomic number, the number of protons in the nucleus. The atomic mass of an atom is the sum of its protons and neutrons.

1. Why is Dalton's concept of the atom considered to be a theory while Democritus's concept of the atom is not?
2. Why is an atom defined as the smallest particle of an element even though an atom is made up of even smaller particles?
3. Calculate the average atomic mass of gallium, which consists of a mixture of 60.1% gallium-69 and 39.9% gallium-71.
4. Make a table that compares the masses and charges of electrons, protons, and neutrons.
5. In the current atomic model, protons and neutrons are composed of smaller particles called quarks. This model does not propose that electrons are made of smaller particles. What could cause scientists to change this model to include smaller particles that make up the electron?

Lesson 2

Essential Question

How Are Elements Organized?

Suppose that you download a large collection of songs. How can you make it easy to find a particular song when you want it or know what musician performed a particular song? What if later you found new songs you wanted. When downloaded, you found that the songs did not fit with the organization you had developed for your songs. What would you do? Would you delete the songs that didn't fit? Would you develop a better organization system? Chemists faced that problem as they discovered more and more elements with a wide variety of chemical and physical properties.

Objectives

- Explain how the periodic table is organized.
- Distinguish between a group and a period on the periodic table.

The Periodic Table Explain

Currently about 118 known elements are known to scientists. The number is not exact because a few elements that have been synthesized must be confirmed before scientists agree that these elements do exist. Other elements may be discovered by the time you read this paragraph. About 90 of the elements occur in nature, and the rest have been made by physicists in laboratories.

Before the eighteenth century, only 13 elements were known. These elements included copper, gold, and sulfur. As scientists studied matter and broke compounds into their components, they discovered many more elements. By the middle of the nineteenth century, more than 60 elements were known, including gases such as oxygen, nitrogen, and chlorine. Each element has its own properties. People wondered how the elements could be organized to keep track of the similarities and differences. Think of categories into which elements could be grouped. What categories seem to provide the best organization? How could organizing this information help us to better understand the elements and make it easier to find uses for them?

The periodic table organizes the elements in a chart.

How would organizing the elements by their properties be useful?

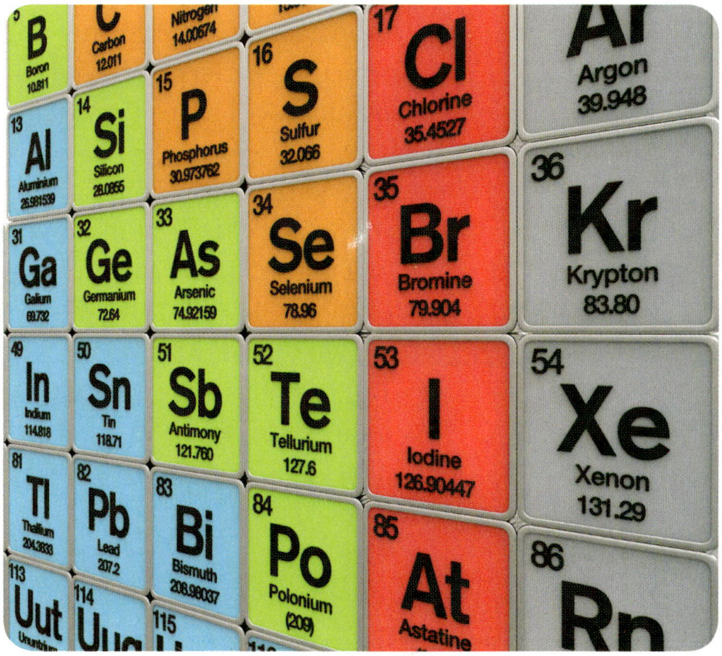

Using Atomic Masses

In the mid-1800s, a Russian scientist named Dmitri Mendeleev looked for a way to organize the 63 elements that were known at that time. Mendeleev made up a set of cards, one for each known element, that included the chemical and physical properties of the element. Although protons and neutrons were not yet known, the atomic masses of the elements had been determined experimentally, so those values were included on the cards. Mendeleev worked with different arrangements of the cards, sorting the cards into rows and columns.

Explore-a-Lab

Structured Inquiry

What organization methods would you use?

Examine the cards that your teacher provides to you. Notice differences and similarities among the cards. Now work out an arrangement that organizes the cards in an orderly way in both columns and rows. Write a short description of your organization of the cards and why you decided to use that organization. Describe how the cards are related within rows and within columns. What organizing principle did you use to classify the cards? How is this activity similar to Mendeleev's research? How is it different?

When Mendeleev arranged the elements according to increasing atomic masses, he noticed that those that had similar properties occurred in a repeating pattern. This *periodic table of elements* is a chart in which all the elements are arranged in columns and rows so that the properties of the elements occur in a repeating pattern. To keep the repeating pattern in his periodic table, Mendeleev had to leave some blank spaces in his arrangement.

Mendeleev made a rather bold prediction: new elements would be discovered that would fit into those blank spaces. He even predicted the properties of those unknown elements. Many scientists were skeptical and questioned the ability to predict the properties of an unknown element. However, the value of the chart as an organizational tool was shown within a few years. Mendeleev developed his chart between 1868 and 1870. In 1875, the new element gallium was discovered, followed by scandium in 1879 and germanium in 1886. The properties of each of these new elements matched Mendeleev's predictions. Do you think these discoveries prove that Mendeleev's chart of the elements was the best way to organize them? Could there be another way of organizing them that would be useful?

Using Atomic Numbers

Mendeleev did have one problem. A few elements did not fit the repeating pattern that he had established in his periodic table. Recall that Mendeleev had organized his periodic table based on the atomic masses of the elements. He had to switch the order of certain elements so that the one with the greater atomic mass appeared first on his periodic table. Only then did the properties of these elements fit the repeating pattern. Mendeleev thought that experiments would later reveal that the atomic masses of these elements had been wrongly calculated. Why would Mendeleev have reached that conclusion?

Because protons had not yet been discovered, Mendeleev did not know about atomic numbers. In 1914, a British scientist named Henry Moseley suggested that the elements be arranged in a periodic table based on increasing atomic numbers, not on increasing mass numbers. Once this was done, all the elements, including the ones identified since Mendeleev's time, were arranged so that their repeating properties were evident. The modern periodic table, shown in this lesson, is a chart that displays all of the elements, arranged in order of atomic number.

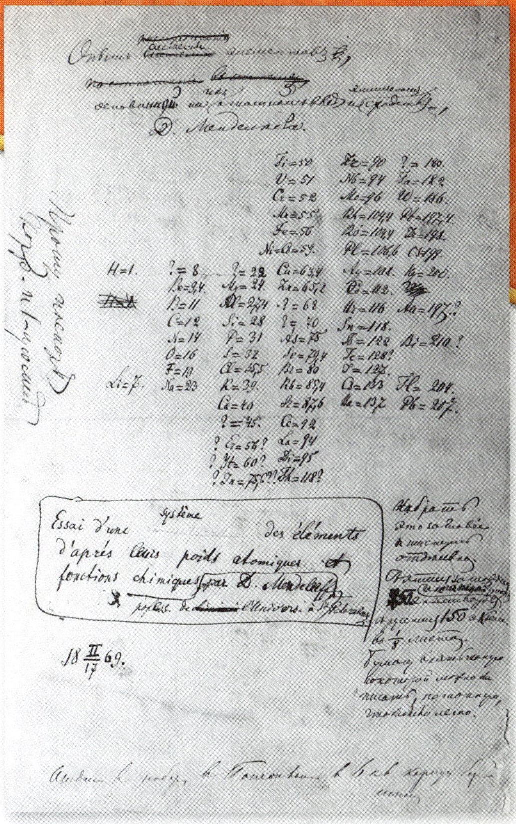

Mendeleev's original notes for the periodic table of elements

Why do you think the information about a few elements contains question marks?

Reading the Periodic Table Explain

Each element in the periodic table is represented by its own box, or element key, on the periodic table. In the version of the periodic table used in this book, the text of the element key contains four basic pieces of information about every element: symbol, name, atomic number, and atomic mass. Some periodic tables provide additional information about each element. The symbol for an element is made up of one to three letters. A one-letter symbol, such as H for hydrogen, is capitalized. For a two-letter symbol, such as He for helium, only the first letter is capitalized. Some of the symbols come from Latin names for the elements, such as sodium (Na—*natrium*), iron (Fe—*ferrum*), and lead (Pb—*plumbum*). Why do you think scientists did not just choose the first two letters of each element name as its symbol?

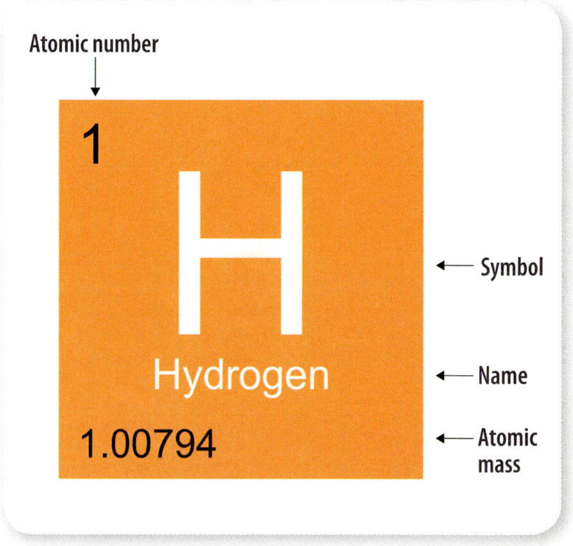

Periodic Table of the Elements

Group → / Period ↓

Key: 1 ← Atomic number; H ← Symbol; Hydrogen ← Name; 1.00794 ← Atomic mass

Period	1 I A	2 II A	3 III B	4 IV B	5 V B	6 VI B	7 VII B	8 VIII B	9 VIII B
1	1 H Hydrogen 1.00794								
2	3 Li Lithium 6.941	4 Be Beryllium 9.012182							
3	11 Na Sodium 22.98976928	12 Mg Magnesium 24.305							
4	19 K Potassium 39.0983	20 Ca Calcium 40.078	21 Sc Scandium 44.9559	22 Ti Titanium 47.867	23 V Vanadium 50.9415	24 Cr Chromium 51.9961	25 Mn Manganese 54.938045	26 Fe Iron 55.845	Atomic number 27 Co Cobalt 58.933195
5	37 Rb Rubidium 85.4678	38 Sr Strontium 87.62	39 Y Yttrium 88.90585	40 Zr Zirconium 91.224	41 Nb Niobium 92.9063	42 Mo Molybdenum 95.96	43 Tc Technetium [98]	44 Ru Ruthenium 101.07	45 Rh Rhodium 102.9055
6	55 Cs Caesium 132.9054519	56 Ba Barium 137.327	57-71 Lanthanoids (Lanthanide)	72 Hf Hafnium 178.49	73 Ta Tantalum 180.94788	74 W Tungsten 183.84	75 Re Rhenium 186.207	76 Os Osmium 190.23	77 Ir Iridium 192.217
7	87 Fr Francium [223]	88 Ra Radium [226]	89-103 Actinoids (Actinide)	104 Rf Rutherfordium [267]	105 Db Dubnium [268]	106 Sg Seaborgium [271]	107 Bh Bohrium [272]	108 Hs Hassium [270]	109 Mt Meitnerium [276]

Lanthanoids (Lanthanide)	57 La Lanthanum 138.90547	58 Ce Cerium 140.116	59 Pr Praseodymium 140.90765	60 Nd Deodymium 144.242	61 Pm Promethium [145]	62 Sm Samarium 150.36
Actinoids (Actinide)	89 Ac Actinium [227]	90 Th Thorium 232.03806	91 Pa Protactinium 231.03588	92 U Uranium 238.02891	93 Np Neptunium [237]	94 Pu Plutonium [244]

Color Key

Alkali metals	Alkaline earth metals	Transition metals	Post-transition metals	Metaloids
Other nonmetals	Halogens	Noble gases	Lanthanide	Actinide
Radioactive element	Synthetic element	H Gas	Hg Liquid	Li Solid

10 VIII B	11 I B	12 II B	13 III A	14 IV A	15 V A	16 VI A	17 VII A	18 VIII A
								2 He Helium 4.002602
			5 B Boron 10.811	6 C Carbon 12.0107	7 N Nitrogen 14.0067	8 O Oxygen 15.9994	9 F Fluorine 18.9984032	10 Ne Neon 20.1797
			13 Al Aluminium 26.9815386	14 Si Silicon 28.0855	15 P Phosphorus 30.973762	16 S Sulfur 32.065	17 Cl Chlorine 35.453	18 Ar Argon 39.948
28 Ni Nickel 58.6934	29 Cu Copper 63.546	30 Zn Zinc 65.38	31 Ga Gallium 69.729	32 Ge Germanium 72.64	33 As Arsenic 74.9216	34 Se Selenium 78.96	35 Br Bromine 79.904	36 Kr Krypton 83.798
46 Pd Palladium 106.42	47 ← Symbol Ag Silver Name 107.8682	48 Cd Cadmium 112.411	49 In Indium 114.818	50 Sn Tin 118.71	51 Sb Antimony 121.76	52 Te Tellurium 127.6	53 I Iodine 126.90447	54 Xe Xenon 131.293
78 Pt Platinum 195.084	79 ← Atomic mass Au Gold 196.966569	80 Hg Mercury 200.59	81 Tl Thallium 204.3833	82 Pb Lead 207.2	83 Bi Bismuth 208.9804	84 Po Polonium [209]	85 At Astatine [210]	86 Rn Radon [222]
110 Ds Darmstadtium [281]	111 Rg Roentgenium [280]	112 Cn Copernicium [285]	113 Uut Ununtrium [286]	114 Fl Flerovium [289]	115 Uup Ununpentium [288]	116 Lv Livermorium [293]	117 Uus Ununseptium [294]	118 Uuo Ununoctium [294]

63 Eu Europium 151.964	64 Gd Gadolinium 157.25	65 Tb Terbium 158.9253	66 Dy Dysprosium 162.5	67 Ho Holmium 164.93032	68 Er Erbium 167.259	69 Tm Thulium 168.93421	70 Yb Ytterbium 173.054	71 Lu Lutetium 174.9668
95 Am Americium [243]	96 Cm Curium [247]	97 Bk Berkelium [247]	98 Cf Californium [251]	99 Es Einsteinium [252]	100 Fm Fermium [257]	101 Md Mendelevium [258]	102 No Nobelium [262]	103 Lr Lawrencium [262]

Structured Inquiry

Discover

Record your work for this inquiry. Your teacher may also assign the related Guided Inquiry.

Science Journal

The Mendeleev Lab of 1869

How can you use the periodic table to organize elements?

Materials

- set of known element cards

Procedure

1. Inspect the properties of the known elements.
2. Arrange the cards of the known elements in a crude representation of the periodic table.
3. Once the known elements are in place, inspect the properties of the unknowns to see where their properties would best "fit" the trends of the elements of each group.
4. Assign the proper element name to each of the unknowns. **Record** the name and symbol for each of the "unknowns" in your ***Science Journal***.

Notes:

- The unknown elements are from the following groups in the periodic table. Each group listed below contains at least one unknown.

  1 2 11 13 14 17 18
- None of the known elements serves as one of the nine unknown elements.
- No radioactive elements are used during this experiment. The relevant radioactive elements include Fr, Ra, At, and Rn.
- You may not use your textbook or other reference materials. You have been provided with enough information to determine each of the unknown elements.

Analyze Results

Compare your table with those of other class members. Discuss which properties you used to determine the identities of the unknown elements.

Create Explanations

1. How can you use the periodic table to organize elements?
2. What trends can you observe in the rows and columns of the periodic table?
3. How can you use the table to predict the properties of an element that you have never encountered?
4. Why are no two elements exactly alike?

Periods

You may recall that each horizontal row of elements from left to right on the periodic table is called a *period*. As you move across a period, the properties of the elements follow a repeating pattern.

As you learned earlier, the atomic number is the number of protons in the nucleus of an element's atoms. Because atoms are neutral, the number of electrons in an atom is the same as its atomic number. These electrons are arranged around the nucleus in different orbitals. The number of electrons that can be in a particular orbital is limited. The periodic table provides information about the locations of electrons in an atom of an element based on the period in which the element appears. There are seven periods in the periodic table. The two rows at the bottom of the table are actually parts of Periods 6 and 7. Why do you think they are placed below the table? Examine the atomic numbers of these elements. Where do these segments fit?

The lowest energy level, or first orbital, of an atom can only have two electrons. You can see that the first period of the table has only two elements—hydrogen and helium. Hydrogen has one electron and helium has two electrons. Lithium (atomic number 3) has three electrons. Because the first orbital is filled by the first two electrons, the third electron is located in the second orbital. From left to right across the second period, each element has one additional electron in the second orbital. At the right side of the period, neon has 10 electrons, 2 in the first orbital and 8 in the second orbital. At the beginning of Period 3, sodium has one electron in the third orbital. How many electrons would potassium, located below sodium, have in its fourth orbital? The electrons in the highest orbital of an atom are much more important in determining the chemical properties than are the electrons in lower orbitals.

Check out your ***Science Journal*** to model Rutherford's experiment.
Extend

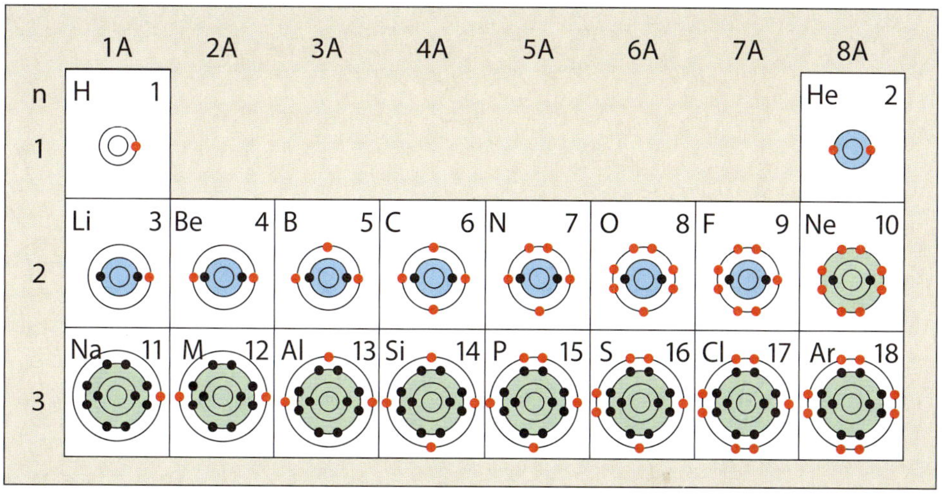

This table shows how electrons are arranged in the atoms of the elements of the first three periods.

What patterns do you see in electron arrangements in the rows and in the columns?

This pattern continues for these groups in all of the periods. Notice that the illustration does not include the elements in Groups 3–12. These elements have either one or two electrons in their highest orbital. Their electron structures are more complicated, because they have additional electrons in lower orbitals.

Groups

The periodic table arranges elements into rows and columns. A *group*, you may remember, consists of the elements in a particular column of the periodic table. The elements in a group have similar chemical and physical properties. All of the elements in each group have the same arrangement of electrons in their highest orbitals. Which electrons do you think have the greatest influence on the properties of the element? At the top of the periodic table, the groups are labeled with the numbers 1 through 18. A second set of labels shows an alternative group-labeling system that uses Roman numerals I–VIII and the letters A and B, but we will not be using that system.

Concept Check Assess/Reflect

Summary: How are elements organized? Scientists have observed that there are groups of chemical elements that have similar properties. The periodic table of the elements, based on a chart developed by Dmitri Mendeleev, arranges the elements into similar groups. The table can be used to predict properties of elements based on those of related elements. The elements are organized in the table in periods (rows) and groups (columns). Within a group, the elements have similar chemical and physical properties. The properties of the elements across a period vary in a repeating pattern.

1. How is the modern periodic table different from Mendeleev's original periodic table?
2. How do you know that all of the elements in the first seven periods of the periodic table have already been discovered or made?
3. The element whose atomic number is 119 has not yet been discovered. If it is discovered, in which group should this element be placed?
4. How will the pattern of properties of elements in Period 8 of the periodic table compare with those elements that are already on the table?
5. Suppose that you could put another element on one row of the periodic table. What would you call it? What would be its properties? Has this method ever helped scientists discover other elements?

Lesson 3

Essential Question

What Are Metals?

Recall that the properties of the elements follow a repeating pattern in the periodic table as you move across a period or down a group. Think about elements that you have seen that you would call metals. Examples of these elements are copper, aluminum, iron, and gold. What do these elements have in common? There are other metal elements that you are not likely to have seen in pure form. Examine the photo of sodium in water. When sodium is added to water, a vigorous reaction occurs. How does this compare with adding copper or aluminum to water? Both copper and sodium are metals; why do you think they react so differently? What other elements do you think would react in the same way as sodium when they are added to water?

Objectives

- Identify the general properties of metals.
- Describe the alkali metals.
- Describe the alkaline earth metals.
- Describe the transition metals.

Vocabulary

malleability
ductility
alkali metal
alkaline earth metal
transition metal

Metals Explain

Both sodium and potassium are classified as metals. You know that a *metal* is an element that is a good conductor of electricity and heat. Most of the elements on the periodic table are metals. Where do you think the most reactive metals are placed on the periodic table? Why are they placed there?

Recall that most metals are hard, shiny solids with high melting points. One exception is mercury, which is a liquid at room temperature. Metallic elements are generally somewhat flexible and bendable. They can be hammered into thin sheets or other shapes, a property known as **malleability**. Most metals can also be drawn into long, thin, flexible wires, which is a property known as **ductility**. Why is mercury classified as a metal, even though it is not a solid at room temperature and therefore is not malleable or ductile?

Sodium and water react very rapidly when they come together.

Based on the location of sodium on the periodic table, can you predict another element that would react in a similar way?

Explore-a-Lab

Structured Inquiry

Which metal is more malleable—aluminum, iron, or copper?

Gather aluminum foil (10 cm × 10 cm), copper wire (thin, uninsulated), sheet metal (heavy gauge, 10 cm × 10 cm), steel wool, disposable gloves, and a hammer. Put on safety goggles. Fold the aluminum foil square in half as many times as you can. Measure the dimensions of the square. Place the folded aluminum square on the sheet metal plate. Carefully use the hammer to pound on the aluminum square to get it as flat as possible. Measure and record the dimensions of the aluminum square. Put on the disposable gloves. Take enough steel wool to form a compact ball about the size of a large marble. Place the steel-wool ball on the sheet-metal plate. Use the hammer to pound on the steel-wool ball to make it into a flat sheet. Measure and record the dimensions of the steel-wool square. Roll the copper wire into a ball and pound it flat. Which is more malleable—aluminum, iron, or copper? Which of the three metals tears more easily? Which of the metals stretches the best? Which of the three metals do you think conducts electricity the best? Explain how you would determine this.

The chemical properties of metals vary widely. Some metals, such as sodium and potassium, react with oxygen or water vapor so rapidly that they must be stored in ways that keep them away from the air. Others, such as gold, react very slowly with oxygen or water vapor. A gold object can remain bright and shiny for thousands of years. Between these extremes are metals such as iron and copper that gradually rust or tarnish. Examine a small object that appears to be made of a single substance. Determine whether the object is a metal or not a metal. What characteristics did you use to make your determination?

Pure sodium is soft enough that you can cut it with a knife. The fresh cut is very shiny.

Why is the fresh cut face shiny while the rest of the piece of sodium has a dull white surface?

Alkali Metals Explain

The elements in Group 1 of the periodic table are the alkali metals. The atoms of the **alkali metals** have one electron in the highest orbital. That electron is easily removed in chemical reactions, which makes the alkali metals extremely reactive compared with other metallic elements. They are so reactive that they are never found in pure form in nature. They must be stored in airtight containers or covered with a layer of oil to isolate them from oxygen and water vapor in the air. Otherwise, the metals would react rapidly and violently with these substances. The alkali metals are very similar to one another and have very similar chemical and physical properties. What characteristic do the alkali metals share that would explain their tendency to have violent reactions with oxygen?

The melting points of the alkali metals are very low compared with those of most other metals. Sodium, for example, melts at 98°C, lower than the boiling point of water. Cesium melts at just over 28°C, the temperature of a warm room. As pure elements, the alkali metals are much softer than most other metals. Cutting a block of an alkali metal in half with a sharp knife does not take much effort. What can you conclude about hardness being a physical property of metals?

The most common alkali metals are sodium and potassium. Although these metals are not found in pure form, they are very common in nature. For example, sodium combines with the element chlorine to form sodium chloride, which is familiar table salt. The oceans of Earth contain a large amount of dissolved sodium chloride. Potassium, in the form of potassium compounds, is an important part of your diet and a necessary nutrient for plants. Many fruits, including bananas, are good sources of potassium. How would the properties of potassium chloride compare to those of sodium chloride?

Alkaline Earth Metals Explain

The elements in Group 2 of the periodic table are the **alkaline earth metals**. As with the alkali metals, the alkaline earth elements all have similar properties. They are very reactive metals, although not nearly as reactive as the alkali metals. Why do you think this is so? Most of the alkaline earths react with water, but not as vigorously as the alkali metals. The pure metals can burn in air if they are exposed to a flame. The melting points of the alkaline earth metals are much higher than those of corresponding alkali metals but lower than those of many other metals. Magnesium, for example, melts at 649°C compared to 98°C for sodium. How would the melting point of calcium compare with the melting point of potassium?

Calcium and magnesium are the most familiar alkaline earth metals. Calcium is a very important part of the human diet because calcium is a major component of bones and teeth. Magnesium is mixed with other metals to form alloys that are very light but stronger than steel. High-performance bicycles are built using an alloy of aluminum and magnesium. Limestone, a common form of rock used in buildings and roads, is made of calcium or magnesium combined with carbon and oxygen.

Small particles of magnesium burn in air, causing the sparkler to send out starry flashes.

How does the flammability of magnesium compare to that of metals that you use daily?

Mercury expands and contracts evenly, making it an ideal material for making thermometers and barometers.

Is mercury more or less reactive than calcium?

Transition Metals Explain

The elements in Groups 3–12 of the periodic table are the **transition metals**. Although they make up 10 groups of the periodic table, the transition metals are often treated as a single large group. Most of the metals that are familiar from your daily life are transition metals. These include iron, copper, silver, and gold. Examine several different metal objects, such as an iron nail, a silver spoon, and a piece of copper wire. How are these transition metals similar to one another? How are they different from one another? What properties of the metals make them particularly useful to people?

While many of the properties of transition metals are similar, each metal has specific properties that make it unique. All of the elements are good conductors of electric current, but silver and copper are better conductors than most other metals. Because copper is less expensive and more available than silver, copper is usually used for the electrical wiring in buildings. Iron, nickel, and titanium are especially strong among the metals, so they are commonly used as structural materials in buildings and vehicles. Gold and platinum are extremely unreactive and do not tarnish when exposed to air and water vapor. These metals are used in jewelry and as electrical components in sensitive equipment where this resistance to chemical reactions makes them very useful.

Most of the transition metals are hard and shiny. They are much less reactive than the alkali metals and alkaline earth metals. In general, the transition metals have rather high melting and boiling points compared with other elements. Mercury, however, is a liquid at room temperature even though it is a transition metal. What properties would you predict for mercury if it were cooled below its melting point?

Structured Inquiry

Discover

Record your work for this inquiry. Your teacher may also assign the related Guided Inquiry.

Science Journal

What's Your Family?

What properties are used to classify elements?

Procedure

Materials
- conductivity tester
- element samples
- graduated cylinder (10-mL)
- hammer
- hydrochloric acid (1.0 M)
- sandpaper
- 7 test tubes
- test tube rack

1. Put on your safety goggles and apron.
2. Label the test tubes. Label the first one *Sample 1*, the second *Sample 2*, and so on.
3. Select one of the samples and observe its appearance. In the data table, **record** the sample's color and whether it has luster (shiny appearance) or no luster (dull appearance). You may need to clean the surface of the sample with sandpaper to determine luster.
4. Place the sample on a hard surface. Gently tap the sample with a hammer to **determine** whether the sample is malleable or brittle. Malleable samples flatten when tapped with a hammer; brittle samples shatter. Record your **observations**.
5. Test whether the sample is a conductor. Use the conductivity tester to test this property. If the bulb lights up, the sample conducts electricity. Record your observations.
6. Take a small sample of the element. Use the graduated cylinder to add 5 mL of 1.0 M hydrochloric acid to a test tube. Test the sample for its reactivity to acid by placing it in one of the test tubes containing hydrochloric acid. If the sample bubbles, it is reacting to the acid. Strong bubbling indicates the sample is very reactive. Weak bubbling indicates the sample is weakly reactive. No bubbles indicate the sample is nonreactive to acid. Record your observations.
7. Repeat Steps 3–6 for the remaining samples.

Analyze Results

How did your observations **compare** with those of your classmates?

Create Explanations

1. What properties are used to classify elements?
2. Which elements are nonmetals? What properties do they have in common?
3. Are there any elements that seem to have properties of both metals and nonmetals? If so, what are they?
4. Of the elements you tested, which one was most reactive to acid?

Lanthanides and Actinides

Included in the transition metals are the two partial rows at the bottom of the table. These elements, known as the lanthanides and actinides, are often called rare earth metals. Uranium and plutonium have been used as fuel for nuclear power since the middle of the twenthieth century. Many of the lanthanides have become very important in the last few decades because they have properties that make them useful in parts for computers and communication devices and in manufacturing powerful magnets.

Posttransition Metals Explain

There are a few metals located in Groups 13–16. These metals are generally not very reactive and have properties that are similar to those of the transition metals. Except for aluminum and gallium in Group 13, these metals are found in the lower part of a column. Familiar posttransition metals are aluminum, tin, and lead. Aluminum is often used to construct the bodies of airplanes. Aluminum atoms have less mass than most other metals. What properties is aluminum most likely to have that make it useful for this application?

Concept Check Assess/Reflect

Summary: What are metals? Most of the elements on the periodic table are metals. Metals are characterized by being good conductors of heat and electric current. In general, metals are hard, malleable, ductile, and have shiny luster. Alkali metals, in Group 1 of the periodic table, are very reactive. Alkaline earth metals are somewhat less reactive than alkali metals. The most familiar metals are among the transition metals. In general, these metals are hard and not extremely reactive.

1. How are properties such as malleability and ductility important in the common uses of metals?
2. Think about the electron structure of metal atoms described in the previous lesson. How might that structure account for the ability of metals to conduct electric current?
3. Potassium is a good conductor of electric current. Why is potassium never used in household electrical wiring?
4. For what types of applications would metals not be suitable materials?

Lesson 4

Essential Question

What Are Nonmetals and Metalloids?

Objectives

- Identify the general properties of nonmetals.
- Describe the halogens.
- Describe the noble gases.
- Describe the metalloids and their applications.

Vocabulary

halogen

salt

Although carbon is a solid like most metals, carbon does not conduct electricity. Although argon belongs to the same period as the alkali metal sodium, argon does not react with other substances. Carbon and argon belong to another category of elements on the periodic table. These elements do not have the properties that you examined as characteristics of metals. Think about the properties of metals. What properties would you predict for nonmetal elements?

Nonmetal Elements Explain

Both carbon and argon are examples of nonmetals. Recall that *nonmetals* are elements that are poor conductors of electricity and heat. In the solid state, nonmetal elements are brittle and tend to fracture when struck. These elements cannot be hammered or pulled into shape. They are not shiny. The nonmetal elements are shown on the periodic table in this book in boxes that are orange, red, or gray toward the right side of the periodic table. Which groups include nonmetals?

Based on the photo, can you predict how the melting points of halogens change from the top to the bottom of the group?

Research the melting points of the alkali metals and the halogens. How do the melting point trends in each group of elements compare?

Many of the nonmetals, including nitrogen, oxygen, argon, and sulfur, are common in nature as pure elements. Eleven of the 20 nonmetals are gases. One nonmetal element, bromine, is a liquid at room temperature. Bromine belongs to a group of elements on the periodic table that includes only nonmetals and has a special name—the halogens.

Halogens

The **halogens** are the nonmetal elements in Group 17 of the periodic table. The name *halogen* means "salt-forming" because the halogens combine with metals to form salts. Recall that sodium, an alkali metal, combines with chlorine, a halogen, to form table salt. A **salt** is an ionic compound produced by the reaction of an acid with a base. Many salts are a combination of a reactive metal element and a reactive nonmetal element. What other salts could be included in your diet?

Halogens are very reactive elements, and so they can be dangerous. Breathing fluorine or chlorine gases can cause severe burns to the throat and lungs. Fluorine is so reactive that it cannot be stored in a glass bottle because the fluorine reacts rapidly with the glass. What kind of container might you use to store fluorine? Chlorine is used in low concentrations as a disinfectant to kill bacteria in water.

Although their chemical properties are very similar, the physical properties of the halogens are quite different. For example, chlorine is a greenish gas, bromine is a dark red liquid, and iodine is a dark gray solid. Iodine is the least reactive of the common halogens and the least hazardous. Solutions of iodine are sometimes used to disinfect cuts and abrasions.

Noble Gases

You learned earlier in your study of science that the elements in Group 18 are known as the *noble gases*. These nonmetal elements are unusual in that they are very unreactive. In fact, there are no known substances that contain a noble gas that is naturally combined with another element. However, scientists have been able to combine a noble gas with other elements in the laboratory. How do you think they accomplished this?

An interesting use for noble gases is the colored light tubes used in making bright signs. Even though the occupied energy levels of the atoms are filled, electrons can be moved up into empty higher levels by adding energy from an electric current. When an electron moves back to a lower level, the atom emits light. The color of the light that is emitted depends on the particular transition between energy levels, so different colors of lights can be made by using different gases. Why would a gas that is not reactive likely be better for this application than another gas?

Scripture Spotlight

Read **Colossians 3:14** to see what Paul says is the "bond of perfection." Physically, two or more atoms joined together create a molecule. Spiritually, what might be joined together by this bond of perfection?

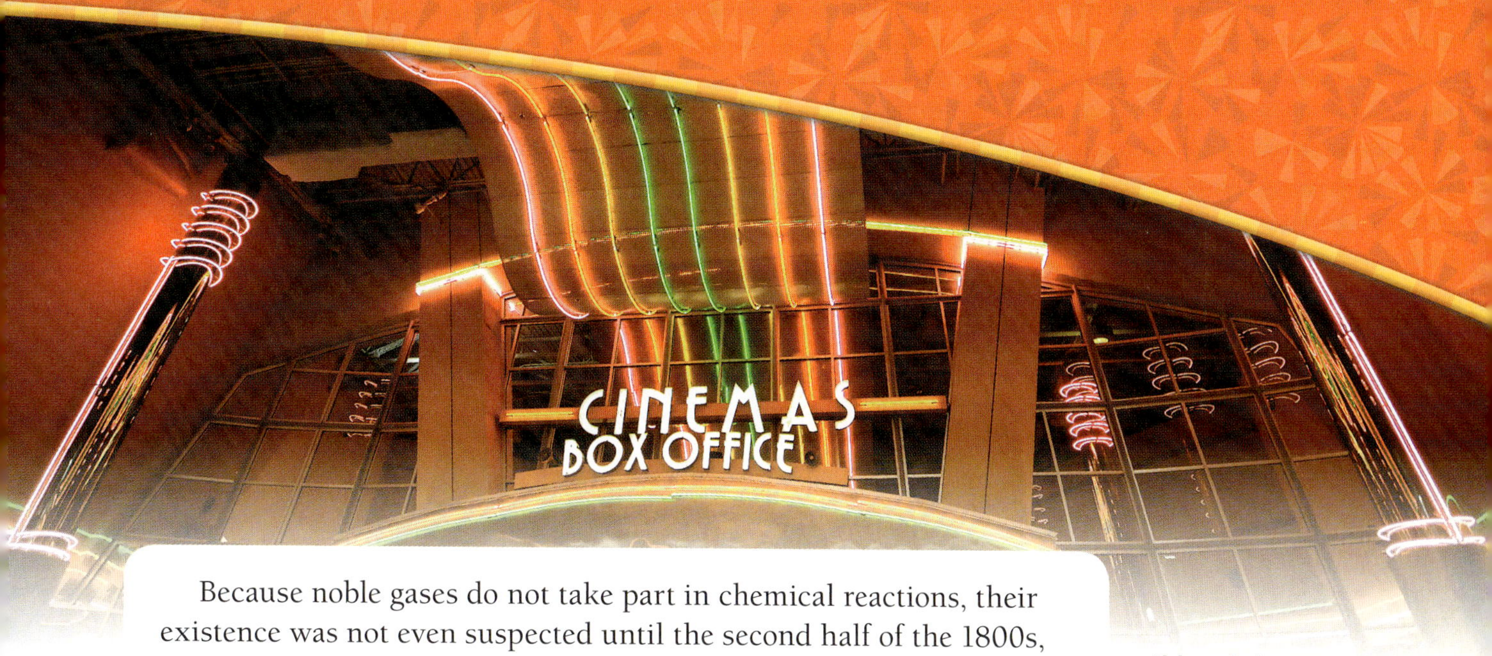

Although they are called *neon lights*, colored light tubes use various pure and mixed noble gases to produce different colors.

Because noble gases do not take part in chemical reactions, their existence was not even suspected until the second half of the 1800s, years after Mendeleev developed his periodic table. The simplest noble gas, helium, was named after the Sun because that is where it was discovered. Helium was later found in small amounts in the atmosphere. Another noble gas, argon, is actually quite common on Earth. It makes up almost 1% of the atmosphere that surrounds us. How could the property of not reacting with other materials make noble gases useful?

Other Nonmetals

Hydrogen is located in an unusual part of the periodic table for a nonmetal. It is found at the top of the alkali metal column because a hydrogen atom has only one electron in its orbital. However, hydrogen is not a metal. Hydrogen is a nonmetal element that really does not fit in as part of any group.

While Groups 17 and 18 contain only nonmetals, Groups 13–16 contain both nonmetals and metals. Group 13 contains only one nonmetal—boron. Group 14 contains one nonmetal (carbon) and two metals (tin and lead). Group 15 contains two nonmetals (nitrogen and phosphorus) and one metal (bismuth). Group 16 contains three nonmetals (oxygen, sulfur, and selenium) and one metal (livermorium). Oxygen is a fairly reactive nonmetal, although it is much less reactive than the halogens. Combustion is the chemical process by which flammable materials react with oxygen, releasing large amounts of energy as heat and light. Sulfur is used to make a substance called sulfuric acid, which is the most widely used substance in the chemical industry. Livermorium does not occur naturally. It is named after the Lawrence Livermore Lab in California, the place where it was first made. Two other elements in Group 16 will be introduced in the next section.

Bismuth is usually alloyed with other metals for use in electrical fuses and automatic fire-sprinkler heads.

Structured Inquiry

Discover

Record your work for this inquiry. Your teacher may also assign the related Guided Inquiry.

Science Journal

A Family of Elements

What are the chemical properties of halogen?

Procedure

1. Label the test tubes *A–H*. Add about 5 mL of each of these halide solutions: tube A—sodium chloride, tube B—sodium bromide, tube C—sodium iodide, tube D—ammonium chloride, tube E—ammonium bromide, tube F—potassium chloride, tube G—potassium bromide, tube H—potassium iodide.
2. On the data table, record the compound in each test tube and identify the halogen that is a component of the compound.
3. Add 5 drops of silver nitrate solution to test tube A and **record** your **observations**.
4. Repeat Step 3 for test tube B.
5. **Make a hypothesis** about the reactions of elements in the halogen family with silver nitrate and record it in your ***Science Journal***.
6. Repeat Step 3 for test tube C through test tube H.

Materials

- 0.1 M aqueous silver nitrate solution
- 0.1 M aqueous sodium chloride solution
- 0.1 M aqueous sodium bromide solution
- 0.1 M aqueous sodium iodide solution
- 0.1 M aqueous ammonium chloride solution
- 0.1 M aqueous ammonium bromide solution
- 0.1 M aqueous potassium chloride solution
- 0.1 M aqueous potassium bromide solution
- 0.1 M aqueous potassium iodide solution
- test tubes
- dropper

Analyze Results

Describe a chemical property of the halogen family.

Create Explanations

1. What are the chemical properties of halogens?
2. What part of the silver nitrate solution reacted with the halogens?
3. What is the name of the product of the reaction in test tube A?
4. **Predict** what would happen if you combined solutions of sodium fluoride and silver nitrate.

Metalloids **Explain**

As you know, an element that exhibits the properties of both a metal and nonmetal is classified as a *metalloid*. The metalloids are elements that exhibit metallic properties under some circumstances and nonmetallic properties under other circumstances. The metalloids are solids at room temperature and tend to be hard and brittle. There are seven metalloids in the periodic table, located in Groups 13–16. Tellurium and polonium are the metalloids in Group 16. They are silvery-white solids that have properties of both metals and nonmetals. Can you identify the other five metalloids in addition to tellurium?

Since the mid-1900s, metalloids have become very important because they function as semiconductors. Recall that a *semiconductor* is a substance that acts as an electrical conductor under some conditions but does not conduct electricity under other conditions. Because of this property, metalloids such as silicon are used to make the computer chips and transistors for modern processors and communications systems. Why do you think the properties of metalloids make them so useful for electronic applications?

Thin wafers of the element silicon are used to make tiny microchips used in computers.

What property of metalloids do you think would make them useful for building electronic components?

Lesson Activity

Use the Internet to find out what elements are in a number of common minerals. Use the periodic table to identify the elements and classify them as metals or nonmetals. Are most minerals made up of metals, nonmetals, or a combination of metals and nonmetal elements? Did you find any minerals that are also pure elements?

What elements are found in minerals?

Concept Check **Assess/Reflect**

Summary: What are nonmetals and metalloids? Although most of the elements are metals, many nonmetals are very important to living organisms. The nonmetal elements are poor conductors of heat and electric current. In the solid state, nonmetal elements are brittle and cannot be hammered or drawn into shape. Most nonmetals are gases, although some are solids and one, bromine, is a liquid. The properties of nonmetals are generally the opposite of the properties of metals. Metalloids are elements that have some properties of nonmetals and some properties of metals. Metalloids are important in electronics applications.

1. How do you know that all of the elements in the atmosphere, such as oxygen and nitrogen, are nonmetals?
2. Very reactive solids are sometimes stored in containers filled with a noble gas. Why are noble gases useful for this application?
3. Sulfur is one of the few elements that occur naturally in an elemental form. What can you conclude about the chemical properties of sulfur?
4. How could you predict whether an element is a metal or a nonmetal based on the arrangement of its electrons?

Lesson 5

Essential Question

What Are Radioactive Elements?

Have you ever discovered something by accident? Scientists often make discoveries that they did not expect to find. In 1896, for example, a French scientist named Henri Becquerel conducted an experiment using a fluorescent mineral. Fluorescent minerals glow when they are exposed to light. What might cause the elements in fluorescent minerals to emit light? What experiment could you do to try to answer this question?

Fluorite crystals from certain locations fluoresce due to impurities in the crystal.

Becquerel's Discovery Explain

Becquerel thought that fluorescent minerals may also give off X-rays. So he wrapped a photographic plate in paper and placed a fluorescent mineral on top of it. He then placed it in bright sunlight. When Becquerel later removed the paper wrapping, he saw an image of the mineral on the photographic plate. How did he know that the exposure came from X-rays and not from light emitted by the glowing mineral?

Objectives

- Distinguish between radiation and radioactivity.
- Compare and contrast the forms of radiation.
- Define transmutation.
- Define the half-life of a radioactive element.
- Distinguish between nuclear fission and nuclear fusion.

Vocabulary

radiation
radioactivity
radioactive isotope
alpha decay
transmutation
beta decay
gamma decay
nuclear fission
nuclear fusion

The exposed plate showed that energy had interacted with the film.

Why was Becquerel surprised to see an image on the film?

Becquerel tried to repeat his experiment, but it turned out to be a cloudy day. So he put all his materials in a drawer, where he left them for a few days. When he finally took out the materials from the drawer, he decided to develop the photographic plate. He was quite surprised. What did Becquerel see according to the illustration below?

Radiation Explain

What Becquerel had discovered accidentally was that the mineral gave off energy that passed through the paper and created the image on the photographic plate. He concluded that this energy came from an element that was in the mineral. This element was uranium. The energy that uranium gives off is called radiation. **Radiation** is the energy that an atom gives off when changes occur in its nucleus. The process by which nuclei give off energy is called **radioactivity**.

Recall that the number of neutrons in the nucleus of an atom can vary. Forces that act among all of the protons and neutrons hold the nucleus of the atom together. If these forces and the forces that push protons away from one another are balanced, then the nucleus of the atom is stable. One force that occurs among neutrons and protons is called the *strong force*. Why do you think scientists chose that name?

Nuclear Instability

In some isotopes, either there are not enough neutrons or there are too many neutrons for the forces to stay in balance. Radiation is emitted as the protons and neutrons rearrange to form a more stable grouping. Uranium is one of many elements that exhibit radioactivity. These elements are known as **radioactive isotopes**. Almost all of the naturally occurring radioactive elements are in the seventh period of the periodic table. Why do you think most of the naturally occurring radioactive isotopes involve elements with fairly large atomic numbers?

Radioactive Decay **Explain**

Radiation is also called *radioactive decay*. This term comes from the fact that an unstable nucleus gives off particles or decays to become more stable. In the process, the radioactive isotope gives off energy. There are three main types of radiation produced by the decay of a nucleus.

Alpha Decay

One type of radioactive decay is known as alpha decay. In an **alpha decay**, a radioactive isotope emits an alpha particle. An alpha particle has two protons and two neutrons. What element atom has this same combination of two protons and two neutrons? Because of its mass and charge, an alpha particle does not travel very far through matter. In fact, a sheet of paper blocks most alpha particles. Human skin that is exposed to a stream of alpha particles, however, can develop a painful burn-like wound.

A uranium atom has 92 protons in its nucleus. After its nucleus releases an alpha particle, only 90 protons remain in the nucleus. Because each element has a specific number of protons, an atom that undergoes alpha decay is no longer a uranium atom. What element has it become? The process by which one element changes into a different element as a result of radioactive decay is called **transmutation**. The diagram shows the transmutation of uranium-235 (92 protons, 143 neutrons). How many neutrons will be present in an atom that forms as a result of this decay?

Alpha Decay

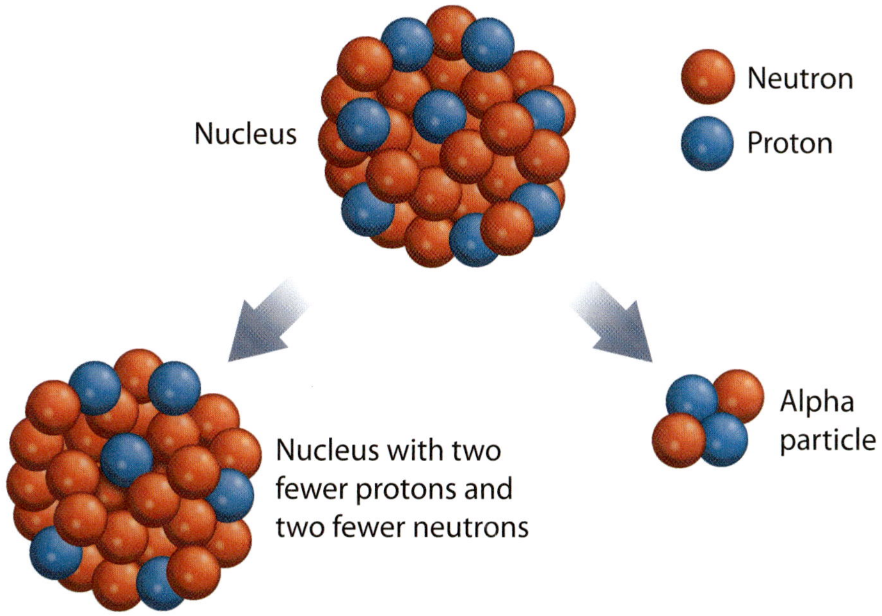

When the nucleus emits an alpha particle, it changes to a different element.

Why is it not necessary to show the electrons in illustrations of radioactivity?

Beta Decay

A second type of radioactive decay is called beta decay. **Beta decay** involves the release of a beta particle from the nucleus of a radioactive isotope. A beta particle is an electron. What mass and charge does a beta particle have? Most beta particles will travel right through a sheet of paper. A stream of beta particles can be stopped by a piece of wood that is 2–3 cm thick. Why is a beta particle more penetrating than an alpha particle? Why does this kind of radiation cause more serious injury than alpha radiation?

Beta Decay

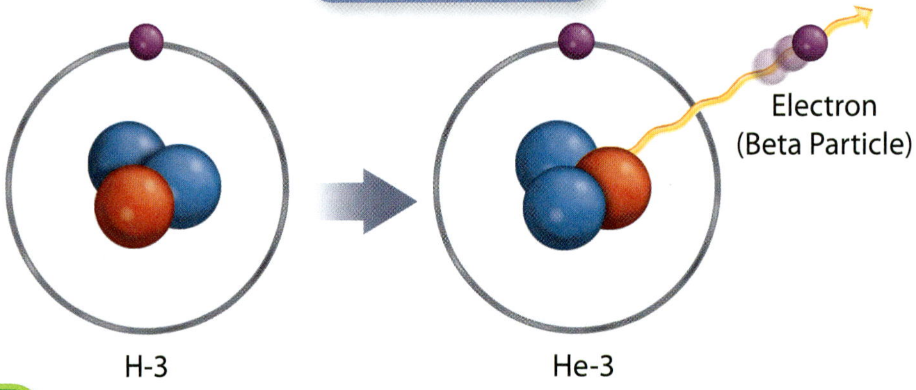

During beta radiation, a nucleus emits a high-energy electron. At the same time, one of the neutrons transforms into a proton.

Explore-a-Lab

Structured Inquiry

How long does it take for radioactive atoms to become stable?

Fold a piece of paper in half. On the top half, write down the day of the month you were born. This is just a random number for the purpose of starting this activity.

Multiply your birthday number by 5000. Write down this number just below your birthday and label it *Beginning Amount*. This number will represent the initial population of unstable or radioactive atoms of your nuclide or isotope.

When the teacher calls "half-life," divide that number by 2. Round to the nearest whole number. Write your new answer below the fold line of your paper. This will divide your original population of atoms in half.

Now tear this whole sheet of paper in half along the fold line. Place the piece of paper that was the top half on the desk in front of you. This represents the first "half-life" and half the population of nuclide atoms that are now "stable" and are no longer radioactive.

Keep the torn-off bottom sheet that represents the "unstable" nuclide atoms and get ready to repeat the process over and over: Divide by 2. Record. Tear in half. Make sure to record your stable atoms' "half-life" data on the piece you tear off and set in front of you. Go through seven more half-lives, for a total of eight half-lives. What stayed the same during the activity? What changed during the activity? How many half-lives do you think it would take to get to zero radioactive atoms?

Gamma Decay

A third type of radioactive decay is called gamma decay. **Gamma decay** involves the release of gamma rays from the nucleus of an atom. Gamma rays have no mass or charge. Does gamma decay result in a transmutation? Why or why not?

Gamma rays have enough energy to pass though wood. It takes a piece of lead several centimeters thick or a concrete slab about 1 m thick to stop gamma rays. Because gamma rays carry so much energy and can pass though flesh, they are very dangerous.

Comparing Radiation

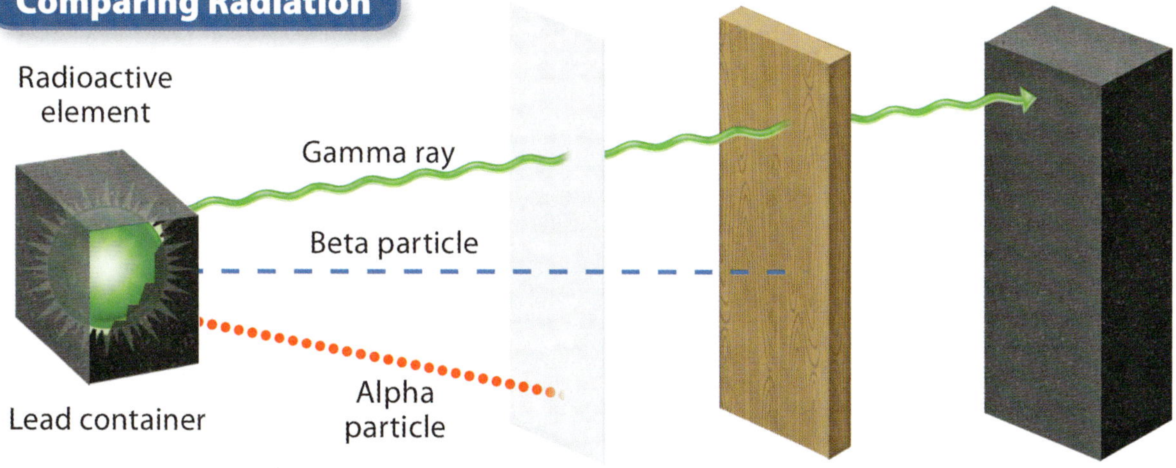

The three types of radiation have very different amounts of energy.

What is the purpose of the lead container in the design of this experiment?

Half-Life **Explain**

Nuclei of radioisotopes decay at a predictable rate. Although it is impossible to predict exactly when a particular nucleus will emit radiation, it is possible to predict the percentage of the total number of nuclei in a sample that will decay in a given period. The amount of time during which half the nuclei in a sample will emit radiation is called the *half-life* of the radioisotope. The type of radiation produced and the half-life are characteristics of each radioactive isotope.

Every sample of a particular radioactive isotope has the same half-life. For example, chlorine-39 has a half-life of 55.7 minutes. In slightly less than an hour, half of the chlorine-39 atoms on Earth will have emitted a beta particle and become argon-39. Uranium-235, on the other hand, has a half-life of about 70,000,000 years. Only a very tiny fraction of the existing uranium-235 will decay over the same one-hour period. Some radioisotopes made by humans have half-lives measured in fractions of a second.

During one half-life, one-half of the nuclei will decay, no matter how many were there at the beginning of that time. During the next half-life, one-half of the remaining nuclei will decay, leaving only one-fourth of the original amount of the radioisotope. Another half-life reduces the amount to one-eighth, and so on. Is there a limit to the number of half-lives that a radioisotope can exist?

Even though the decay of a particular nucleus cannot be predicted, the decay rate of a group of nuclei is very predictable.

How could you use the ratio of parent and daughter isotopes to calculate how long a sample has been decaying?

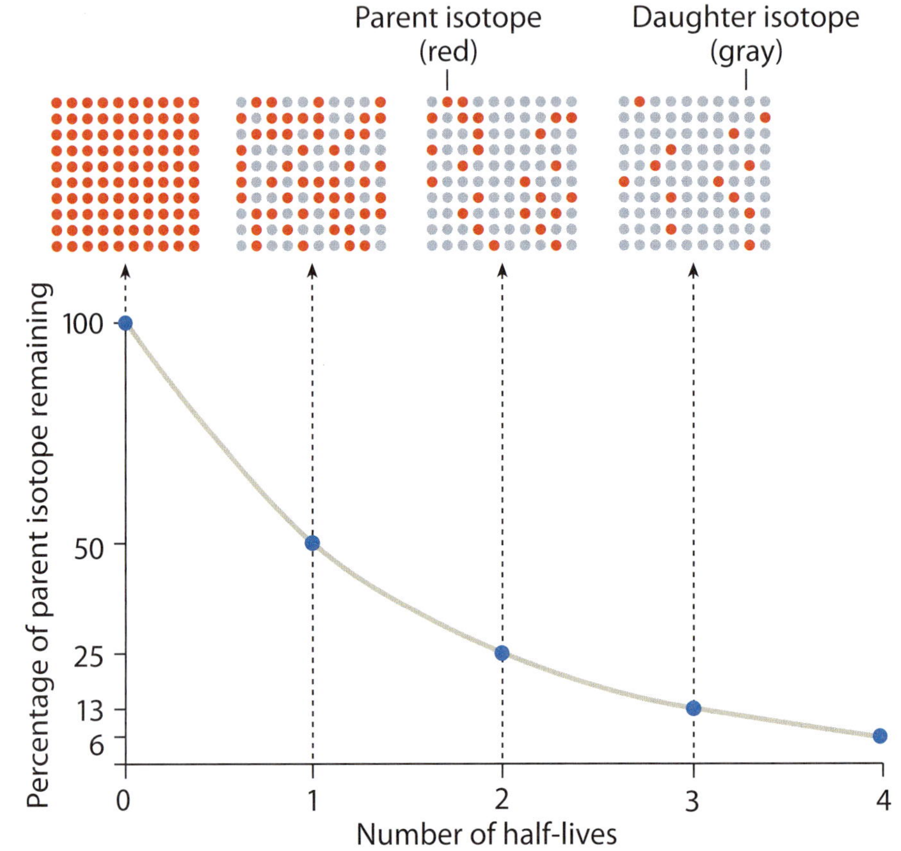

Math in Science

The ratio of isotope that decays and its product can be used to determine how long a sample has been undergoing radioactive decay. For example, iron-59 undergoes beta decay to form cobalt-59. The half-life of this decay is 44.5 days. If a sample that was known to be pure iron-59 initially is analyzed and found to be 50% iron and 50% cobalt, how old is the sample? One-half of the original iron has decayed, so the sample is one half-life old, or 44.5 days. If the sample consists of 87.5% cobalt and 12.5% iron, it is 3 half-lives old, or 133.5 days. How old is a sample consisting of 75% cobalt and 25% iron? Is it possible to know for sure how old a sample is, based on this method of dating? What assumptions are involved?

Structured Inquiry

Discover

Record your work for this inquiry. Your teacher may also assign the related Guided Inquiry.

Science Journal

Candied Half-Life

How does radiometric decay occur?

Materials

- candy-coated chocolate bites (large package)
- plastic cups (large)

Procedure

1. Label one cup *R* for "radioactive" and the other cup *D* for "decayed," or no longer radioactive.
2. Count your candies and **record** the number.
3. Place all of your candies in cup R.
4. Shake the candies and pour them on your desktop. Count all of the candies with the marks facing up. Record the number on the data table. Return these candies to cup R.
5. Place the rest of the candies in cup D.
6. Repeat Steps 4–5 until there are no more candies to place back into cup R.
7. For the second and third rounds, repeat Steps 3–6 and record your data on the corresponding row.
8. **Calculate** the average values for the three trials and record the average values on the bottom row of the data table.

Analyze Results

Graph the average values.

Create Explanations

1. How does radiometric decay occur?
2. For each trial, what is the chance that a particular candy will decay?
3. If half the candies decay in each trial, why aren't all of the candies gone after the second trial?
4. **Predict** how many more trials would be needed to complete the decay sequence if you started with twice as many candies.

Radioactive Isotope Dating Explain

Faith Connection

While many scientists accept the ages that are calculated using radioactive isotope dating, other scientists point to possible weaknesses of this method of dating. Many Christians believe that the long ages that are approximated in this way are not consistent with Scripture. What are some things that could influence this process and cast doubt on its accuracy?

Scientists can use the half-lives of radioactive isotopes to determine the age of materials. Archaeologists use carbon-14 dating to find the ages of materials that were once living. Carbon-14 is a radioactive isotope of carbon. It undergoes beta decay to form nitrogen-14 with a half-life of 5730 years. Because all of the isotopes of carbon have the same chemical properties, they all act the same way in living things. Carbon-14 is produced in the atmosphere by radiation from space, so atmospheric carbon dioxide has a known ratio of carbon-14. The percentage of carbon in a growing plant is the same as that found in the atmosphere. The same percentage occurs in animals that eat the plant. When an organism dies, though, it no longer takes in carbon. Then the amount of carbon-14 is fixed.

Recall, however, that carbon-14 is radioactive. That means the amount of carbon-14 in the organism begins to decrease as soon as the organism dies. How much of the original carbon-14 remains after 5730 years? By comparing the amount of carbon-14 remaining in an animal fossil with the amount of carbon-12, scientists can calculate when the organism died. This dating method works for samples that are less than 50,000–60,000 years old. After that, not enough carbon-14 is left to work with.

For older samples and samples that were never alive, scientists use the decay of other radioactive isotopes. Many combinations of isotopes are useful, depending on the age and type of sample. The decay of uranium into lead provides information about the age of very old rock formations. Uranium-235 passes through a series of radioactive decay, eventually forming lead-207 with a half-life of about 700 million years. Uranium-238 decays to form lead-206 with a half-life of about 4.5 billion years. By studying the ratios of uranium and lead isotopes in a sample, scientists can approximate the age of different rocks.

Radioactive isotope dating puts the age of the fossil ammonites at more than 65 million years old.

Which radioactive isotopes could not be used to date the fossil? Why?

Energy From the Nucleus Explain

In the first half of the 1900s, Albert Einstein proposed that during nuclear changes, some mass is converted into energy. His famous equation $E = mc^2$ indicates that a small amount of mass (m) is equivalent to a very large amount of energy (E). Scientists discovered that during radioactive decay, a very small amount of mass is converted to energy.

Nuclear Fission

The conversion of mass into energy occurs in a process known as nuclear fission. **Nuclear fission** is the process by which a large radioactive nucleus splits into smaller nuclei and releases energy. An example of a radioactive isotope that undergoes nuclear fission is uranium-235. The nucleus of this atom splits to form two smaller atoms and emits three neutrons. If you added all the masses of these three neutrons and the two smaller nuclei that were formed, you would find that their combined masses are slightly less than the mass of the protons and neutrons that were in the original uranium-235 nucleus. This mass was not lost but rather was converted into energy as predicted by Einstein's equation.

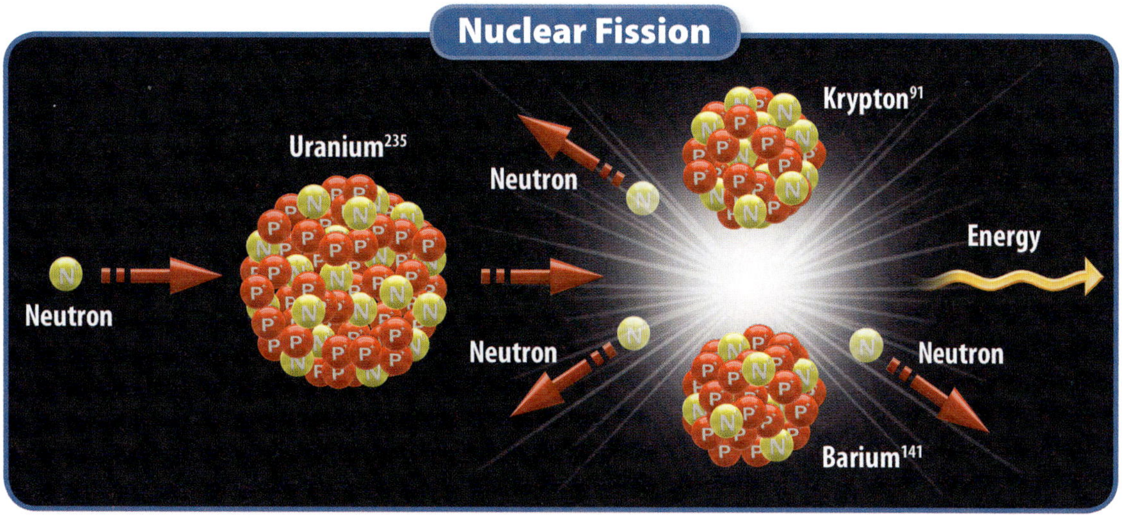

The single neutron striking the uranium nucleus can start a chain reaction that results in a large amount of energy.

Why is it necessary to remove some of the neutrons in a nuclear power plant?

The nuclear fission of uranium-235 begins when it is bombarded by a neutron, as shown above. Think about the products of the nuclear fission of uranium-235. In addition to two atomic nuclei, three neutrons are produced. If each of these neutrons strikes another uranium nucleus, three more fission reactions will occur, producing energy and nine neutrons. These 9 neutrons can produce 27 neutrons. As this *chain reaction* proceeds, the number of neutrons increases rapidly: 81, 243, 729, 2187, and so on. Each level of the series of reactions also produces three times as much energy as the previous level. What will happen if there is a lot of uranium and the reaction is not controlled?

In a nuclear power reactor, the reaction is controlled by using materials that can absorb some of the neutrons. The energy produced is used to make steam that runs a generator, producing electricity.

Nuclear Fusion

Another process in which mass can be converted into energy is nuclear fusion. **Nuclear fusion** is the process of combining small nuclei to form a larger nucleus. A lot of energy is needed to bring the nuclei together in a fusion reaction, but then the reaction releases a lot more energy. Nuclear fusion is the process that powers the Sun and other stars as hydrogen fuses into helium, releasing energy.

Nuclear fusion only occurs at very high temperatures. Engineers have studied ways to control the fusion process to produce usable energy in a fusion power plant. However, such a plant would require higher pressures and temperatures than we are currently able to control. Nuclear fusion power may someday be possible, but not yet. In what ways could nuclear fusion be a better source of power than those that we currently use?

During this nuclear fusion reaction, four hydrogen nuclei fuse to form one helium nucleus. The neutrinos that are emitted have no charge and very little mass.

Based on the changes in matter and charge, what can you deduce about positrons?

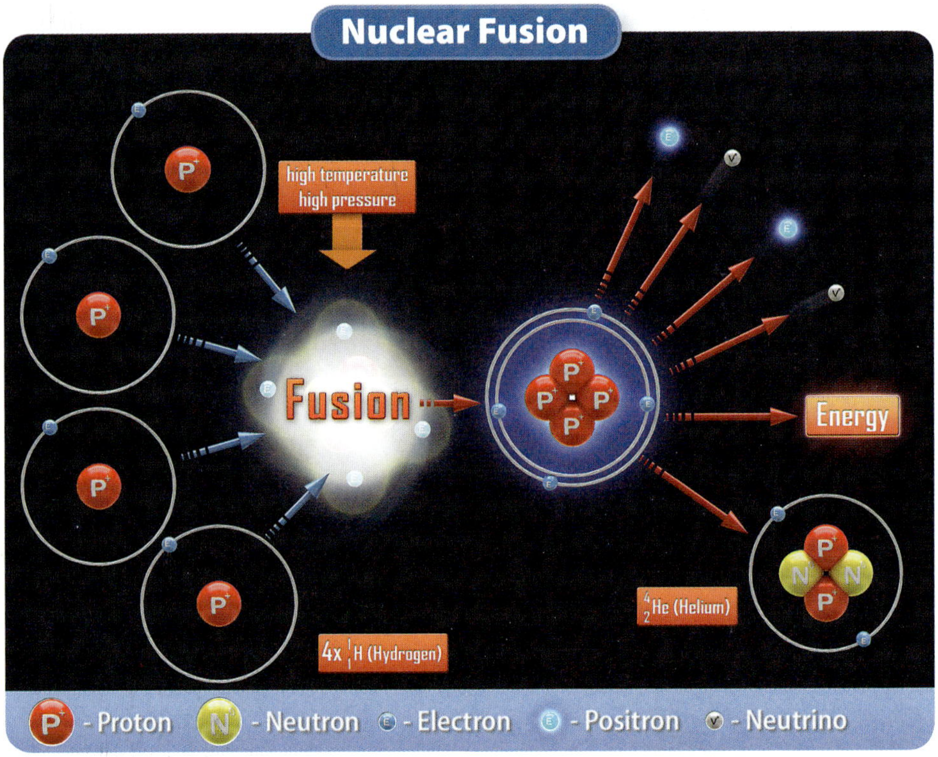

Explore-a-Lab

Structured Inquiry

How are the processes of fusion and fission different?

Use a graduated cylinder to carefully measure 2 mL of dish detergent, 5 mL of glycerin, and 6 mL of water into an aluminum tart pan. Use a wood splint to stir the liquid until it is of a uniform consistency. Make two bubble wands from two 25-cm pieces of thin wire. Take each piece of wire and form a 3- to 4-cm circle at the center of the length. Twist the two ends together to form the handle. Hold one wire frame in each hand. Dip the two circular wire frames in the solution. Gently blow through each wire frame to create a bubble with a diameter a little larger than the frame and catch the bubble on the frame. Bring the frames and the bubbles together. Let the bubbles press against each other until they form one large bubble. This illustrates the fusion process. Stretch the bubble by pulling the two frames farther apart until the bubble separates into two bubbles, one in each frame. This demonstrates the fission process. When this is done a little faster, a small bubble may be released. What is illustrated by this bubble?

Concept Check **Assess/Reflect**

Summary: What are radioactive elements? Some atomic nuclei are not stable due to the ratio of protons and neutrons. These nuclei emit particles and energy as they rearrange themselves into a more stable form in a process known as radioactivity. The energy and particles given off during this process are called radiation. The most common types of radiation are alpha particles, beta particles, and gamma rays. Alpha particles have the least amount of energy. Gamma rays have the most energy. Emission of alpha or beta particles changes the nucleus to that of a different element, a process known as transmutation. Radioactive decay occurs at a predictable rate that is a characteristic of a radioactive isotope. During one half-life, one-half of the nuclei undergo decay. Scientists use the half-lives of elements in substances to estimate the age of the substance. Nuclear fission and nuclear fusion are processes that change nuclei, forming new elements and releasing large amounts of energy. In nuclear fission, large nuclei break apart to form two or more smaller nuclei. In nuclear fusion, smaller nuclei combine to form a larger nucleus.

1. How does the radioactive decay of a nucleus make it more stable?
2. Uranium isotopes are separated from one another to make fuel for nuclear power plants. Why would it be very difficult to determine when a sample of uranium-238 was processed by using radioactive dating?
3. Nickel-67 has a half-life of 20 sec. What fraction of a sample of nickel-67 would remain unchanged after one minute?

People in Science

Extend

Get to Know
Robert Boyle

The Shannon Portrait of the Hon. Robert Boyle F. R. S.
by Johann Kerseboom, 1689. CHF Collections. Photograph by Will Brown.

Robert Boyle was born in 1627. He was not religious as a child or as a young man. But during his studies in Europe, he observed God's hand in nature when he experienced a severe thunderstorm. From then on, he found God's presence to be connected to the natural scientific world he loved to observe.

Robert Boyle was a man ahead of his time. First, he developed the scientific method. You can thank Robert Boyle for the consistent and acceptable results you get each time you conduct an experiment. He believed in something called the "new philosophy." This stressed that observation, experiment, and logical thinking were all necessary to formulate accurate scientific understanding. Boyle was a founding member of the Royal Society, which was established to help this new way of thinking continue. It is the oldest scientific society in existence today.

Boyle also used the newly invented microscope to study the various excretions from the bodies of humans and animals. He became a strong supporter of *corpuscularism*. This hypothesis stated that substances were made of particles and that the particles moved about. Boyle understood that these particles joined together to form new materials but was unable to accurately explain how or why this occurred. Today we call this the atomic theory.

Robert Boyle is most famous for his studies with air and its properties. He made a vacuum pump so he could conduct experiments in an airless environment. The results allowed him to state an important relationship of gases—their volume varies inversely with pressure. That is, if you increase pressure, the volume of the gas decreases. Today this important relationship is known as Boyle's law.

Finally, Boyle was a humanitarian. Throughout his life he worked to improve the lives of people. For example, he wanted to improve farming methods. He discussed the possibility of making freshwater from saltwater. He wrote about making medicines safer and more effective. He studied the improvement of medical practices. He even wrote about the possibility of preserving food in vacuum packaging, much like we have today in canned and jarred foods. Robert Boyle truly was a man ahead of his time.

Concept Check

1. Which of Boyle's works made him a "man ahead of his time"?
2. How might people's lives have changed if Boyle had been successful in some of his humanitarian efforts?

Science and Technology

Extend

Macromolecules

Imagine a day without Styrofoam™ or plastic. That is what life was like before 1839 when polystyrene, or Styrofoam, as it is commonly called, was invented. Since that time, more and more plastic-like materials have been invented as scientists have learned to work with materials we call polymers.

You have learned that all things are made of atoms and that atoms bond together to make molecules. When individual molecules bond together, they become larger and more complex. When many molecules link together, macromolecules are formed. And, when many macromolecules link together, polymers are formed. Some polymers form naturally, such as cellulose in plants. They can also be formed synthetically. These are used to make plastics, Styrofoam, and nylon.

Recall that atoms are in constant motion. That means that polymers are also constantly moving. The larger the polymers, the slower the atoms in the molecules move. This results in polymers that are harder and more difficult to break apart. They can also become more difficult to break down.

You are surrounded by materials made of macromolecules every day. And because recycling is common in many communities worldwide, we have become familiar with the many different types of polymers that are made from these macromolecules. Consider the recycling symbol on the bottom of different types of plastic containers. Polyethylene terephthalate, or PET, is used to make soda bottles and laundry detergent bottles. High-density polyethylene, or HDPE, is used to make milk jugs and shampoo bottles. Low-density polyethylene, or LDPE, is used to make garbage bags, tape, and disposable diapers. Polyvinyl chloride, or PVC, is used to make pipes and shower curtains. Polypropylene, or PP, is used for chip and cookie bags and Tupperware® containers.

Concept Check

1. Describe the differences between materials made from HDPE and LDPE.
2. Relate the speed of atomic movement in macromolecules to that of gases versus solids.

Chapter 12 Review Assess/Reflect

Study Guide

Lesson 1

1. Greek philosopher Democritus called the smallest piece of matter *atomos*. In the 1800s, John Dalton described atoms as unbreakable spheres. In the nineteenth century, J. J. Thomson proposed that atoms consist of a positively charged mass embedded with negatively charged electrons. Ernest Rutherford concluded that the nucleus contained positively charged particles called protons. James Chadwick discovered the neutron. According to Niels Bohr, electrons move around the nucleus in specific orbits called energy levels. Scientists now describe the atom as having a central nucleus with the probability of electrons being in a particular location around the nucleus, called electron clouds.
2. Energy is absorbed when an electron moves to a higher energy level and emitted when an electron moves to a lower energy level.
3. The atomic number of an element is the number of protons in the nucleus. The atomic mass of an atom is the sum of protons and neutrons in the nucleus. The average atomic mass of an element is the weighted atomic masses of its atoms.
4. Isotopes are atoms of an element with different numbers of neutrons. Two isotopes of an element have the same chemical properties but different atomic masses.

Lesson 2

1. Elements are arranged on the periodic table in rows and columns. The rows, or periods, of the table order elements based on increasing atomic number. The columns, or groups, of the table include elements with similar physical and chemical properties.
2. Across a period, elements have increasing numbers of outer orbital electrons. Within a group, elements have increasing numbers of occupied orbitals.

Lesson 3

1. Metals are good conductors of heat and electric current. General properties of metals include ductility, malleability, and a shiny luster.
2. The alkali metals form Group 1 of the periodic table. Alkali metals are very reactive with water and air. The alkalis are soft metals with low melting points compared with most other metals.
3. The alkaline earth metals form Group 2 of the periodic table. They are very reactive, although less so than the alkali metals.
4. Transition metals are the elements in the central part of the periodic table. They generally react very slowly with air and water. The hardness of the transition metals makes them useful for structural and building materials.

Lesson 4

1. Nonmetals are poor conductors of heat and electric current. Many nonmetals are gases at room temperature. In the solid state, nonmetals are brittle with a dull luster. They are not malleable or ductile.
2. The halogens form Group 17 of the periodic table. They are very reactive nonmetal elements.
3. Noble gases form Group 18 of the periodic table. These gaseous elements are the most nonreactive of all elements. Because they are so inert, the noble gases had not even been discovered at the time that Mendeleev designed the first periodic table.

4. Metalloids are elements that have some properties of metals and some properties of nonmetals. Because of these properties, metalloids are important for electronic applications.

Lesson 5

1. Radiation is energy emitted from an atomic nucleus when unstable nuclei rearrange to a more stable form. Radioactivity is the process by which the energy is emitted.
2. The three major types of radiation are alpha particles, beta particles, and gamma radiation. Alpha particles are identical to a helium nucleus. Beta particles are identical to an electron. Gamma radiation is a form of electromagnetic energy, similar to high-energy X-rays.
3. Transmutation is the change of one element to another element due to a change in the number of protons in the nucleus. Transmutation occurs during alpha and beta radiation.
4. Half-life is the period during which one-half of the nuclei of a radioisotope decay and form a new substance. Half-life is a characteristic of a radioisotope.
5. Nuclear fission is the process of splitting a large nucleus into smaller nuclei, releasing energy. Nuclear fusion is the process of combining smaller nuclei to produce a larger nucleus and energy.

Vocabulary Check

Explain how each pair of items is related.

1. energy level—orbital
2. period—group
3. metal—conduction
4. nonmetal—noble gas
5. nuclear fusion—nuclear fission

Multiple Choice

Choose the best answer.

6. What particles determine the atomic mass of an atom?
   A. protons only
   B. electrons only
   C. protons and electrons
   D. protons and neutrons
7. Which property applies to all metals?
   A. completely filled outer orbital
   B. conduct electric current
   C. more protons than electrons
   D. solid at room temperature
8. What is another way to describe an alpha particle?
   A. electron
   B. helium nucleus
   C. hydrogen nucleus
   D. neutron

Check Point

Answer the following questions.

9. How do the numbers of electrons compare in two isotopes of an element?
10. What structural feature of their atoms is characteristic of elements in a group of the periodic table?
11. If additional elements are discovered, in what period are they almost certain to belong?
12. Which group of metals is likely to have the most vigorous reactions with the halogen chlorine?
13. Why are the halogens more toxic than the noble gases?
14. Nuclear reactors have rods made of materials that absorb neutrons that strike them but do not release additional neutrons. What is the purpose of these rods?

Chapter 13

Types of Chemical Bonds

Scripture Spotlight

All of the matter God created on Earth comes from the elements. God's Design allows these elements to combine and interact in different ways to give us the great variety of matter you see every day. By studying how these elements combine, you can better understand God's Design. In this chapter, you will read the following passages.

Ephesians 4:3 (p. 479)
Colossians 3:14 (p. 489)

This student is studying the small molecules God created and how they are held together by chemical bonds.

The Big Idea

God's natural laws that govern the planets and galaxies also govern the atoms and the bonds they form with one another.

How do you think the strength of a chemical bond might affect the state of matter?

Inquiry Kick-Off Engage

Elements are the alphabet in the language of molecules. They form the words, sentences, and paragraphs that make up the matter around you. Why do some elements join and others don't? If elements join, can they be separated? What role does energy play in the joining of elements? In your *Science Journal,* you will observe two elements join together to make a new substance.

Science Journal

Lesson 1

Objectives

- Explain how atoms form chemical bonds.
- Compare and contrast the formation and features of ionic and metallic bonds.

Vocabulary

cation

anion

ionic bond

metallic bond

Essential Question

How Do Chemical Bonds Form?

Why do fireworks produce colors such as reds, blues, and greens? Why does a stretched rubber band return to its original shape when released? Why does silver tarnish? The answers to these questions have to do with what happens when atoms form chemical bonds. Each of these actions takes place for the same reason. In each case, the result is something that has less energy than the initial state and is therefore more stable. Releasing energy to achieve stability is a driving force in nature. In the case of a chemical bond, the atoms have less energy and are more stable after they have joined. How would you separate these bonded atoms so that they are no longer bonded?

Ionic and Metallic Bonds Explain

You already know that *chemical bonding* is an attractive force between atoms or groups of atoms with complementary electronic characteristics. Recall that atoms have electrons that orbit the nucleus in various energy levels. Those electrons in energy levels closer to the nucleus have less energy than those farther away from the nucleus. Which electrons in an atom would have the most energy? Also recall that those electrons that are in the outermost energy level are known as *valence electrons*. Valence electrons determine how elements react with other elements and how they bond together. Bonding involves the interactions of valence electrons to form atoms that become more stable.

Energy is released during a chemical reaction in these glowsticks. A fluorescent dye accepts the energy and converts it to light.

Atoms react so that they have the same number of valence electrons as the noble gas nearest to them in the periodic table. When potassium (K) reacts, the atom of which noble gas would have the same number of valence electrons as the potassium atom?

Ionic Bonds

One way atoms react is by gaining or losing one or more electrons. Consider what can happen to a sodium (Na) atom as it forms an ionic bond. Study the periodic table (on the inside cover of this book) and the diagram of a Na atom below. Notice that Na has one valence electron. The nearest noble gas to sodium is neon (Ne), which has one less proton and one less electron. If the sodium atom loses its valence electron, the atom would have the same number of valence electrons as neon, which would make the Na atom more stable. When it loses the electron, however, the atom has 11 protons and only 10 electrons. Recall that an *ion* is an atom or group of atoms that has an electrical charge because it gained or lost electrons. What happens to the charge on the sodium atom?

Scripture Spotlight

Imagine Christians as atoms being joined together in a chemical bond. What bonding agent is responsible for the spiritual bond described in **Ephesians 4:3**?

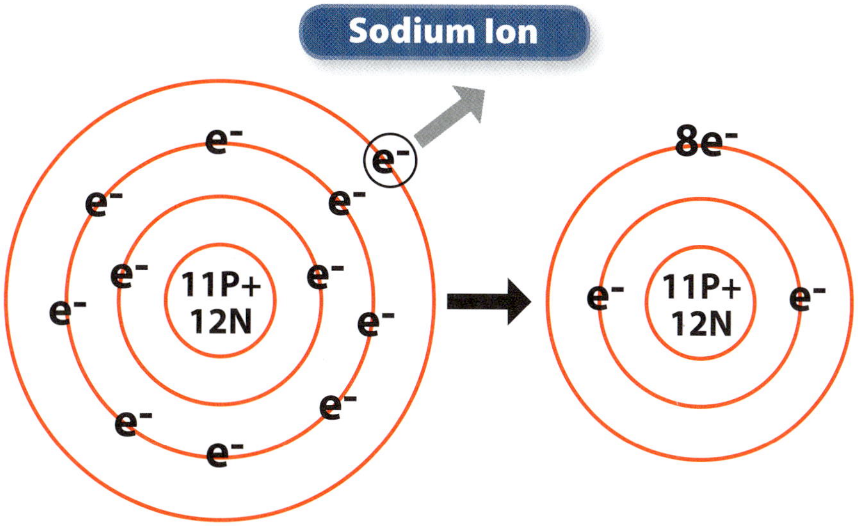

Sodium (Na)

Sodium ion (Na^+)

The sodium atom gains the outer energy configuration of neon by losing its valence electron.

Why doesn't sodium gain one electron when it forms a bond?

The electrons lost by one atom can be gained by another atom. Chlorine (Cl) has seven electrons in its outer energy level. If a chlorine atom gains an electron lost by a sodium atom, it becomes an ion with a charge of −1. What causes this −1 charge on the chlorine ion?

Any ion with a positive charge is called a **cation**. An ion with a negative charge is called an **anion**. After the transfer of electrons, the oppositely charged sodium and chloride ions attract one another. The force of attraction that holds the ions together is called an **ionic bond**. What elements other than sodium and chlorine do you think form ionic bonds? Why?

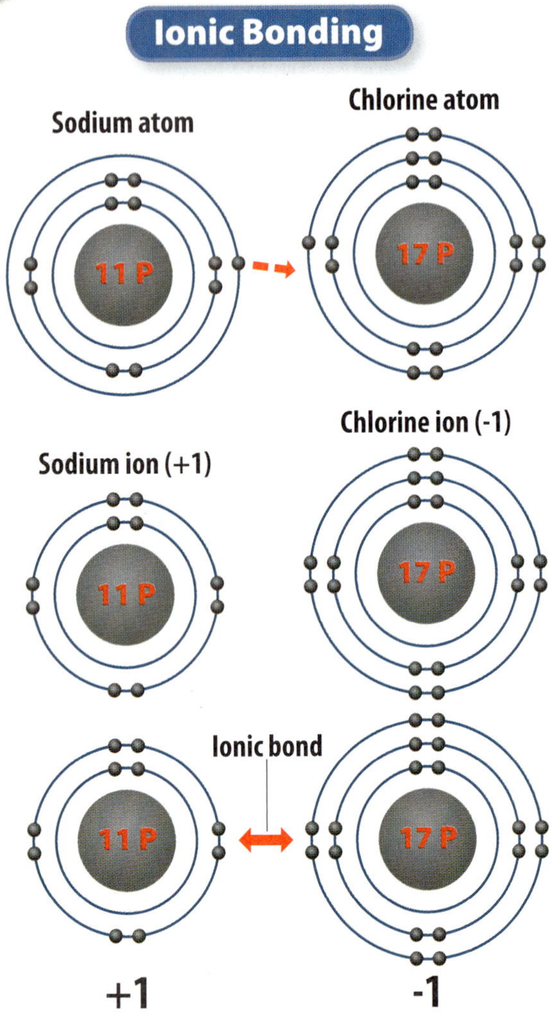

Ionic bonds hold the sodium chloride crystal together.

All Group 1 elements except hydrogen can form similar ionic bonds with all Group 7 elements.

Why do you think hydrogen atoms might not form an ionic bond such as this?

The compound that forms as a result of the transfer of electrons between the sodium and chloride ions is sodium chloride—common table salt. Its chemical formula is written as NaCl to show that the ratio of sodium and chlorine atoms is 1:1. Another substance that is formed by an ionic bond is $MgCl_2$. What is the ratio of magnesium atoms (Mg) to chlorine atoms (Cl) in this case? Why are two chlorine atoms needed for this compound?

When sodium combines with chlorine, many atoms of chlorine become cations, and many atoms of sodium become anions at the same time. When ionic bonds form between these ions, the familiar salt crystal *lattice* (a regular repeating internal unit of structure) results. Therefore, a salt crystal does not consist of a cation and an anion that have formed an ionic bond. Rather, the crystal consists of a vast number of ions that bond with one another to form a lattice structure. How does the lattice structure affect the shape of the salt crystal? Do all solids with a lattice structure have the same kind of structure and shape? Why or why not?

Math in Science

Complete the following table.

Atom	Change	Charge
Sodium (Na)	Loses one electron	
Aluminum (Al)	Loses three electrons	
Oxygen (O)	Gains two electrons	
Iodine (I)	Gains one electron	
Copper (Cu)	Loses either one or two electrons	

Structured Inquiry

Discover

Record your work for this inquiry. Your teacher may also assign the related Guided Inquiry.

Science Journal

Investigate an Ionic Bond

How does an ionic bond form?

Procedure

Materials (for each group)
- 8 small foam balls
- 8 large foam balls
- box of toothpicks
- black marking pen

1. Work in pairs to **model** the formation of a sodium chloride crystal. Use the marker to put a "–" on the large balls, which represent chloride ions. Put a "+" on the small balls, which represent sodium ions.
2. Use a toothpick to connect the centers of a small and a large ball to represent sodium chloride (NaCl). Do the same thing with the other small and large balls. Connect all the NaCls you made to form a sodium chloride crystal. Do this by referring to the illustration below.
3. Working with other groups, connect the sodium chloride crystals you made so that they form an even larger crystal.
4. **Observe** the model crystal you made. Draw a picture of the crystal structure in your ***Science Journal***.

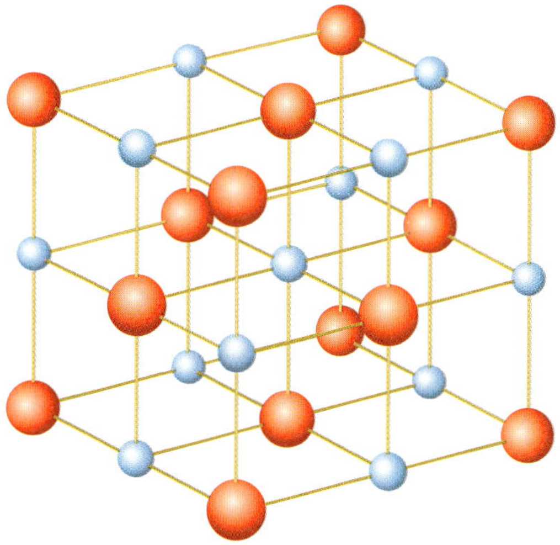

Analyze Results

Review the model crystal. What surrounds each sodium and chloride ion? Why do you think a sodium chloride crystal is brittle?

Create Explanations

1. How does an ionic bond form?
2. Recall which ions the two balls represent. What gave each ion its respective charges?
3. When you stacked all the models together to form a sodium chloride crystal, what shape do they create? Based on the way the ions bond together, why does the salt crystal assume this shape?

Explore-a-Lab

Guided Inquiry

How do sodium and oxygen form a bond?

An oxygen atom needs to gain two electrons to become stable. Recall that a sodium atom loses only one electron to achieve stability. Yet sodium and oxygen do form an ionic bond. Using small and large foam balls and toothpicks, show how sodium and oxygen form an ionic bond. Use the foam balls to form ionic compounds between the following elements: K and Br; Ca and I; Mg and O; and Mg and S.

Other Ionic Compounds

Ionic bonds tend to form between many metals and nonmetals, but the number of electrons that one atom gains does not need to be the same as the number of electrons that one other atom loses. For example, calcium (Ca) is in Group 2 of the periodic table, so it has two valence electrons. It can reach the electron configuration of argon by losing the two electrons. The diagram below shows how it might lose those electrons to a Group 7 element. What charge does the calcium ion have? What charge does the chlorine atom have? Why?

When atoms form ionic compounds, the total charges of the positive and negative ions must equal zero. This compound is written $CaCl_2$—a ratio of one calcium atom to two chlorine atoms. How would you write the formula for aluminum chloride if an aluminum atom loses three electrons?

Ionic Compound

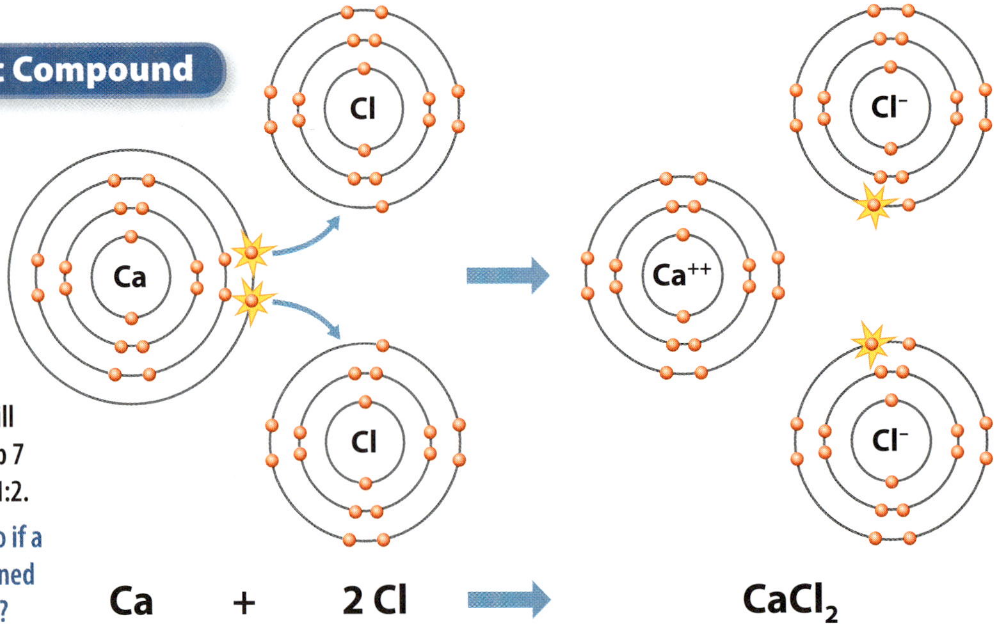

Any Group 2 element will combine with any Group 7 element in the ratio of 1:2.

What would be the ratio if a Group 1 element combined with a Group 6 element?

Properties of Ionic Compounds

In general, ionic compounds have high melting points because the attractive forces between ions are very strong. A lot of energy is needed to break the ionic bonds and cause an ionic solid to melt. How would the melting point of table salt compare with the melting point of sugar, which is not an ionic compound?

Even though they are made of electrically charged ions, ionic solids are very poor conductors of electricity because the ions are not free to move. Dissolved or melted ionic solids tend to be good conductors of electricity because the crystal lattice breaks down and the ions separate and are free to move. How does this action improve electrical conduction?

Explore-a-Lab

Guided Inquiry

How does the amount of salt in a solution affect electrical conductivity?

If dissolved ionic solids are good conductors of electricity, a solution containing dissolved NaCl should conduct electricity and cause a light bulb to light. Make a conductivity tester as shown in this picture. Use distilled water as the liquid.

What happens when the beaker is filled with 300 mL of distilled water? What do you think is the least amount of NaCl that you can add to 300 mL of water to make the set-up conduct electricity? Design an experiment to use the least amount of salt in the water to conduct electricity. What happens if you change the amount of water? Only change one variable at a time and record all of your observations. Why do you think the amounts of water and salt make a difference in creating a conductor?

Metallic Bonds Explain

Sodium is classified as a metal. Chlorine is classified as a nonmetal. Sodium chloride is therefore an example of a metal losing one or more electrons to a nonmental, which gains one or more electrons, to form an ionic bond. But what kind of bond holds together atoms in a metal, such as aluminum, iron, or silver?

Metallic Bonding

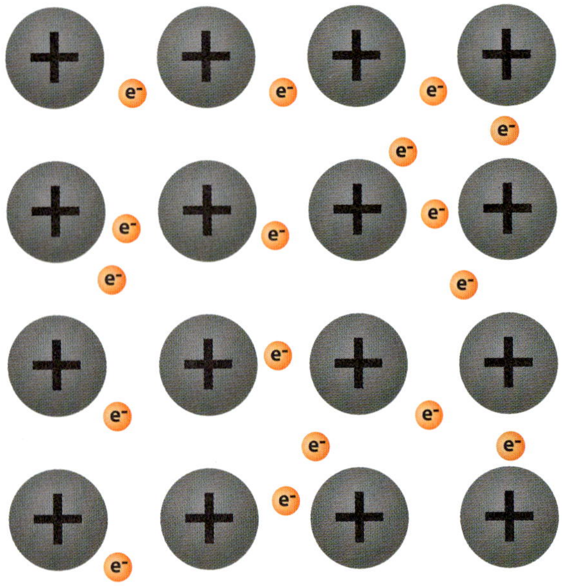

Freely moving electrons in a metal are temporarily attracted to the closest positive ions.

Why do the atoms in a metal not have a charge?

Metals easily lose valence electrons because they are not held strongly. Therefore, the atoms in a metal form a metallic bond. In a **metallic bond**, each atom donates its valence electrons to a "sea of electrons" that move freely between the positive ions. The electrons are constantly in motion and do not "belong" to any one atom. Various metals, such as iron (Fe), lithium (Li), potassium (K), gold (Au), and silver (Ag), are examples of metallic bonding. How are metallic bonds different from ionic bonds? How does this kind of bonding affect the properties of the materials? How do these properties compare with properties of ionic compounds? Why is there a difference? Which do you think is stronger—an ionic bond or a metallic bond? Why?

Because their electrons are free to move, metals possess certain properties. Metals are usually excellent conductors of heat and electricity because their free electrons act as charge carriers. Metals have high melting and boiling points and are strong solids. Metals are less fragile than ionic crystals because there is no rigid lattice structure. This allows us to flatten metals without breaking them. As you know, this property is known as *malleability*. Metals also have the ability to be drawn into fine wires. This property, you will recall, is called *ductility*. Another feature of metals is that they may have a metallic luster and opaqueness.

Alloys, mixtures of two metals, are examples of metallic bonding too. Brass, an alloy of copper and nickel, is an example of a metallic alloy that exhibits metallic bonding.

Explore-a-Lab

Structured Inquiry

How do metallic bonds affect heat conduction?

Fill a pot with hot tap water. To get the water as hot as possible, you may need to let the water run a minute before filling your pot. *Use caution not to splash anyone or get burned.* Place three large spoons—one made of wood, one of plastic, and one of metal—in the pot so that only their handles are in the water. Make sure that the spoons do not touch each other. Add a dab of butter to the bowl of each spoon that is extended above the water. Record your observations. Explain your results.

Concept Check Assess/Reflect

Summary: How do chemical bonds form? Bonding occurs when valence electrons are attracted by positively charged nuclei of nearby atoms. An ion is an atom or group of atoms that has an electrical charge because it has gained or lost electrons. Any ion with a positive charge is called a cation. An ion with a negative charge is called an anion. Cations and anions can combine, with the force of attraction that holds them together being an ionic bond. In a metallic bond, electrons are free to move about. This moveability enables metals to be good conductors of electricity and heat. Metals are less fragile than ionic solids because they have no rigid structure (crystal lattice). This property enables us to stretch and hammer metals into shapes without breaking them.

1. Explain why elements form chemical bonds.
2. Describe the three steps in the formation of an ionic bond.
3. Imagine that some atoms of lithium (Li) and oxygen (O) form an ionic bond.
   - **a.** How many lithium and oxygen atoms are required to form one unit of this compound?
   - **b.** What are the charges on the Li and O ions?
   - **c.** What other elements would form compounds with the same ratio of positive and negative charges?
4. Explain why copper is a good conductor of electricity as a solid, but sodium chloride only conducts electricity when it is melted or dissolved in water.
5. Draw illustrations that show how the following ionic bonds: Ca and I; Be and O; Li and S.
6. Draw a diagram to illustrate the metallic bonds that are present in a gold bar.

Lesson 2

Essential Question

How Do Covalent Bonds Form?

Objectives

- Explain covalent bonding.
- Explain polar and nonpolar molecules.
- Describe how a macromolecule forms.

Vocabulary

covalent bond
electronegativity
polar
polar covalent bond
nonpolar
nonpolar covalent bond
macromolecule

Suppose that you and a friend are collecting the U.S. Commemorative State Quarters. Your friend is missing the Utah coin to complete her collection of all 50 states. You not only have a complete set, but you also have an extra coin of Utah. You give it to your friend to complete her collection. As a result, you both have a complete set. How is this similar to what happens when an ionic bond forms? Now imagine that your friend has a complete set and that you are missing Utah. How can you both have a complete set, even if you both cannot have a complete set at all times? In this case, having a complete set is similar to what happens when another type of bond is formed between atoms. What do you think happens in this case?

Covalent Bonds Explain

What happens when two hydrogen atoms are near each other? Each hydrogen atom has only one electron, which is found in the first energy level. This energy level can hold a maximum of two electrons to become stable. Why doesn't one hydrogen atom simply give up its one electron to the other atom? If it did, what would the hydrogen atom become?

Rather than give up their one electron, both hydrogen atoms can achieve stability by having two electrons in their first energy level. If two hydrogen atoms share their electrons, they each have the configuration of a helium atom.

The strongest chemical bond that forms between atoms when they share electrons is called a **covalent bond**. The electrons shared by the atoms are attracted to the nuclei of both atoms. The atoms must stay together to complete their outer energy levels. You know that atoms that are held together by covalent bonds are called *molecules*. If Na and Cl are held together by bonds, then why don't they form a molecule? What are they called instead?

Covalent Bonding

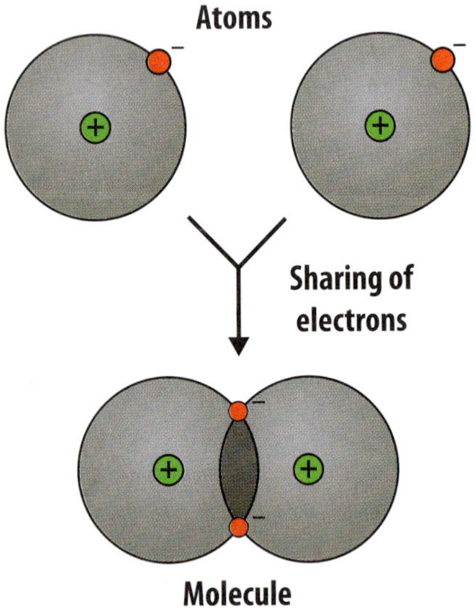

Compounds formed by covalent bonding are called molecular compounds.

What is the main difference between an ionic bond and a covalent bond?

Electron Dot Diagrams

The top part of the diagram below shows all the energy levels of two chlorine atoms that form a chlorine molecule. In chemistry, the inner energy levels are often omitted from drawings because they do not change when the elements bond. Notice that in the bottom diagram, the valence electrons are shown as dots around the chemical symbol. These simplified drawings are called *electron dot diagrams*. The shared pair of electrons is shown as a single straight line that connects the two atoms in the covalent bond. After bonding, how many valence electrons will each chlorine atom have?

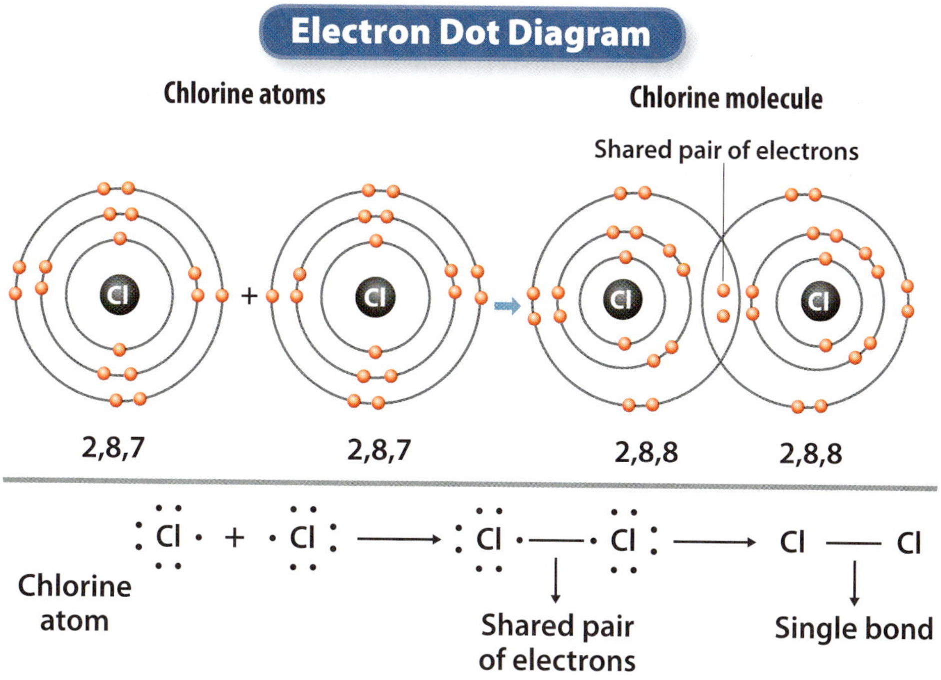

The atoms involved in a covalent bond do not have to be of the same element.

How would you use an electron dot diagram to show the formation of a covalent bond between hydrogen (H) and chlorine (Cl)?

Double and Triple Covalent Bonds

Atoms can share more than one electron. For example, carbon (C) has four valence electrons. It needs four more to complete its outer energy level. Oxygen (O) needs two electrons to complete its outer energy level. Both atoms will be complete if carbon shares a pair of electrons with each of two oxygen atoms to form carbon dioxide (CO_2). This type of bond is called a *double covalent bond*. The two double covalent bonds in carbon dioxidecan be written C=O=C.

Nitrogen (N) has five valence electrons, so an additional three would complete its outer energy level. When two nitrogen atoms combine with one another, they share three pairs of electrons. This is called a *triple covalent bond*, which can be written as N≡N.

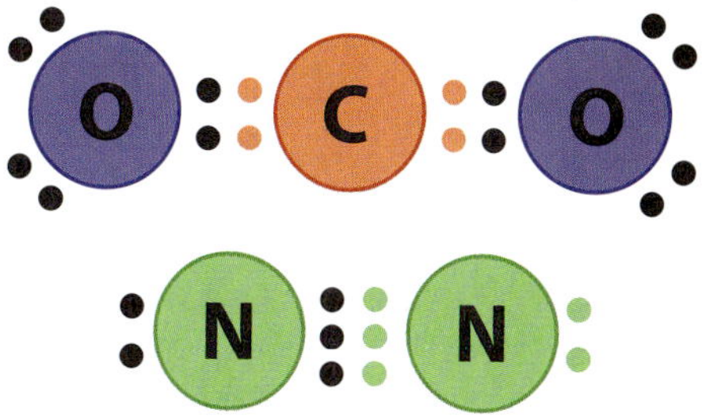

Count the number of dots (electrons) around each atom to see how the atoms gain a noble gas configuration by sharing electrons.

What type of bond would form between two oxygen (O) atoms?

Structured Inquiry

Discover

Record your work for this inquiry. Your teacher may also assign the related Guided Inquiry.

Science Journal

Build Models of Molecules

How can you model molecular elements and compounds?

Procedure

1. In your ***Science Journal***, draw the electron dot diagrams for each of the following molecules: hydrogen (H_2), chlorine (Cl_2), fluorine (F_2), oxygen (O_2), water (H_2O), carbon dioxide (CO_2), ammonia (NH_2), methane (CH_2), and methyl chloride (CH_2Cl).
2. Use the materials provided to build **models** of each of these molecules. Use a different-colored gumdrop to represent each element and toothpicks to represent covalent bonds.
3. **Communicate**: Draw pictures of the models in your ***Science Journal***. Provide a key to indicate which color represents each element.

Materials

- gumdrops of 5 different colors
- toothpicks
- periodic table

Analyze Results

How are your models similar? How are they different?

Create Explanations

1. How can you model molecular elements and compounds?
2. Why can't two atoms form a quadruple covalent bond?
3. Do you think there is a limit to the size of the molecules you could build in this way? Explain your answer.
4. What has to happen to molecules of hydrogen (H_2) and oxygen (O_2) for water (H_2O) to form?

Polar and Nonpolar Molecules **Explain**

Some elements have a greater affinity for valence electrons than others. This affinity for valence electrons is called **electronegativity**. Electronegativity is not a measured value but a description of the attraction of the nucleus for valence electrons. An atom with an electronegativity of 4 has twice the attraction as an atom with an electronegativity of 2.

Electronegativity

		H 2.2					He
Li 1.0	Be 1.6	B 2.0	C 2.6	N 3.0	O 3.4	F 4.0	Ne
Na 0.93	Mg 1.3	Al 1.6	Si 1.9	P 2.2	S 2.6	Cl 3.2	Ar
K 0.82	Ca 1.3	Ga 1.6	Ge 2.0	As 2.2	Se 2.6	Br 3.0	Kr
Rb 0.82	Sr 0.95	In 1.8	Sn 2.0	Sb 2.1	Te 2.1	I 2.7	Xe
Cs 0.79	Ba 0.89	Tl 2.0	Pb 2.3	Bi 2.0	Po 2.0	At	Rn

Electronegativity increases from bottom to top in a column and from left to right across a group.

Why are no electronegativity values assigned to the noble gases?

Key:
- 0.1–0.99
- 1.0–1.99
- 2.0–2.99
- 3.0–3.99
- 4.0

You may think of electronegativity as somewhat like a tug-of-war game between two atoms. If you have two atoms with the same electronegativity, the valence electrons will not be pulled in any one direction more than another, so they will remain between the two atoms. When atoms of the same element form a covalent bond, the attraction of both nuclei for the shared pair of electrons is the same. What happens to the valence electrons of these two atoms when they form a molecule? Why?

As in a tug-of-war with contestants of different strengths, if you have an atom with higher electronegativity, it will pull the shared electrons toward itself. When atoms of different electronegativity form a covalent bond, they share the electrons unequally. This is often the case when atoms of two different elements form a covalent bond. For example, hydrogen has an electronegativity of 2.1, and chlorine's value is 3.5. Do you think the electrons are pulled closer to the hydrogen nucleus or the chlorine nucleus when a covalent bond forms? Why?

Scripture Spotlight

With the definition of a molecule in mind, read **Colossians 3:14** to see what Paul says is the "bond of perfection." Physically, two or more atoms joined together create a molecule. Spiritually, what might be joined together by this bond of perfection?

Because the electrons are shared unequally by the hydrogen and chlorine atoms in a hydrogen chloride molecule, the chlorine atom gains a slightly negative charge and the hydrogen atom a slightly positive charge. These are not ions, however, because the charges are partial. The partial charges at the ends of the molecule can attract or repel other molecules, depending on the partial charges on those other molecules.

A molecule in which electrons are unequally shared is said to be **polar**. Water is an example of a polar molecule. One or more bonds in a polar molecule are **polar covalent bonds**. As the difference in electronegativity between two atoms joined by a polar covalent bond increases, the bond becomes more polar. A **nonpolar** molecule is one in which the electrons are equally shared. A bond in which the electrons are equally shared is a **nonpolar covalent bond**.

When the difference in electronegativity between two atoms is 1.7 or greater, the bond between them is usually ionic. When it is less than 1.7 but greater than 0.5, the bond is generally covalent. Differences of less than 0.5 are considered to be nonpolar.

The blue regions around the nuclei represent locations in which there is a high probability of finding electrons.

Which would you expect to have the greatest attractions for one another: the ions in an ionic compound or the atoms or groups of atoms in a polar covalent compound? Why?

Electronegativity Differences

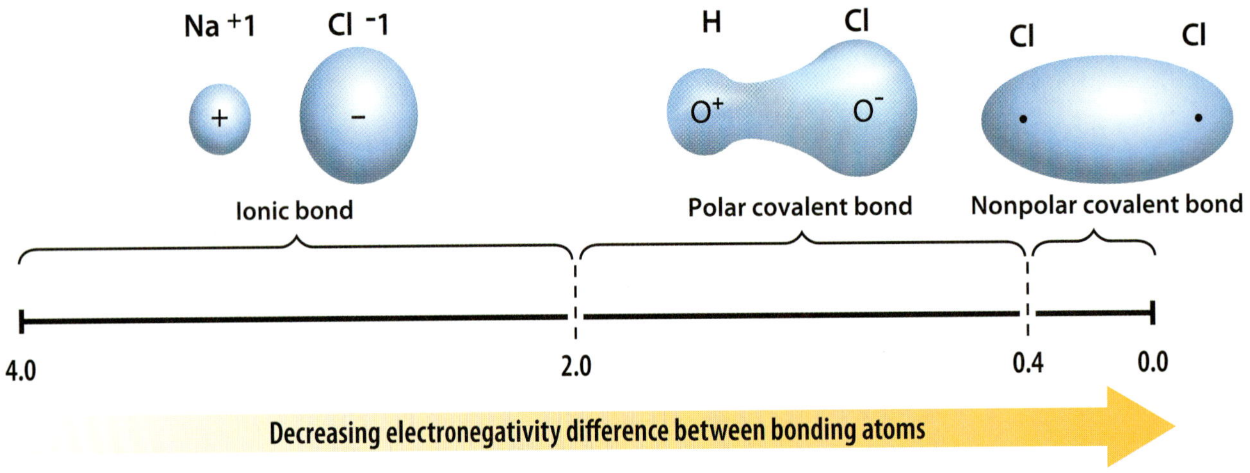

Molecular Shapes and Properties

The bonds in a molecule can be polar, while the entire molecule is nonpolar. Notice that the three atoms in a carbon dioxide molecule are held in a straight line by the double bonds.

Carbon Dioxide

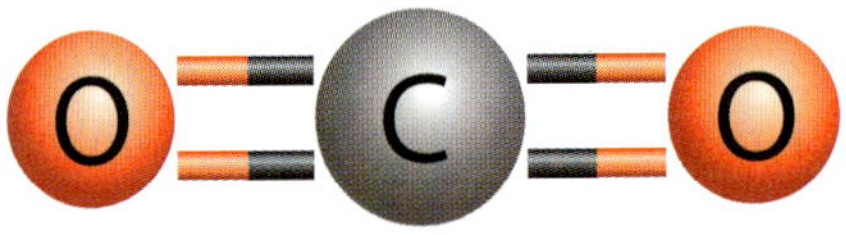

CO_2 molecules have no particular attraction for one another, so the molecules remain separated as a gas at room temperature.

Oxygen has an electronegativity of 3.5, while carbon's electronegativity is only 2.5. The shared pairs of electrons are both pulled toward the oxygen atoms at the opposite ends of the molecule. Is this carbon dioxide molecule polar or nonpolar?

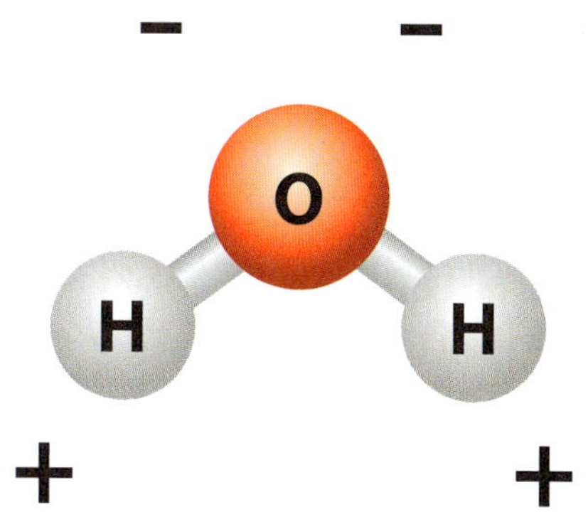

How could two water molecules attract each other?

Unlike carbon dioxide, the three atoms in a water molecule are not arranged in a straight line. Rather, they form an angle because valence electrons around the oxygen atom cause the two hydrogen atoms to move apart. Because of its bent shape, a water molecule is polar. The region where the oxygen atom is located has a negative charge at times. The regions where the two hydrogen atoms are located have a positive charge at times. These two regions with their opposite charges account for the polarity of the water molecule.

Water molecules attract one another and tend to remain closer to one another than do molecules of carbon dioxide. Why do you think this is so? As a result of this attraction, water molecules form a liquid at room temperature. Changing this liquid to a gas requires a lot of energy in the form of heat to break apart the attractive forces between water molecules. As a result, H_2O boils at 100°C. In contrast, CO_2 boils at –57°C. Why does CO_2 have such a low boiling point even though this molecule consists of three atoms, just like water?

Check out your *Science Journal* for a Structured Inquiry exploring covalent and ionic bonding.

Extend

Explore-a-Lab

Structured Inquiry

How does the polarity of molecules affect the surface tension of water?

Fill a clean bowl with water. Work with a partner to see how many paperclips you can float on the water. Once you have several paperclips floating on the surface, add a drop of dish soap to the water. Observe what happens. Try floating a paperclip in the water with the soap. Explain what happens in terms of the polarity of the molecules.

Macromolecules **Explain**

A **macromolecule** is a molecule with many atoms arranged in long chains, rings, or other complex structures. Many of these molecules are based on carbon atoms, each of which can bond with four other atoms. The compounds can have many single, double, and triple bonds between hundreds of different atoms. Macromolecules are often made up of smaller molecules bonded together. For example, protein molecules are made up of smaller molecules called amino acids.

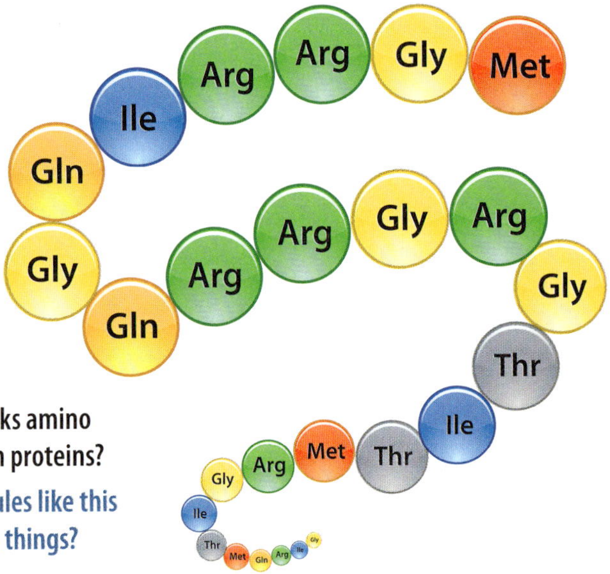

What type of bond links amino acids together to form proteins?

Why are macromolecules like this so important to living things?

Concept Check **Assess/Reflect**

Summary: How do covalent bonds form? The chemical bond that forms between nonmetal atoms when they share electrons is called a covalent bond. A molecule in which electrons are unevenly shared is said to be polar. A nonpolar molecule is one in which the electrons are evenly shared. Macromolecules are molecules that contain a large number of atoms.

1. How do ionic and covalent bonds differ?
2. What do you need to know to predict whether a molecule will be polar or nonpolar?
3. Draw an electron dot diagram for the carbon tetrachloride (CCl_4) molecule. Then use lines to represent the bonds.
4. Predict what would happen if water were a linear molecule, just like carbon dioxide.

Lesson 3

Essential Question

How Are Chemical Formulas Helpful?

As you have studied science, you have learned that when studying living organisms, scientific names are preferable to common names. Why are scientific names useful? Now think about substances as common as salt and sugar. How might using the common names for these substances be confusing? How could this confusion cause problems? How do scientists avoid this confusion?

Objectives

- Explain the significance of a chemical formula.
- Distinguish between molecular and structural formulas.
- Given a chemical formula, name the compound.
- Identify isomers of a compound.

Vocabulary

chemical formula
molecular formula
structural formula
organic chemistry
isomer

Chemical Formulas Explain

The chemical formulas for compounds are much more specific than common names, and they tell you which elements the compounds contain. A **chemical formula** is a shorthand notation that identifies the elements that make up the compound and the ratio in which they are present. For example, the formula for water is H_2O, the formula for table salt is NaCl, and the formula for table sugar is $C_{12}H_{22}O_{11}$. Based on its chemical formula, the chemical name for water would be *dihydrogen monoxide*. How is the chemical name more descriptive than its common name, *water*?

You know that a *chemical symbol* is an abbreviation used to represent an element. Examples include Al (aluminum), Fe (iron), B (boron), and Br (bromine.) The periodic table of the elements shows all the symbols of the elements.

When elements combine, we can represent the compounds they form by chemical formulas. The symbol for an element itself represents one atom of the element. If more than one atom of the element is present in the compound, the number is written as a subscript, such as the *11* after oxygen in the formula for sugar.

Baking soda ($NaHCO_3$) is an ionic compound. Because ionic compounds do not have molecules, the formula represents the simplest ratio of atoms in the compound. In baking soda, the ratio is 1Na:1H:1C:3O.

The chemical forumla for table sugar is $C_{12}H_{22}O_{11}$.

What elements make up table sugar? How might this formula be different for other sugars?

Molecular Formulas

There are several types of chemical formulas. Each provides different descriptive information. For example, one type of chemical formula is known as a molecular formula. A **molecular formula** shows the various elements that are present and the ratio in which they are found in a molecule. For example, the molecular formula for methane is written as CH_4. This molecular formula shows the number of atoms of each element in one molecule of methane—one atom of carbon and four atoms of hydrogen. Why is NaCl not considered a molecular formula?

Molecular Formulas for Ethyl Alcohol

C_2H_6O

C_2H_5OH

CH_3CH_2OH

Notice that there are three ways in which the molecular formula for ethanol can be written. What additional information about the structure of the compound is provided by each formula as you proceed down the list?

Faith Connection

In a similar way, think of symbols to represent several characteristics of God that you appreciate most. Put those symbols together to create a formula that describes God.

Structural Formulas

Another type of formula that provides specific details about how the atoms are arranged in a molecule is called a structural formula. A **structural formula** shows the types and numbers of atoms present in a molecule and the arrangement of the atoms. Why do you think these formulas are called structural?

Structural Formula for Ethyl Alcohol

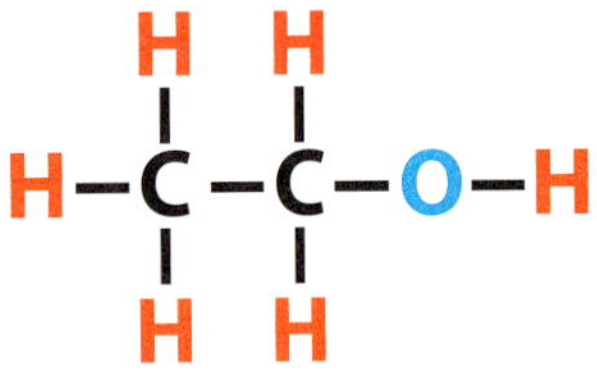

A structural formula shows how the C, H, and O atoms are arranged in an ethanol molecule.

How does the molecular formula CH_3CH_2OH compare with the structural formula for ethanol?

Lesson Activity

Write the molecular formulas for the following compounds that you might find in a home:

- Hydrogen peroxide, in which the hydrogen and oxygen atoms are present in a 2:2 ratio
- Isopropyl alcohol (rubbing alcohol), in which the carbon, hydrogen, and oxygen atoms are present in a 3:8:1 ratio
- Acetic acid (vinegar), in which the carbon, hydrogen, and oxygen atoms are present in a 2:4:2 ratio
- Caffeine, in which the carbon, hydrogen, nitrogen, and oxygen atoms are present in an 8:10:4:2 ratio

Now create your own compound and give it a name.

What information do you need to write molecular formulas?

Structured Inquiry

Discover

Record your work for this inquiry. Your teacher may also assign the related Guided Inquiry.

Science Journal

Build and Name Ionic Compounds

How can you build a model of an ionic compound, write its formula, and name it?

Materials

- element squares
- paperclips
- periodic table

Procedure

1. **Observe** the element squares. Divide them among the members of your group. Each square shows the atomic symbol and number for an element, as well as its energy levels. Electrons in the outer energy level are represented by black dots. Valence electrons in Groups 1 and 2 are represented by paperclips. Valence electrons in Groups 5, 6, and 7 are represented by black dots. Empty holes represent electrons needed to complete the outer energy level and to give the atom an inert gas configuration.
2. Combine a metallic element with a nonmetallic element by transferring "electrons" (paperclips) from the metal to the nonmetal to fill the outer energy level. You may need more than one "atom" of one or both elements to do this.
3. Write the formula for the ionic compound you made. Then name it. (Example: NaCl, sodium chloride). **Record** your work in the ***Science Journal***.
4. Repeat Steps 2 and 3 to make as many ionic compounds as you can with the element squares you have. Hint: Every metal can be combined with every nonmetal to form a different compound.

Analyze Results

Communicate: How many different ionic compounds would you have been able to build using the element squares?

Create Explanations

1. Locate your elements on the periodic table. What other elements that were not part of your element squares will form a compound with fluorine in which the compound has the same ratio as sodium fluoride?
2. What is the formula for a compound that forms between any Group 2 element (X) and any Group 6 element (Y). Why?
3. Could you have built ionic compounds containing three elements using your element squares? Explain.

Naming Compounds **Explain**

It may seem as though there is a bewildering array of ways to name chemical compounds. But chemists have devised rules for naming compounds, whether they are ionic or molecular, so that there is no uncertainty about the compounds under discussion.

Ionic Compounds

To name ionic compounds that consist of only two elements, follow these simple rules:

1. Name the first element in the compound. The first element is the metal that is located toward the left side of the periodic table.
2. Change the name of the second element so that it ends in the suffix *-ide*. The second element is the nonmetal that is located toward the right side of the periodic table.

For example, NaCl is sodium chloride. K_2O is potassium oxide. The number of atoms of each element in the ionic compound does not matter in the name. What do you think the name for $AlBr_3$ is?

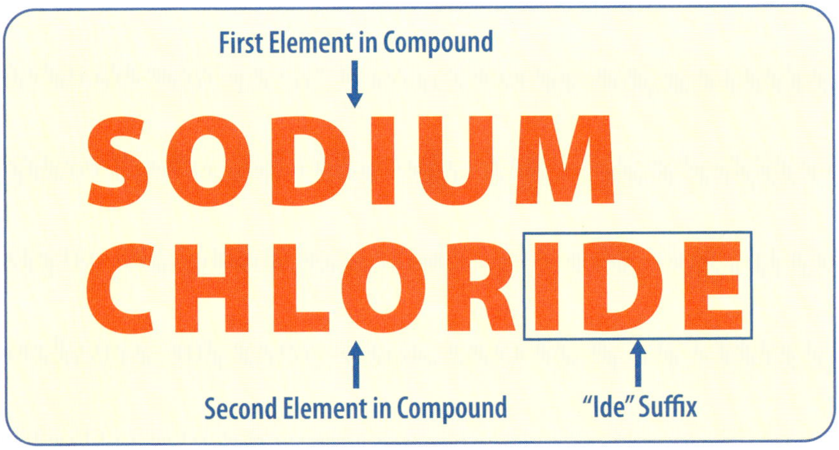

Salt can be obtained through mining and solar evaporation, which is shown in this photo.

Molecular compounds

Most simple molecular compounds only have two words in their names. The name carbon monoxide (CO) tells you that you have one carbon (C) atom and one oxygen (O) atom. (You can also use the prefix *mono-* for one atom.) Notice the prefix *mono-* is used only for the second element. Also remember that the second word ends in *-ide*.

(1) carbon + (1) oxygen =carbon monoxide (CO)

The table below lists the prefixes used to name molecular compounds. Use this table to name the following molecular compounds: SO_2, NI_3, CCl_4, SF_6, and H_2O.

Number of Atoms	Prefix
1	*mono-*
2	*di-*
3	*tri-*
4	*tetra-*
5	*penta-*
6	*hexa-*
7	*hepta-*
8	*octa-*
9	*nona-*
10	*deca-*

Lesson Activity

Work in small groups to make a card game of the elements. Make cards for the elements in Groups 1, 2, 16, and 17. Place the name of each element on a separate card and indicate the number of valence electrons available for bonding. Shuffle the cards together and deal the cards to each player. The object of the game is to lay cards out to form compounds and to write the correct chemical formula and name of the chemical compound. Be sure to form both ionic and molecular compounds. Develop a scoring system and rules for play. Share your completed card game with the class.

How can you practice naming chemical compounds?

Organic Compounds

Most molecular compounds are based on carbon. The chemistry of carbon compounds is called **organic chemistry** because life on Earth is based on carbon. What chemical processes did you learn about in life science that involve carbon? How does life depend on these processes?

Many molecular compounds share certain structures or arrangements of atoms to form a series. For example, compounds composed only of carbon and hydrogen are called *hydrocarbons*. A hydrocarbon might have one, two, three, four, five, or more carbon atoms. The prefix in the names of these compounds tells the number of carbon atoms. The names for all simple hydrocarbons held together by only single bonds end in the suffix *-ane*. Hydrocarbons have the general formula C_nH_{2n+2}, where *n* is the number of carbon atoms. These compounds are all good fuels because they combine easily with oxygen.

Hydrocarbons	CH_4	C_2H_6	C_3H_8	C_4H_{10}	C_5H_{12}
	Methane	Ethane	Propane	Butane	Pentane

Alcohols

Methanol

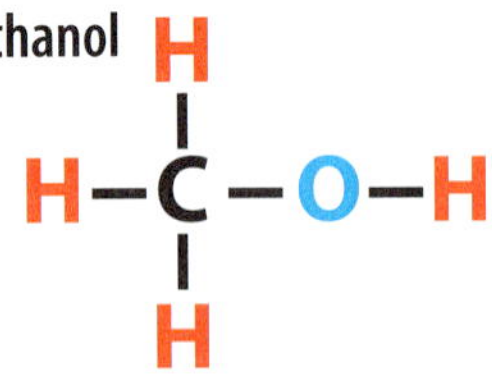

Ethanol

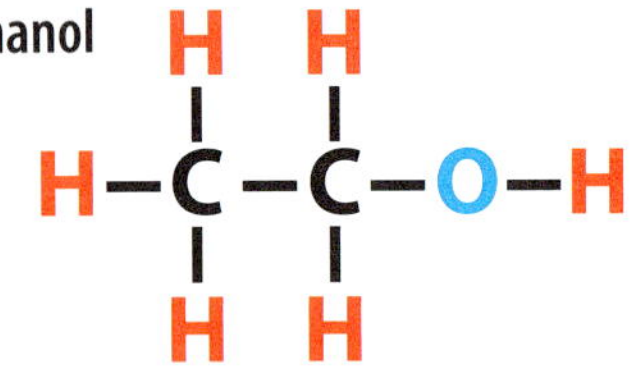

Propanol

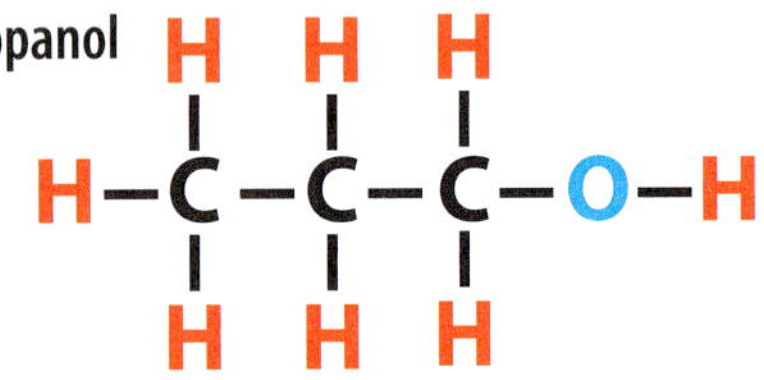

The suffix *-ol* stands for the OH group.

Another series of compounds with similar properties is the alcohols. In an alcohol, one or more of the hydrogen atoms in a hydrocarbon is replaced by an OH group, as shown in the diagram. The same prefixes are used to describe the number of carbon atoms, but all simple alcohols end with the suffix *-ol*. How would you write the general formula for an alcohol?

Alcohols

CH_3OH	C_2H_5OH	C_3H_7OH	C_4H_9OH	$C_5H_{11}OH$
Methanol	Ethanol	Propanol	Butanol	Pentanol

Isomers

Once a molecule contains sufficient atoms, it is possible to arrange the same atoms in different ways. Compounds with the same molecular formula but different structural formulas are called structural **isomers**. To understand isomers, you might think of the Creator as a baker and these isomer molecules as His dough. The recipe for the dough is the same, but it can be baked into loaves, rolls, or twisted and made into many shapes in three dimensions. What is the difference between molecules A and C shown below?

A molecule with the molecular formula C_3H_8O could be arranged in several different ways to form different compounds with different properties. Look at the structural formulas below. Molecules A and C both contain the OH group, so they are both alcohols. How do you think the different shapes affect the properties of the isomers? Isomer A is propanol, and isomer C is isopropanol, commonly called rubbing alcohol.

Isomer B is not an alcohol because the oxygen is bonded to two carbons. Although tens of thousands of different carbon compounds exist, naming them is easy once you know the prefix for the number of carbon atoms and the suffixes for different types of compounds.

Isomers

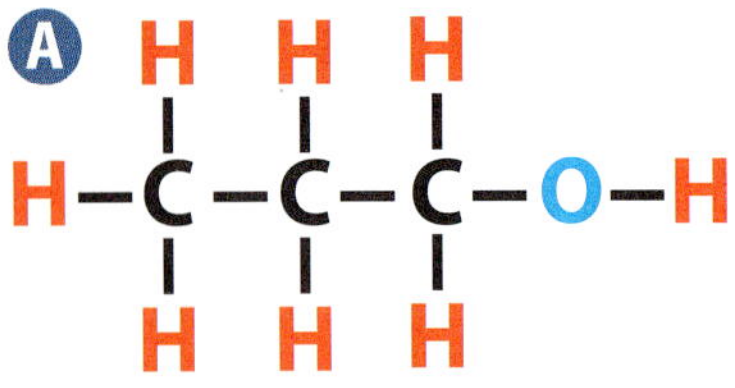

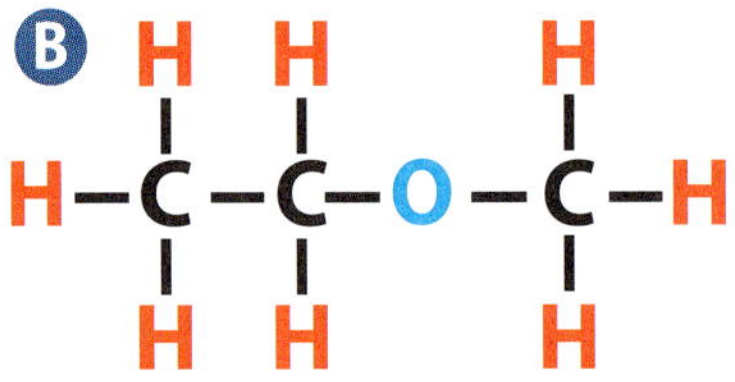

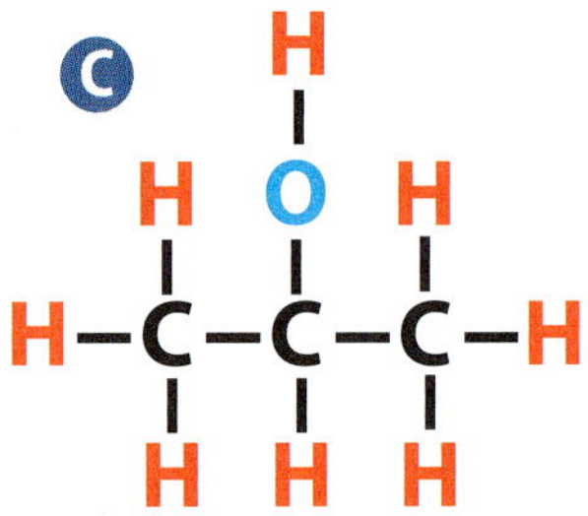

The structural formulas show three possible ways that C_3H_8O can be arranged.

If you put the OH in isomer A on the left carbon rather than the right carbon in the chain, would it be a different isomer? Explain.

Why do you think so many different isomers exist in nature? Isomers have important roles in biological systems. Because of their complexity, every important macromolecule is made up of many different types of atoms (carbon, hydrogen, nitrogen, phosphorous, and many others) containing chemical bonds. The proteins that make up our bodies are isomers, as are the carbohydrates that make up many of our basic foods. For example, starch and cellulose are isomers of each other. While one makes up many foodstuffs, the other is an almost indigestible part of many plant structures. The different properties of these isomers help them function. Moreover, the Creator appears to have favored different isomers for different purposes over the many different forms that exist. It all depends on the arrangement of the atoms and their bonds in space.

Explore-a-Lab

Structured Inquiry

How can you model isomers?

Use color gumdrops and toothpicks to build four isomers of $C_4H_{10}O$.

Concept Check

Assess/Reflect

Summary: How are chemical formulas helpful? A chemical symbol represents an element. The chemical symbols for all of the elements are found in the periodic table. Chemical names for compounds are a shorthand way to identify the elements in a compound. A chemical formula can give information about the structure of the molecule. A molecular formula indicates both the elements that are present and the ratio in which they are present. A structural formula indicates the elements that are present in a compound and the way in which the elements are arranged or bonded to each other. Compounds with the same molecular formula but different structural arrangements are called structural isomers.

1. What is the name of the ionic compound Al_2S_3?
2. What is the name of the molecular compound S_2O_3?
3. Which of these general formulas represents an ionic compound between a Group 1 element (X) and a Group 16 element (Y)?

   A. XY   **C.** XY_2

   B. X_2Y **D.** X_2Y
4. Draw a structural formula for CH_3CH_2Cl.
5. Draw at least three different structural isomers for C_5H_{12}. Be sure that every carbon forms four bonds.

Science and Technology

Extend

Get to Know

Cryogenics

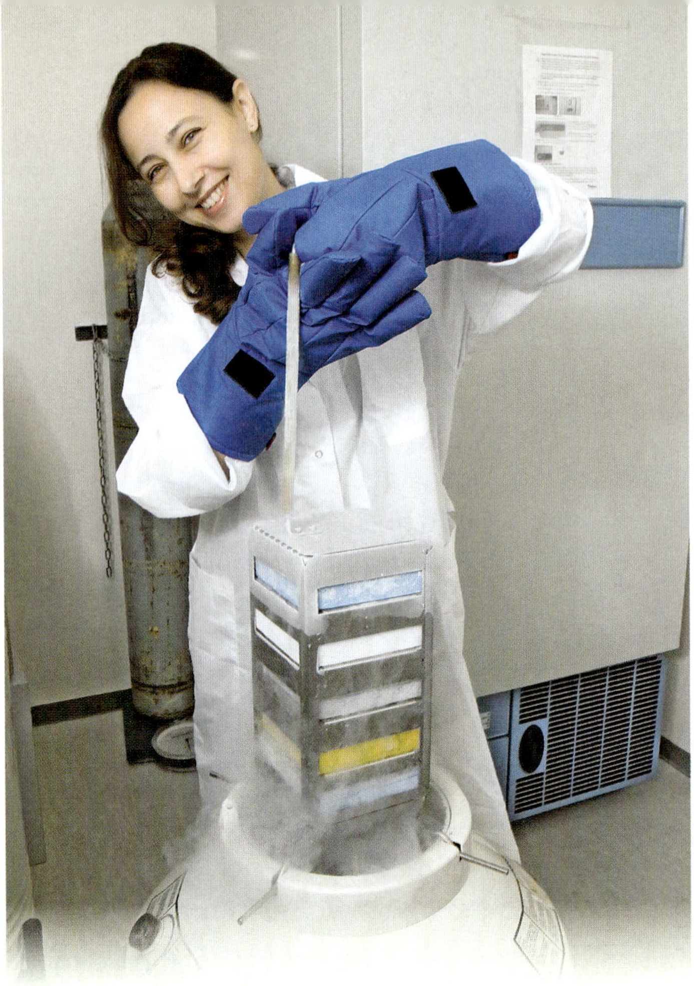

The science of cryogenics deals with the production of very low temperatures and the study of the effects of those temperatures. The word *cryogenics* comes from two Greek words—*kryos*, meaning "frost," and *genic*, meaning "to produce."

A temperature range for cryogenics has been defined but is not precise. Generally, it refers to temperatures below –150°C. At these temperatures, certain materials take on irregular properties. Some metals become superconductive, while liquid helium becomes resistant to flowing. The importance of cryogenics comes from these unusual properties.

Cryogenics has many purposes. Cryogenic liquids, such as oxygen, nitrogen, and argon, are frequently used in medical technologies and in industry. MRI scanners depend on cryogenics to help their superconducting magnets stay cold, allowing the MRI scanner to see more clearly and produce a sharper image. If you have ever had a wart removed, the doctors may have used the cryoprobe, a needlelike instrument that freezes and destroys certain kinds of tissues, such as warts, skin tumors, and eye tumors. In addition, biological substances such as whole blood and bone marrow can be preserved using cryogenic temperatures.

One of the best-known industrial uses for cryogenics is the process of cooling the air to low temperatures to liquefy gases such as oxygen and nitrogen. Once liquefied, the gases occupy less space in liquid form than in gaseous form, which improves storage and transportation. These same liquefied gases can be used as coolants in various industrial processes, including to quickly freeze foods. The frozen pizza you eat may have been frozen using liquid nitrogen.

While advances in the use of cryogenics have been made in the medical and industrial fields, its use is also expanding into the fields of energy production and space exploration. How do you think this technology could be used in those fields?

Concept Check

1. Explain what occurs in certain materials when they are exposed to very low temperatures within the defined range of cryogenics.
2. Identify four ways cryogenics is used in the medical and industrial fields.

Plasma in Homes and Businesses

Plasma, an ionized gas, is the fourth state of matter. Plasmas are the most common state of matter in the Universe, more common than solids, liquids, or gases. They are even common here on Earth.

Science and technology have learned to harness the abundance of plasmas. They are in many common items we use every day. Look around your home. Do you have a flat-panel television, a computer, a fluorescent lamp, or synthetic cloth on your furniture or in your clothing? If you do, you are benefiting from the use of plasma.

Plasma is used commercially in various industries. Many businesses use high-efficiency plasma lighting in retail, manufacturing, and warehousing facilities. Diesel engine turbocharger housings are covered with a high-temperature-resistance surface coating using plasma spraying. Robotic arms are used in the high-temperature plasma spraying process. Plasma is also used in large semiconductors to conduct electricity. NASA uses plasma thrusters on spacecraft.

Plasma systems are also used for surface cleaning and sterilization. The plasma systems kill bacteria on various surfaces, destroying viruses, fungi, and spores in seconds. These systems are environmentally friendly because they are less harmful than other detergents and chemicals used for pretreating and cleaning surfaces. The medical and food industries both make use of plasma systems for surface cleaning and sterilization.

Plasma is important in the electronics, automobile, optics, textile, and machine and toolmaking industries, as well as in energy, environmental, and medical technologies. The benefits of plasma technology stem from its amazing potential for flexible, ecological, environmentally friendly, and energy-saving processes.

Concept Check

1. Identify three plasma-containing items commonly found in homes.
2. Explain the benefits of plasma systems used for cleaning.

Chapter 13 Review Assess/Reflect

Study Guide

Lesson 1

1. Atoms form chemical bonds when attractive forces exist between atoms or groups of atoms. Bonding involves the interaction of valence electrons to form atoms that are more stable.

2. In ionic bonds, cations and anions are held together by their attraction for each other. Vast numbers of the ions bond to form a crystal lattice, or crystal. Ionic bonds tend to form between metals and nonmetals. In a metallic bond, each atom of a metal donates its valence electrons to a sea of electrons that moves freely between the positive ions.

Lesson 2

1. Elements form chemical bonds by sharing electrons. This chemical bond is called a covalent bond.

2. When atoms held together by a covalent bond share the electrons equally, the covalent bond is nonpolar. A polar covalent bond forms when the electrons are not shared equally.

3. Complex molecules that contain many atoms and are found in plants and animals are called macromolecules. Macromolecules form when many atoms combine in long chains, rings, or other complex structures. Many of these molecules are based on carbon atoms, each of which can bond with four other atoms. The compounds can have many single, double, and triple bonds between hundreds of different atoms.

Lesson 3

1. A chemical formula indicates the kind and number of atoms in one unit of a compound.

2. A molecular formula shows the elements in the compound and the ratio in which their atoms are present. A structural formula shows how the atoms are arranged to form the compound.

3. To name ionic compounds that consist of only two elements, follow these simple rules: **1.** Name the first element in the compound. The first element is the metal that is located toward the left side of the periodic table. **2.** Change the name of the second element so that it ends in the suffix *-ide*. The second element is the nonmetal that is located toward the right side of the periodic table.

4. A system of prefixes is used to name organic compounds, compounds based on carbon.

5. Molecules with the same molecular formula but different structural formulas are called isomers.

Vocabulary Check

Compare the terms in each pair.

1. anion—cation
2. polar covalent bond—covalent bond
3. ionic bond—metallic bond
4. isomer—chemical formula
5. symbol—formula

Multiple Choice

Choose the best answer.

6. What information does the formula for an ionic compound provide?
   A. number of atoms in the molecule
   B. simplest ratio of atoms in the compound
   C. arrangement of atoms in the compound
   D. type of bonding in the molecule

7. What is the name of NO_3?
   A. nitrogen oxide
   B. trinitrogen trioxide
   C. nitrogen trioxide
   D. trinitrogen dioxide

8. What is the formula for pentane?
   A. C_2H_6
   B. C_5H_{12}
   C. C_3H_8
   D. C_4H_{10}

9. Which is true about the electrons in a nonpolar covalent bond?
   A. shared equally
   B. shared unequally
   C. gained
   D. lost

10. What does the 2 in the formula N_2O describe?
   A. nitrogen atoms in the molecule
   B. oxygen atoms in the molecule
   C. ions in the compound
   D. molecules in the compound

11. What are compounds with the same molecular formula but different structures called?
   A. ions
   B. isomers
   C. molecules
   D. diatomic

12. What is the main reason atoms form chemical bonds?
   A. gain electrons
   B. lose electrons
   C. form molecules
   D. become more stable

Check Point

Answer the following questions.

13. Copper is a good conductor of electricity. It can be hammered into different shapes and pulled into wires. Name and describe the type of chemical bond that gives copper these properties.

Use the **model** to answer Question 14. In the model, black represents carbon (C), aqua represents hydrogen (H), and red represents oxygen (O).

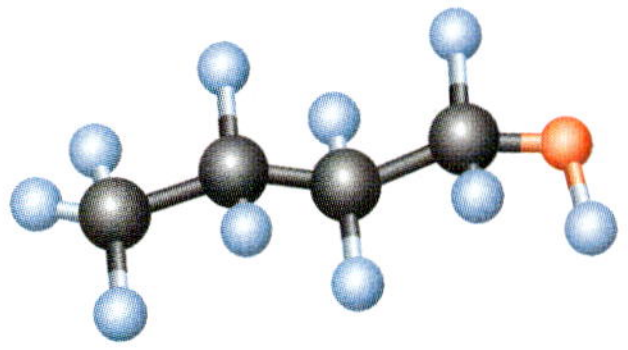

14. A. What is the molecular formula for the compound?
   B. What is the name of the compound?
   C. Draw the structural formula represented by this model.
   D. Describe the types of bonds in the compound.
   E. Draw two other isomers of the compound.

15. The electronegativity of hydrogen is 2.1, and the electronegativity of sulfur is 2.5.
   A. Describe the type of bond that will form between hydrogen and sulfur.
   B. Write the molecular formula for the compound.
   C. Name the compound.
   D. **Predict** the state of matter of the compound at room temperature (solid, liquid, or gas). Explain.

16. Sodium and chlorine combine to form sodium chloride (NaCl).
   A. Explain why sodium atoms and chlorine atoms form chemical bonds.
   B. Explain why sodium bonds with chlorine in a 1:1 ratio, while magnesium and chlorine bond in a 2:1 ratio.

Chapter 14

Chemical Reactions

Scripture Spotlight

You know that God created all matter and that His laws allow that matter to react in different ways. By studying chemical reactions you can see God's creativity and attention to design. In this chapter, you will read the following passages.

Exodus 19:16–18 (p. 508)
2 Corinthians 5:13–17 (p. 513)
2 Peter 3:9–13 (p. 522)
Matthew 6:10–20 (p. 523)

The Big Idea

New substances can form when electrons from different atoms interact. These interactions happen according to physical laws that God established at Creation.

What changes are taking place in this picture that suggest new substances are forming?

Fireworks are a dazzling use of chemical reactions, but as you will learn, not all chemical reactions emit light.

Inquiry Kick-Off

Engage

What chemical reactions do you experience every day? How would life be different if you didn't experience chemical reactions? In your ***Science Journal***, you will experiment with a reaction that takes place in your body.

Science Journal

Lesson 1

Objectives

- Define chemical change.
- Identify indicators of chemical change.
- Compare chemical change to physical change.

Vocabulary

combustion reaction

precipitate

Essential Question

How Can We Observe Chemical Changes?

Think of the last time you saw a cooked egg. What happened to the white of the egg? Why? Now think about the rust that forms on iron if it is left outside in the weather. What material combines with the iron to make the rust? Look at the apple in the picture below. Have you seen this happen before? Why does this change to the apple occur? How can you prevent this from happening? Is this a chemical or a physical change? How do you know?

Chemical Change Explain

Why do bananas turn brown as they ripen? Why do apples get soft, brown spots if they are dropped or mishandled? As you will see, chemical changes are responsible for many of the changes that you observe every day. Although many of these changes can be observed with your own eyes, a chemical change is a process that occurs on a molecular scale. You know that a *chemical change* is a process that:

- alters the composition of atoms or molecules.
- makes new substances that have different properties from the original substances.

How can you tell that a chemical change has taken place on the cut surface of this apple?

Chemical Changes Form New Substances

Is baking a cake a chemical change? A chemical change alters the composition of one or more substances to form new substances. Changing the number of atoms or the type of atoms in a molecule is an example of forming a substance. Changing a substance to this extent involves breaking and making chemical bonds.

Explore-a-Lab

Structured Inquiry

How can you identify a chemical change?

You will need four 250-mL beakers, a plastic spoon, 2 g sodium chloride (salt), 120 mL vinegar, 120 mL distilled water, 60 mL detergent, and 20 dirty pennies. Label the four beakers *1, 2, 3, 4*. In beaker 1, mix 60 mL of detergent with 60 mL of water. In beaker 2, put 60 mL of vinegar. In beaker 3, mix 1 g of salt and 60 mL of water. And in beaker 4, mix 60 mL of vinegar with 1 g of salt. Place 5 pennies in beaker 1. Use the spoon to stir them for 10 sec. Then wait 1 min and use the spoon to remove the pennies. Place the pennies on a paper towel folded in half and pat them dry with another paper towel. Record your observations. Repeat this process with the three other beakers. Explain why the reactions you observed happened. In what beaker do you think a chemical change occurred? Why?

For example, a new substance can be made by combining sodium (Na) with chlorine gas (Cl) to make the compound sodium chloride (NaCl). In this case, new chemical bonds are formed between the sodium atoms and the chlorine atoms. New substances can also be made by breaking bonds. For example, mixing sodium chloride (NaCl) with vinegar ($HC_2H_3CO_2$) breaks the bonds between the sodium and chlorine atoms and makes a solution of hydrochloric acid (HCl) and sodium acetate ($NaCH_3O_2$).

Chemical Reactions

Recall that the process of making or breaking chemical bonds and rearranging atoms in molecules is called a *chemical reaction*. A chemical change is the result of a chemical reaction. Here are some examples of chemical reactions, or chemical changes, that you can easily observe:

- Spaghetti cooks in boiling water.
- Wood burns in a campfire.
- Milk spoils without refrigeration.
- Rust forms on an old bicycle.

Think about your everyday activities. What chemical reactions have you observed since arriving at school today? Can you think of more examples of chemical reactions that you observe every day?

Chemical changes, or chemical reactions, make all parts of your day possible. A common chemical reaction is a combustion reaction. In a **combustion reaction**, molecules of a fuel source combine with oxygen to give off carbon dioxide and water. Combustion reactions, such as burning fossil fuels, heat water to make large amounts of steam to run generators that give us electricity. The cells in your body carry out combustion reactions so that you get energy from the foods you eat.

In this combustion reaction, the candle wax changes into carbon dioxide and water.

Can you think of other examples of combustion reactions?

A combustion reaction is a chemical change that can be easy to observe. A burning candle is an example of a combustion reaction that gives off both heat and light. The candle wax is reacting with the oxygen in the air. During the reaction, the bonds between the atoms in the wax are broken. The carbon and hydrogen atoms in the wax combine with molecules of oxygen in the air to become molecules of carbon dioxide and water.

Chemical Changes and Properties of Substances

What is the same about all chemical changes? They are all reactions that form new substances. This means that during a chemical change, the properties of the substances that form are different from those of the substances you started with.

Can you think of some physical properties of substances? Every pure substance has both physical properties and chemical properties. As you know, *physical properties* describe how and when a substances changes physically. For example, the boiling point of a substance describes the temperature at which it changes from liquid to gas. Also recall that a *chemical property* describes how and when a substance changes chemically. For example, a chemical property such as reactivity to water describes how a substance changes when exposed to water. Other chemical properties include reactivity to oxygen and reactivity to acids and bases.

Rust forming on an old automobile is due to a chemical change in iron. During this chemical change, the iron atoms in the metal parts combine with the oxygen molecules in carbon dioxide and water in the air to form iron oxide. You know iron oxide as the red-brown, flaky substance called rust.

Scripture Spotlight

Read **Exodus 19:16–18**. What chemical reactions happened as God came down to Mount Sinai?

What properties of the metal change when the iron forms rust?

Structured Inquiry

Discover

Record your work for this inquiry. Your teacher may also assign the related Guided Inquiry.

Science Journal

Making Changes

How can you identify a chemical change?

Materials

- 12 test tubes
- iodine solution in dropper bottle
- 3 beakers
- graduated cylinder
- 30 mL of vinegar
- 30 mL of milk
- 30 mL of orange juice
- 30 mL of cornstarch solution
- 30 mL of clean tap water
- 40 g of baking soda

Procedure

1. Label one of the beakers *Iodine*, one beaker *Vinegar*, and the last beaker *Baking soda*.
2. Label three test tubes *Milk*, three test tubes *Orange juice*, three test tubes *Water*, and three test tubes *Starch solution*.
3. Add 10 mL of milk, orange juice, water, and cornstarch solution to one set of test tubes. Place these test tubes in the beaker labeled *Iodine*.
4. Add a few drops of iodine to each of the four labeled test tubes. **Record** your **observations** in the data table in your ***Science Journal***.
5. Repeat Step 3. Place these test tubes in the beaker labeled *Vinegar*.
6. Add 10 drops of vinegar to each of the four test tubes. Record your observations.
7. Repeat Step 3. Place these test tubes in the beaker labeled *Baking soda*. Add 1.0 g of baking soda to each of the four test tubes. Record your observations.

Analyze Results

Review the results in your data table. Which combinations of liquid and testing compound produced a chemical change?

Create Explanations

1. How can you identify a chemical change?
2. Describe how you would design an experiment to find out whether an unknown white powder is cornstarch or baking soda. What results would you expect, and how would you use the results to determine the identity of the white powder?
3. Color change is an indicator of a chemical change. Is adding lime-flavored drink mix to water and making the water green in color a chemical change? Explain your reasoning.

Look at the table of substances and properties. The individual elements sodium and chlorine have properties that are different from the properties of a compound that combines the two elements together, such as sodium chloride. How are the properties of carbon and oxygen different from the properties of carbon dioxide?

Properties of Substances

Substance	Appearance at Room Temp.	Melting Point	Reactivity to Water	Reactivity to Oxygen
Sodium	Silvery white solid	97.72°C	Very reactive, can ignite, forms $NaOH$	Very reactive, forms Na_2O
Chlorine	Yellowish-green gas	−100.98°C	Reactive in some conditions, can form HCl	Unreactive
Carbon	Dark gray solid (graphite)	3550°C	Unreactive in normal conditions	Reactive only at high temperatures, forms CO_2
Oxygen	Colorless gas	−218.79°C	Unreactive	Unreactive
Sodium chloride	White solid	801°C	Dissolves into Na^+ Cl^- ions	Unreactive
Carbon dioxide	Colorless gas	−55.6°C	Reactive, forms a weak acid H_2CO_3	Unreactive

Science Journal

Check out your ***Science Journal*** for a Structured Inquiry that investigates chemical reactions.

Extend

Indicators of Chemical Change **Explain**

You know a chemical change is occurring if there are indicators of a chemical reaction. Many chemical reactions can give off heat, light, or gas. Chemical reactions can also form solids out of a solution or cause colors to change.

Color Change

Do you think dying a white T-shirt blue is a chemical change? How about painting a white wall red? Color change is often the result of a chemical reaction. However, some color changes are not from chemical reactions. For example, paint is different from dye. Paint contains a mix of pigments that make up a specific color. Paint simply covers an old color with a new color, so it doesn't require a chemical reaction. Dye is a chemical color that bonds to a surface and requires a chemical reaction. Dye must be carefully applied to a surface so that it properly reacts and bonds. Many of the colors in textiles are made by treating the surface of the fabric with a chemical dye.

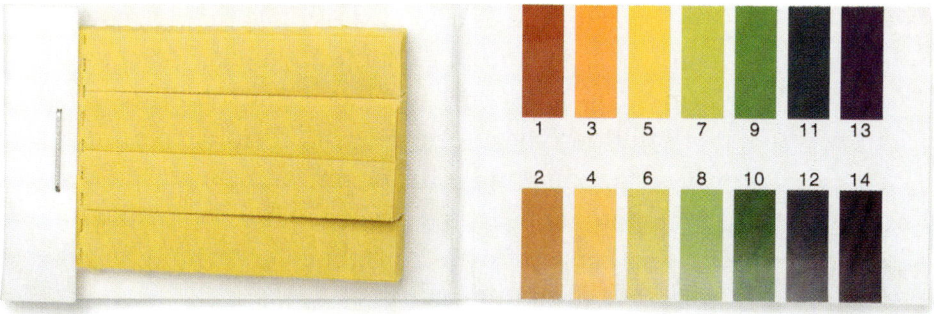

A change in pH is an indication of a chemical change. A pH indicator strip can be used to measure the pH of a solution.

Can you think of some other examples of when observing a color change can be useful?

Some chemical reactions change the color of the starting substances as new substances are formed. Often scientists take advantage of color change by using the substance as an indicator dye. For example, the acidity of a solution can be measured with an indicator dye that changes color depending on the acids or bases in a solution. An indicator dye must change color in a predictable way. This means that the dye changes color only when a specific substance is present. Do you think an indicator dye changes color when it is added to a solution that it is not specific for? Why or why not?

Explore-a-Lab

Structured Inquiry

Is tie-dying a physical or a chemical change?

You need a white T-shirt, assorted powdered drink packages (different flavors/colors), vinegar, small plastic bowls, plastic gloves, and rubber bands. For every color you wish to use, put 1 package of powdered drink and 30 mL vinegar in individual plastic bowls. Mix until dissolved. Pull and twist your T-shirt into different shapes and secure with rubber bands. Dip the areas gathered by the rubber bands into the bowls of color. Make sure you wear gloves so your hands don't get stained. The longer you leave the parts you are dying in the color, the darker the colors will be. Keep the rubber bands on and set your T-shirt in the sun to dry. When dry, remove all of the rubber bands. If you want to make the colors more permanent, iron the dry shirt on medium heat, using a cloth or rag between the shirt and the iron. Let the shirt set for 24 hours before washing. To be safe, wash separately the first time. Are the changes to the shirt physical changes or chemical changes? Could they be both physical changes and chemical changes? Explain your reasoning.

Formation of a Precipitate

A **precipitate** is a solid substance that forms in a solution as a result of a chemical change. Precipitates form because the new substance is insoluble in the solution. The appearance of precipitates can vary depending upon the chemical change. Some precipitates appear as a cloudy substance in solution, while others appear as a bright color forming in the bottom of a solution.

When a solution of potassium iodide is added into a solution of lead nitrate, a chemical reaction takes place, and a new yellow substance, lead iodide, forms. The lead iodide is a precipitate.

Why do you think the lead iodide is unable to disperse in the solution?

Explore-a-Lab

Structured Inquiry

How do precipitates form?

You need 20 mL of baking soda water in a 100-mL beaker and 20 mL of calcium chloride water in a 50-mL beaker. Pour the calcium chloride water into the baking soda water. Observe the solutions as they combine. Record your observations. What results do you have that demonstrate that a chemical change occurred? How would you design an experiment to demonstrate that the substance formed from combining the solutions is not either of the starting substances (baking soda or calcium chloride)?

Gas, Light, and Heat

If you put a clean iron nail in a solution of acid, you can observe tiny bubbles forming on the surface of the nail. Iron reacts with acids and gives off hydrogen gas. The tiny bubbles are hydrogen gas. Bubbles are an indication that a chemical reaction is taking place. What are other examples of chemical reactions that form bubbles?

Light and heat are also indicators of chemical change. When you see a substance ignite, burn, or glow, it is a sign of a chemical change because some types of chemical reactions produce light or heat. Have you ever used a glowstick or a hot pack? Glowsticks contain substances that do not glow until they react with one another. When you snap a glowstick, the substances combine and the stick begins to glow. The stick glows because light is released as part of the chemical reaction between the substances. Hot packs are similar, but when the substances combine, they release heat instead of light. Do you think burning wood in a fireplace is an example of a chemical change? Why or why not?

Explore-a-Lab

Structured Inquiry

How can you make a hand warmer?

You will need a pencil, pencil sharpener, zip-top sandwich bag, salt, water, vermiculite, iron powder, and pulverized activated charcoal. Sharpen your pencil until you have about 40 g of shavings. In the zip-top bag, mix the shavings, 5 g of salt, and 30 g each of vermiculite, iron powder, and charcoal. Seal the bag and shake to mix the contents as evenly as possible. Open the bag and add the water. Reseal the bag, making sure to capture air in the bag. Shake the bag again and observe the reaction. How long did the reaction take to start? What happened? What ingredients caused the reaction?

Scripture Spotlight

Read **2 Corinthians 5:13–17**. Would you describe allowing your heart to be with Christ as being more like a physical or chemical reaction? Explain.

Comparing Physical and Chemical Changes Explain

Imagine you are heating water on a gas stove. A tea kettle full of water sits on the flames of the stove. The water starts to boil and makes steam. Describe what physical and chemical changes are occurring.

Remember that in a physical change, the material or substance does not change into a different substance. Examples of physical change include changes of state, such as boiling, melting, and freezing of water. During a change of state, the atoms in the substance remain in the same chemical bonds in molecules. Physical changes to do not change the arrangement of atoms in molecules, and the substances do not change their chemical properties. In a chemical change, the chemical properties of substances change, and in many chemical changes the arrangement of atoms in molecules changes to form entirely different substances.

The top illustration on the right shows atoms and molecules in a physical change. The bottom illustration shows atoms and molecules in a chemical change. How would you explain the difference between physical and chemical changes using the illustrations?

Physical Change

Solid (ice)	Liquid (water)	Gas (steam)
The water molecules are locked into rigid positions and are close together.	The water molecules are still close together, but can move around to some extent.	The water molecules are far apart and move randomly.

Chemical Change

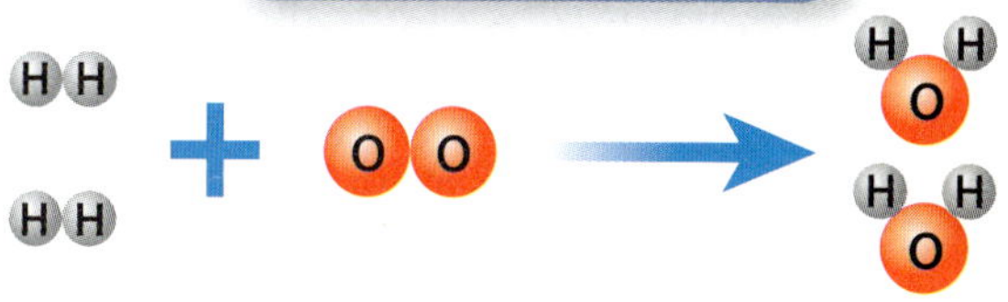

Explore-a-Lab

How can you model physical and chemical changes?

Imagine you have a pot of boiling water over a flame on a gas stovetop. The boiling water is an example of a physical change and the burning methane gas on the stovetop is an example of a chemical change. You will use gumdrops and toothpicks to model what is happening with the molecules in each example. Use white gumdrops for oxygen atoms, yellow gumdrops for hydrogen, and red gumdrops for carbon. Use the gumdrops and toothpicks to make one molecule of methane (CH_4) and two molecules of oxygen (O_2).

Draw the molecules for the example with boiling water before any change occurs.

Draw the molecules for the example with the flame on the stove before any change occurs. Now, using the model molecules, demonstrate the change that occurs in each example. You may or may not have to rearrange the atoms that make up the molecules in the examples. Draw the molecules in each example after each change. Which example models a physical change? Which example models a chemical change? Why? What physical and chemical changes occur when you make pancakes?

Concept Check Assess/Reflect

Summary: How can we observe chemical changes? Chemical changes tell you how and when a substance changes its chemical properties. Because a chemical change is the result of a chemical reaction, it makes new substances out of the starting substances. You can recognize a chemical change by observing indicators of a chemical reaction, such as change in color, the formation of a precipitate, or the formation of gas, light, or heat. Chemical changes are different from physical changes because chemical changes alter the arrangement and bonds of atoms in molecules, while physical changes do not alter the composition of substances.

1. How are new substances made in a chemical change?
2. How could you use the properties of a substance to show that a chemical change has occurred?
3. Why is the formation of gas bubbles in solution an indicator of a chemical change?
4. How is a chemical change different from a physical change?
5. How could you design an experiment to test whether mixing powdered lemonade in water results in a chemical change? Include an experimental set-up, the observations you would record, and the evidence that would be needed to show that a chemical change does or does not occur.

Lesson 2

Essential Question

How Can We Describe Chemical Reactions?

Objectives

- Explain the purpose of chemical equations.
- Describe the law of conservation of mass.
- Describe chemical reactions as endothermic or exothermic.
- Describe common types of chemical reactions.

How would you describe making pancakes? Would you use the words *double-replacement reaction* or *endothermic* in your description? A chemist could describe making pancakes with chemical reactions and describe the changes in matter and energy involved in the reactions. As you will learn, chemical reactions can be described by chemical equations and by how substances and energy change during a reaction.

Chemical Equations Explain

Chemical reactions change the composition of a substance. Scientists describe chemical reactions using chemical equations. A *chemical equation* is a shorthand way to communicate how substances change in a chemical reaction. Similar to the math equations you are familiar with, in chemistry, a chemical equation is called an equation because it shows that there are two sides to every reaction. Chemical equations are balanced between these two sides, just as math equations are.

What ingredients go into making pancakes? What do you get as a result of mixing and cooking these ingredients? Just like the process of making pancakes, chemical reactions can also be described by the starting ingredients and the end result. Recall that the *reactants* are the substances that go into and are used up in the reaction. The *products* are the substances that are formed as a result of the reaction.

Making pancakes involves a chemical reaction.

What are the reactants? What is the product?

Scientists write chemical equations based on the reactants and products in the reaction. Since the reactants are the starting substances, scientists write the reactants on the left side of a chemical equation. This way, you read the reactants before you read the products. Scientists write the products on the right side of a chemical equation to show that the products form as a result of the reaction. In the example below, methane and oxygen gas are the reactants, and carbon dioxide and water are the products.

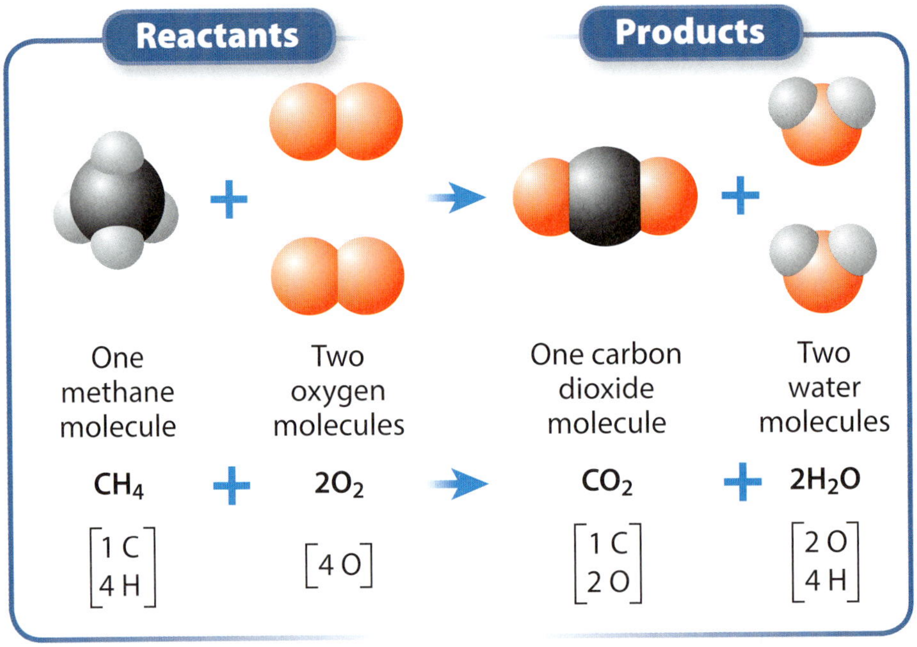

Do you know why there is a large number 2 in front of the oxygen and the water molecules in the equation? The number 2 is a *coefficient*. In chemical equations, the coefficient tells you how many molecules of each substance are in the reaction. You can use the coefficient to find the number of atoms of each element by multiplying the coefficient by the subscript of each atom. Recall that if there is no subscript, this means there is one atom of that element in the molecule.

As you can see, chemical equations are written with the chemical symbols for each element and the chemical formula for each compound in the reaction. The arrow between the reactants and the products means "yields," and a + sign on the left side means "reacts with," while a + sign on the right side means "and." The physical state of a substance is often written next to the substance in parentheses: solids (s), liquids (l), and gas (g). Read the chemical equation below.

$$2Na\ (s) + Cl_2\ (g) \rightarrow 2NaCl\ (s)$$

How could you use gumdrops and toothpicks to model the chemical reaction of sodium with chlorine?

Math in Science

Look at the three chemical equations below. How many atoms of each element are in the reactants? How many atoms of each element are in the products?

$$2H_2 + O_2 \rightarrow 2H_2O$$

$$2Fe_2O_3 \rightarrow 4Fe + 3O_2$$

$$Cl_2 + 2NaBr \rightarrow 2NaCl + Br_2$$

Conservation of Mass **Explain**

In the reaction of methane and oxygen gas, how do the numbers of each kind of atom compare in the reactants and products? You learned earlier that in chemical reactions, the law of conservation of mass states that no matter is created or destroyed in the reaction. Scientists show conservation of mass by writing balanced chemical equations for reactions. A balanced chemical equation shows that the number of atoms for each element is equal on both sides of the equation.

Balancing Chemical Equations

How would you write the chemical equation for the reaction between aluminum metal and oxygen gas to produce aluminum oxide? The chemical symbol for aluminum is Al, the chemical symbol for oxygen is O, and the compound aluminum oxide is written as Al_2O_3. So you might be tempted to write the following equation:

$$Al\ (s) + O_2\ (g) \rightarrow Al_2O_3\ (s)$$

Is this equation balanced? Count the number of atoms for each element on both sides of the equation. Is there an equal number of atoms for each element on both sides? The correct balanced equation is:

$$4Al\ (s) + 3O_2\ (g) \rightarrow 2Al_2O_3\ (s)$$

Remember that coefficients show the number of molecules of each substance in a reaction. Scientists use coefficients to make the number of atoms of each element on both sides of the equation equal. How many atoms of oxygen are on each side of the balanced chemical equation for aluminum oxide?

Using what you have learned about coefficients and chemical equations, look at the three chemical equations below. Which chemical equations are balanced?

A. $C_2H_4\ (g) + O_2\ (g) \rightarrow CO_2\ (g) + H_2O\ (l)$

B. $2C_2H_6\ (g) + 7O_2\ (g) \rightarrow 4CO_2\ (g) + 6H_2O\ (l)$

C. $PCl_3 + 3H_2O \rightarrow H_3PO_3 + 3HCl$

It might help you to make a chart to count the number of atoms of each element on each side of the equation. Below is an example chart for equation A.

$C_2H_4 + O_2 \rightarrow CO_2 + H_2O$		
Type of Atom	**Reactants**	**Products**
C	2	1
H	4	2
O	2	3

You can see that there are unequal numbers of atoms of each element in the reactants and products, which means that equation A is not balanced. Complete a chart to see whether equations B and C are balanced.

Energy and Chemical Reactions **Explain**

How would you describe a moving object? Would you use energy to describe a boulder sitting on top of a large hill? Just as scientists use energy to describe moving objects, scientists also describe chemical reactions in terms of energy. In a chemical reaction, the bonds between the atoms in molecules are broken or made, and new molecules are formed. Making and breaking bonds requires energy. This means that in all chemical reactions, some amount of energy changes from reactants to products. This energy is usually in the form of heat. What chemical reactions release heat and help you cook food?

Exothermic and Endothermic Reactions

Similar to how total mass is conserved in chemical reactions, the total energy between a reaction and the surrounding environment is also conserved. Chemical reactions can either release heat into the environment or absorb heat from the environment. Whether a particular reaction releases or absorbs heat depends on the energy difference between the reactants and the products.

If the reactants have more energy than the resulting products, it is an *exothermic reaction*. These reactions release energy in the form of heat. Exothermic reactions feel hot to the touch because heat is released into the environment. If the reactants have less energy than the resulting products, it is an *endothermic reaction*. These reactions absorb energy in the form of heat. Endothermic reactions feel cold to the touch because heat is absorbed from the surrounding environment to use in the chemical reaction.

Structured Inquiry
Discover

Record your work for this inquiry. Your teacher may also assign the related Guided Inquiry.

Science Journal

Where Did It Go?

How can you verify the law of conservation of mass?

Procedure

Trial 1

1. Pour 30 mL of vinegar into the beaker. Place about 12 g of baking soda into about a 1/3 section of the tissue and fold the contents into a neat little packet. Use the balance to determine the mass of the beaker, vinegar, and baking soda. **Record** the mass of each in your ***Science Journal.***

2. Place the packet into the beaker and allow the reaction to complete. Wait three minutes. Measure and record the mass of the remaining products.

Trial 2

3. Pour 30 mL of vinegar into the 2-L bottle, using the funnel (to avoid getting the neck of the container wet). Prepare another packet of baking soda in a tissue as in Step 1. **Measure** and record the total mass of these reactants and the 2-L bottle.

4. Tilt the bottle (as much as possible without spilling the contents), insert the packet of baking soda into its neck, and tightly secure the cap.

5. Tip the bottle upright and allow the contents to fully react. Wait three minutes. Measure and record the mass of the container with products.

Materials

- 2-L bottle
- baking soda
- balance
- balloon (12 in)
- beaker (600 mL)
- Erlenmeyer flask (250 mL)
- funnel
- graduated cylinder
- safety goggles
- tissue paper
- vinegar
- weighing dish

Analyze Results

Compare the results of the two trials. Which trial has results that demonstrate the law of conservation of mass? How do you know?

Create Explanations

1. How can you verify the law of conservation of mass?

2. Why was there a difference between the mass of the reactants and the mass of the products in Trial 1?

3. Why do you suppose the mass of the products compared as they did with the mass of the reactants in Trial 2?

4. How would you design an experiment that could confirm the conservation of mass with a smaller margin of error?

The transfer of heat from a reaction to the environment can be measured as a temperature change in the environment. Would you expect the temperature of the surrounding environment to increase or decrease with an endothermic reaction?

Look at the diagrams below that show the energy change during exothermic and endothermic reactions. Given what you know, which diagram would best show the combustion of candle wax?

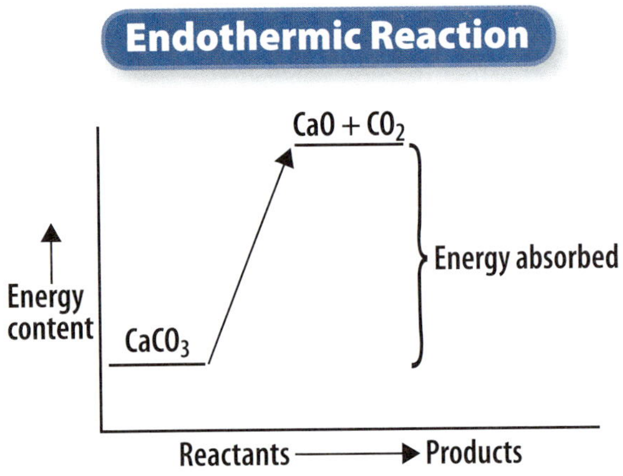

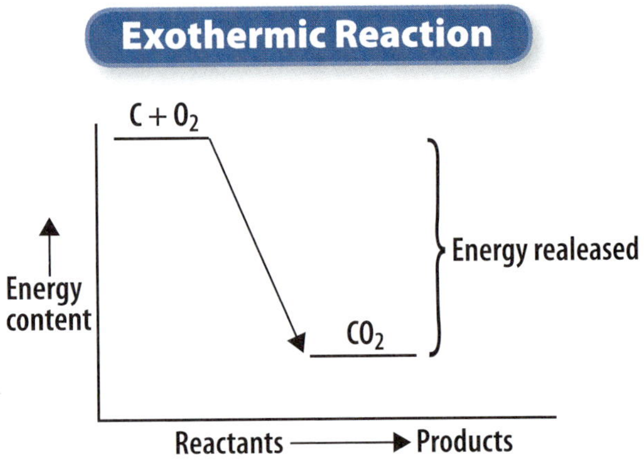

Explore-a-Lab

Structured Inquiry

How do we identify endothermic and exothermic reactions?

You will need two insulated-foam cups, a thermometer, potassium chloride (KCl), calcium chloride ($CaCl_2$), glass stirring rods, and a 100-mL beaker of water. Make a data table to record the starting and ending temperatures of the two reactions that will be tested.

Wear gloves and safety goggles while working with these chemicals. Work with a partner and add 50 mL of water to a plastic foam cup. Hold the thermometer in the cup and record the starting temperature of the water. Remove the thermometer and add 35 g of potassium chloride to the cup. Use the stirring rod to mix the solution. Place the thermometer in the cup. Observe and record the temperature of the solution. Record the temperature every minute for 5 min. Repeat the procedure using a new cup with 50 mL of water and 35 g of calcium chloride. Find the change in temperature for each reaction. Which reaction is exothermic? Which reaction is endothermic? Make a graph to show the temperature change over time for each reaction.

Types of Chemical Reactions **Explain**

How would you describe the weather today? Would you describe how sunny it is or what the temperature is outside? As you know, there are many ways to describe the weather. Similarly, there are many ways to describe how substances change in chemical reactions. In general, though, scientists describe the way substances change in two ways. One way is to describe how the atoms change their bonds with other atoms. A second way is to describe how electrons are exchanged between atoms. The names *combustion*, *synthesis*, *decomposition*, and *single- or double-replacement reactions* all describe how atoms of reactants change bonds. The name *oxidation-reduction reaction* describes how electrons are exchanged between atoms of reactants. As you learn about types of reactions, keep in mind that many reactions can be described both by how atoms change bonds with one another and by how electrons are exchanged.

What is the fuel source in this combustion reaction?

Combustion Reactions

When you observe something burning, you are actually observing a combustion reaction. Some of the reactions that break down sugar in living cells are also combustion reactions. Can you name a combustion reaction that occurs in plants?

Combustion reactions have three requirements: a fuel source, oxygen, and heat. Fuel sources can be hydrocarbons, or molecules that are made up of many carbon and hydrogen atoms, such as methane or propane.

Explore-a-Lab

Structured Inquiry

How do you think the way a substance combusts is related to chemical energy?

You need a candle, matches, a dissection pick (or long pin or needle), a small marshmallow, a peanut, and a glass dish or metal tray to work over. You will see how long each food burns by lighting it in the flame of the candle. Before you do this, predict which food will burn faster and explain your prediction. Then put the marshmallow on the end of the pick. Use the matches to light the candle. Hold the marshmallow in the flame until it is lit and remove it from the flame. Hold the burning marshmallow over the glass dish or metal tray so that no burning parts fall onto a flammable surface. Repeat the procedure with the peanut. Note how the peanut burns differently from the marshmallow. Do your observations agree or disagree with your predictions? Why do you think the marshmallow and the peanut burn differently?

Fuel sources can also be substances such as glucose, which, as you know, is a simple sugar found in many types of cells. The reactants of the reaction are the fuel source and oxygen. The products of the reaction are carbon dioxide and water. Here are some examples of chemical equations for combustion reactions.

Combustion Reactions

Combustion of propane: $C_3H_8 + 5O_2 \rightarrow 3CO_2 + 4H_2O$

Combustion of glucose: $C_6H_{12}O_6 + 9O_2 \rightarrow 6CO_2 + 6H_2O$

Scripture Spotlight

As described in **2 Peter 3:9–13**, what elements would burn if Earth was set on fire?

Synthesis and Decomposition Reactions

In synthesis reactions, two elements or compounds react to form a single product. The formation of sodium chloride (NaCl) and the formation of water are examples of synthesis reactions.

Synthesis Reactions

$2Na\ (s) + Cl_2\ (g) \rightarrow 2NaCl\ (s)$

$2H_2\ (g) + O_2\ (g) \rightarrow 2H_2O\ (l)$

A decomposition reaction is the opposite of a synthesis reaction; it occurs when a compound breaks into two or more simpler substances. There is usually only one reactant in decomposition reactions. For example, when you heat magnesium chloride, it breaks down into magnesium and chlorine.

Explore-a-Lab

Structured Inquiry

How can you observe a synthesis reaction and a decomposition reaction?

You need two evaporating dishes, tweezers, a dropper or pipette, steel wool, 10 mL of hydrogen peroxide (10%) in a small beaker, 50 mL of vinegar in a 100-mL beaker, and potato slices. Make a marble-sized ball of steel wool and soak it in the beaker of vinegar for a few seconds. Use the tweezers to remove the steel wool and place it in one of the evaporating dishes. Record any changes in the steel wool over the next half hour. Start a different reaction by adding a potato slice to the other evaporating dish. Use a dropper or pipette to add a few drops of hydrogen peroxide on the top of the potato slice. Record any indicators of a chemical reaction that you can observe. Steel wool is iron. What do you think the steel wool is reacting with? Hydrogen peroxide (H_2O_2) is very light sensitive. What do you think happens when it is exposed to light? Which reaction do you think is the synthesis, and which do you think is the decomposition? Explain how your observations helped you make each conclusion.

Decomposition Reaction

$$MgCl_2\ (s) \rightarrow Mg\ (s) + Cl_2\ (g)$$

Single- and Double-Replacement Reactions

In single-replacement reactions, a single element replaces another element of a compound. There are usually two substances as reactants: an element and an ionic compound. For example, chlorine can replace sodium in sodium bromide to form sodium chloride and bromine.

Single-Replacement Reaction

$$Cl_2 + 2NaBr \rightarrow 2NaCl + Br_2$$

In order for a single-replacement reaction to occur, the single element must be more reactive than the element that is in the compound. Chemists have charts that list elements by their reactivity, and these charts help them determine whether a single element can replace another element in a compound. For example, copper (Cu) is more reactive than silver (Ag). Because it is more reactive than silver, a copper atom can replace the silver atom in the compound silver nitrate, $AgNO_3$.

Scripture Spotlight

What does **Matthew 6:10–20** say happens to treasures gathered on Earth? What types of "treasures" might be better to accumulate on Earth?

Single-Replacement Reaction

$$Cu + 2AgNO_3 \rightarrow 2Ag + Cu(NO_3)_2$$

In double-replacement reactions, two compounds exchange elements. Most of these reactions occur in aqueous solutions, which means that the compounds are dissolved in water. Chemists use the symbol (aq) to show that the compounds in the reaction are aqueous. Since these reactions are in water-based solutions, if a product is insoluble it often is a precipitate. The following example shows that the precipitate silver chloride forms in a double-replacement reaction.

Double-Replacement Reaction

$$AgNO_3\ (aq) + NaCl\ (aq) \rightarrow AgCl\ (s) + NaNO_3\ (aq)$$

When copper wire is placed in aqueous silver nitrate, the product is silver and copper nitrate.

What type of chemical reaction is this?

Explore-a-Lab

Structured Inquiry

Can you identify the substances from two different chemical reactions?

You will need two test tubes, gloves, two droppers, a stir stick, safety goggles, a small piece of steel wool (Fe), a solution of potassium iodide (0.1 M), and a solution of copper II sulfate (0.1 M). Put on your gloves and safety goggles. Label the test tubes *Fe* and *KI*. Use a dropper to add copper sulfate solution to each test tube so that there is about 3 cm of solution in each tube. Add a small piece of steel wool to the "Fe" test tube, using the stir stick to help move the steel wool down the tube if necessary. Use a clean dropper to add 2–3 droppers of the solution of potassium iodide to the "KI" test tube. Allow both test tubes to sit for five minutes, and record all of your observations for both test tubes. What substances do you think formed in each of the solutions? Which reaction do you think was a single replacement, and which reaction do you think was a double replacement? Explain how you reached your conclusions.

Concept Check

Assess/Reflect

Summary: How can we describe chemical reactions? Chemical reactions can be described by matter and energy changes. Some descriptions are true for all chemical reactions. Matter is conserved within all chemical reactions, and energy is conserved within the overall reaction and its surrounding environment. Chemical reactions can be described by the energy, or heat, change that occurs. All chemical reactions transfer energy between the reaction and its surrounding environment. Exothermic reactions release heat into the environment, while endothermic reactions absorb heat from the environment. You can identify heat changes in a reaction by measuring the change in temperature surrounding a reaction. Chemical reactions can also be described by how matter changes, or the type of reaction that occurs. Common types of chemical reactions include combustion, synthesis, decomposition, single replacement, and double replacement.

1. How could you describe, in three different ways, the reaction that occurs when heating magnesium and oxygen to form magnesium oxide?
2. How does balancing chemical equations support the law of conservation of mass?
3. How would you set up an experiment, measure, and make calculations to determine whether a reaction is an exothermic reaction?
4. How can you distinguish between a single-replacement and a synthesis reaction?

Lesson 3

Essential Question

How Can We Control Chemical Reactions?

What do you think happens when you throw two bouncy balls around in a small room? Do you think the balls would collide with one another? Do you think you could avoid being hit by the balls? What if you throw 20 bouncy balls around the same room? How do you think colliding bouncy balls could model molecules starting a chemical reaction? For a chemical reaction to occur, the particles of reactants need to collide. Particles of reactants can be atoms, molecules, or ions. Anything that changes the chances of reactant particles colliding can affect how the reaction proceeds.

Objectives

- Identify the factors that affect the rate of chemical reactions.
- Explain the effect of surface area, temperature, and concentration on the rate of chemical reactions.

Vocabulary

rate of reaction

concentration

activation energy

You can imagine that many collisions are happening in this room as the balls bounce off of one another.

How are these bouncing balls like combining the atoms of two liquids together?

Factors That Affect Chemical Reactions Explain

All reactions need energy to start because energy is needed to make or break the chemical bonds between atoms. Much of the energy used to start changing chemical bonds is in the form of kinetic energy. You might recall that kinetic energy is the energy of motion. The kinetic energy to start a chemical reaction comes from the motion of the particles of the reactants. The faster the particles of the reactants move, the greater the kinetic energy of the reactants. Can you think of a way to increase the kinetic energy of a reactant?

The Start of a Chemical Reaction

Not every collision results in a chemical reaction. Two things need to happen in order to start a chemical reaction. First, the reactant particles must collide with enough energy to break or make chemical bonds. The amount of energy needed depends on the properties of the elements and compounds in the reaction. Second, the reactant particles must come together with the correct orientation. This means the atoms or molecules must collide so that their reactive parts come together.

Reactants and Energy

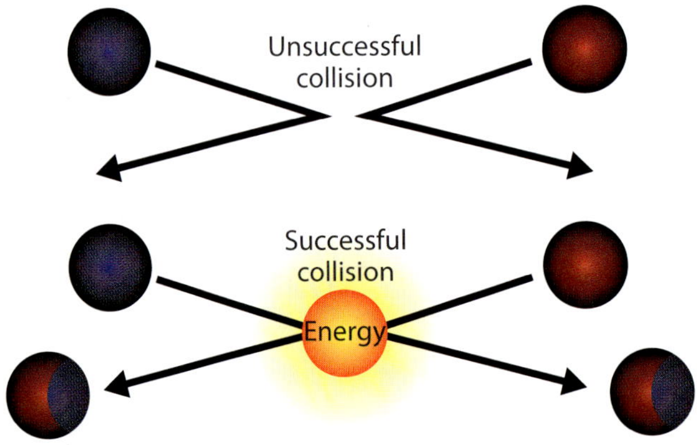

Why does only the second situation in the diagram result in a chemical reaction?

Rate of Reaction

How long does it take you to walk 1 kilometer? How can you measure your walking speed? You can find the speed that you walk by measuring your walking rate, or the time it takes you to walk a certain distance. Speed can also be used to describe chemical reactions. The speed of a chemical reaction can be measured by the rate at which reactants change into products. Chemical reactions occur at different speeds. Some reactions take longer than others to start and finish.

Scientists are often interested in the time that it takes to start reactions and to complete reactions. This is because scientists want to control how fast the products of a reaction are formed. Scientist refer to the speed of a chemical reaction as the **rate of reaction**.

If you want to control chemical reactions, it is important to understand what factors affect the rate of reaction. Any factor that changes the likelihood of collision between the reactants changes the rate of reaction. Four general factors affect the rate of reaction: temperature, concentration, surface area, and catalysts. How might scientists use these factors to control chemical reactions?

Temperature

Do strawberries ripen faster in warmer temperatures or in cooler temperatures? How does temperature affect the ripening of fruit? Recall that heat energy affects the kinetic energy of molecules. This means that adding or removing heat from a chemical reaction affects the rate of reaction. Here is how it happens: When you add heat energy to a substance, you increase the kinetic energy of the particles. This results in an increase in the frequency of collisions between the particles. Increasing the frequency of collisions increases the chances of molecules colliding in a way that starts a chemical reaction. How could you control the chemical reaction that spoils milk?

Explore-a-Lab

Structured Inquiry

How does temperature affect the rate of reaction?

You will need two effervescent tablets, a cup of cold tap water and a cup of hot (but not boiling) water, a thermometer, and a stopwatch or a clock. Label the cups *cold* and *hot*. You will observe and measure how the tablet reacts with water at two different temperatures. Make a data table to record the temperature of water, time to react, and other observations for each temperature of water tested. Place the thermometer in the cold water and record the temperature. Remove the thermometer from the cup. Add the tablet to the cold water and start the stopwatch or look at a clock. Observe the reaction until it appears to be complete. Record the time it took to complete the reaction. Repeat this procedure with the second tablet and the cup of hot water. Which temperature had the faster reaction rate? Explain how increasing or decreasing the temperature of the water changes the rate of reaction.

Concentration

Think again of the bouncy balls in a small room. When do more collisions occur—when there are 2 balls or 20 balls bouncing around the room? **Concentration** is the amount of substance in a given volume. In chemical reactions involving solutions, the likelihood of molecules colliding with one another is related to the concentration of the reactants in solution. When more molecules of reactants are in a given volume, the chance of the reactants colliding is greater. Look at the illustration below. Would you increase or decrease the concentration of a solution to slow down the rate of reaction?

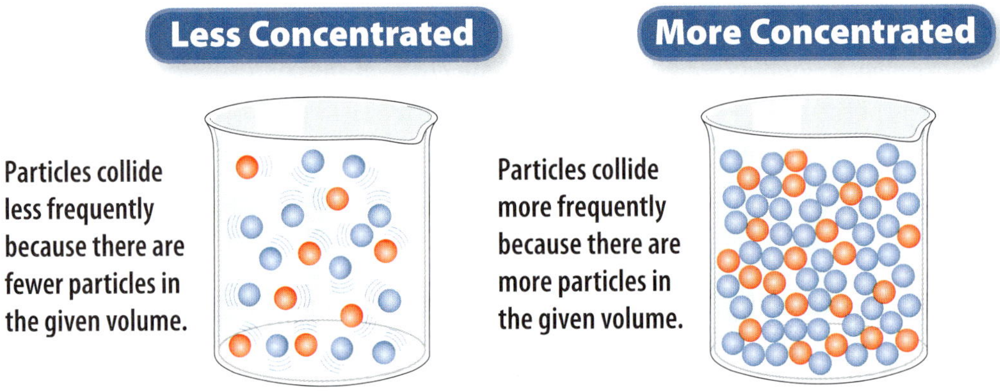

Surface Area

Why does chopping vegetables into smaller pieces make them cook faster? Another way to increase the rate of reaction is to increase the surface area of the reactants. Surface area applies to reactants in solid form. When more surface of a solid reactant is exposed, more collisions are possible. Increasing the frequency of collisions increases the rate of reaction. The illustration below shows a solid reactant and a reactant in solution. When the solid reactant is in a single clump, it has a limited surface area exposed for the molecules of reactant in solution to collide with. When the solid reactant is in smaller pieces, there is more surface area exposed for molecules of reactant in solution to collide with.

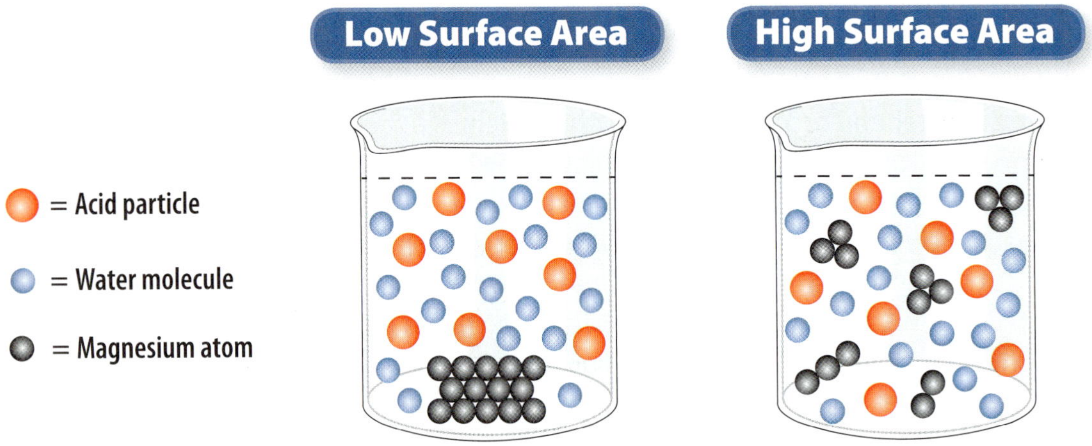

Structured Inquiry

Discover

Record your work for this inquiry. Your teacher may also assign the related Guided Inquiry.

Science Journal

Concentrate on Concentrations

How does concentration affect the rate of reaction?

Procedure

1. Label the cups *100*, *50*, *25*, and *water*. You will test the reaction of baking soda and vinegar in four different concentrations of vinegar solution to determine how concentration affects the rate of reaction.
2. Make a table in your ***Science Journal*** to **record** the time it takes to complete the reaction for each concentration of vinegar solution.
3. Write a **prediction** for how which concentrations of vinegar will result in a faster reaction.
4. **Measure** 5 g of baking soda in a small cup. Repeat this so you have 4 cups with 5 g of baking soda in each.
5. Fill each clear plastic cup to half full with the appropriate vinegar solution. Line up the cups and check that all of the cups have an equal volume of solution.
6. Add 5 g of baking soda to the plastic cup with 100% vinegar and start the stopwatch or watch the time on a clock.
7. **Observe** the reaction. When it stops bubbling, record the total time it took to complete the reaction.
8. Repeat the procedure with each cup of vinegar solution.

Materials

- 4 cups, clear plastic
- balance
- 4 cups, small
- stopwatch or clock
- 50 mL 100% vinegar solution
- 50 mL 50% vinegar solution
- 50 mL 25% vinegar solution
- 50 mL of tap water
- 20 g baking soda

Analyze Results

Compare the results for each concentration of vinegar in your data table. How is concentration related to the rate of the reaction you tested? Was your prediction correct or incorrect?

Creating Explanations

1. How does concentration affect the rate of reaction?
2. How can changing the frequency of collisions change the rate of reaction?
3. Imagine a reaction is happening so fast that you can't accurately observe or measure it. Explain how you could make the reaction more easily observable.

Explore-a-Lab

Structured Inquiry

How does surface area affect the rate of reaction?

You will need two antacid tablets, two small beakers, 100 mL of hydrochloric acid (HCl), a stopwatch, a mortar and pestle, gloves, and goggles. Put on the gloves and goggles. Pour 50 mL of HCl into each small beaker. Add a whole antacid tablet to one beaker of acid and start the stopwatch. Observe the reaction, and record the time it takes to complete the reaction. Place the other antacid tablet in the mortar and use the pestle to crush the tablet as best you can. Tap the powder out of the mortar onto a small piece of paper. Pick up the sides of the paper to form a chute and tip the paper over the second beaker so that the powder slides into the beaker of acid. Observe the reaction, and record the time it takes to complete the reaction. How does crushing the antacid tablet affect the time it takes to react with the acid? Explain how surface area affects the rate of reaction.

Activation Energy

To start a chemical reaction, reactant particles need to collide with enough energy to break or make new chemical bonds. If reactants collide without enough energy, they just bump into one another and no reaction occurs. The energy needed to start a chemical reaction is called the **activation energy**. Scientists use diagrams with the activation energy to show how the energy of a reaction changes over the course of the reaction. The diagram below is an example of the energy of a reaction as the reaction moves from reactants to products.

Energy of Activation

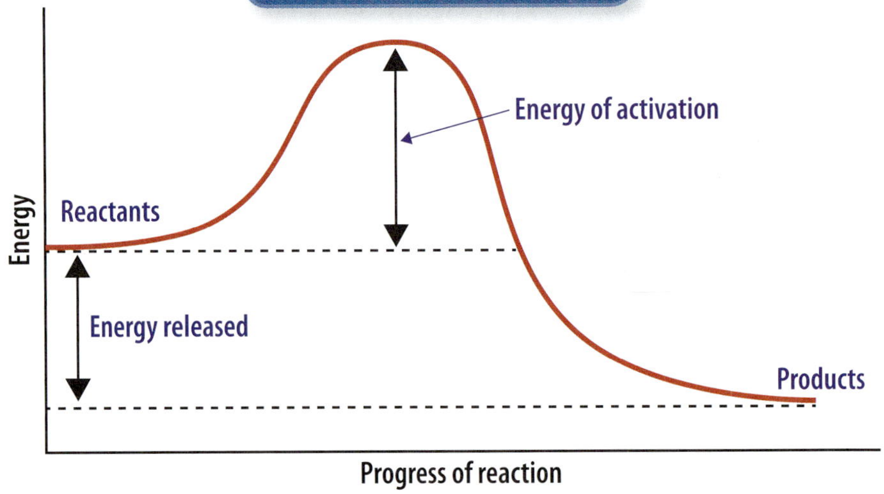

In the diagram, the activation energy looks like a hill where the reactants come together. For the reaction to proceed, the reactants must have enough energy to make or break the chemical bonds that allow the products to form. This diagram is for an exothermic reaction. You can see that the reactants have more energy than the products. The remaining energy from the reaction is released as heat. Draw a diagram that shows the energy of an endothermic reaction over the course of a reaction.

Explore-a-Lab

Structured Inquiry

How is dropping a popper like starting a chemical reaction?

Work in groups and get a popper and a meterstick. Predict the height the popper will bounce if you drop it straight down. Decide on a starting height to drop the popper from every time you drop it. Then drop the popper without turning it inside out and observe the height at which it returns. Repeat the drop a few times and then estimate the average height it reaches. Turn the popper inside out. Predict the height you think it will reach. Drop the popper with the bulge pointing upward. Repeat the drop a few times and estimate the average height it reaches. How did you observe the difference in the amount of energy released? If the popper represents a chemical reaction, how does the height the popper reaches represent the activation energy?

Catalysts

Recall that a *catalyst* is anything that increases the rate of reaction without being changed by the reaction. Catalysts are not reactants or products in the reaction, which means they can be used over and over in the same reaction. Remember at the start of a chemical reaction, there must be enough energy in the collision of reactants to start a chemical reaction. Catalysts work by allowing the reactants to collide in a way that requires less energy to start the chemical reaction.

Catalysts are very important to many industries, including the manufacture of chemicals, the preparation of fuel sources, and the design of new medicines. Catalysts in living systems are important to many of the reactions that take place in cells. The catalytic converter in cars is an example of a metal catalyst; the element platinum is often the catalyst in this reaction. Platinum is often coated onto metal plates and used in reactions that involve hydrogen or oxygen gases.

The catalytic converter in a car changes carbon monoxide into carbon dioxide by passing carbon monoxide and oxygen gas through a platinum-coated honeycomb. The catalyst works by holding both reactants on the plate. Because of the weak bond to the platinum, the oxygen molecules break into two oxygen atoms. These atoms can then move across the plate and collide with carbon monoxide, forming carbon dioxide. The new molecule of carbon dioxide releases from the platinum because it is held by only a weak bond. Why is the platinum reusable after the reaction takes place?

Without the platinum catalyst, the reactant molecules of oxygen gas and carbon monoxide would need to collide with a much greater energy to break the bonds between the oxygen atoms. The catalyst allows this reaction to occur with a lower-energy collision, and this is how it increases the rate of reaction.

Catalytic Converter

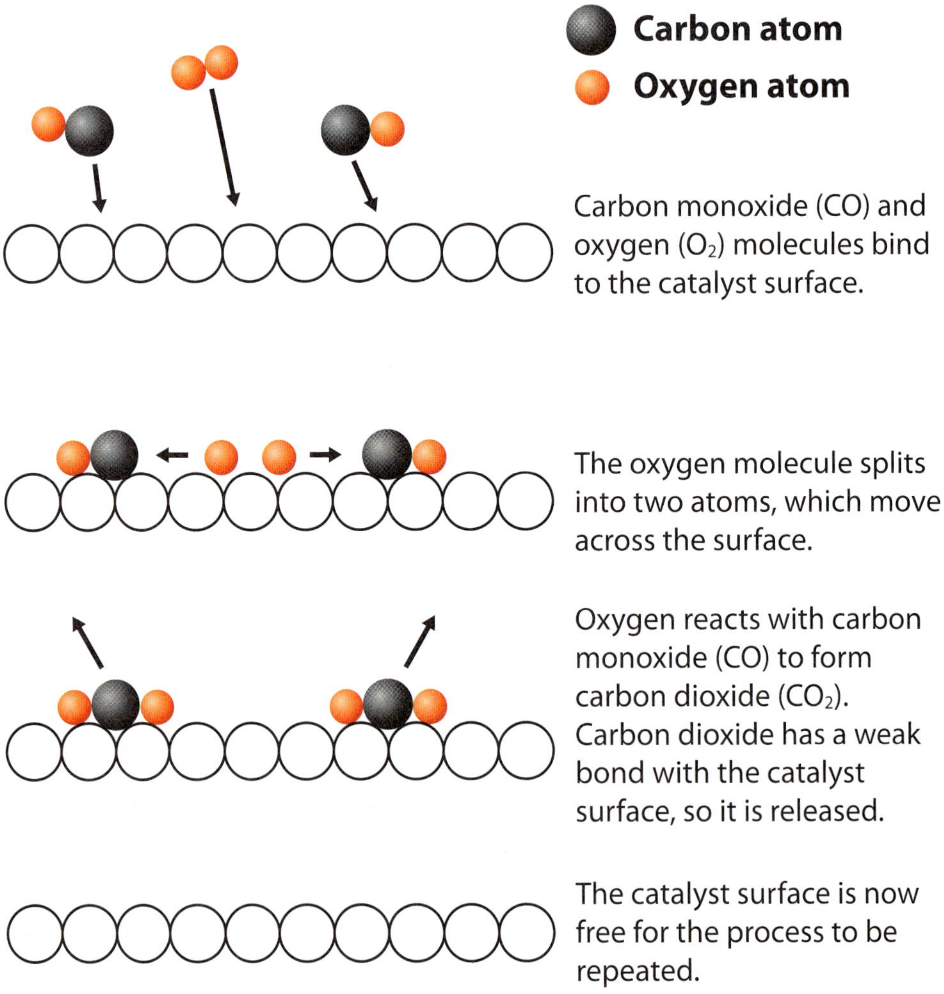

Explore-a-Lab

Structured Inquiry

How do catalysts affect chemical reactions?

Make two balls of loosely formed steel wool about the size of a large marble. Place both balls of steel wool in a beaker of vinegar and allow them to soak for about 1 min. Use tweezers to remove the balls of steel wool and shake them dry over a paper towel. Place a piece of folded paper towel in the bottom of a Petri dish. Place one of the dried steel wool balls on the folded paper towel in the Petri dish. Place the other steel wool ball in a beaker of clean tap water. Let the balls set undisturbed for 24 hr. At the end of 24 hr, compare the appearance of the two balls. Record your observations. Do the balls look the same? Why or why not? Write the chemical equation for the reaction that has occurred. A catalyst is at work in one of the reactions. Which one? What is the catalyst that helped the reaction to occur?

Concept Check Assess/Reflect

Summary: How do we control chemical reactions? You can control chemical reactions by changing the frequency with which reactant particles collide. Increasing the frequency of collisions increases the chances that a collision makes a chemical reaction and so increases the rate of reaction. There are three main factors that can change the frequency of reactants colliding with each other: temperature, concentration, and surface area. Another factor that can affect the rate of reaction is a catalyst. Catalysts work by allowing a chemical reaction to occur with less energy than is usual for the collision of reactants. Scientists use these factors to control the rates of reaction for many reasons, including in laboratory experiments and in designing new chemicals or medicines.

1. How and when do chemical reactions occur?
2. How does increasing the temperature surrounding a reaction change the frequency of collisions between reactant particles and increase the rate of reaction?
3. What are some ways you could make a reaction between solid sodium chloride and bromine gas proceed at a faster rate?
4. What are some ways you could slow down a reaction between a solution of citric acid and a solution of sodium bicarbonate?
5. How is it possible that a catalyst can be used over and over in the same reaction?

People in Science

Extend

Get to Know
Donald B. DeYoung

Dr. Donald B. DeYoung is a creationist scientist specializing in solid state and nuclear science and astronomy. He is chairman of the Department of Physical Science at Grace College and Seminary in Indiana, where he has taught since 1972.

DeYoung earned a doctorate degree in physics from Iowa State University, and both a master's degree and a bachelor's degree in physics from Michigan Technological University. He is well known for his books and articles supporting creationist views, but this was not always the case.

DeYoung was not a six-day creationist at the start of his physics training. His studies focused on the big bang origin of the Universe and our animal ancestry. He was not aware of the creationist view. A colleague challenged him to consider the possibility of supernatural creation, and it has been the center of his attention ever since.

Called to Serve

Dr. DeYoung's research and writings have helped people open their minds to the possibilities of the creationist point of view. His research in a wide range of scientific fields has convinced him and provides evidence supporting his belief in a literal six-day Creation.

The two most basic laws in science, the thermodynamic laws of nature, are the basis for DeYoung's scientific belief in creationism. The first law of thermodynamics states that energy is conserved or constant at all times; energy can be neither created nor destroyed. The second law of thermodynamics states that everything deteriorates, breaks down, and becomes less ordered over time. From these laws, DeYoung discerned that energy did not spontaneously begin, and it will not last indefinitely. He points out that secular and natural science offer no acceptable explanations. Instead, these laws are consistent with the biblical six-day Creation.

Dr. DeYoung continues to chair the Department of Physical Science at Grace College, teach, and write. He has authored many articles for periodicals such as the *Journal of Chemical Physics* and *Creation Research Society Quarterly*. He has also written several books on Bible science topics, including one for children. Some of his most popular books are *Weather and the Bible, In Six Days: Why Fifty Scientists Choose to Believe in Creation, Astronomy and the Bible,* and *The Moon: Its Creation, Form and Significance*.

Concept Check

1. What event opened DeYoung's mind to the creationist view?
2. Identify and explain the two scientific laws that are the basis for DeYoung's work.

Science and Technology

Extend

Nuclear Power Plants

Each year since 1990, about 20% of the electricity in the United States is generated by nuclear power plants. Thirty-one states around the country currently have nuclear power plants. Thirty-six of the 100 plants currently operating have two or more reactors.

Two types of nuclear reactors in the United States are pressurized water reactors (PWRs) and boiling water reactors (BWRs). The differences between the PWR and the BWR are:

- the way steam is produced,
- the pressure in the system, and
- the process the steam goes through after being produced.

Perhaps the most important benefit of nuclear energy is that it is the world's largest source of emission-free energy. Nuclear power plants do not produce greenhouse gases or any controlled air pollutants, such as sulfur and particulates. The use of nuclear energy in place of other energy sources helps keep the air clean, preserve Earth's climate, avoid ground-level ozone formation, and prevent acid rain. Nuclear energy has the lowest impact on water, land, air, habitats, and species. Nuclear energy is the most eco-efficient of all energy sources because it produces the most electricity relative to its environmental impact.

The greatest risk associated with nuclear reactors is the effects of radiation on people and the environment. An accident could result in the release of dangerous levels of radiation that could affect the health and safety of the people living near the nuclear power plant. The radiation contained in a nuclear reactor can go through the human body, where it can damage biological cells and cause cancer. If radiation strikes sex cells, it can cause genetic diseases.

Concept Check

1. Describe the differences between the two types of nuclear reactors.
2. Discuss one benefit and one risk to people associated with nuclear power plants.

Chapter 14 Review Assess/Reflect

Study Guide

Lesson 1

1. A chemical change is any process that alters the composition of atoms or molecules and forms new substances that have different properties from the original substances.
2. Indicators of chemical change include the formation of light, heat, or gas, as well as the formation of a precipitate or a change in color.
3. Chemical changes alter the arrangement of atoms in molecules and form new substances as a result. Physical changes do not alter the arrangement of atoms in molecules and do not make new substances.

Lesson 2

1. Chemical equations communicate the reactants, products, and rearrangement of atoms in a chemical reaction.
2. Chemical reactions follow the law of conservation of mass. Scientists show this by balancing chemical equations so that both sides of an equation have equal numbers of atoms for each element.
3. Chemical reactions can be described as endothermic or exothermic. In endothermic reactions, the reactants absorb heat from the environment. They feel cool or cold to the touch. In exothermic reactions, the reactants release heat into the environment. They feel warm or hot to the touch.
4. Common types of chemical reactions include combustion (burning), synthesis, decomposition, single replacement, and double replacement.

Lesson 3

1. A chemical reaction occurs only when particles of reactants collide with enough energy at the correct orientation. Factors that affect the rate of a chemical reaction include temperature, concentration, and surface area.
2. Temperature affects the rate of reaction by changing the frequency of collisions between particles of reactant. Concentration affects the rate of reaction by changing the number of particles in a solution. Surface area affects rate of reaction by changing the amount of exposed surface that can collide with other particles. Catalysts change the rate of reaction by allowing a reaction to occur with less energy than usual from the collision of reactant particles.

Vocabulary Check

Use the vocabulary terms to complete the sentences.

1. A(n) __________ is a substance formed as part of a chemical reaction.
2. If the energy of the reactants is less than the energy of the products, it is a(n) __________ reaction.
3. You observe small bubbles forming in a beaker where you just mixed two substances. This is a good indication that a(n) __________ has occurred.
4. If you increase the __________ of a solution, you will increase the rate of reaction.
5. The __________ is the amount of energy needed to start a chemical reaction.
6. If you lower the temperature of a solution in a chemical reaction, you will most likely also lower the __________.

Multiple Choice.

Choose the best answer.

7. Which does not describe chemical change?
   A. It alters the composition of atoms and molecules.
   B. It is the result of a chemical reaction.
   C. It is unaffected by the temperature and concentration of substances.
   D. It forms new substances with different properties from the original substances.

8. What does the term *exothermic reaction* describe?
   A. the energy change from reactants to products in a chemical reaction
   B. the rearrangement of atoms in molecules
   C. the type of change that occurs in a decomposition reaction
   D. the types of elements that are consumed and produced in the reaction

9. Which cannot be used to control the rate of reaction?
   A. the temperature of the environment surrounding a reaction
   B. the amount of reactant in a given volume of solution
   C. the amount of heat applied to a reaction
   D. the time a reaction is allowed to proceed

10. Which situation would increase the rate of reaction between solid and liquid reactants?
    A. increasing the volume of water in the reactant solution
    B. decreasing the temperature of the solution
    C. increasing the surface area of the solid reactant
    D. decreasing the surface area of the solid reactant

11. Which describes a single-replacement reaction?
    A. Two compounds exchange elements to form a precipitate in an aqueous solution.
    B. A more reactive element replaces a less reactive element in a compound.
    C. Ions of a metal in solution gain electrons and change to atoms of the same metal.
    D. A single reactant breaks into two different compounds.

Check Point

Answer the following questions.

12. How could you demonstrate the difference between physical and chemical changes? Describe a demonstration you could do in front of the class.

13. How does lowering temperature decrease the rate of reaction? Use the words *heat*, *kinetic energy*, and *collision* in your explanation.

14. How can you describe the reaction $2C_2H_6 + 7O_2 \rightarrow 4CO_2 + 6H_2O$ + heat?

15. Explain three ways that controlling chemical reactions helps us store, cook, or prepare food.

16. Describe an experiment that would test the rate of reaction in the ripening of bananas. What set-up, measurements, calculations, and results would you need to make?

Periodic Table of the Elements

Period ↓ / Group →	1 I A	2 II A	3 III B	4 IV B	5 V B	6 VI B	7 VII B	8 VIII B	9 VIII B
1	1 H Hydrogen 1.00794								
2	3 Li Lithium 6.941	4 Be Beryllium 9.012182							
3	11 Na Sodium 22.98976928	12 Mg Magnesium 24.305							
4	19 K Potassium 39.0983	20 Ca Calcium 40.078	21 Sc Scandium 44.9559	22 Ti Titanium 47.867	23 V Vanadium 50.9415	24 Cr Chromium 51.9961	25 Mn Manganese 54.938045	26 Fe Iron 55.845	27 Co Cobalt 58.933195
5	37 Rb Rubidium 85.4678	38 Sr Strontium 87.62	39 Y Yttrium 88.90585	40 Zr Zirconium 91.224	41 Nb Niobium 92.9063	42 Mo Molybdenum 95.96	43 Tc Technetium [98]	44 Ru Ruthenium 101.07	45 Rh Rhodium 102.9055
6	55 Cs Caesium 132.9054519	56 Ba Barium 137.327	57-71 Lanthanoids (Lanthanide)	72 Hf Hafnium 178.49	73 Ta Tantalum 180.94788	74 W Tungsten 183.84	75 Re Rhenium 186.207	76 Os Osmium 190.23	77 Ir Iridium 192.217
7	87 Fr Francium [223]	88 Ra Radium [226]	89-103 Actinoids (Actinide)	104 Rf Rutherfordium [267]	105 Db Dubnium [268]	106 Sg Seaborgium [271]	107 Bh Bohrium [272]	108 Hs Hassium [270]	109 Mt Meitnerium [276]

Lanthanoids (Lanthanide)	57 La Lanthanum 138.90547	58 Ce Cerium 140.116	59 Pr Praseodymium 140.90765	60 Nd Deodymium 144.242	61 Pm Promethium [145]	62 Sm Samarium 150.36
Actinoids (Actinide)	89 Ac Actinium [227]	90 Th Thorium 232.03806	91 Pa Protactinium 231.03588	92 U Uranium 238.02891	93 Np Neptunium [237]	94 Pu Plutonium [244]

Color Key

Alkali metals	Alkaline earth metals	Transition metals	Post-transition metals	Metaloids
Other nonmetals	Halogens	Noble gases	Lanthanide	Actinide
Radioactive element	Synthetic element	H Gas	Hg Liquid	Li Solid

10 VIII B	11 I B	12 II B	13 III A	14 IV A	15 V A	16 VI A	17 VII A	18 VIII A
								2 He Helium 4.002602
			5 B Boron 10.811	6 C Carbon 12.0107	7 N Nitrogen 14.0067	8 O Oxygen 15.9994	9 F Fluorine 18.9984032	10 Ne Neon 20.1797
			13 Al Aluminium 26.9815386	14 Si Silicon 28.0855	15 P Phosphorus 30.973762	16 S Sulfur 32.065	17 Cl Chlorine 35.453	18 Ar Argon 39.948
28 Ni Nickel 58.6934	29 Cu Copper 63.546	30 Zn Zinc 65.38	31 Ga Gallium 69.729	32 Ge Germanium 72.64	33 As Arsenic 74.9216	34 Se Selenium 78.96	35 Br Bromine 79.904	36 Kr Krypton 83.798
46 Pd Palladium 106.42	47 Ag Silver 107.8682	48 Cd Cadmium 112.411	49 In Indium 114.818	50 Sn Tin 118.71	51 Sb Antimony 121.76	52 Te Tellurium 127.6	53 I Iodine 126.90447	54 Xe Xenon 131.293
78 Pt Platinum 195.084	79 Au Gold 196.966569	80 Hg Mercury 200.59	81 Tl Thallium 204.3833	82 Pb Lead 207.2	83 Bi Bismuth 208.9804	84 Po Polonium [209]	85 At Astatine [210]	86 Rn Radon [222]
110 Ds Darmstadtium [281]	111 Rg Roentgenium [280]	112 Cn Copernicium [285]	113 Uut Ununtrium [286]	114 Fl Flerovium [289]	115 Uup Ununpentium [288]	116 Lv Livermorium [293]	117 Uus Ununseptium [294]	118 Uuo Ununoctium [294]

63 Eu Europium 151.964	64 Gd Gadolinium 157.25	65 Tb Terbium 158.9253	66 Dy Dysprosium 162.5	67 Ho Holmium 164.93032	68 Er Erbium 167.259	69 Tm Thulium 168.93421	70 Yb Ytterbium 173.054	71 Lu Lutetium 174.9668
95 Am Americium [243]	96 Cm Curium [247]	97 Bk Berkelium [247]	98 Cf Californium [251]	99 Es Einsteinium [252]	100 Fm Fermium [257]	101 Md Mendelevium [258]	102 No Nobelium [262]	103 Lr Lawrencium [262]

Using the Glossary

The Glossary is like a small dictionary. It lists important science terms from your book. The words and terms are listed in alphabetical order. Each word is followed by its meaning. Some words have more than one meaning. This Glossary gives the scientific meaning. That is how the word is used in this book. After the meaning is a page number that tells on what page the word is first used.

A

abiogenesis The theory that life came from nonliving things. (p. 17)

abrasion When bits of sediment carried by the wind strike other rocks, slowly weathering and reshaping them. (p. 371)

absolute age The geologic age of a fossil geologic structure or event expressed in units of time, usually in years. (p. 378)

acid A substance that releases a hydrogen ion (H^+) when dissolved in water. Acids are sour to taste. (p. 419)

acid-base indicator A substance that will change to a specific color when exposed to an acid or base. (p. 421)

activation energy The energy needed to start a chemical reaction. (p. 530)

addiction The condition of needing to regularly have something (tobacco or alcohol) or do something (gamble). (p. 206)

aerobic exercise A type of physical activity that uses large muscles and is rhythmic and repetitive. (p. 226)

alkali metal An element in Group 1 of the periodic table. They are very reactive metals that do not occur freely in nature. (p. 450)

alkaline earth metal An element in Group 2 of the periodic table. (p. 451)

alpha decay A type of radioactive decay in which a radioactive isotope emits an alpha particle made up of two protons and two neutrons. (p. 463)

alternation of generations Two-phased reproductive process in all plants, where the plant alternates between a spore-producing plant (the sporophyte) and a gamete-producing plant (the gametophyte). (p. 55)

altitude The height above sea level. (p. 276)

amorphous solid A solid in which each atom has a specific position where it remains, but the atoms do not have the ordered arrangement of a crystalline material. (p. 408)

anabolic steroid An artificially produced hormone that promotes the growth of muscle mass. (p. 235)

anaerobic exercise A type of physical activity made up of brief, strength-related activities, such as sprinting or weightlifting, that build muscle and physical strength. (p. 226)

analogous structures Different structures that have the same function such as a bird wing, butterfly wing, and bat wing. (p. 31)

angiosperm A vascular plant that produces seeds in the ovaries of flowers. (p. 55)

anion An ion with a negative charge. (p. 479)

anorexia nervosa An eating disorder that involves starving oneself and that is characterized by an irrational fear of gaining weight and a distorted perception of one's body. (p. 235)

anther The structure at the end of the stamen that contains structures that produce the spores that develop into pollen. (p. 77)

artificial selection The process of selecting plants or animals with desirable characteristics to become parents of the next generation in order to perpetuate the characteristic in offspring. (p. 21)

atherosclerosis A condition caused when plaque builds up on the inside walls of the arteries, narrowing the arteries and restricting or reducing blood flow, which can lead to heart attack or stroke. (p. 176)

average atomic mass The weighted average of the atomic masses of an element's isotopes. (p. 438)

B

bandwagon An advertising technique where groups of attractive people are shown using a product. (p. 261)

basal cell cancer The most common type of skin cancer. It can usually be treated by surgically removing the cancerous cells. (p. 185)

base A substance that releases a hydroxide ion (OH^-) when dissolved in water. Bases slippery to feel. (p. 419)

Batesian mimicry A type of mimicry that refers to two or more animals that appear similar, but only one of them is dangerous to predators. (p. 132)

bedding The very distinct boundaries of sedimentary rock layers found within the geologic column that do not blend gradually into one another. (p. 384)

benign tumor A noncancerous tumor. (p. 184)

beta decay A type of radioactive decay in which a radioactive isotope releases a beta particle from its nucleus. Beta particles consist of high energy electrons. (p. 464)

biopsy A procedure, which involves taking a small sample of cells for testing, used to determine whether a tumor is benign or malignant. (p. 184)

bioturbation The continual burrowing activity of creatures that live within sediment. (p. 384)

blizzard A winter storm characterized by low temperatures, winds that gust or blow at 56 km/hr or more, and low visibility because of blowing snow that lasts for at least three hours. (p. 307)

body composition The ratio of body fat to lean body tissue. (p. 228)

bulimia An eating disorder that involves eating big meals and then vomiting. Also referred to as "binging and purging." (p. 235)

C

Cambrian explosion The abrupt appearance of most animal phyla in the Cambrian layers. (p. 390)

carbon dating A process in which scientists compare the ratio of carbon-14 to the ratio of carbon-12 to estimate the age of a fossil. Also known as radiocarbon dating. (p. 388)

carcinogen A substance that can cause cancer. (p. 184)

cardiovascular disease Heart disease that involves a structural or functional abnormality of the heart or the blood vessels. Leading cause of disease in the United States. (p. 176)

CAT scan (computerized axial tomography scan) A specialized X-ray technique that provides many detailed cross-sections of an area, producing a clearer image than a regular X-ray. (p. 191)

cation An ion with a positive charge. (p. 479)

chemical formula A shorthand notation that identifies the elements that make up a compound and the ratio in which they are present. (p. 493)

chlorofluorocarbon (CFC) A chemical ingredient used in the manufacture of aerosol sprays, packing materials, and refrigerants that is a major contributor to the destruction of the ozone layer in Earth's atmosphere. (p. 355)

cirrhosis A disease of the liver caused by excessive drinking that can damage liver cells and cause scar tissue to accumulate. Cirrhosis can lead to death. (p. 213)

cladistics Another name for the process of phylogenetic analysis. (p. 43)

cladogram Another name for a phylogenetic tree. (p. 40)

classical conditioning A learned behavior to associate one stimulus with another, unrelated one. (p. 136)

climate zone A region that shares similar climate characteristics. (p. 313)

colloid A mixture in which the particles are larger than those in a solution but smaller than those in a suspension. (p. 418)

color blindness A genetic disorder that causes portions of the spectrum to be undetectable. Also called color vision deficiency. (p. 169)

combustion reaction A chemical reaction in which molecules of a fuel source combine with oxygen to give off carbon dioxide and water. (p. 507)

competition The struggle among organisms to obtain the necessities for survival and reproduction. (p. 120)

compression A property in which a substance can be forced to occupy a smaller volume. (p. 409)

concentrated solution A solution where a large amount of solute is dissolved in a solvent. (p. 417)

concentration The amount of substance in a given volume. (p. 528)

condensation nuclei Tiny airborne particles of dust, pollen, and other materials that attract water droplets and then join to form a cloud. (p. 283)

conifer A cone-bearing plant with needlelike or scalelike leaves. (p. 70)

conservation of design The idea that an intelligent Creator designed a flexible general plan, which could be adapted for different animals. (p. 32)

contour plowing An agricultural practice of plowing along the natural contours of the land to reduce overly disturbed land and soil erosion and runoff. (p. 337)

convection cell The entire rising-and-sinking, cyclical movement of air. (p. 290)

cork cambium A meristem located outside the phloem that produces cork tissue and is responsible for secondary growth in stems. (p. 99)

cotyledon Large food storage area in a seed that functions as the first "seed leaf" of a plant embryo. (p. 74)

countershading A form of camouflage in an organism, where its dorsal surfaces are darker and its ventral surfaces are lighter. (p. 131)

covalent bond The chemical bond that forms between atoms when they share electrons. (p. 486)

creationism The theory that God created the Universe and all life. (p. 17)

crop rotation The process of planting different crops in a field to improve the quality of the soil. (p. 337)

cryptic coloration The use of color as a form of camouflage to blend into the background. (p. 131)

crystalline solid A solid whose atoms are arranged in a very regular, three-dimensional pattern. (p. 408)

cyanobacteria Photosynthetic organisms that produce oxygen. According to the big bang theory these organisms introduced oxygen into the Earth's atmosphere, allowing evolution to occur. (singular: cyanobacterium) (p. 275)

cystic fibrosis A recessive, single gene in which the body produces very thick, sticky mucus that interferes with breathing and food absorption. (p. 168)

D

dead zone Area of water with very low levels of dissolved oxygen. (p. 148)

deciduous A plant that drops its leaves in the fall. (p. 67)

deflation When wind carries bits of sediment as it blows across land. (p. 371)

deforestation The removal or clear-cutting of trees from an area. (p. 139)

denitrification The process in which bacteria convert nitrates back into nitrogen gas by removing the oxygen atoms. (p. 127)

diabetes A disease resulting from the body's inability to regulate the amount of sugar in the blood due to problems with the production of the hormone insulin. (p. 179)

dietary supplement A product taken orally that contains one or more ingredients that are intended to add to your diet and that are not considered food. (p. 254)

dilute solution A solution where a small amount of solute is dissolved in a solvent. (p. 417)

dioecious Plants that have the male and female flowers on different plants. (p. 78)

disclaimer A legal statement that attempts to absolve a company or other source from responsibility for any consequences resulting from taking the company's product or advice. (p. 264)

disruptive coloration A form of camouflage (such as spots or stripes) that breaks up the shape and outline of an animal, making it harder to detect. (p. 131)

Down syndrome A genetic disorder in which a person has an extra copy of chromosome 21, resulting in characteristic facial features, and delays in physical, mental and social development. (p. 170)

drug Any substance, other than food, that changes the way your body or mind works. (p. 202)

ductility The ability to draw metals into a wire. (p. 449)

E

eating disorder Any of a group of conditions in which people focus entirely on food and their weight while their health suffers. (p. 235)

electron cloud The area of probability of an electron being in a particular location around the nucleus. (p. 437)

electronegativity An affinity for valence electrons. (p. 489)

emphysema A disease caused by tobacco smoke and certain enviromental factors that damages the tiny sacs (alveoli) in the lungs where gas exchange occurs. (p. 214)

endothermic A type of change in which an object absorbs energy while it changes its state. (p. 410)

endothermic reaction A chemical reaction in which the reactants have less energy that the resulting products. (p. 518)

endurance activity An activity that raises your heart rate and breathing, such as walking, running, or swimming. (p. 225)

energy level Possible locations around the nucleus where elections with specific energy values can be found. (p. 436)

energy resource A source of energy that is used to produce other types of energy. (p. 325)

estivation A dormant state that helps animals survive a lack of resources in hot, dry weather. (p. 135)

estuary The area where the freshwater of a river meets the saltwater of the ocean and mixing occurs. (p. 148)

evergreen A plant that retains its leaves all year long. (p. 67)

exogenesis The theory that life began somewhere else. (p. 19)

exothermic A type of change in which an object gives off energy while it changes its state. (p. 411)

exothermic reaction A chemical reaction in which the reactants have more energy that the resulting products. (p. 518)

exotic species Non-native species that may be introduced by natural processes or by human activity. (p. 111)

extensive property A physical property of matter that depends on the amount of matter present, such as mass and volume. (p. 404)

extinct When all members of a species have died off. (p. 109)

F

fixity of species The idea that plants and animals never change. (p. 21)

Food and Drug Administration (FDA) A U.S. government agency that has the task of ensuring that fruits, vegetables, and shell eggs sold in the United States are safe and healthful. (p. 251)

fracking The process of drilling and injecting fluid into the ground at high pressure to break apart shale rocks and release natural gas resources. Also known as hydraulic fracturing. (p. 144)

frond The leaf of a fern. (p. 57)

G

gamete A male (sperm) or female (egg) reproductive cell. (p. 55)

gametophyte The form of a plant that develops from a spore in the second part of the two-part reproduction cycle and that produces gametes (egg and sperm). (p. 55)

gamma decay A type of radioactive decay in which an atom releases gamma rays from its nucleus. (p. 465)

genetic disorder A health problem caused partly or totally by a defect in one or more genes or chromosomes. (p. 165)

geologic time scale The dates associated with the geologic column that relate life forms to geologic events in Earth's history. (p. 380)

geotropism The response of a plant to the effects of gravity. (p. 104)

gymnosperm A vascular plant that produces seeds that are not enclosed in fleshy fruits. (p. 54)

H

habituation A learned behavior in which an animal learns to stop responding to a particular stimulus over time. (p. 136)

half-life The time it takes for one half of the amount of an unstable isotope, or parent isotope, to decay into daughter isotopes. (p 380)

halogen A nonmetal element in Group 17 of the periodic table. Means "salt-forming." (p. 456)

Health Canada A Canadian government agency that uses scientific research to modernize the health-care system, regulate food safety, promote the health of Canadians, and communicate the importance of disease prevention. (p. 251)

heliotropism The movement of a plant as it follows the Sun throughout the day. Also called solar tracking. (p. 103)

heterogeneous mixture A mixture that has components that are not uniform or that has regions with different properties. (p. 413)

hominid An apelike animal that walked on two legs. (p. 393)

homogeneous mixture A mixture in which the components are uniformly distributed throughout the mixture. (p. 414)

homologous structures Structures that are similar in construction but that have different functions such as a human hand, whale flipper, and bat wing. (p. 31)

I

imprinting A learned behavior that can occur only during a "critical period" shortly after birth. (p. 138)

informed interventionism The concept that acknowledges God's intervention at certain times in history. (p. 17)

Intelligent Design The theory that certain features of the Universe and of living things are the result of an intelligent cause. (p. 26)

intensive property A physical property of matter that does not depend on the amount of matter present, such as color, density, hardness, and temperature. (p. 404)

intoxication When a person's mental and physical abilities are impaired because of drug use. (p. 211)

intrusion The rock that is formed as magma is pushed up into cracks or spaces in the existing rock. (p. 373)

invasive species An exotic species that harms the environment. (p. 111)

ionic bond Chemical bonding that involves the complete transfer of electrons between atoms. It produces two oppositely charged ions. The force of attraction that holds ions together. (p. 479)

isomer One of two or more compounds with the same molecular formula but different structural formulas. (p. 498)

isotope One of two or more atoms of the same element that have the same number of protons but different numbers of neutrons. (p. 378)

J

jet stream A fast-flowing wind between 10 km and 15 km above Earth's surface. (p. 293)

L

lamination A sedimentary rock structure in which thin layers, or laminae, are recognizable. (p. 386)

local winds Winds that generally move short distances and that result from pressure differences caused by the differences in how quickly matter on Earth heats up. (p. 293)

M

macroevolution The theory that all present life forms descended from a simple, one-celled organism. Macroevolution has not been shown to take place. (p. 17)

macromolecule A molecule with many atoms arranged in long chains, rings, or other complex structures. (p. 492)

malignant tumor A cancerous tumor. (p. 184)

malleability The ability to flatten metals into thin sheets or other shapes without breaking them. (p. 449)

mammogram A special X-ray of breast tissue. (p. 191)

mass movement Landslides, mudslides, and rock falls. (p. 371)

mass number The sum of the number of protons and the number of neutrons in an atom's nucleus. (p. 438)

material resource A natural resource that is used to make other products. (p. 325)

melanoma A serious form of skin cancer that can be life threatening if it metastasizes. (p. 185)

meristem A growth region of a plant. (p. 98)

metallic bond A type of bond where each atom donates its valence electrons to a "sea of electrons" that move freely between the positive ions. (p. 484)

metastasis The spread of cancerous cells from their origin to other parts of the body. (p. 184)

microevolution Small changes in a species over time. (p. 23)

mine reclamation The process of restoring land that has been mined to a natural or usable state. (p. 341)

molecular formula A type of chemical formula that shows how much of each element is present in a molecule. (p. 494)

monoecious A plant that has perfect flowers or imperfect flowers of both sexes. (p. 78)

morphological intermediate A creature that shares characteristics with different animal groups. (p. 392)

morphology What an animal looks like in terms of its structure or shape. (p. 392)

mountain breeze Local winds in mountainous regions that form at night when the air on the mountaintop cools more quickly than the air on the valley floor and then moves down the mountainside into the valley. (p. 296)

MRI (magnetic resonance imaging) A specialized diagnostic test that uses magnets, radio waves, and a computer to create detailed images of internal tissues and organs. (p. 193)

Müllerian mimicry A type of mimicry that refers to two or more animals that look alike, and both have strong defenses against predators. (p. 132)

multifactorial disease A disease that results when several factors, such as a genetic factor and an environmental factor, work together to cause the disease. (p. 174)

mutagen An environmental factor that causes a change in the DNA of an organism. (p. 165)

mutation A sudden change that occurs when a gene is altered, damaged, or lost. (p. 24)

N

native species Plants that live naturally in an environment. (p. 111)

natural selection The process whereby organisms better adapted to their environment tend to survive and produce more offspring. (p. 22)

nicotine A chemical in tobacco that makes the central nervous system, heart, and other organs work faster. (p. 214)

nitrogen fixation The process of changing nitrogen gas from the air into nitrate compounds that plants can use. (p. 127)

no-till farming A technique of growing crops from year to year without disturbing the soil through plowing and tilling. (p. 337)

nonpoint-source pollution Pollution that is difficult to trace to a single source. (p. 349)

nonpolar A molecule in which the electrons are equally shared. (p. 490)

nonpolar covalent bond A bond in which the electrons are equally shared. (p. 490)

nonvascular plant A plant that lacks vascular tissue and absorbs water and nutrients directly into its cells, transporting these materials throughout the plant through diffusion and osmosis. (p. 53)

nuclear fission The process by which a large radioactive nucleus splits into smaller nuclei and releases energy. (p. 469)

nuclear fusion The process of combining small nuclei to form a larger nucleus. (p. 470)

nutrient A chemical in food that the body uses for energy, growth, and cell repair. (p. 220)

O

obesity The condition of being overweight. (p. 179)

oncogene A gene that is predisposed to causing cancer. (p. 184)

operant conditioning A learned behavior to modify future behavior based on past events. (p. 136)

orbital A specific level of electron energy that an electron cloud corresponds to. (p. 437)

organic chemistry The chemistry of carbon compounds. (p. 497)

organic farm A farm that uses only naturally occurring substances to fertilize its crops and that uses natural substances instead of toxic pesticides to protect its crops from weeds and insect invaders. (p. 337)

origins Beginnings. (p. 16)

ovary One of the three structures that make up the female part of the flower and where the eggs form. (p. 78)

over-the-counter (OTC) medicine Medication that can be purchased without a prescription. (p. 202)

P

panspermia The idea that life is actually widespread throughout the Universe. (p. 19)

parsimony The selection of one cladogram out of many that requires the fewest evolutionary changes. (p. 41)

particulate matter Tiny pieces of solid matter, including dust, ash, pollen, and salts, in Earth's atmosphere. (p. 276)

PET scan (positron emission tomography scan) A specialized diagnostic test that produces a three-dimensional, color image of a body part to show how it is functioning rather than just how it looks. (p. 194)

pH scale A measure of the hydrogen ion concentration in a solution, thereby identifying the strength of an acid or a base. (p. 424)

phloem Cells or tissues in the vascular system that transport the food the plant makes to all cells in the plant. (p. 53)

phototropism The directional response of plants to light. (p. 103)

phyletic gradualism The process in which organisms with new body plans would appear gradually in the fossil record. (p. 389)

phylogenetic analysis The process of grouping organisms based on multiple shared characteristics or traits that were not shared with organisms thought to have been their ancestors. (p. 39)

phylogenetic tree Graphic of the phylogenetic analysis used to see how animals are related. (p. 39)

physical dependence Where the user has a chemical need for the drug. (p. 206)

pistil The female part of a flower. It includes the stigma, style, and ovary. (p. 77)

point-source pollution Pollution that comes from a single point. (p. 348)

polar A molecule in which the electrons are unequally shared. (p. 490)

polar covalent bond A bond in a polar molecule. (p. 490)

polygamous A plant that has both perfect flowers and imperfect flowers of both sexes. (p. 78)

precipitate A solid substance that forms in a solution as a result of a chemical change. (p. 512)

prescription An instruction that a physician or other authorized medical professional writes, specifying the type, dosage, and frequency of a medication for a patient. (p. 202)

primordial soup The carbon dioxide, methane, ammonia, and water that filled the oceans at some point in Earth's past. (p. 18)

psychological dependence Where the user believes he or she needs a particular substance (drug) or activity (gambling) to feel good or even just to feel normal. (p. 206)

Q

quark One of many smaller particles that make up the protons and neutrons of an atom. (p. 438)

R

radiation The energy that an atom gives off when changes occur in its nucleus. (p. 462)

radioactive decay The process in which unstable isotopes, or parent isotopes, break down into other isotopes. (p. 380)

radioactive isotope An element that exhibits radioactivity. (p. 462)

radioactivity The process by which nuclei give off energy. (p. 462)

rate of reaction The speed of a chemical reaction. (p. 527)

refusal skill A strategy that can help an individual resist negative peer pressure. (p. 248)

rhizoid A rootlike structure of a nonvascular plant that anchors the plant to a surface and that enables the plant to absorb water and nutrients directly into its cells. (p. 59)

rhizome An underground, horizontal stem of ferns and other plants that contains vascular tissue for transporting water and nutrients. (p. 61)

role model A person looked to by others as an example to be imitated. (p. 249)

S

salt An ionic compound produced by the reaction of an acid with a base. (p. 456)

saturated solution A solution where the concentration of the solute reaches the point where no more solute can be dissolved in the solvent. (p. 417)

secondhand smoke The smoke from burning cigarettes and from the smoke-polluted air smokers exhale. (p. 217)

sickle-cell anemia A genetic disorder that causes red blood cells to form a sickle, or crescent-moon, shape that impairs blood flow. (p. 166)

simple sugar A sugar that cannot be broken down into smaller carbohydrate units. (p. 91)

smog A yellow-brown haze formed when sunlight reacts with pollutants that are released into the air from the burning of fossil fuels. (p. 355)

sori The reproductive structures of the sporophyte often found on the underside of the fern frond. (singular: sorus) (p. 57)

spontaneous generation The theory that life can arise from nonliving materials. (p. 17)

sporangia Structures on each sorus that produce spores. (singular: sporangium) (p. 57)

spore A reproductive cell that can develop into a new individual without being fertilized in ideal conditions. (p. 55)

sporophyte The form of a plant that produces spores during one part of the two-part reproduction cycle. (p. 55)

stamen The male part of a flower. (p. 77)

stasis No change for long periods. (p. 393)

stewardship The act of accepting responsibility and taking action as an individual to improve the quality of the environment. (p. 343)

stigma The top of the style that is covered with hairs or a sticky material that traps pollen and is one of the three structures that make up the female part of a flower. (p. 78)

stomata Tiny openings in the surface of plant leaves that allow gases to enter and exit during photosynthesis. (singular: stoma) (p. 77)

storm surge An abnormal rise of water generated by a storm that can be pushed on shore by strong winds. (p. 307)

strip farming An agricultural practice of growing alternating rows of a cultivated plant, such as corn, and a sod-forming plant, such as alfalfa, to reduce soil erosion. (p. 337)

structural formula A type of notation that shows the types and numbers of atoms present in a molecule and the arrangement of the atoms. (p. 494)

style The hollow stalk that rises up from the ovary and one of the three structures that make up the female part of a flower. (p. 78)

sublimation A process in which ice or snow evaporates directly into water vapor without first becoming a liquid. (p. 412)

supersaturated solution A solution that has more solute dissolved in it than could be dissolved by the solvent under normal conditions. (p. 417)

support network A group of people who share your values and who will support you when you decide not to participate in dangerous or questionable activities. (p. 248)

surface tension The attractive force that exists between the particles on the surface of a liquid. (p. 409)

suspension A mixture in which the particles in the solvent are so large that they will settle out unless the mixture is constantly stirred. (p. 417)

sustainable agriculture Farming methods that help keep the land productive. (p. 140)

sustainable logging A practice in which loggers balance the use of the forests for wood production with protecting the ecosystem and the environment. (p. 342)

systematics The classification system used presently to classify organisms. (p. 37)

T

taxis An innate behavior that is the movement toward or away from a stimulus. (p. 133)

tectonic plate One of several slabs in Earth's solid lithosphere that floats on top of the molten asthenosphere. (p. 367)

testimonial An advertising technique where a satisfied customer or a famous person praises a product. (p. 261)

thigmotropism A plant's movement or growth response to touch. (p. 106)

threatened species A species that is "likely to become an endangered species within the foreseeable future." (p. 113)

topography The features of an area described in terms of elevation and slope. (p. 312)

transition metal An element in Groups 3–12 of the periodic table. (p. 452)

transmutation The process by which one element changes into a different element as a result of radioactive decay. (p. 463)

transpiration Loss of water from a plant through the stomata of the plant's leaves, triggering water to be pulled up through the xylem of the plant. (p. 97)

trophic level The position an organism occupies in a food web, based on its source of energy. (p. 123)

tropism Any movement in a plant that occurs in response to a stimulus. (p. 102)

tumor A mass of cells that form as a result of uncontrolled cell division. (p. 182)

U

unconformity An event where the process of deposition in an area stops for a period of time and the layers of rock that had been formed are eroded before deposition starts again, leaving a gap in the geologic record. (p. 373)

urban sprawl An effect of urbanization in which urban areas expand into surrounding countryside. (p. 334)

urbanization The expansion of urban areas into rural areas. (p. 333)

V

valley breeze Local winds in mountainous regions that form during the day when the Sun rapidly heats up the valley air, which then rises up the mountainside. (p. 296)

vascular cambium The stem layer that produces new xylem and phloem cells. (p. 96)

vascular system A structure inside vascular plants that moves water and nutrients from one part of the plant to another by the processes of diffusion and osmosis. (p. 53)

vascular tissue The tissue in plants that makes up the vascular system, consisting of phloem, which carries food, and xylem, which carries water and minerals. (p. 96)

vestigial structure A structure that does not appear to have any use and that is often presumed to be a "leftover." (p. 29)

viscosity A liquid's resistance to flow. (p. 409)

visibility The greatest distance at which objects can be seen. (p. 307)

W

withdrawal Body symptoms that a person experiences when he or she stops using a drug that he or she has a physical dependence on. (p. 206)

X

xylem Cells or tissues in the vascular system that transport water and inorganic nutrients from the roots to the rest of the plant. (p. 53)

Z

zygote A cell that forms when two gametes (egg and sperm) unite. (p. 55)

Index

F

G

H

I

N

O

P

T

U

V

W

X

Y

Z

Photo Credits

CVR Rob Marmion/Shutterstock, (Background) Radu Razvan/Thinkstock, (Banner) (1) Fotorich01/Shutterstock, (2) Janprchal/Shutterstock, (3) Courtesy of Larry Blackmer, (4) Steshkin Yevgeniy/Shutterstock, (5) Lucarelli Temistocle/Shutterstock, (6) NASA Images/NASA, (7) Matej Pavlansky/Shutterstock Barnett/Shutterstock; **SS 0** IStockphoto/Thinkstock; **SS 5** Bruce Rolff/Shutterstock; **SS 6–7** iStockphoto/Thinkstock; **SS 8** Frangipani/Shutterstock; **9** (l) iStockphoto/Thinkstock, (r) iStockphoto/Thinkstock; **13–14** Stockbyte/Thinkstock; **14–15** Wasu Watcharadachaphong/Shutterstock; **16–17** Johan Swanepoel/Shutterstock; **19** Petarg/Shutterstock; **22** Julia Remezova/Shutterstock; **23** (l) Steve McWilliam/Shutterstock, (r) Martin Fowler/Shutterstock; **24** Eric Isselee/Shutterstock; **27** Johan Swanepoel/Shutterstock; **28** Dorling Kindersley/Thinkstock; **30** Eric Isselee/Shutterstock; **32** (l to r) iStockphoto/Thinkstock, (2) iStockphoto/Thinkstock, (3) ANP/Shutterstock, (4) Jasenka Luksa/Shutterstock, (5) iStockphoto/Thinkstock; (6) iStockphoto/Thinkstock, (7) ANP/Shutterstock, (8) Bob Orsillo/Shutterstock; **34** Tatiana Makotra/Shutterstock; **35** Liew Weng Keong/Shutterstock; **37** Adisa/Shutterstock; **40** (l) iStockphoto/Thinkstock, (cl) iStockphoto/Thinkstock, (cr) Sergii Votit/Shutterstock, (r) Alta Oosthuizen/Shutterstock; **42** (t to b) karakotsya/Shutterstock, (2) Meoita/Shutterstock, (3) stihii/Shutterstock, (4) kavalenkava volha/Shutterstock, (5) stihii/Shutterstock; **44** Dr. Raúl Esperante/Geoscience Research Institute; **45** Paul Maguire/Shutterstock; **48–49** Mayovskyy Andrew/Shutterstock; **50** (l) Lee Prince/Shutterstock; **50–51** (r) Melinda Fawver/Shutterstock; **53** iStockphoto/Thinkstock; **54** (t) Todd Boland/Shutterstock, (b) Stephen Farhall/Shutterstock; 57 Dorling Kindersley RF/Thinkstock; **59** photong/Shutterstock; **60** Madlen/Shutterstock; **61** Repina Valeriya/Shutterstock; **62** iStockphoto/Thinkstock; **64** scphoto60/Shutterstock; **66** iStockphoto/Thinkstock; **67** iStockphoto/Thinkstock; **69** (t) Jorge Salcedo/Shutterstock, (b) Jody./Shutterstock; **70** iStockphoto/Thinkstock; **73** CreativeNature.nl/Shutterstock; **75** (t) Richard Griffin/Shutterstock, (b) Krasowit/Shutterstock; **78** BlueRingMedia/Shutterstock; **81** Hkannn/Shutterstock; **83** iStockphoto/Thinkstock; **84** littleny/Shutterstock; **85** Shots Studio/Shutterstock; **88–89** iStockphoto/Thinkstock **90** Designua/Shutterstock; **96** Luka Skywalker/Shutterstock; **97** Dimarion/Shutterstock; **99** (t) Tom Oliveira/Shutterstock, (b)iStockphoto/Thinkstock; **100–101** Fuse/Thinkstock; **102** Noppasin/Shutterstock; **104** Stephane Bidouze/Shutterstock; **106** Marco Uliana/Shutterstock; **107** iStockphoto/Thinkstock; **108** Anton Foltin/Shutterstock; **109** Aspen Photo/Shutterstock; **110** iStockphoto/Thinkstock; **111** Seroff/Shutterstock; **113** Dan Daggett/Shutterstock; **114** meunierd/Shutterstock; **115** Szasz-Fabian Ilka Erika/Shutterstock; **118–119** Howard Sandler/Shutterstock; **120** Kletr/Shutterstock; **126** photoiconix/Shutterstock; **130** Decha Thapanya/Shutterstock; **131** (l to r)Ryan M. Bolton/Shutterstock, (2) Elmarie Dreyer/Shutterstock, (3) EcoPrint/Shutterstock, (4) iStockphoto/Thinkstock; **132** Ellen Hui/Shutterstock; **133** Steve Byland/Shutterstock; **135** (t) ken_cheung/Shutterstock, (b) Tom Reichner/Shutterstock; **136** Michelle D. Milliman/Shutterstock; **139** Dr. Morley Read/Shutterstock; **143** Designua/Shutterstock; **144** Hemera/Thinkstock; **145** Leonid Ikan/Shutterstock; **147** Lone Wolf Photos/Shutterstock; **148–149** Lone Wolf Photos/Shutterstock; **156** Bibliotheca Scarensis; **157** Tom Reichner/Shutterstock; **160** Fuse/Thinkstock; **162–163** iStockphoto/Thinkstock; **164** Geanina Bechea/Shutterstock; **165** Getty Images News/Thinkstock; **166** Alila Medical Images/Shutterstock; **169** Portokalis/Shutterstock; **170** Alila Medical Images/Shutterstock; **171** Alila Medical Images/Shutterstock; **172** PathDoc/Shutterstock; **174** Wavebreakmedia Ltd/Thinkstock; **179** Dmitry Lobanov/Shutterstock; **181** Fly_dragonfly/Shutterstock; **182** Designua/Shutterstock; **184** Spencer Freeman, Florida Hospital; **185** iStockphoto/Thinkstock; **187** dotshock/Shutterstock; **189** emin kuliyev/Shutterstock; **190** Hemera/Thinkstock; **191** Fuse/Thinkstock; **193** Stockbyte/Thinkstock; **194** Triff/Shutterstock; **196** Will Hughes/Shutterstock; **197** Courtesy Cami Wenger; **199** Alila Medical Images/Shutterstock; **200–201** Fuse/Thinkstock; **202–203** Fabiana Ponzi/Shutterstock; **205** Royalty-Free/Masterfile; **210** Ingram Publishing /Thinkstock; **213** Alila Medical Media/Shutterstock; **217** iStockphoto/

Thinkstock; **219** Maridav/Shutterstock; **220** iStockphoto/Thinkstock; **221** Courtesy of USDA; **225** Jacek Chabraszewski/Shutterstock; **226** Jacek Chabraszewski/Shutterstock; **229** Monkey Business Images/ Shutterstock; **230** Edyta Pawlowska/Shutterstock; **232** hin255/Shutterstock; **236** prudkov/Shutterstock; **238** Viorel Sima/Shutterstock; **239** iStockphoto/Thinkstock; **242** Creatas/Thinkstock; **244** stester/ Shutterstock; **246** wavebreakmedia/Shutterstock; **248** Pressmaster/Shutterstock; **249** iStockphoto/ Thinkstock; **251** Monika Wisniewska/Shutterstock; **254** Tish1/Shutterstock; **255** Tshirt Designs/ Shutterstock; **258** iStockphoto/Thinkstock; **261** RetroClipArt/Shutterstock; **263** iStockphoto/Thinkstock; **264** Dmitry Lobanov/Shutterstock; **267** Fuse/Thinkstock; **266** Tyler Olson/Shutterstock; **270–271** Markus Gann/Shutterstock; **272–273** razlomov/Shutterstock; **274–275** Manamana/Shutterstock; **277** Kotenko Oleksandr/Shutterstock; **279** Designua/Shutterstock; **282** Zern Liew/Shutterstock; **285** peresanz/Shutterstock; **286** MarcelClemens/Shutterstock; **289** Phil MacD Photography/ Shutterstock; **290** Dorling Kindersley RF/Thinkstock; **291** Vladislav Gurfinkel/Shutterstock; **292** Dorling Kindersley RF/Thinkstock; **293** leonello calvetti/Shutterstock; **296** Dorling Kindersley RF/Thinkstock; **297** Praisaeng/Shutterstock; **302** Patryk Kosmider/Shutterstock; **303** Dorling Kindersley RF/Thinkstock; **305** lafoto/Shutterstock; **306–307** Leonard Zhukovsky/Shutterstock; **309** Photobank gallery/ Shutterstock; **311** Peter Hermes Furian/Shutterstock; **312** Dorling Kindersley RF/Thinkstock; **314** outdoorsman/Shutterstock; **315** BMJ/Shutterstock; **316** Getty Images News/Thinkstock; **317** SMART Radar, University of Oklahoma. Photo by Keli Pirtle, NOAA National Severe Storms Lab; **321** iStockphoto/Thinkstock; **322–323** anweber/Shutterstock; **324** Epidote/Shutterstock; **326** Greg Randles/Shutterstock; **327** PhotographyByMK/Shutterstock; **329** Lefteris Papaulakis/Shutterstock; **331** Robin Lund/Shutterstock; **334** donsimon/Shutterstock; **335** Photodisc/Thinkstock; **336** ermess/ Shutterstock; **339** Pabkov/Shutterstock; **340** (l) Jason Maehl/Shutterstock, (r) A.S. Zain/Shutterstock; **341** Jiri Foltyn/Shutterstock; **343** S.Borisov/Shutterstock; **344** Marcin-linfernum/Shutterstock; **346** iStock Editorial/Thinkstock; **355** trekandshoot/Shutterstock; **359** Alexandr Makarov/Shutterstock; **360** iStockphoto/Thinkstock; **361** Miks Mihails Ignats/Shutterstock; **364–365** Thomas Barrat/ Shutterstock; **366** Johan Swanepoel/Shutterstock; **367** Dorling Kindersley RF/Thinkstock; **368** Dorling Kindersley RF/Thinkstock; **371** VLADJ55/Shutterstock; **372** (t) James Steinberg/Science Source, (b) holbox/Shutterstock; **374** (bkgd) Thomas & Pat Leeson/Science Source, (inset) AridOcean/ Shutterstock; **375** Dan Schreiber/Shutterstock; **376** iStockphoto/Thinkstock; **378** MarcelClemens/ Shutterstock; **384** thoron/Shutterstock; **385** Courtesy Leonard Brand; **391** John A. Anderson/ Shutterstock; **392** Lorraine Hudgins/Shutterstock; **393** Dorling Kindersley/Thinkstock; **396** Courtesy Mary Baumgardner; **397** Earth Observatory, NASA; **400–401** Troy Wegman/Shutterstock; **402–403** zhu difeng/Shutterstock; **404** Trudy Wilkerson/Shutterstock; **407** Gabor Havasi/Shutterstock; **409** Madlen/Shutterstock; **410** Alix Kreil/Shutterstock; **413** Svetlana Foote/Shutterstock; **414** (l) Gyvafoto/Shutterstock, (r) iStockphoto/Thinkstock; **416** discpicture/Shutterstock; **418** Dan Kosmayer/Shutterstock; **420** rockstar_images/Shutterstock; **421** Charles D. Winters/Science Source; **422** Shawn Hempel/Shutterstock; **426** Photos.com/Thinkstock; **427** Robert Red/Shutterstock; **430–431** iStockphoto/Thinkstock; **432** Igor Plotnikov/Shutterstock; **437** Faraways/Shutterstock; **441** Nerthuz/Shutterstock; **443** RIA Novosti/Science Source; **449** Albert Russ/Shutterstock; **450** Andraž Cerar/Shutterstock; **451** Corepics VOF/Shutterstock; **452** MarcelClemens/Shutterstock; **455** Andrew Lambert Photography/Science Source; **457** (t) Ivan Cholakov/Shutterstock, (b) mitzy/ Shutterstock; **459** magnetix/Shutterstock; **461** Albert Russ/Shutterstock; **468** MarcelClemens/ Shutterstock; **470** general-fmv/Shutterstock; **472** Georgios Kollidas/Shutterstock; **473** nikkytok/ Shutterstock; **476–477** Purestock/Thinkstock; **478** bluesnote/Shutterstock; **493** Garsya/Shutterstock; **496** Palo_ok/Shutterstock; **500** Yuri Bathan (yuri10b)/Shutterstock; **501** iStockphoto/Thinkstock; **504–505** R. Gino Santa Maria/Shutterstock; **506** Seqoya/Shutterstock; **508** (t) Aksenova Natalya/ Shutterstock, (b) Arangan Ananth/Shutterstock; **511** Coprid/Shutterstock; **512** Lindsey Moore/ Shutterstock; **515** AlenaNex/Shutterstock; **521** alexsvirid/Shutterstock; **523** Charles D. Winters/Science Source; **525** andrea crisante/Shutterstock; **534** Courtesy Don B. DeYoung; **535** iStockphoto/Thinkstock.